USING MASS SPECTROMETRY FOR DRUG METABOLISM STUDIES

Second Edition

CRC Press
Taylor & Francis Group
6000 Broken Sound Parkway NW, Suite 300
Boca Raton, FL 33487-2742

Printed in the United States of America on acid-free paper
10 9 8 7 6 5 4 3 2 1

International Standard Book Number: 978-1-4200-9220-2 (Hardback)

Library of Congress Cataloging-in-Publication Data

Using mass spectrometry for drug metabolism studies / editor, Walter A. Korfmacher. -- 2nd ed.
p. cm.
Includes bibliographical references and index.
ISBN 978-1-4200-9220-2 (hardcover : alk. paper)
1. Drugs--Metabolism. 2. Drugs--Spectra. 3. Mass spectrometry. I. Korfmacher, Walter A. II. Title.

RM301.U85 2010
615'.7--dc22 2009034840

Visit the Taylor & Francis Web site at
http://www.taylorandfrancis.com

and the CRC Press Web site at
http://www.crcpress.com

This book is dedicated to the most important people in my life:

Madeleine Korfmacher
Joseph Korfmacher
Mary McCabe
Michael McCabe
Brian McCabe

Contents

Preface

The reason for writing this second edition is simple—science marches on and there are new topics and areas of interest that did not exist in 2003, when the first edition of this book was written. We now have UPLC and Orbitraps as well as DESI and DART to name a few of the new instruments and techniques that are now part of our toolset for using mass spectrometry for drug metabolism studies. This second edition is a completely new book with 14 new chapters that were written solely for this edition. While some of the topics are the same as in the first edition, the newly written chapters provide the latest thinking on how best to use mass spectrometry in a drug metabolism setting.

This book is designed to be a reference book for professionals in both mass spectrometry and various drug metabolism areas. It will also be a useful reference book for medicinal chemists working in the area of new drug discovery who want to learn more about drug metabolism and how it is used for participation in lead optimization efforts. The chapters are written by scientists who are experts in the topic and provide not only a summary of the current best practices but also an extensive review of recent scientific literature, including many references for further reading. Each chapter has been written so that it can be read separately from the other chapters, but, together, the 14 chapters cover a wealth of information on various topics that relate to mass spectrometry and drug metabolism studies.

Some of the chapters are written to cover general topics, while other chapters cover a specific area of interest. For example, there is a chapter on metabolite identification as well as a chapter on UPLC. There was also an effort on my part to include newer areas of interest to drug metabolism scientists. As one example of a new topical area, a chapter on biomarkers has been added in this second edition.

I would like to thank all the contributing authors for their efforts to make this second edition complete. I also thank CRC Press for supporting this effort. In addition, I would like to thank the management of Schering-Plough Research Institute for supporting my efforts to make this book a reality. Finally, I would like to thank my family for all their support, with special thanks to my wife, Madeleine.

Walter A. Korfmacher

Editor

Dr. Walter A. Korfmacher is a distinguished fellow of exploratory drug metabolism at Schering-Plough Research Institute in Kenilworth, New Jersey. He received his BS in chemistry from St. Louis University in 1973. He then went on to obtain his MS in chemistry in 1975 and his PhD in chemistry in 1978, both from the University of Illinois in Urbana. In 1978, he joined the Food and Drug Administration (FDA) and was employed at the National Center for Toxicological Research (NCTR) in Jefferson, Arkansas. While at the NCTR, he also held adjunct associate professor positions in the College of Pharmacy at the University of Tennessee (Memphis) and in the Department of Toxicology at the University of Arkansas for Medical Sciences (Little Rock). After 13 years at the NCTR, he joined Schering-Plough Research Institute as a principal scientist in October 1991.

Dr. Korfmacher is currently a distinguished fellow and the leader for a group of 20 scientists. His research interests include the application of mass spectrometry to the analysis of various sample types, particularly metabolite identification and trace organic quantitative methodology. His most recent applications are in the use of high-performance liquid chromatography (HPLC) combined with atmospheric pressure ionization mass spectrometry and tandem mass spectrometry for both metabolite identification as well as nanogram per milliliter quantitative assay development for various pharmaceutical molecules in plasma. He is also a leader in the field of developing strategies for the application of new mass spectrometry (MS) techniques for drug metabolism participation in new drug discovery and is frequently invited to speak at scientific conferences.

In 1999–2000, Dr. Korfmacher was the chairperson of the North Jersey Mass Spectrometry Discussion Group and in 2002 he received the New Jersey Regional Award for Achievements in Mass Spectrometry. Dr. Korfmacher is a member of the editorial boards of three journals: *Rapid Communications in Mass Spectrometry*, *Current Drug Metabolism*, and *Drug Metabolism Letters*. He is also an associate editor of pharmaceutical sciences for the *Journal of Mass Spectrometry*. Dr. Korfmacher has edited a book entitled *Using Mass Spectrometry for Drug Metabolism Studies* (CRC Press, 2005) and has written chapters in four additional books. He has more than 125 publications in the scientific literature and has made more than 75 presentations at various scientific forums.

Contributors

Richard M. Caprioli
Department of Biochemistry
Mass Spectrometry Research Center
Vanderbilt University Medical Center
Nashville, Tennessee

and

Department of Biochemistry
Vanderbilt University
Nashville, Tennessee

Swapan Chowdhury
Department of Pharmaceutical Sciences and Drug Metabolism
Schering-Plough Research Institute
Kenilworth, New Jersey

Inhou Chu
Exploratory Drug Metabolism
Schering-Plough Research Institute
Kenilworth, New Jersey

S. Nilgün Çömezoğlu
Department of Biotransformation
Pharmaceutical Candidate Optimization
Bristol-Myers Squibb Company
Princeton, New Jersey

Lan Gao
Department of Pharmaceutical Sciences and Drug Metabolism
Schering-Plough Research Institute
Kenilworth, New Jersey

Gérard Hopfgartner
School of Pharmaceutical Sciences
Life Sciences Mass Spectrometry
University of Geneva and University of Lausanne
Geneva, Switzerland

Yunsheng Hsieh
Exploratory Drug Metabolism
Schering-Plough Research Institute
Kenilworth, New Jersey

W. Griffith Humphreys
Department of Biotransformation
Pharmaceutical Candidate Optimization
Bristol-Myers Squibb Company
Princeton, New Jersey

Steven E. Klohr
Analytical Research & Development
Bristol-Myers Squibb Company
New Brunswick, New Jersey

Walter A. Korfmacher
Exploratory Drug Metabolism
Schering-Plough Research Institute
Kenilworth, New Jersey

Mike S. Lee
Milestone Development Services
Newtown, Pennsylvania

Fangbiao Li
Exploratory Drug Metabolism
Schering-Plough Research Institute
Kenilworth, New Jersey

Hong Mei
Exploratory Drug Metabolism
Schering-Plough Research Institute
Kenilworth, New Jersey

Richard A. Morrison
Exploratory Drug Metabolism
Schering-Plough Research Institute
Kenilworth, New Jersey

Joanna R. Pols
Department of Pharmaceutical Sciences
and Drug Metabolism
Schering-Plough Research Institute
Kenilworth, New Jersey

Ragu Ramanathan
Department of Biotransformation
Pharmaceutical Candidate Optimization
Bristol-Myers Squibb Company
Princeton, New Jersey

Michelle L. Reyzer
Department of Biochemistry
Mass Spectrometry Research Center
Vanderbilt University Medical Center
Nashville, Tennessee

and

Department of Biochemistry
Vanderbilt University
Nashville, Tennessee

Sam Wainhaus
Exploratory Drug Metabolism
Schering-Plough Research Institute
Kenilworth, New Jersey

Jing-Tao Wu
Department of Drug Metabolism
and Pharmacokinetics
Millennium Pharmaceuticals Inc.
Cambridge, Massachusetts

Xiaoying Xu
China Novartis Institutes for
BioMedical Research Co. Ltd.
Shanghai, China

1 Strategies and Techniques for Bioanalytical Assays as Part of New Drug Discovery

Walter A. Korfmacher

CONTENTS

1.1 INTRODUCTION

The task of new drug discovery remains a formidable undertaking. Current estimates of the cost of bringing a new drug to the market are in the range of \$1.2–\$1.5 billion. There is also a significant time commitment—typically it takes 10–14 years to bring a compound from initial discovery to being an approved drug in the market. One of the reasons it takes so long and costs so much is that there is a lot of attrition along the way (most compounds fail). The challenge of new drug discovery is to sort through millions of compounds in a compound library to find a few initial lead compounds and then sift through thousands of new compounds as part of the lead optimization phase with the goal of getting 10–20 compounds that are suitable for development. As shown in Figure 1.1, it takes a total of 14 compounds selected for development in order to reach the goal of 1 compound that becomes a new drug on the market (for more on this topic see Chapter 2). It is this critical lead optimization phase during which there is a continuous need for getting thousands of compounds assayed that the capabilities of a higher throughput bioanalytical scientist are required.

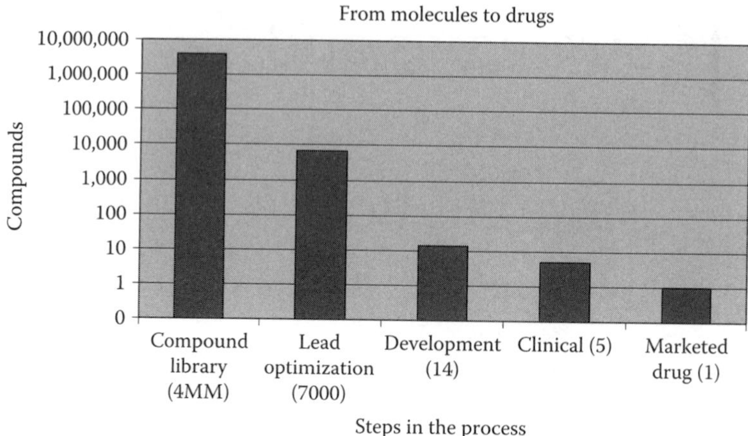

FIGURE 1.1 Schematic chart showing the effect of compound attrition in the drug discovery process from compound libraries to drug approval. The *x*-axis is the stage or point in the process. The *y*-axis is the number of compounds at that point.

The last 20 years have produced enormous changes in how new drug discovery is performed. Before 1990, most major pharmaceutical companies had little need for drug metabolism expertise before a compound was recommended for development. As long as a new compound showed efficacy in some *in vivo* model, it could be a candidate for nomination [1,2]. What changed the landscape was a study that showed that 40% of clinical compounds failed due to human pharmacokinetics [3]. This finding led major pharmaceutical companies to set up exploratory drug metabolism (EDM) departments. The goal of the EDM departments has been to reduce the attrition rate due to pharmacokinetic (PK) issues for new compounds in the clinic [4–6]. As documented by Kola and Landis [7], the desired results have been achieved; currently, less than 10% of new compounds in the clinic fail due to PK reasons. This change in strategy led to the need to develop higher throughput bioanalytical assays in a new drug discovery environment. During the last two decades, another significant change occurred—the high-performance liquid chromatography–tandem mass spectrometer (HPLC–MS/MS) became the instrument of choice for almost all *in vitro* and *in vivo* assays that are routinely performed by bioanalytical scientists in these EDM departments [6,8–22]. This new technology has continued to evolve. As shown in Figure 1.2, in addition to the triple quadrupole mass spectrometer, we now have access to new analytical tools such as the Exactive MS, the LTQ-Orbitrap MS, and the LTQ-FTMS system. The new analytical tools as well as other MS systems have become essential for both quantitative and qualitative assays that are now routinely performed in EDM departments as part of new drug discovery. The reader can find discussions on the utility of these MS systems as well as other types of mass spectrometers in various chapters of this book. This chapter will focus on strategies for quantitative bioanalytical assays to support various *in vivo* PK studies that are now routinely used in new drug discovery.

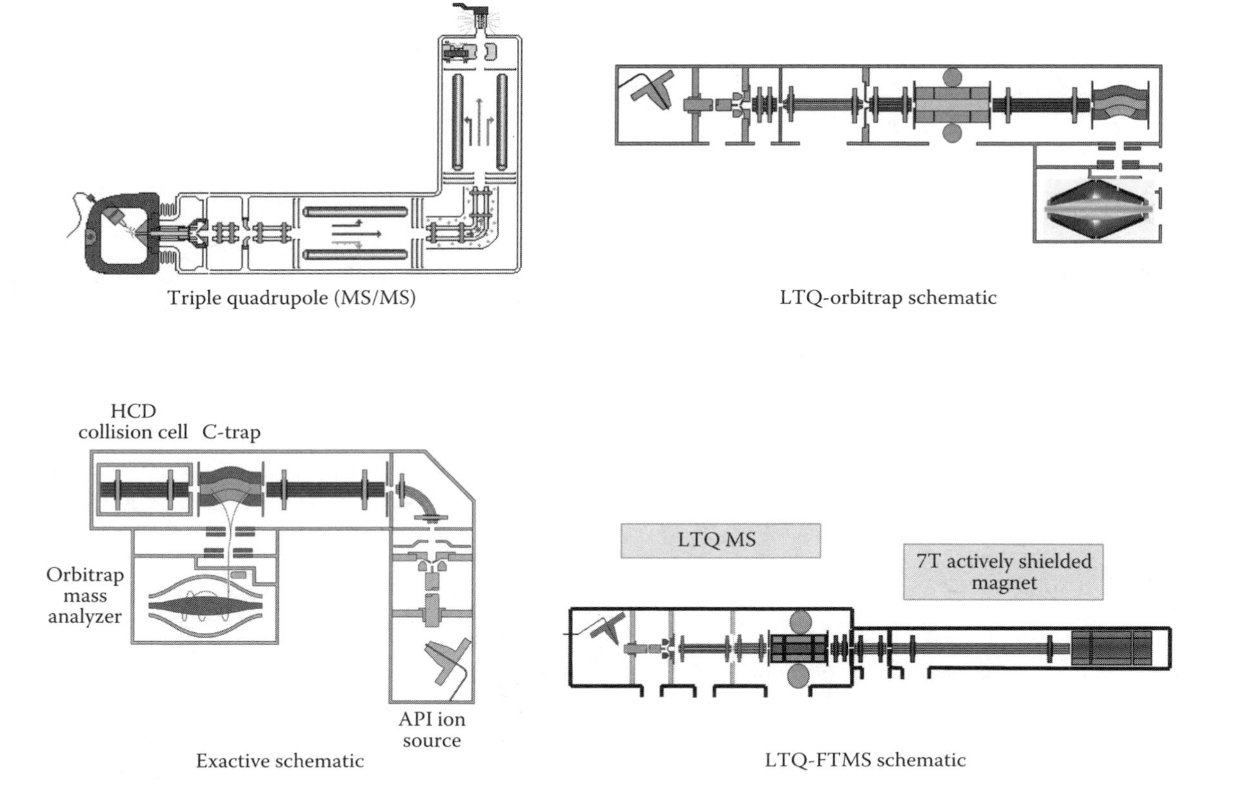

FIGURE 1.2 Four types of mass spectrometers that are used for various drug metabolism assays. (From Pappas, J., Thermo Fisher Scientific Inc., Waltham, MA. With permission.)

1.2 REVIEW OF RECENT LITERATURE

In the past several years, multiple reviews and book chapters have been published on topics related to discovery bioanalysis [10,15,19,20,22–27]. Most of the issues that are cited deal with one of four topic areas: sample preparation, faster assay run times, use of robotics, and matrix effects. Most of the recent papers related to discovery bioanalysis also discuss ways to deal with one or more of these issues.

There are significant challenges in working in discovery bioanalysis. One challenge is that one must develop methods for many new compounds on a daily basis. In the PK screening stage, for example, one analyst might be expected to develop assays for as many as 20 different compounds in one week and then use these assays to measure 5–10 plasma samples for each compound and issue the results to various discovery teams or upload them to a corporate database. The second challenge is that speed is important. The expectation for getting results quickly has become tougher. Five years ago, a two-week turnaround time was acceptable; currently, a one-week turnaround is considered routine. For high-priority studies, discovery teams demand results in a few days or less!

The third challenge is that pharmaceutical companies are developing more potent compounds which means that the dose used in preclinical efficacy studies is lower than 5 years ago which also means that the bioanalytical scientist needs to develop assays with lower limits of quantitations (LOQs) than were needed in the past. This environment is challenging and has forced the bioanalytical scientist to develop generic assays that work most of the time for most of the compounds while looking for ways to get lower LOQs. Fortunately, the MS instrument manufacturers have developed improved MS systems and software tools that have helped the bioanalytical scientists in both of these areas. In addition, new or improved types of chromatography have also helped to make assays based on HPLC–MS/MS systems both faster and more sensitive.

One example of a combined approach for increasing throughput is provided by De Nardi and Bonelli [28]. In their article, they describe reducing the HPLC–MS/MS sample runtime from 5 to 2 min by changing from a 50 × 4.6 mm HPLC column to a 30 × 2.1 mm HPLC column. The HPLC gradient was termed ballistic due to the very short time (1 min) for switching from A (aqueous) to B (organic) mobile phase. The use of these ballistic gradients is now common and is one way to reduce the HPLC–MS/MS assay runtime [22,29–32]. De Nardi and Bonelli [28] also compared a standard acetonitrile protein precipitation procedure (1:3::plasma:acetonitrile) with a protein precipitation procedure that combined acetonitrile and DMSO. In their assay, the standard acetonitrile protein precipitation procedure provided better results. The authors also compared the analysis of a rat PK study using their previous 5 min gradient with an analysis using the 2 min gradient; these two methods were shown to give essentially identical results.

Briem et al. [33] described a combined approach for the higher throughput sample preparation and analysis of plasma samples from discovery preclinical PK studies. Their process combines robotic sample preparation with fast HPLC–MS/MS analysis to speed up the assay plus they add additional steps based on understanding the PK properties of the compound and checking for matrix effects to ensure that the

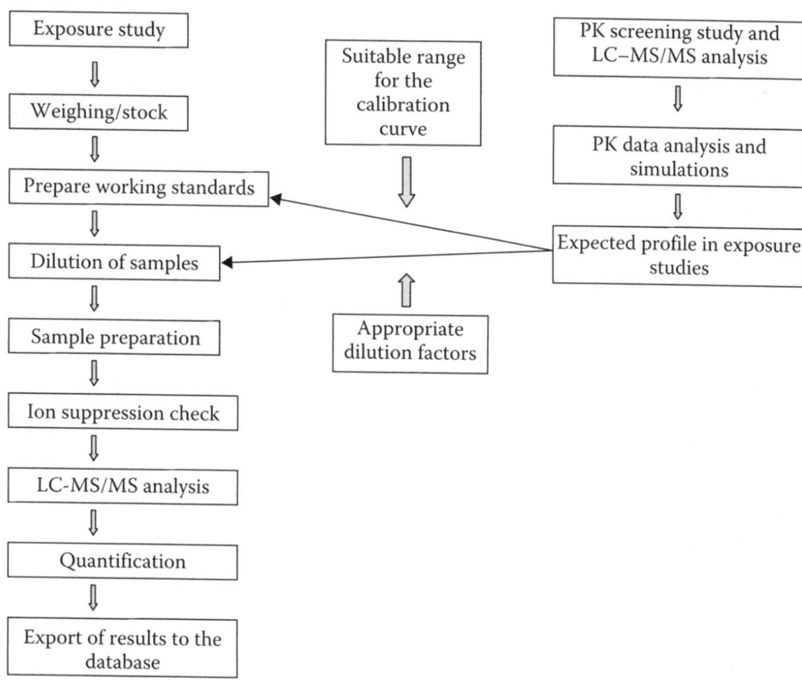

FIGURE 1.3 Workflow for discovery PK studies using the expected PK profile to guide the assay setup. (Adapted from Briem, S. et al., *Rapid Commun. Mass. Spectrom.*, 21, 1965, 2007. With permission.)

data is accurate. Figure 1.3 shows their proposed workflow for the analysis of plasma samples from discovery PK studies. In their scheme, they make use of initial rat PK screening data to calculate various PK parameters for a new chemical entity (NCE) and then these PK parameters are used to predict the likely plasma concentrations of the NCE in subsequent PK studies. The primary value of this procedure was to guide the analyst regarding which study samples should be diluted before the sample analysis was performed, thereby reducing the times that samples would have to be reassayed because they were above the limit of quantitation. This type of advanced planning for a discovery PK study is reasonable and is a good example of how a little extra planning can save time and effort in the long run.

Briem et al. [33] also describe the use of a liquid handler to perform the sample preparation steps in an automated manner. The sample preparation consisted in taking a 25 μL aliquot of the plasma sample and adding 150 μL of acetonitrile (ACN) containing the internal standard (IS) and then following typical protein precipitation (PPT) procedure steps. The authors noted that the 1:6 ratio of plasma to ACN worked well for the liquid handler and exceeded the 1:3 criterion reported by Polson et al. [34] as the minimum ratio of plasma to ACN needed to remove at least 96% of the proteins from the plasma sample. The authors also noted that use of EDTA (instead of heparin) as the anticoagulant helped to avoid clots in plasma as has been reported previously by other authors [35–37]. The authors also stated that the robotic method provided a fourfold improvement over the manual method in terms of the sample preparation time.

Several papers were published on the use of solid phase extraction (SPE) as an alternative procedure for sample preparation of plasma samples from PK studies [38–41]. For example, Mallet et al. [39] described the use of "an ultra-low" elution volume 96-well SPE plate for use in drug discovery applications including plasma assays. The authors described the use of a special low-volume 96-well SPE plate that was designed for working with small volumes of plasma. Each well of the plate was packed with 2 mg of a high-capacity SPE sorbent that could be used for plasma volumes as small as 50 μL and as large as 750 μL. One big advantage of the plate is that the elution step required only 25 μL of solvent, thereby providing a potential concentrating effect. A second advantage is that due to the low volume of the elution step, there is no need for an evaporation and reconstitution as is typically needed in most SPE applications. For drug discovery applications, 50 μL of plasma was diluted 1:1 with the IS aqueous solution and is then loaded onto the preconditioned SPE (either generic reversed-phase or mixed-mode) plate. The 25 μL eluant was captured in a clean 96-well plate and then diluted with 50 μL of water. For comparison purposes, 50 μL of plasma was diluted 1:1 with the IS aqueous solution and is then precipitated with 1 mL of ACN. After vortexing and centrifugation, the supernatant was evaporated to dryness and then the residue was reconstituted with 25 μL of the eluting solvent and then diluted with 50 μL of water. When the authors used propranolol as a test compound for this sample preparation comparison, the results were dramatic. As can be seen in Figure 1.4, while the signal for the sample from the PPT

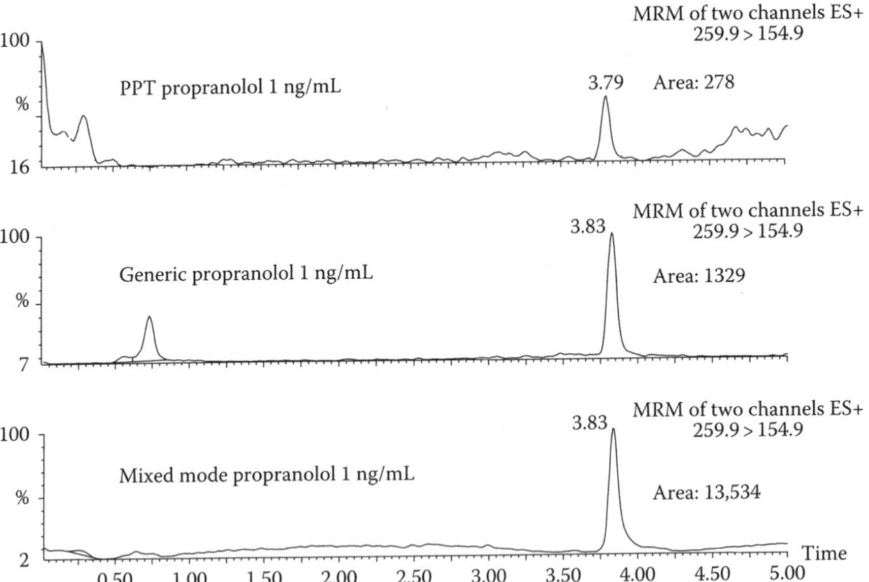

FIGURE 1.4 A comparison of three extraction protocols for the analysis of rat plasma samples for propranolol. (Reprinted from Mallet, C.R. et al., *Rapid Commun. Mass Spectrom.*, 17, 163, 2003. With permission.)

was acceptable, the signal from the generic SPE method was fourfold higher and the mixed-mode SPE method produced an impressive 40× higher signal from the same concentration. The authors also showed that their mixed mode SPE system was suitable for assaying a drug and at least two of its metabolites. Figure 1.5 shows the HPLC–MS/MS analysis of terfenadine and two of its metabolites at a concentration of 0.5 ng/mL in plasma; all three compounds were easy to detect at this concentration using this sample cleanup procedure.

Xu et al. [42] described the use of low-volume plasma sample precipitation procedure that was well suited for drug discovery PK assay applications. In this method, only 10 µL of plasma was used, and it was subjected to PPT using 60 µL of ACN (spiked with the IS) and then assayed using a generic HPLC–MS/MS procedure. In this report, the comparison was made between a "standard" PPT procedure and the proposed low-volume PPT procedure. In the standard-volume (1:3) PPT procedure, a 50 µL plasma sample was precipitated with 150 µL of ACN and the supernatant was transferred to another 96-well plate and a 5 µL aliquot was injected into the HPLC–MS/MS system. In the low-volume (1:6) PPT procedure, a 10 µL plasma sample was precipitated with 60 µL of ACN and the supernatant was transferred to another 96-well plate and a 5 µL aliquot was injected into the HPLC–MS/MS system. In each case the samples were assayed for a discovery test compound using a fast 1.5 min gradient assay. In the first comparison, blank plasma was assayed using either the low- or high-volume procedure. As shown in Figure 1.6, the low-volume procedure produced a significantly lower background signal than did the high-volume procedure. These data show the value of effectively injecting less matrix material into the HPLC–MS/MS system. The authors then showed the results of assaying

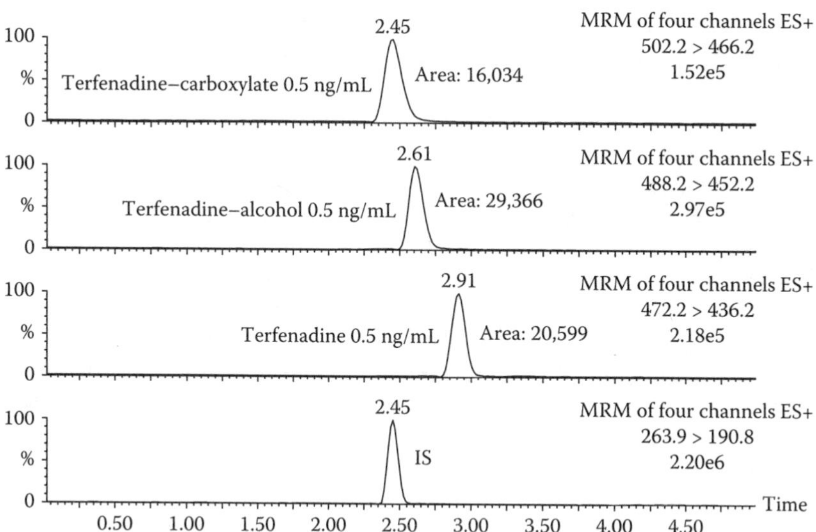

FIGURE 1.5 HPLC–MS/MS assay result for terfenadine and metabolites using a mixed-mode extraction protocol. (Reprinted from Mallet, C.R. et al., *Rapid Commun. Mass Spectrom.*, 17, 163, 2003. With permission.)

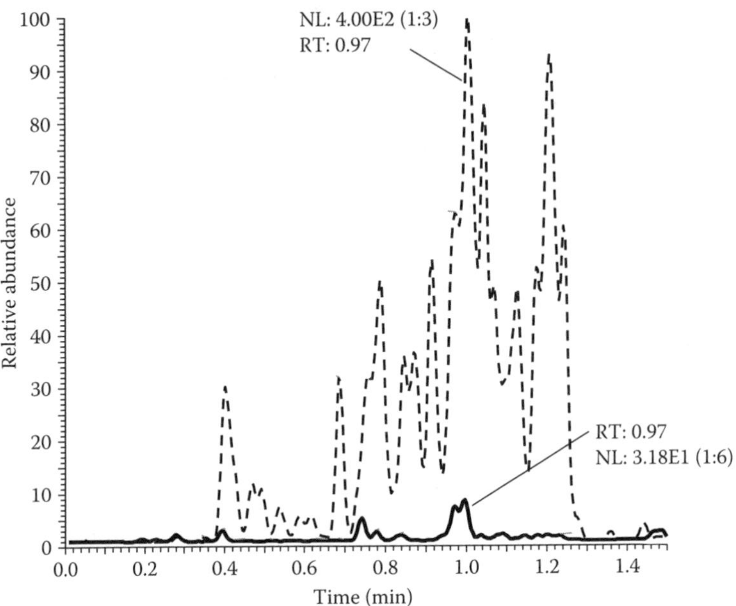

FIGURE 1.6 Normalized mass chromatograms of blank control plasma obtained using PPT ratios of 1:3 (dotted line) or 1:6 (solid line). (Reprinted from Xu, X. et al., *Rapid Commun. Mass Spectrom.*, 19, 2131, 2005. With permission.)

the low standards (0.1 and 0.5 ng/mL) using the two sample preparation methods. The results for the 0.1 ng/mL standard are shown in Figure 1.7 and the results for the 0.5 ng/mL standard are shown in Figure 1.8. The data shown in Figure 1.7 are quite impressive—while the absolute peak height for the 1:3 PPT analyte peak is higher than that for the 1:6 PPT analyte peak (as would be expected), the signal-to-noise ratio (SNR) is clearly much better for the 1:6 PPT sample; indeed, the LOQ for the 1:3 PPT method would be greater than 0.1 ng/mL, while the assay LOQ for the 1:6 PPT method might well be less than 0.1 ng/mL. When the two sample preparation procedures were compared for the 0.5 ng/mL standard (Figure 1.8), both gave acceptable SNR levels for a discovery bioanalytical assay, but the 1:6 PPT sample preparation procedure was clearly the better result. Xu et al. [42] then compared the assay results in a rat PK study and demonstrated that both the standard method and the low-volume method provided essentially the same results. Figure 1.9 shows the results of a concentration time profile for the test compound after it was dosed orally in a rat—it can be seen in this figure that the two sample preparation methods produced essentially identical results. The main advantages of the low-volume method were that smaller sample volumes were needed (this is often important in discovery rodent PK studies) and that (at least in this case) a better SNR was achieved.

Xu et al. [43] published a report on a process for rapid HPLC–MS/MS method development in a drug discovery setting. In this report, the authors provided a three-step process for rapid bioanalytical method development for HPLC–MS/MS methods suitable for drug discovery PK applications. Figure 1.10 shows the flowchart

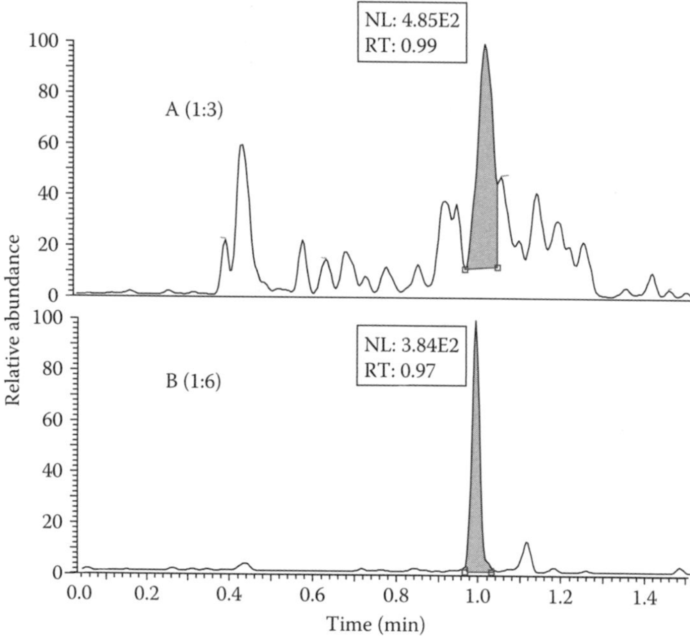

FIGURE 1.7 Normalized mass chromatograms of a test compound spiked into rat plasma (0.1 ng/mL) after using PPT ratios of 1:3 (A) or 1:6 (B). (Reprinted from Xu, X. et al., *Rapid Commun. Mass Spectrom.*, 19, 2131, 2005. With permission.)

that Xu et al. [43] used to describe the rapid method development procedure. The first step is to develop the selected reaction monitoring (SRM) method for the test compound and then to inject the test compound onto the HPLC–MS/MS system and use the generic HPLC conditions. If the generic conditions provide a suitable HPLC peak, then the first checkpoint is to test for matrix effects by using the post-column infusion technique [44–46]. In the postcolumn infusion technique, a solution of the test compound is infused into the HPLC–MS/MS system between the HPLC column and the MS source during the analysis of a sample extract from a blank sample following PPT and the MS is set to monitor the SRM transition for the test compound. In this experiment, typically one will observe a drop in the MS signal during the void volume of the HPLC column, but the signal should return to the level it was before the sample was injected (the dip is caused by matrix effects from the sample extract). The goal is to observe little or no matrix effects during the elution time of the test compound. If a matrix-effects problem is observed, then one needs to go into the "nonroutine assay options box" for a fix. The suggested fixes include revising the chromatography or changing to either liquid–liquid extraction (LLE) or SPE. If a matrix-effects problem is not observed, then one can proceed to the second checkpoint—the interference test. Even though HPLC–MS/MS systems are very selective, it is still possible to have some interference at the LOQ of an assay. Typically, discovery bioanalytical PK assays would have a range of 1–5000 ng/mL; in this case, the LOQ standard would be 1 ng/mL. Therefore, the interference test

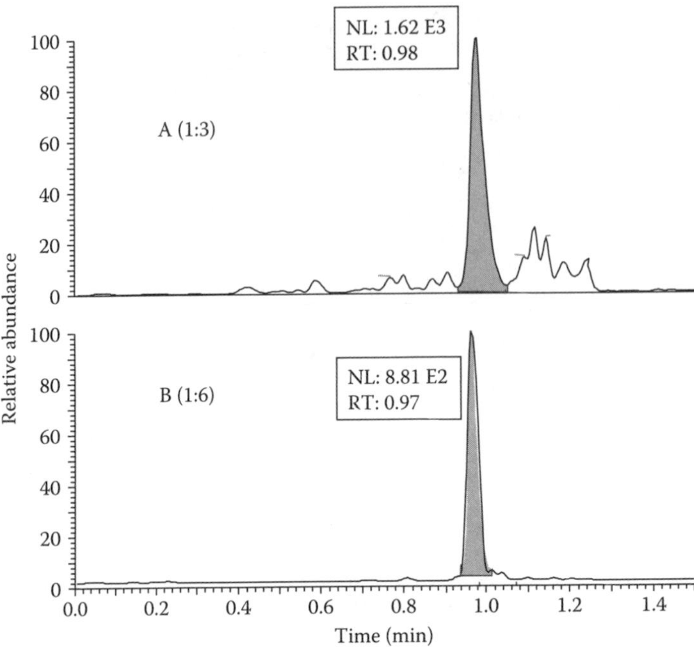

FIGURE 1.8 Normalized mass chromatograms of a test compound spiked into rat plasma (0.5 ng/mL) after using PPT ratios of 1:3 (A) or 1:6 (B). (Reprinted from Xu, X. et al., *Rapid Commun. Mass Spectrom.*, 19, 2131, 2005. With permission.)

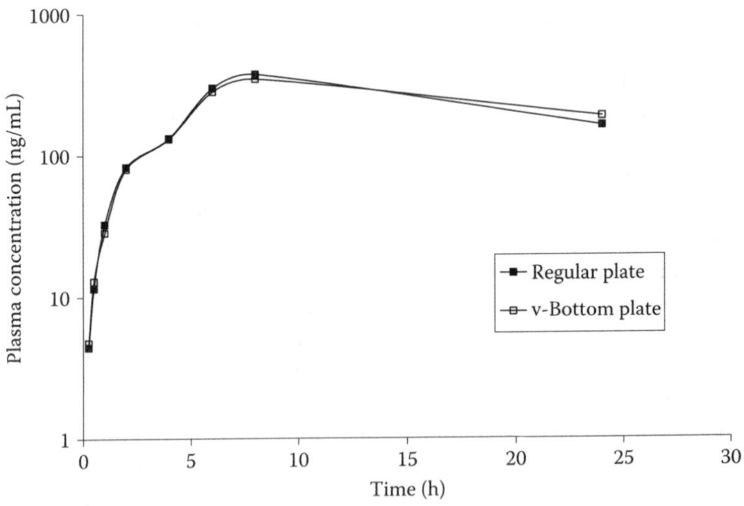

FIGURE 1.9 Representative plasma concentration time profile resulting from a po dose of a test compound into a rat and comparing the standard method results with the low-volume method results. (Reprinted from Xu, X. et al., *Rapid Commun. Mass Spectrom.*, 19, 2131, 2005. With permission.)

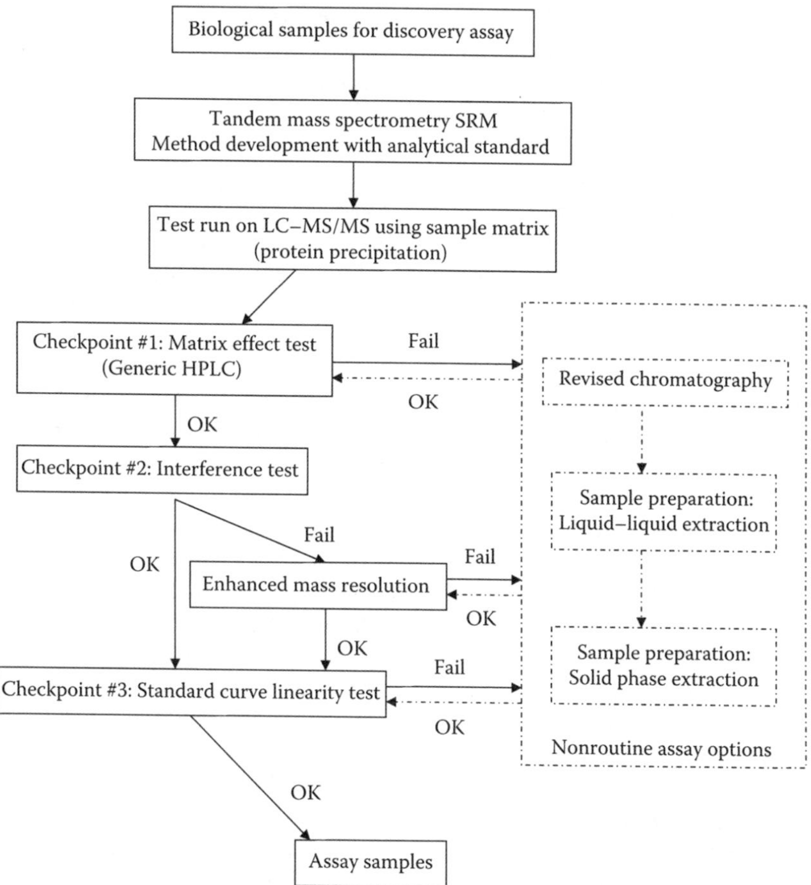

FIGURE 1.10 Fast method development paradigm for use in drug discovery applications. (Reprinted from Xu, X. et al., *Anal. Chem.*, 77, 389 A, 2005. With permission.)

would be to extract and inject the LOQ standard as well as a zero standard to check for any interference with the assay. If an interference is seen, two solution paths are suggested. The faster path (if available) is to increase the mass resolution on the MS/MS system. In a typical triple quadrupole system, both Q1 and Q3 are set to "unit mass resolution," which is normally defined as 0.7 Da full width half maximum (FWHM) [23]. In most triple quadrupole systems, if one increases the mass resolution setting, then one would observe a significant loss in sensitivity. For some triple quadrupole systems, one can increase the mass resolution from 0.7 to 0.2 Da FWHM with only a minor loss in sensitivity [47–49]. Thus, if one has the option, then the faster possible solution is to use higher mass resolution; Xu et al. [43] recommended the use of 0.2 Da FWHM for Q1 and 0.7 Da FWHM for Q3. If this path does not work (or is not available), then one should go to the "nonroutine assay options box" for a fix. The third checkpoint is the standard curve linearity test. The test is to simply extract and inject the standard curve in order to check that it is linear. While standard curves over the range of 1–5000 ng/mL may not always be linear, they should be able to be fit by using weighted linear regression with a power curve fit or a quadratic fit [10,43,50]. If the standard curve does not meet the assay acceptance criteria, then one is again referred to the "nonroutine assay options box" for a fix. Xu et al. [43] end the article with a case study of an unusual standard curve issue and how it was solved by methodical experimentation.

For HPLC–MS/MS assays, the mobile phase is an important consideration. For reversed-phase HPLC systems (the most common), mobile phase A is water plus one or more modifiers while mobile phase B is usually either acetonitrile or methanol with one or more modifiers. Modifiers have to be volatile for HPLC–MS/MS assays. Typical modifiers are acetic acid, formic acid, and ammonium acetate. Formic acid is so popular that it is now available as a premixed HPLC solvent (0.1% in water or acetonitrile). Most other modifiers are not recommended. Triethylamine (TEA) and trifluoroacetic acid (TFA) are a problem as they are known to cause ion-suppression problems. For some special HPLC–MS/MS assays that need ion-pairing reagents, hexylamine and heptafluorobutyric acid have been found to be successful [51–53]. Gao et al. [54] recently evaluated a series of ion-pairing reagents in terms of their suitability for an HPLC–MS/MS assay.

Bakhtiar and Majumdar [55] published a well-written article on possible problems and solutions when assaying small molecule drugs in biological fluids using HPLC–MS/MS. One of their topics is the importance of mobile phase modifiers. For example, they state that the pH of the mobile phase has been reported to affect the sensitivity of the method. A recent example of the importance of mobile phase pH can be found in Delatour and Leclercq [56] and the importance of mobile phase additives can be seen in the report by Gao et al. [57]. Bakhatiar and Majumdar [55] have a long section of their report devoted to matrix effects (also referred to as ion suppression). Matrix effects can simply be described as any significant reduction or enhancement of an analyte's signal intensity due to some component in the sample matrix [58]. In one example, Bakhatiar and Majumdar [55] show how matrix effects can lead to nonreproducible results for standard curves. The plot shown in Figure 1.11 is the results from a mouse plasma assay where the assay followed the common practice of front and back standard curves with the assay samples injected in between the two standard curves (this is a good practice as it is one way to ensure that the samples were assayed properly). In this example, it was clear that there was a problem in the assay because the two standard curves were divergent (they should, of course, coplot). The authors attributed the problem to matrix effects from the mouse plasma samples and solved the problem by switching from PPT to LLE as a method for sample preparation.

Phospholipids in plasma have been shown to cause significant matrix effects [59–61]. The authors suggested that an improved sample cleanup should try to remove phospholipids and noted that Shen et al. [62] have described a strong cation exchange SPE cleanup that was effective in removing some of the endogenous phospholipids from plasma. Wu et al. [63] described the utility of colloidal silica polyanions in combination with divalent and trivalent cations for the removal of phospholipids from plasma. It would be highly desirable to have a simple generic cleanup step that would remove phospholipids from plasma while not having any effect on the analytes

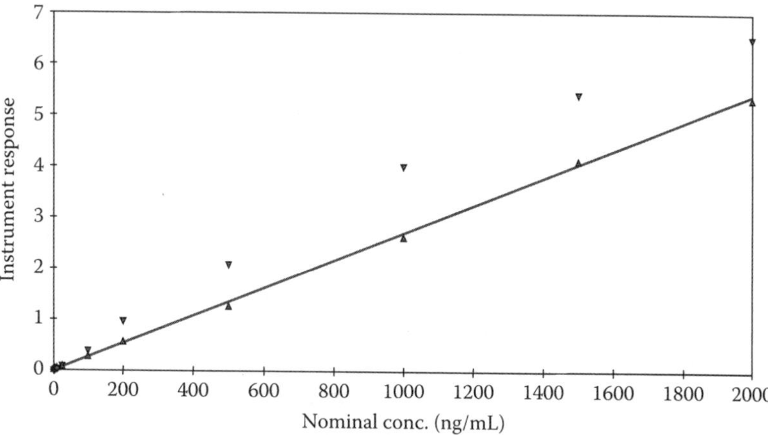

FIGURE 1.11 Plot of two separate standard curves obtained using PPT of mouse plasma samples spike with a test compound and using a structural analog as the IS. One set of standards was injected before the injection of mouse PK plasma samples while the second standard curve set was injected after the mouse PK samples. The assay was based on HPLC–ESI–MS/MS. (Reprinted from Bakhtiar, R. and Majumdar, T.K., *J. Pharmacol. Toxicol. Methods*, 55, 262, 2007. With permission.)

of interest in the plasma sample. The authors tried various cleanup procedures and settled on one that worked well for a 100 μL aliquot of plasma. The process removed about 90% of the targeted phospholipids, while having a minimal effect on the recovery of the eight model test compounds.

Marchi et al. [64] reported on the utility of various sample cleanup procedures for reducing the matrix effects that are caused by various plasma constituents. They found that the best sample preparation procedure was to use PPT followed by an online SPE system. The authors also stated that with this sample preparation procedure, atmospheric pressure photoionization (APPI) was the least affected by matrix effects, followed by atmospheric pressure chemical ionization (APCI) and then electrospray ionization (ESI).

Bakhtiar and Majumdar [55] have a very long section on the potential issue of the possible interference that can occur from metabolites of the dosed compound in a PK study. This is an important issue that is sometimes overlooked in a discovery bioanalytical setting. The potential for getting false positive signals from a metabolite has increased as scientists concentrate on shorter HPLC–MS/MS sample runtimes (a.k.a. cycle times). In recent years, a few reports of metabolite interference with HPLC–MS/MS assays have been published [65–67]. Conjugated metabolites (including acyl glucuronides, sulfate conjugates, as well as *O*-glucuronides and *N*-glucuronides) can potentially lead to metabolite interferences [68]. For example, acyl glucuronides of a test compound are an issue because they are known to be highly susceptible to in-source fragmentation back to the test compound; therefore if these metabolites are not separated from the test compound, they can add to the signal for the test compound (false positive) [65,67,69–71].

Another issue covered by Bakhtiar and Majumdar [55] is the problem of analyte carryover. They cite multiple references that discuss various carryover issues as a potential problem for HPLC–MS/MS assays [29,72,73]. The authors note that carryover can be traced to various steps in the process, but is often due to either the liquid handler (sample preparation step) or the autoinjector. They then list 10 specific suggestions for reducing carryover. While carryover can still be a major issue in some cases, the newer autosamplers are generally very good at keeping carryover to an acceptable level as long as appropriate wash solutions are utilized. Chang et al. [74] have proposed a fairly sophisticated process for monitoring sample preparation carryover (a.k.a. cross-contamination of wells). In their report, they spike additional analytes into every other well and then monitor cross-contamination of each well by assaying these additional analytes along with the analyte of interest. This technique might be useful if one is looking for a way to test a sample handling process to ensure that it does not lead to a significant level of cross-contamination.

Zeng et al. [73] provided a mathematical evaluation of carryover in HPLC–MS/MS assays. Their report suggested that the carryover effect should be viewed as a constant percentage of the previous sample and the significance (if any) depends on the concentration of the following sample. Therefore, the carryover effect is relative and is highly dependent on the concentration ratio of the previous sample to the following sample. This report supports the common practice of placing standard curve samples in order (low to high) and then following the high standard with one or two blank samples before assaying the study samples in a logical sequence in order to minimize the effect of carryover on the results.

Another potential source of matrix effects in discovery bioanalytical assays is the dosing formulation that is used in the PK study [75–77]. Xu et al. [75] reported on an extensive study of the potential for various commonly used dosing formulation components to cause matrix effects. Xu et al. [75] found that methyl cellulose (MC) and hydroxypropyl-β-cyclodextrin (HPBCD) did not lead to significant matrix effects when used in oral (po) or intravenous (iv) dosing formulations. On the other hand, both PEG 400 and Tween 80 caused significant time-dependent matrix effects when used as part of po or iv dosing formulations. In addition, these matrix effects were not limited to ESI assays, but were also observed in some APCI assays. An example of the time-dependent nature of the matrix effects can be seen by looking at the data shown in Figure 1.12 [75].

Leverence et al. [78] describe the potential for signal enhancement matrix effects from various over-the-counter drugs that might be taken along with the test compound (in clinical studies). In this report they demonstrated the possibility of two- to threefold signal enhancement from either ibuprofen or naproxen. Their solution to the issue was to improve the chromatography for the assay.

While stable isotope-labeled ISs are generally considered "safe" in terms of matrix effects, Wang et al. [79] reported on an assay where matrix effects were an issue despite the use of a deuterium-labeled IS. In this case the reason that was given for the problem was that the chromatography was too good—there was some separation between the analyte and its IS and this led to the matrix-effect problem.

Chambers et al. [80] describe a systematic and comprehensive strategy for reducing the potential for matrix effects in HPLC–MS/MS assays. They studied various

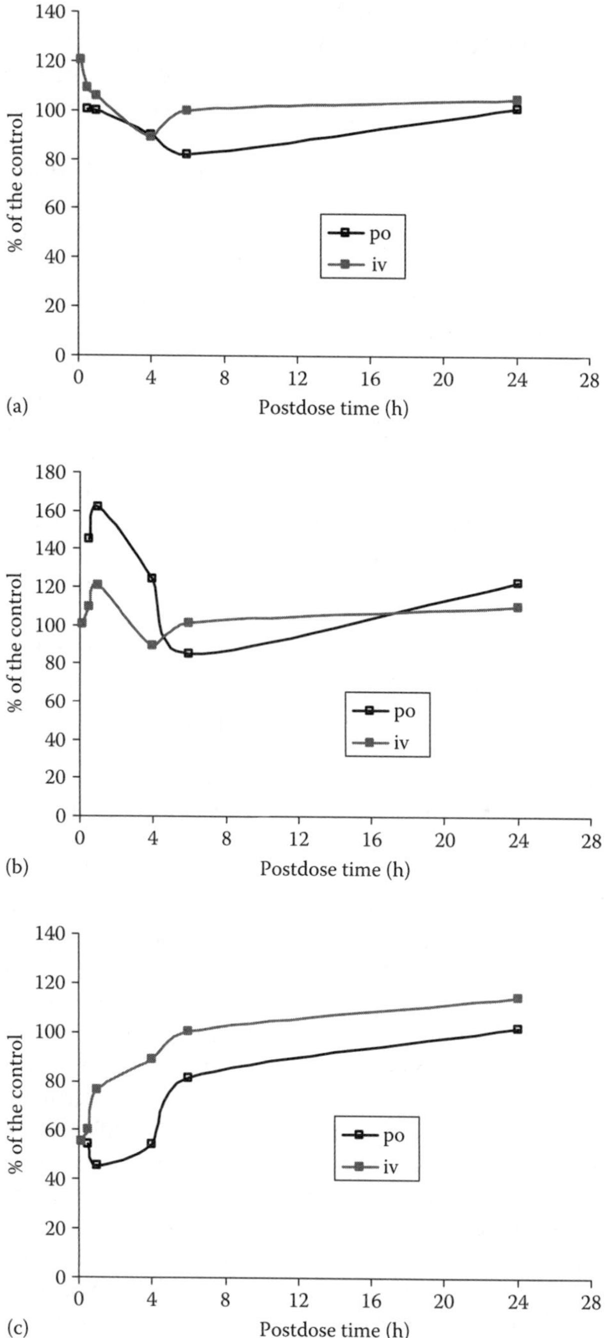

FIGURE 1.12 Propranolol time-dependent MS response after 0.1% Tween 80 was used as in the formulation in iv and po dose routes. HPLC–ESI–MS/MS results are shown for (a) Thermo-Finnigan, (b) ABI Sciex, and (c) waters micromass. (Reprinted from Xu, X. et al., *Rapid Commun. Mass Spectrom.*, 19, 2643, 2005. With permission.)

sample preparation techniques and found that PPT was less effective than either SPE or LLE. They also reported that mobile phase pH could be used to separate analytes from matrix effects in some cases. They also reported that ultra-performance liquid chromatography–tandem mass spectrometry (UPLC–MS/MS) was better than HPLC–MS/MS in dealing with matrix effects.

Wang et al. [81] discussed the utility of a generic HPLC–ESI–MS/MS assay along with a simple PPT-based sample preparation for supporting discovery PK studies. In this work, the plasma sample volumes were set to 20 μL. The authors reported that good linearity was achieved over a range of 2–25,000 ng/mL in favorable cases. Wang et al. [81] also stated that keeping the amount of formic acid at a low concentration (0.015%) in their water/methanol mobile phases helped to reduce the observed matrix effects.

In a review article by Xu et al. [17], the authors describe the utility of UPLC–MS/MS for increasing the throughput of discovery bioanalytical assays when compared to HPLC–MS/MS. They reported that UPLC–MS/MS showed sensitivity increases of up to tenfold over HPLC–MS/MS while providing for shorter runtimes (up to fivefold faster). Another advantage of UPLC–MS/MS is that it provides for better chromatographic resolution than what can be obtained using typical HPLC–MS/MS systems [82–87]. For these reasons, UPLC–MS/MS has been well received by the bioanalytical community and has already been utilized for multiple pharmaceutical applications [53,88–95]. For example, Yu et al. [91] described the utility of UPLC–MS/MS for the rapid analysis of various model compounds. As Yu et al. [91] noted, it is important to have a fast scanning mass spectrometer when using UPLC due to the very sharp chromatographic peaks that are obtained. Typically the current models of triple quadrupole MS systems are able to scan fast enough to work with UPLC separations. As shown in Figure 1.13, UPLC was able to separate four test drugs in less than 30 s and the MS system was able to collect approximately 15 data points per peak [91]. For more on UPLC, see Chapter 8.

Xu et al. [17] also discussed the potential for high-field asymmetric waveform ion mobility spectrometry (FAIMS) to provide additional specificity to HPLC–MS/MS assays. FAIMS is added to an HPLC–MS/MS system by inserting it in between the atmospheric pressure ionization (API) source and the entrance to the MS system. The transmission of ions through the FAIMS device is controlled (in part) by the compensation voltage that is applied. The compensation voltage can be adjusted such that the mobility of ions through the FAIMS can be very selective. The selectivity is based on the shape of the molecules rather than their molecular weight. In this way, compounds that are isobaric may be separated by FAIMS. Thus FAIMS has the potential to reduce interferences from endogenous compounds that are isobaric with the analyte of interest. In one example of the potential utility of FAIMS, Kapron et al. [96] described how FAIMS was used to remove an interference of an *N*-oxide metabolite from the assay for the dosed compound. Recently, Wu et al. [97] reported on a study of mobile phase composition effects when dealing with FAIMS and concluded that FAIMS was not affected by changes in typical mobile phase components. At this time, FAIMS is still a potentially useful tool for adding to HPLC–MS/MS and may prove useful in certain assays.

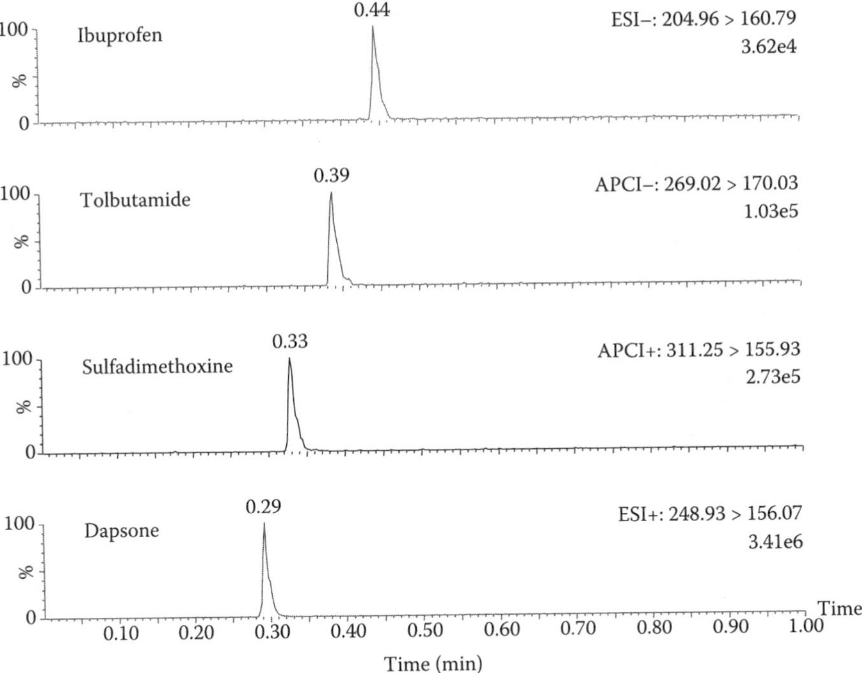

FIGURE 1.13 SRM chromatograms obtained from a four compound mixture analysis based on a single injection assay. In this case, the mass spectrometer was switching between the four ionization modes (+/– ESI and APCI) rapidly. Approximately 15 data points were obtained for each peak. (Reprinted from Yu, K. et al., *Rapid Commun. Mass Spectrom.*, 21, 893, 2007. With permission.)

Finally, rapid rodent oral PK screening approaches have continued to generate publications. While cassette dosing seems to have fallen out of favor [98], there are still some laboratories that follow this practice for *in vivo* PK screening [99–102]. An alternative approach for *in vivo* oral PK screening is the cassette-accelerated rapid rat screen (CARRS) described by Korfmacher et al. [103]. In the CARRS approach, two rats are dosed with one test compound and the test compounds are organized into cassettes of six compounds in order to provide a systematic approach for the oral PK screen. To date, the CARRS approach has been used to evaluate over 15,000 compounds. Mei et al. [104] described a retrospective study of the CARRS assay in which the results for 100 compounds that were assayed by both CARRS and the conventional "full PK" approach were compared. The results suggested that the CARRS assay was a reliable rat oral PK screen that could be used to filter out compounds with poor rat PK in an efficient manner. Liu et al. [105] reported on the utility of "snapshot PK" as an oral rodent PK screen. The "snapshot PK" approach is very similar to the CARRS approach, with the primary difference that it is scaled down (in terms of compound needed) because it is designed as a mouse oral PK screen. Han et al. [106] described the utility of "rapid rat" oral PK screening by using an approach based on the CARRS-paradigm. Han et al. [106] noted that sample pooling across compounds could also be used (with some exceptions) as a way to further reduce the number of samples that need to be assayed by HPLC–MS/MS.

1.3 CURRENT PRACTICES

As shown in Figure 1.14, the flowchart for discovery PK assays is still very much the same as it was 5 years ago, what has changed in that time is some of the technology that is used for the various steps. Currently, PPT is still the preferred way to do sample preparation. PPT has the advantage that it is easy to implement, it is fast (it can also be readily automated using various robotic devices), and it requires no method development. The major change in this area is the trend toward lower plasma volumes for sample analysis. Five years ago, it was common to use 40–50 μL of plasma for the assay; currently, 10–20 μL plasma samples are becoming routine. In addition,

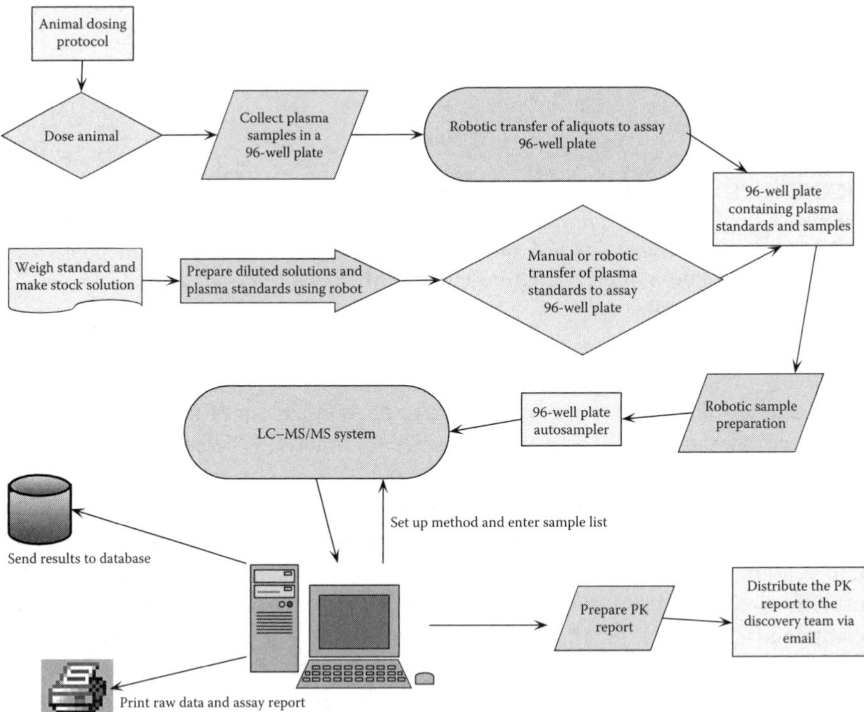

FIGURE 1.14 Discovery PK analysis flowchart showing the multiple steps that are involved from the animal dosing to the HPLC–MS/MS assay followed by the report preparation and the electronic delivery of the report to the discovery team. (Reprinted from Korfmacher, W., Bioanalytical assays in a drug discovery environment, in *Using Mass Spectrometry for Drug Metabolism Studies*, Korfmacher, W. (ed.), CRC Press, Boca Raton, FL, 2005, p. 1. With permission.)

the PPT ratio has been changing; in the past it was usually 1:3 (plasma:acetonitrile), while current practice is to use 1:4 or 1:6 as standard PPT ratios. This change has been driven by several factors: (1) the increased sensitivity of the triple quadrupole mass spectrometers means that a more diluted sample can be assayed and still have the analyte be detected; (2) by injecting a more diluted sample, one can reduce the likelihood of having matrix-effects issues; and (3) by injecting a more diluted sample, one may have a better LOQ due to a reduction in background interferences as described by Xu et al. [42].

A second major change in the last 5 years is the implementation of various fast HPLC technologies that have allowed most discovery PK assays to be routinely performed with 1–2 min cycle times [28–30,94,107]. While this change has been due to multiple advances in chromatography, the biggest change came with the introduction of UPLC, which was rapidly accepted by the bioanalytical community and has led to the widespread use of UPLC–MS/MS. This one change has allowed analysts to operate differently in some cases. With 5 min cycle times, a typical discovery PK study with 90 samples (test samples plus standards) takes 450 min or about 8 h to complete; this amount of time is a good match for an overnight assay. With a 1.5 min cycle time (injection to injection), the same 90 samples can be assayed in less than 2.5 h—so this could be performed during an 8 h workday. This leads to the possibility of shorter turnaround times plus it allows the analyst to see the results of an assay before leaving work for that day.

The disadvantage of fast chromatography combined with PPT is that it has a greater potential for matrix effects than would an approach based on more extensive sample cleanup or a longer chromatographic runtime. While various ways to detect matrix effects have been reported, Matuszewski et al. [108] have provided a clear way to measure the matrix effect (ME) for an analyte in an assay as well as the recovery (RE) from the extraction procedure plus the overall process efficiency (PE) for the procedure. Their method is to prepare three sets of samples and then to assay them using the planned HPLC–MS/MS method. The first set is the neat solution standards diluted into the mobile phase before injection—this provides the A results. The second set is the analyte spiked into the blank plasma extract (after the extraction step)—this provides the B results. The third set is the analyte spiked into the blank plasma before the extraction step; these are then extracted and assayed along with the two other sets; set three data are the C results. Using these three sets of data, then allows for the following calculations:

$$\mathrm{ME}(\%) = \mathrm{B}/\mathrm{A} \times 100$$

$$\mathrm{RE}(\%) = \mathrm{C}/\mathrm{B} \times 100$$

$$\mathrm{PE}(\%) = \mathrm{C}/\mathrm{A} \times 100 = (\mathrm{ME} \times \mathrm{RE})/100$$

The utility of this procedure is that it allows one to identify the source of the problem if an assay shows poor process efficiency. The main disadvantage is that it is a lot of effort, which can be justified for a validated assay, but would not likely be used for most discovery PK assays except as a tool for understanding the source of a problem when an assay is not working properly.

One of the simplest ways to deal with most matrix effects is dilution. If the samples are diluted 1:10 with the same matrix that is used for the standard curve preparation, then if there are any matrix effects, they should be the same for the samples and the standards. This "dilution solution" is a very good fix for matrix effects that are caused by certain dosing formulation constituents [58,75,109], a problem that has become more common as medicinal chemists look for ways to solubilize some of their highly insoluble test compounds by using formulation additives like PEG 400 or Tween 80. The reason the "dilution solution" can be utilized is that the newest generation of triple quadrupole mass spectrometers have significantly better sensitivity than the triple quadrupole mass spectrometers that were available 5 years ago. This increased sensitivity has also led to a change in the standard injection volume for a typical HPLC–MS/MS assay. In the past, it was common to inject 25–30 μL of the sample extract onto the HPLC–MS/MS system, currently 5–10 μL would be the standard injection volume. With the current generation of autosamplers, it is now possible to inject sample volumes as low as 1 μL in an HPLC–MS/MS assay. The advantage of the lower injection volume is that one is injecting less matrix than with the larger sample volumes. Less matrix means less chance of matrix effects. Again, the reason that this can be done is that the current generation of triple quadrupole mass spectrometers are all very sensitive so that less analyte is needed for quantitation.

While a typical standard curve for a discovery PK study would still be in the 1–10,000 ng/mL range, it is becoming more likely that one would need an assay in the 0.1–1000 ng/mL range. The reason is that test compounds are becoming more potent so that lower doses are being given (in some cases) to test animals in various efficacy assays. As stated above, this has not been an issue for the bioanalytical community because the triple quadrupole mass spectrometers have become more sensitive.

Carryover is still an issue in some assays. While the new generation of autosamplers is generally better in this regard than their predecessors, carryover is still a problem for many compounds. In our laboratory, we use the relative carryover method to assess carryover. In this method, one can assess the level of carryover by having a blank sample (blank plasma or solvent blank) injected after the high standard. The relative analyte peak areas of the high standard versus the blank provide the relative carryover assessment. When this assessment ratio is expressed as a percentage, we would use 0.1% as the normal cutoff for carryover. The logic is simple. As long as one is careful to arrange the standards from low to high and then follow the standards with at least two blank samples and then arrange the samples in order of sample collection for each animal with a blank in between each animal, then carryover should not cause any significant error in the sample analysis. This concept of relative carryover was described recently by Zeng et al. [73]. The important concept is that what matters is the relative concentrations of adjacent (in injection time) samples. This is best understood by an example. If a sample has an analyte concentration of 1000 ng/mL and the amount of carryover is 0.1%, then the carryover to the next sample (in this example) would be 1 ng/mL. So, if the

1000 ng/mL is followed by a sample that had an analyte concentration of 100 ng/mL (tenfold change in concentration), then the carryover would add 1 ng/mL, so this sample would be measured at 101 ng/mL—this would amount to a 1% error. Even if the 1000 ng/mL sample was followed by a 10 ng/mL sample, it would only lead to a 10% error (still acceptable in a discovery PK assay). In most PK studies, the relative change from one timepoint to the next is less than tenfold, so a 0.1% carryover is acceptable for these assays. If the carryover is more than 0.1% (up to 0.5%), it is still possible to perform the assay, but one would have to evaluate the samples more carefully to ensure that the carryover did not have a significant effect on the sample results.

The major vendors for triple quadrupole mass spectrometers all provide software tools that can be used for getting the precursor to product ion transitions as well as collision cell voltages that are optimized for the test compound. In most cases this can be performed in a batch mode where the standards are submitted in a 96-well plate format for unattended MS/MS method development. This is important. In rapid *in vivo* PK assays, it would not be unusual for an analyst to have to develop MS/MS methods for 20–30 new compounds each week. Ideally, the transition and MS optimization data would be saved in a database and be available in case the compounds were assayed again at some future date. In addition to MS method development, the MS systems all have software that is well suited to automated peak area measurements and quantitative analysis using linear or nonlinear (power curve or quadratic) regression analysis including various weighting regimes (typical weighting is $1/x$ or $1/x^2$). In some cases, this is all that is needed. When PK parameters are needed, then one can use WATSON (Thermo-Finnigan) LIMS for both the regression analysis and the calculation of standard PK parameters. In either case, the actual data report is generally delivered electronically—either as an email attachment or as a data report that has been made available by uploading it to a corporate database.

An important part of discovery bioanalysis is the ability to distinguish the assay requirements for an early screen versus a late-stage compound that is in the final PK studies that will determine if the NCE can be recommended for development. As shown in Table 1.1, we have divided the various stages of discovery and development into four levels in terms of assay requirements. Level I is for early screening

TABLE 1.1
Rules for Discovery (Non-GLP) Assays (Level I)

Drug Stage	Assay Type	Summary of Major Rules	GLP?
Screening	Level I	Use a two-point standard curve	No
Lead optimization	Level II	Use a multipoint standard curve but no QCs	No
Lead qualification	Level III	Use a multipoint standard curve plus QCs	No
Development	Level IV	GLP rules	Yes

Source: Adapted from Korfmacher, W., Bioanalytical assays in a drug discovery environment, in *Using Mass Spectrometry for Drug Metabolism Studies*, Korfmacher, W. (ed.), CRC Press, Boca Raton, FL, 2005, p. 1. With permission.

assays, such as CARRS or early efficacy screening studies. Level II is for the lead optimization phase when one is performing initial "full PK" studies on various compounds in order to pick the new lead compound. Level III is for the lead qualification phase where one is performing single rising dose studies or multiple dose studies on the lead compound in order to ensure that it can be recommended for development. Level IV is for studies that are carried out using good laboratory practices (GLP).

These rules have been described in detail previously by Korfmacher [10] and are shown in this chapter in three tables. Table 1.2 shows the Level I rules in detail [10].

TABLE 1.2
Rules for Discovery (Non-GLP) "Screen" Assays (Level I)

1. Samples should be assayed using HPLC–MS/MS technology.
2. Sample preparation should consist of PPT using an appropriate IS.
3. Samples should be assayed along with a standard curve in duplicate (at the beginning and end of the sample set).
4. The zero standard is prepared and assayed, but is not included in the calibration curve regression.
5. Standard curve stock solutions are prepared after correcting the standard for the salt factor.
6. The standard curve should be in three levels, typically ranging from 25 to 2500 ng/mL (they can be lower or higher as needed for the program); each standard is 10× the one below (thus, a typical set would be 25, 250, and 2500 ng/mL). The matrix of the calibration curve should be from the same animal species and matrix type as the samples.
7. QC samples are not used and the assay is not validated.
8. After the assay, the proper standard curve range for the samples is selected; this must include only two concentrations in the range that covers the samples. A one order of magnitude range is preferred, but two orders of magnitude is acceptable, if needed to cover the samples.
9. Once the range is selected, at least three of the four assayed standards in the range must be included in the regression analysis. Regression is performed using unweighted linear regression (not forced through zero).
10. All standards included in the regression set must be back calculated to within 27.5% of their nominal values.
11. The LOQ may be set as either the lowest standard in the selected range or 0.4 times the lowest standard in the selected range, but the LOQ must be greater than three times the mean value for the back-calculated value of the two zero (0) standards.
12. Samples below the LOQ are reported as zero (0).
13. If the LOQ is 0.4 times the lowest standard in the selected range, then samples with back-calculated values between the LOQ and the lowest standard in the selected range may be reported as their calculated value, provided the SNR for the analyte is at least three (3).
14. Samples with back-calculated values between 1.0× and 2.0× the highest standard in the selected range are reportable by extending the calibration line up to 2× the high standard.
15. Samples found to have analyte concentrations more than 2× the highest standard in the regression set are not reportable; these samples must be reassayed after dilution or along with a standard curve that has higher concentrations so that the sample is within 2× the highest standard.

Source: Adapted from Korfmacher, W., Bioanalytical assays in a drug discovery environment, in *Using Mass Spectrometry for Drug Metabolism Studies*, Korfmacher, W. (ed.), CRC Press, Boca Raton, FL, 2005, p. 1. With permission.

Table 1.3 shows the Level II rules and Table 1.4 shows the Level III rules [10]. These rules have been used successfully for thousands of discovery compounds and have been found to ensure that the reported results are scientifically correct while minimizing the amount of time that is required to develop the assay and report the

TABLE 1.3
Rules for Discovery (Non-GLP) "Full PK" Assays (Level II)

1. Samples should be assayed using HPLC–MS/MS technology.
2. Sample preparation should consist of PPT using an appropriate IS.
3. Samples should be assayed along with a standard curve in duplicate (at the beginning and end of the sample set).
4. The zero standard is prepared and assayed, but is not included in the calibration curve regression.
5. Standard curve stock solutions are prepared after correcting the standard for the salt factor.
6. The standard curve should be 10–15 levels, typically ranging from 1 to 5,000 or 10,000 ng/mL (or higher as needed). The matrix of the calibration curve should be from the same animal species and matrix type as the samples.
7. QC samples are not used.
8. After the assay, the proper standard curve range for the samples is selected; this must include at least five (consecutive) concentrations.
9. Once the range is selected, at least 75% of the assayed standards in the range must be included in the regression analysis.
10. Regression can be performed using weighted or unweighted linear or smooth curve fitting (e.g., power curve or quadratic), but is not forced through zero.
11. All standards included in the regression set must be back calculated to within 27.5% of their nominal values.
12. The regression r^2 must be 0.94 or larger.
13. The LOQ may be set as either the lowest standard in the selected range or 0.4 times the lowest standard in the selected range, but the LOQ must be greater than three times the mean value for the back-calculated value of the two zero (0) standards.
14. Samples below the LOQ are reported as zero (0).
15. If the LOQ is 0.4 times the lowest standard in the selected range, then samples with back-calculated values between the LOQ and the lowest standard in the selected range may be reported as their calculated value provided the SNR for the analyte is at least three (3).
16. Samples with back-calculated values between 1.0× and 2.0× the highest standard in the selected range are reportable by extending the calibration curve up to 2× the high standard as long as the calibration curve regression was not performed using quadratic regression.
17. Samples found to have analyte concentrations more than 2× the highest standard in the regression set are not reportable; these samples must be reassayed after dilution or along with a standard curve that has higher concentrations so that the sample is within 2× the highest standard.
18. The assay is not validated.
19. The final data does not need to have quality assurance (QA) approval. This is an exploratory, non-GLP study.

Source: Adapted from Korfmacher, W., Bioanalytical assays in a drug discovery environment, in *Using Mass Spectrometry for Drug Metabolism Studies*, Korfmacher, W. (ed.), CRC Press, Boca Raton, FL, 2005, p. 1. With permission.

TABLE 1.4
Additional Rules for Discovery (Non-GLP) PK Assays Requiring QC Samples (Level III)

1. Use all the rules for "Full PK—Level II" assays (except rule 7) plus the following rules.
2. Quality control (QC) standards are required, and a minimum of six QCs at three concentrations (low, middle, and high) are to be used. The QC standards should be frozen at the same freezer temperature as the samples to be assayed.
3. The QC standards need to be traceable to a separate analyte weighing from the one used for the standard curve standards.
4. The standard curve standards should be prepared on the same day the samples are prepared for assay—the standard curve solutions needed for this purpose may be stored in a refrigerator until needed for up to 6 months.
5. At least two-thirds of the QC samples must be within 25% of their prepared (nominal) values.
6. If dilution of one or more samples is required for this assay, then an additional QC at the higher level must be prepared, diluted, and assayed along with the sample(s) needing dilution—this QC should be run in duplicate and at least one of the two assay results must meet the 25% criteria.

Source: Adapted from Korfmacher, W., Bioanalytical assays in a drug discovery environment, in *Using Mass Spectrometry for Drug Metabolism Studies*, Korfmacher, W. (ed.), CRC Press, Boca Raton, FL, 2005, p. 1. With permission.

results to the discovery project team in a timely manner [20]. These rules have also been utilized at various contract research laboratories when we sent some of our samples to these vendors for assay. In general, we have found that when assays meet these criteria, they are good assays and when they fail the criteria, they should be performed again.

In addition to the dosed compound, often one can look for common metabolites when assaying plasma samples from a PK study; this step is often referred to as "metabolite profiling." In our laboratory we have used the QTrap MS/MS system to provide both quantitative analysis and metabolite profiling when assaying plasma samples. The hardware and software tools that are part of the QTrap MS/MS system are well suited to this purpose [110–114]. We have found it useful to use three "rules" for deciding when to report a possible metabolite. The metabolite reporting rules are as follows:

1. Each metabolite must be chromatographically separated from the dosed compound as well as other reportable metabolites in order to be reported.
2. Each metabolite must have an SRM response of at least 1% of the dosed compound in order to be reported.
3. A product ion mass spectrum for each reportable metabolite must be obtained and it must provide sufficient information in order to demonstrate that it is a metabolite of the dosed compound.

These rules have helped to avoid the incorrect reporting of false metabolites as well as the unnecessary reporting of minor metabolites. Typically, we will report the metabolites by showing the relative responses of the metabolite and the dosed compound on the same graph; because the *y*-axis of this graph is labeled relative response (as opposed to concentration units), we alert the recipient that the actual concentration ratios may be different from the response ratios for the dosed compound and the metabolite.

1.4 CONCLUSIONS

While the challenge of working in a discovery bioanalytical group continues to be that the discovery teams always want more studies assayed and faster turnaround times than we can readily deliver to them, the new LC–MS instrumentation and software tools that are becoming available are helping us to meet these challenges. In addition to better tools, performing our work in a systematic manner can be very helpful in meeting the discovery bioanalytical timelines. Our new challenge is to determine what additional information we can provide to the discovery teams while still providing fast turnaround times for the requested information.

REFERENCES

1. MacCoss, M. and Baillie, T.A., Organic chemistry in drug discovery, *Science*, 303(5665), 1810, 2004.
2. Lombardino, J.G. and Lowe, III, J.A., The role of the medicinal chemist in drug discovery—then and now, *Nat. Rev. Drug Discov.*, 3(10), 853, 2004.
3. Kennedy, T., Managing the drug discovery/development interface, *Drug Discov. Today*, 2(10), 436, 1997.
4. Yengi, L.G., Leung, L., and Kao, J., The evolving role of drug metabolism in drug discovery and development, *Pharm. Res.*, 24(5), 842, 2007.
5. Kumar, G.N. and Surapaneni, S., Role of drug metabolism in drug discovery and development, *Med. Res. Rev.*, 21(5), 397, 2001.
6. Korfmacher, W.A., Lead optimization strategies as part of a drug metabolism environment, *Curr. Opin. Drug Discov. Devel.*, 6(4), 481, 2003.
7. Kola, I. and Landis, J., Can the pharmaceutical industry reduce attrition rates? *Nat. Rev. Drug Discov.*, 3(8), 711, 2004.
8. Ackermann, B.L., Berna, M.J., and Murphy, A.T., Recent advances in use of LC/MS/MS for quantitative high-throughput bioanalytical support of drug discovery, *Curr. Top. Med. Chem.*, 2(1), 53, 2002.
9. Korfmacher, W.A. et al., HPLC-API/MS/MS: A powerful tool for integrating drug metabolism into the drug discovery process, *Drug Discov. Today*, 2, 532, 1997.
10. Korfmacher, W., Bioanalytical assays in a drug discovery environment, in *Using Mass Spectrometry for Drug Metabolism Studies*, W. Korfmacher, ed. 2005, CRC Press: Boca Raton, FL. p. 1.
11. Korfmacher, W.A., Principles and applications of LC-MS in new drug discovery, *Drug Discov. Today*, 10(20), 1357, 2005.
12. Ackermann, B. et al., Current applications of liquid chromatography/mass spectrometry in pharmaceutical discovery after a decade of innovation, *Annu. Rev. Anal. Chem.*, 1, 357, 2008.

13. Hsieh, Y., HPLC-MS/MS in drug metabolism and pharmacokinetic screening, *Expert Opin. Drug Metab. Toxicol.*, 4(1), 93, 2008.
14. Espada, A. et al., Application of LC/MS and related techniques to high-throughput drug discovery, *Drug Discov. Today*, 13(9–10), 417, 2008.
15. Korfmacher, W., Strategies and techniques for higher throughput ADME/PK assays, in *High-Throughput Analysis in the Pharmaceutical Industry*, P.G. Wang, ed. 2008, CRC Press: Boca Raton, FL. p. 205.
16. Niessen, W.M., Progress in liquid chromatography-mass spectrometry instrumentation and its impact on high-throughput screening, *J. Chromatogr. A*, 1000(1–2), 413, 2003.
17. Xu, R.N. et al., Recent advances in high-throughput quantitative bioanalysis by LC-MS/MS, *J. Pharm. Biomed. Anal.*, 44(2), 342, 2007.
18. White, R.E., High-throughput screening in drug metabolism and pharmacokinetic support of drug discovery, *Annu. Rev. Pharmacol. Toxicol.*, 40, 133, 2000.
19. Korfmacher, W., Strategies for increasing throughput for PK samples in a drug discovery environment, in *Identification and Quantification of Drugs, Metabolites and Metabolizing Enzymes by LC-MS*, S. Chowdhury, ed. 2005, Elsevier: Amsterdam, the Netherlands. p. 7.
20. Xu, X. et al., Designing LC-MS/MS methods that are good enough to provide bioanalytical support for new drug discovery assays, *Am. Drug Discov.*, 2(3), 6, 2007.
21. Balani, S.K. et al., Strategy of utilizing in vitro and in vivo ADME tools for lead optimization and drug candidate selection, *Curr. Top. Med. Chem.*, 5(11), 1033, 2005.
22. Hsieh, Y. and Korfmacher, W.A., Increasing speed and throughput when using HPLC-MS/MS systems for drug metabolism and pharmacokinetic screening, *Curr. Drug Metab.*, 7(5), 479, 2006.
23. Xu, X., Using higher mass resolution in quantitative assays, in *Using Mass Spectrometry for Drug Metabolism Studies*, W. Korfmacher, ed. 2005, CRC Press: Boca Raton, FL. p. 203.
24. Korfmacher, W., Mass spectrometry in early pharmacokinetic investigations, in *Mass Spectrometry in Medicinal Chemistry*, K. Wanner and G. Hofner, eds. 2007, Wiley-VCH: Weinheim, Federal Republic of Germany. p. 401.
25. Korfmacher, W., New strategies for the implementation and support of bioanalysis in a drug metabolism environment, in *Integrated Strategies for Drug Discovery Using Mass Spectrometry*, M.S. Lee, ed. 2005, John Wiley & Sons, Inc.: Hoboken, NJ. p. 359.
26. Jemal, M. and Xia, Y.Q., LC-MS development strategies for quantitative bioanalysis, *Curr. Drug Metab.*, 7(5), 491, 2006.
27. Gomez-Hens, A. and Aguilar-Caballos, M., Modern analytical approaches to high-throughput drug discovery, *Trends Anal. Chem.*, 26(3), 171, 2007.
28. De Nardi, C. and Bonelli, F., Moving from fast to ballistic gradient in liquid chromatography/tandem mass spectrometry pharmaceutical bioanalysis: Matrix effect and chromatographic evaluations, *Rapid Commun. Mass Spectrom.*, 20(18), 2709, 2006.
29. Dunn-Meynell, K.W., Wainhaus, S., and Korfmacher, W.A., Optimizing an ultrafast generic high-performance liquid chromatography/tandem mass spectrometry method for faster discovery pharmacokinetic sample throughput, *Rapid Commun. Mass Spectrom.*, 19(20), 2905, 2005.
30. Hsieh, Y. et al., Fast mass spectrometry-based methodologies for pharmaceutical analyses, *Comb. Chem. High Throughput Screen.*, 9(1), 3, 2006.
31. Tiller, P.R. and Romanyshyn, L.A., Implications of matrix effects in ultra-fast gradient or fast isocratic liquid chromatography with mass spectrometry in drug discovery, *Rapid Commun. Mass Spectrom.*, 16(2), 92, 2002.
32. Tiller, P.R., Romanyshyn, L.A., and Neue, U.D., Fast LC/MS in the analysis of small molecules, *Anal. Bioanal. Chem.*, 377(5), 788, 2003.

33. Briem, S. et al., Combined approach for high-throughput preparation and analysis of plasma samples from exposure studies, *Rapid Commun. Mass Spectrom.*, 21(13), 1965, 2007.
34. Polson, C. et al., Optimization of protein precipitation based upon effectiveness of protein removal and ionization effect in liquid chromatography-tandem mass spectrometry, *J. Chromatogr. B Analyt. Technol. Biomed. Life Sci.*, 785(2), 263, 2003.
35. Sadagopan, N.P. et al., Investigation of EDTA anticoagulant in plasma to improve the throughput of liquid chromatography/tandem mass spectrometric assays, *Rapid Commun. Mass Spectrom.*, 17(10), 1065, 2003.
36. Berna, M. et al., Collection, storage, and filtration of in vivo study samples using 96-well filter plates to facilitate automated sample preparation and LC/MS/MS analysis, *Anal. Chem.*, 74(5), 1197, 2002.
37. Watt, A.P. et al., Higher throughput bioanalysis by automation of a protein precipitation assay using a 96-well format with detection by LC-MS/MS, *Anal. Chem.*, 72(5), 979, 2000.
38. Gao, L. et al., A generic fast solid-phase extraction high-performance liquid chromatography/mass spectrometry method for high-throughput drug discovery, *Rapid Commun. Mass Spectrom.*, 21(21), 3497, 2007.
39. Mallet, C.R. et al., Performance of an ultra-low elution-volume 96-well plate: Drug discovery and development applications, *Rapid Commun. Mass Spectrom.*, 17(2), 163, 2003.
40. Mallet, C.R. et al., Analysis of a basic drug by on-line solid-phase extraction liquid chromatography/tandem mass spectrometry using a mixed mode sorbent, *Rapid Commun. Mass Spectrom.*, 16(8), 805, 2002.
41. Mallet, C.R., Lu, Z., and Mazzeo, J.R., A study of ion suppression effects in electrospray ionization from mobile phase additives and solid-phase extracts, *Rapid Commun. Mass Spectrom.*, 18(1), 49, 2004.
42. Xu, X., Zhou, Q., and Korfmacher, W.A., Development of a low volume plasma sample precipitation procedure for liquid chromatography/tandem mass spectrometry assays used for drug discovery applications, *Rapid Commun. Mass Spectrom.*, 19(15), 2131, 2005.
43. Xu, X., Lan, J., and Korfmacher, W., Rapid LC/MS/MS method development for drug discovery, *Anal. Chem.*, 77(19), 389 A, 2005.
44. King, R. et al., Mechanistic investigation of ionization suppression in electrospray ionization, *J. Am. Soc. Mass Spectrom.*, 11(11), 942, 2000.
45. Miller-Stein, C. et al., Rapid method development of quantitative LC-MS/MS assays for drug discovery, *Am. Pharm. Rev.*, 3, 54, 2000.
46. Bonfiglio, R. et al., The effects of sample preparation methods on the variability of the electrospray ionization response for model drug compounds, *Rapid Commun. Mass Spectrom.*, 13(12), 1175, 1999.
47. Xu, X., Veals, J., and Korfmacher, W.A., Comparison of conventional and enhanced mass resolution triple-quadrupole mass spectrometers for discovery bioanalytical applications, *Rapid Commun. Mass Spectrom.*, 17(8), 832, 2003.
48. Jemal, M. and Ouyang, Z., Enhanced resolution triple-quadrupole mass spectrometry for fast quantitative bioanalysis using liquid chromatography/tandem mass spectrometry: Investigations of parameters that affect ruggedness, *Rapid Commun. Mass Spectrom.*, 17(1), 24, 2003.
49. Pucci, V. et al., Enhanced mass resolution method development, validation and assay application to support preclinical studies of a new drug candidate, *Rapid Commun. Mass Spectrom.*, 20(8), 1240, 2006.
50. Rossi, D.T. and Sinz, M., *Mass Spectrometry in Drug Discovery*. 2002, Marcel Dekker, Inc: New York.

51. Coulier, L. et al., Simultaneous quantitative analysis of metabolites using ion-pair liquid chromatography-electrospray ionization mass spectrometry, *Anal. Chem.*, 78(18), 6573, 2006.
52. Ariffin, M.M. and Anderson, R.A., LC/MS/MS analysis of quaternary ammonium drugs and herbicides in whole blood, *J. Chromatogr. B Analyt. Technol. Biomed. Life Sci.*, 842(2), 91, 2006.
53. Wang, G. et al., Ultra-performance liquid chromatography/tandem mass spectrometric determination of diastereomers of SCH 503034 in monkey plasma, *J. Chromatogr. B Analyt. Technol. Biomed. Life Sci.*, 852(1–2), 92, 2007.
54. Gao, S. et al., Evaluation of volatile ion-pair reagents for the liquid chromatography-mass spectrometry analysis of polar compounds and its application to the determination of methadone in human plasma, *J. Pharm. Biomed. Anal.*, 40(3), 679, 2006.
55. Bakhtiar, R. and Majumdar, T.K., Tracking problems and possible solutions in the quantitative determination of small molecule drugs and metabolites in biological fluids using liquid chromatography-mass spectrometry, *J. Pharmacol. Toxicol. Methods*, 55(3), 262, 2007.
56. Delatour, C. and Leclercq, L., Positive electrospray liquid chromatography/mass spectrometry using high-pH gradients: A way to combine selectivity and sensitivity for a large variety of drugs, *Rapid Commun. Mass Spectrom.*, 19(10), 1359, 2005.
57. Gao, S., Zhang, Z.P., and Karnes, H.T., Sensitivity enhancement in liquid chromatography/atmospheric pressure ionization mass spectrometry using derivatization and mobile phase additives, *J. Chromatogr. B Analyt. Technol. Biomed. Life Sci.*, 825(2), 98, 2005.
58. Mei, H., Matrix effects: Causes and solutions, in *Using Mass Spectrometry for Drug Metabolism Studies*, W. Korfmacher, ed. 2005, CRC Press: Boca Raton, FL. p. 103.
59. Little, J.L., Wempe, M.F., and Buchanan, C.M., Liquid chromatography-mass spectrometry/mass spectrometry method development for drug metabolism studies: Examining lipid matrix ionization effects in plasma, *J. Chromatogr. B Analyt. Technol. Biomed. Life Sci.*, 833(2), 219, 2006.
60. Du, L. and White, R.L., Improved partition equilibrium model for predicting analyte response in electrospray ionization mass spectrometry, *J. Mass Spectrom.*, 44(2), 222, 2009.
61. Du, L. and White, R.L., Reducing glycerophosphocholine lipid matrix interference effects in biological fluid assays by using high-turbulence liquid chromatography, *Rapid Commun. Mass Spectrom.*, 22(21), 3362, 2008.
62. Shen, J.X. et al., Minimization of ion suppression in LC-MS/MS analysis through the application of strong cation exchange solid-phase extraction (SCX-SPE), *J. Pharm. Biomed. Anal.*, 37(2), 359, 2005.
63. Wu, S.T., Schoener, D., and Jemal, M., Plasma phospholipids implicated in the matrix effect observed in liquid chromatography/tandem mass spectrometry bioanalysis: Evaluation of the use of colloidal silica in combination with divalent or trivalent cations for the selective removal of phospholipids from plasma, *Rapid Commun. Mass Spectrom.*, 22(18), 2873, 2008.
64. Marchi, I. et al., Evaluation of the influence of protein precipitation prior to on-line SPE-LC-API/MS procedures using multivariate data analysis, *J. Chromatogr. B Analyt. Technol. Biomed. Life Sci.*, 845(2), 244, 2007.
65. Xue, Y.J. et al., Separation of a BMS drug candidate and acyl glucuronide from seven glucuronide positional isomers in rat plasma via high-performance liquid chromatography with tandem mass spectrometric detection, *Rapid Commun. Mass Spectrom.*, 20(11), 1776, 2006.

66. Xue, Y.J., Liu, J., and Unger, S., A 96-well single-pot liquid-liquid extraction, hydrophilic interaction liquid chromatography-mass spectrometry method for the determination of muraglitazar in human plasma, *J. Pharm. Biomed. Anal.*, 41(3), 979, 2006.
67. Schwartz, M.S. et al., Determination of a prostaglandin D2 antagonist and its acyl glucuronide metabolite in human plasma by high performance liquid chromatography with tandem mass spectrometric detection—a lack of MS/MS selectivity between a glucuronide conjugate and a phase I metabolite, *J. Chromatogr. B Analyt. Technol. Biomed. Life Sci.*, 837(1–2), 116, 2006.
68. Yan, Z. et al., Cone voltage induced in-source dissociation of glucuronides in electrospray and implications in biological analyses, *Rapid Commun. Mass Spectrom.*, 17(13), 1433, 2003.
69. Wainhaus, S.B., Acyl glucuronides: Assays and issues, in *Using Mass Spectrometry for Drug Metabolism Studies*, W.A. Korfmacher, ed. 2005, CRC Press: Boca Raton, FL. p. 175.
70. Wainhaus, S.B. et al., Semi-quantitation of acyl glucuronides in early drug discovery by LC-MS/MS, *Am. Pharm. Rev.*, 5(2), 86, 2002.
71. Shipkova, M. et al., Acyl glucuronide drug metabolites: Toxicological and analytical implications, *Ther. Drug Monit.*, 25(1), 1, 2003.
72. Vallano, P.T. et al., Elimination of autosampler carryover in a bioanalytical HPLC-MS/MS method: A case study, *J. Pharm. Biomed. Anal.*, 36(5), 1073, 2005.
73. Zeng, W. et al., A new approach for evaluating carryover and its influence on quantitation in high-performance liquid chromatography and tandem mass spectrometry assay, *Rapid Commun. Mass Spectrom.*, 20(4), 635, 2006.
74. Chang, M.S., Kim, E.J., and El-Shourbagy, T.A., A novel approach for in-process monitoring and managing cross-contamination in a high-throughput high-performance liquid chromatography assay with tandem mass spectrometric detection, *Rapid Commun. Mass Spectrom.*, 20(14), 2190, 2006.
75. Xu, X. et al., A study of common discovery dosing formulation components and their potential for causing time-dependent matrix effects in high-performance liquid chromatography tandem mass spectrometry assays, *Rapid Commun. Mass Spectrom.*, 19(18), 2643, 2005.
76. Tong, X.S. et al., Effect of signal interference from dosing excipients on pharmacokinetic screening of drug candidates by liquid chromatography/mass spectrometry, *Anal. Chem.*, 74(24), 6305, 2002.
77. Shou, W.Z. and Naidong, W., Post-column infusion study of the "dosing vehicle effect" in the liquid chromatography/tandem mass spectrometric analysis of discovery pharmacokinetic samples, *Rapid Commun. Mass Spectrom.*, 17(6), 589, 2003.
78. Leverence, R. et al., Signal suppression/enhancement in HPLC-ESI-MS/MS from concomitant medications, *Biomed. Chromatogr.*, 21(11), 1143, 2007.
79. Wang, S., Cyronak, M., and Yang, E., Does a stable isotopically labeled internal standard always correct analyte response? A matrix effect study on a LC/MS/MS method for the determination of carvedilol enantiomers in human plasma, *J. Pharm. Biomed. Anal.*, 43(2), 701, 2007.
80. Chambers, E. et al., Systematic and comprehensive strategy for reducing matrix effects in LC/MS/MS analyses, *J. Chromatogr. B Analyt. Technol. Biomed. Life Sci.*, 852(1–2), 22, 2007.
81. Wang, L. et al., "LC-electrolyte effects" improve the bioanalytical performance of liquid chromatography/tandem mass spectrometric assays in supporting pharmacokinetic study for drug discovery, *Rapid Commun. Mass Spectrom.*, 21(16), 2573, 2007.

82. Churchwell, M.I. et al., Improving LC-MS sensitivity through increases in chromatographic performance: Comparisons of UPLC-ES/MS/MS to HPLC-ES/MS/MS, *J. Chromatogr. B Analyt. Technol. Biomed. Life Sci.*, 825(2), 134, 2005.
83. Plumb, R.S. et al., A rapid screening approach to metabonomics using UPLC and oa-TOF mass spectrometry: Application to age, gender and diurnal variation in normal/Zucker obese rats and black, white and nude mice, *Analyst*, 130(6), 844, 2005.
84. Shen, J.X. et al., Orthogonal extraction/chromatography and UPLC, two powerful new techniques for bioanalytical quantitation of desloratadine and 3-hydroxydesloratadine at 25 pg/mL, *J. Pharm. Biomed. Anal.*, 40(3), 689, 2006.
85. Sherma, J., UPLC: Ultra-performance liquid chromatography, *J. AOAC Int.*, 88(3), 63A, 2005.
86. Wren, S.A., Peak capacity in gradient ultra performance liquid chromatography (UPLC), *J. Pharm. Biomed. Anal.*, 38(2), 337, 2005.
87. O'Connor, D. et al., Ultra-performance liquid chromatography coupled to time-of-flight mass spectrometry for robust, high-throughput quantitative analysis of an automated metabolic stability assay, with simultaneous determination of metabolic data, *Rapid Commun. Mass Spectrom.*, 20(5), 851, 2006.
88. Castro-Perez, J. et al., Increasing throughput and information content for in vitro drug metabolism experiments using ultra-performance liquid chromatography coupled to a quadrupole time-of-flight mass spectrometer, *Rapid Commun. Mass Spectrom.*, 19(6), 843, 2005.
89. Plumb, R. et al., Ultra-performance liquid chromatography coupled to quadrupole-orthogonal time-of-flight mass spectrometry, *Rapid Commun. Mass Spectrom.*, 18(19), 2331, 2004.
90. Wang, G. et al., Ultra-performance liquid chromatography/tandem mass spectrometric determination of testosterone and its metabolites in vitro samples, *Rapid Commun. Mass Spectrom.*, 20(14), 2215, 2006.
91. Yu, K. et al., Ultra-performance liquid chromatography/tandem mass spectrometric quantification of structurally diverse drug mixtures using an ESI-APCI multimode ionization source, *Rapid Commun. Mass Spectrom.*, 21(6), 893, 2007.
92. Yu, K. et al., High-throughput quantification for a drug mixture in rat plasma-a comparison of ultra performance liquid chromatography/tandem mass spectrometry with high-performance liquid chromatography/tandem mass spectrometry, *Rapid Commun. Mass Spectrom.*, 20(4), 544, 2006.
93. Al-Dirbashi, O.Y. et al., Rapid UPLC-MS/MS method for routine analysis of plasma pristanic, phytanic, and very long chain fatty acid markers of peroxisomal disorders, *J. Lipid Res.*, 49(8), 1855, 2008.
94. Hsieh, Y. et al., Comparison of fast liquid chromatography/tandem mass spectrometric methods for simultaneous determination of cladribine and clofarabine in mouse plasma, *J. Pharm. Biomed. Anal.*, 44(2), 492, 2007.
95. Licea-Perez, H. et al., A semi-automated 96-well plate method for the simultaneous determination of oral contraceptives concentrations in human plasma using ultra performance liquid chromatography coupled with tandem mass spectrometry, *J. Chromatogr. B Analyt. Technol. Biomed. Life Sci.*, 852(1–2), 69, 2007.
96. Kapron, J.T. et al., Removal of metabolite interference during liquid chromatography/tandem mass spectrometry using high-field asymmetric waveform ion mobility spectrometry, *Rapid Commun. Mass Spectrom.*, 19(14), 1979, 2005.
97. Wu, S.T., Xia, Y.Q., and Jemal, M., High-field asymmetric waveform ion mobility spectrometry coupled with liquid chromatography/electrospray ionization tandem mass spectrometry (LC/ESI-FAIMS-MS/MS) multi-component bioanalytical method development, performance evaluation and demonstration of the constancy of the compensation voltage with change of mobile phase composition or flow rate, *Rapid Commun. Mass Spectrom.*, 21(22), 3667, 2007.

98. Manitpisitkul, P. and White, R.E., Whatever happened to cassette-dosing pharmacokinetics? *Drug Discov. Today*, 9(15), 652, 2004.
99. Sadagopan, N., Pabst, B., and Cohen, L., Evaluation of online extraction/mass spectrometry for in vivo cassette analysis, *J. Chromatogr. B Analyt. Technol. Biomed. Life Sci.*, 820(1), 59, 2005.
100. Zhang, M.Y. et al., Brain and plasma exposure profiling in early drug discovery using cassette administration and fast liquid chromatography-tandem mass spectrometry, *J Pharm Biomed Anal*, 34(2), 359, 2004.
101. Tong, X.S. et al., High-throughput pharmacokinetics screen of VLA-4 antagonists by LC/MS/MS coupled with automated solid-phase extraction sample preparation, *J. Pharm. Biomed. Anal.*, 35(4), 867, 2004.
102. Smith, N.F. et al., Evaluation of the cassette dosing approach for assessing the pharmacokinetics of geldanamycin analogues in mice, *Cancer Chemother. Pharmacol.*, 54(6), 475, 2004.
103. Korfmacher, W.A. et al., Cassette-accelerated rapid rat screen: A systematic procedure for the dosing and liquid chromatography/atmospheric pressure ionization tandem mass spectrometric analysis of new chemical entities as part of new drug discovery, *Rapid Commun. Mass Spectrom.*, 15(5), 335, 2001.
104. Mei, H., Korfmacher, W., and Morrison, R., Rapid in vivo oral screening in rats: Reliability, acceptance criteria, and filtering efficiency, *AAPS J.*, 8(3), E493, 2006.
105. Liu, B. et al., Snapshot PK: A rapid rodent in vivo preclinical screening approach, *Drug Discov. Today*, 13(7–8), 360, 2008.
106. Han, H.K. et al., An efficient approach for the rapid assessment of oral rat exposures for new chemical entities in drug discovery, *J. Pharm. Sci.*, 95(8), 1684, 2006.
107. Wainhaus, S. et al., Ultra fast liquid chromatography-MS/MS for pharmacokinetic and metabolic profiling within drug discovery, *Am. Drug Discov.*, 2(1), 6, 2007.
108. Matuszewski, B.K., Constanzer, M.L., and Chavez-Eng, C.M., Strategies for the assessment of matrix effect in quantitative bioanalytical methods based on HPLC-MS/MS, *Anal. Chem.*, 75(13), 3019, 2003.
109. Schuhmacher, J. et al., Matrix effects during analysis of plasma samples by electrospray and atmospheric pressure chemical ionization mass spectrometry: Practical approaches to their elimination, *Rapid Commun. Mass Spectrom.*, 17(17), 1950, 2003.
110. Li, A.C. et al., Simultaneously quantifying parent drugs and screening for metabolites in plasma pharmacokinetic samples using selected reaction monitoring information-dependent acquisition on a QTrap instrument, *Rapid Commun. Mass Spectrom.*, 19(14), 1943, 2005.
111. Shou, W.Z. et al., A novel approach to perform metabolite screening during the quantitative LC-MS/MS analyses of in vitro metabolic stability samples using a hybrid triple-quadrupole linear ion trap mass spectrometer, *J. Mass Spectrom.*, 40(10), 1347, 2005.
112. Hopfgartner, G. et al., Triple quadrupole linear ion trap mass spectrometer for the analysis of small molecules and macromolecules, *J. Mass Spectrom.*, 39(8), 845, 2004.
113. Hopfgartner, G. and Zell, M., Q Trap MS: A new tool for metabolite identification, in *Using Mass Spectrometry for Drug Metabolism Studies*, W. Korfmacher, ed. 2005, CRC Press: Boca Raton, FL. p. 277.
114. King, R. and Fernandez-Metzler, C., The use of Qtrap technology in drug metabolism, *Curr. Drug Metab.*, 7(5), 541, 2006.

2 The Drug Discovery Process: From Molecules to Drugs

Mike S. Lee and Steven E. Klohr

CONTENTS

2.1 INTRODUCTION

The pharmaceutical industry faces significant and perhaps unprecedented pressures. Many experts now question whether current business models will be sustainable [1–4]. Blockbuster drugs, defined as drugs with over $1 billion in annual sales, have historically provided the bulk of earnings within the pharmaceutical industry. These earnings support the discovery and development of new therapies. The pending demise of the so-called blockbuster drug (Table 2.1) has resulted in dramatic changes in the industry. All facets of the drug development continuum that range from drug discovery to development to manufacturing to sales and marketing have been impacted. Additionally, acceptable risk/benefit ratios are approaching zero, and thus, the regulatory hurdles for new drugs have never been higher [2–5]. The end result has been a decrease in industry productivity while costs (i.e., R&D expenditures) increase (Figure 2.1). Furthermore, the R&D cost for a new drug now exceeds $1.3 billion in 2005 [6,7].

Mergers, in-licensing, layoffs, codevelopment ventures, and outsourcing/offshoring practices are tangible external signs of an industry in the midst of change. Workflows are constantly evaluated and revised, and, at times, eliminated. The cost to develop a drug that fails before approval now account for 75% of the cumulative costs [8]. Furthermore, only 2 of 10 drugs that reach the market are earning enough money to match or exceed the average R&D cost to develop these new medicines [9]. The process to discover, develop, and receive approval to market a drug is long, complex, and costly (Figure 2.2), thus, tools and approaches that can add knowledge, reduce attrition, and reduce R&D costs are in demand. This trend has

TABLE 2.1
The Top 10 Blockbuster Drugs Based on Yearly Worldwide Sales ($b) as of March 2008

Sales Rank	Product	2008 Sales ($b)
1	**Lipitor—Pfizer**	**13.2**
2	**Plavix—BMS/Sanofi**	**7.4**
3	Nexium—AstraZeneca	7.1
4	**Advair—GSK**	**6.9**
5	**Risperdal—J&J**	**4.7**
6	**Zyprexa—Lilly**	**4.6**
7	**Seroquel—AstraZeneca**	**4.5**
8	**Singulair—Merck**	**4.4**
9	**Effexor—Wyeth**	**4.0**
10	Aranesp—Amgen	3.9

Source: IMS Health, MIDAS, March 2008. With permission.
Note: The eight drugs in the boldface will lose exclusivity by the year 2011.

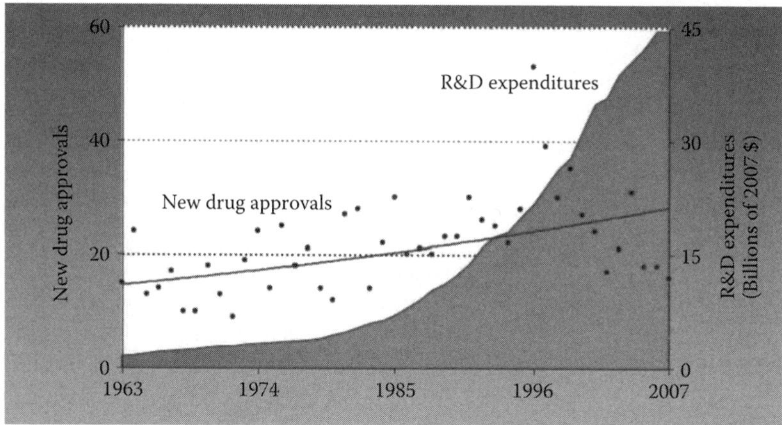

FIGURE 2.1 The new drug approvals from 1963 to 2007 and corresponding R&D expenditures. New drug approvals have not kept pace with rising R&D spending. R&D expenditure adjusted for inflation. (Courtesy of Tufts Center for the Study of Drug Development, Boston, MA. With permission.)

been especially evident in the drug discovery and preclinical stages of drug development. Success prior to clinical development can reduce attrition in development and result in significant savings.

The ability to gather more knowledge faster with smaller quantities of complex samples can support these increased demands on pharmaceutical R&D. The advanced approaches and technologies associated with liquid chromatography–mass spectrometry (LC–MS) continue to play a significant role within drug discovery because of the inherent sensitivity and selectivity of this analytical platform. Improvements with increased throughput and easier-to-use instruments have allowed researchers unprecedented levels of efficiency and productivity.

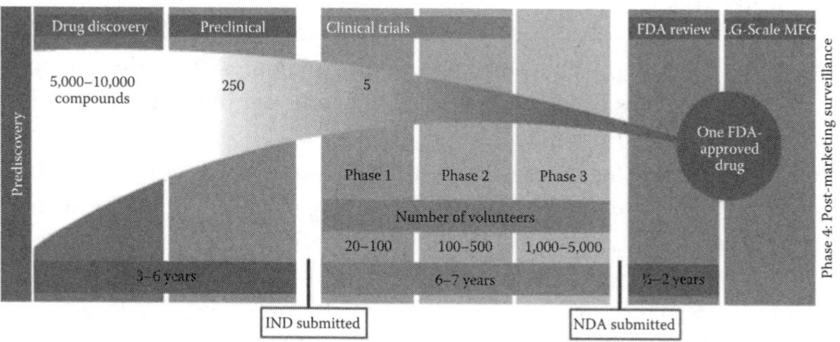

FIGURE 2.2 The long, complex, and costly process of pharmaceutical R&D. (Courtesy of Pharmaceutical Research and Manufacturers of America, *Pharmaceutical Industry Profile 2008*, PhRMA, Washington, D.C., March 2008. With permission.)

So, what is the problem? Why is the pharmaceutical industry experiencing such pain? Why have the recent advances and capabilities not translated into more discovery and more drugs in the market? The current state of the industry suggests that a qualitative change in the overall process, from molecules to drugs, may need to be critically examined or perhaps looked at in a different way.

Drug metabolism plays a significant role in the drug discovery process [10–12]. Furthermore, drug metabolism functions are perhaps most dependent on LC–MS technologies. Absorption, distribution, metabolism, and excretion (ADME) and pharmacokinetics (PK) are the pharmaceutical properties that have been used during drug metabolism studies to define the disposition of a novel drug candidate. In drug discovery, ADME–PK properties are routinely measured using LC–MS techniques. High-throughput LC–MS assays are developed to screen drug candidates and provide the basis for comparative results and quick decisions. The corresponding data obtained from these ADME–PK screens are used to predict a new drug's behavior in humans in an effort to gauge clinical success or failure and distinguish between lead candidates.

This chapter provides an analytical perspective on the drug discovery process, with a specific focus on drug metabolism studies that rely on LC–MS techniques. The purpose is to highlight the critical drug metabolism studies that are used to discern the most promising drug candidates. It is hoped that a critical understanding of these studies in the context of the evolving drug discovery process will provide insight into the analytical requirements for current, and perhaps, future drug discovery strategies. The applications outlined in this chapter are not intended to be comprehensive. Specific chapters in this book as well as several references are recommended for further study to obtain in-depth perspective on analytical applications and preferred uses of LC–MS in drug discovery [13–15].

2.2 DRUG DISCOVERY OVERVIEW

Drug discovery is part of an overall drug development process that involves four distinct stages: drug discovery, preclinical development, clinical development, and manufacturing (Figure 2.3). The drug development process can be conceptually viewed as a series of "go" and "stop" decisions designed to provide focus on the most favorable drug candidates and eliminate weak candidates from the pipeline [13]. Ideally, weak candidates can be culled quickly, before significant development effort and resources have been expended.

The drug discovery stage represents a strategic area of focus comprised of distinct events that are both complicated and dynamic. Drug discovery activities feature four main areas of research: target identification/validation; lead identification; screening; and lead optimization. The work in drug discovery is performed in a nonregulated environment. Thus, the procedures and results are not reported to a regulatory agency such as the Food and Drug Administration (FDA).

Most drug metabolism-based screens are performed during the lead optimization phase of drug discovery. Lead optimization begins once a compound is identified from primary screens and the desired potency, selectivity, and mechanism of

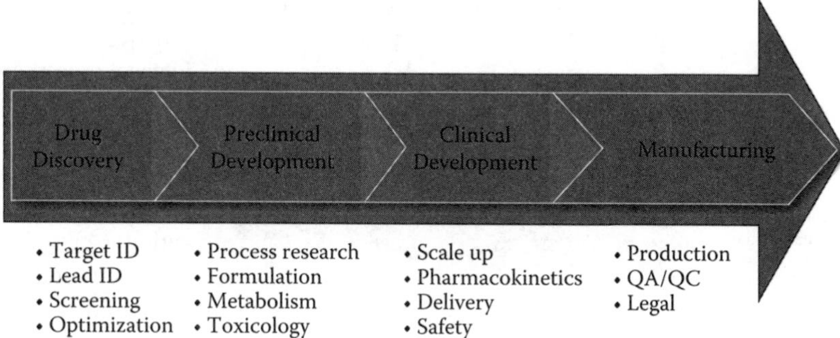

FIGURE 2.3 An overview of the drug development process that illustrates the activities within the four major stages of drug development: drug discovery, preclinical development, clinical development, and manufacturing. (Courtesy of Milestone Development Services, Newtown, PA, 1998. With permission.)

action is demonstrated. Medicinal chemists then modify these lead compounds to improve efficacy, design prodrugs, and determine biologically active metabolites. Animal models are often used as a surrogate to provide correlation with human pharmacodynamics (PD), PK, and ADME properties. This information is fed back to medicinal chemists to assist with the optimization of pharmaceutical properties. During this time, the promising lead compounds are evaluated for toxicity. Also, prospective studies are carried out to determine potential formulation and delivery options.

The lead optimization process is highly iterative, multiparametric, and intensely interdisciplinary. Both qualitative and quantitative information is routinely obtained on analog series of compounds to correlate changes in chemical structure to biological and pharmacological data. The resulting quantitative information provides the basis for the comparison of lead compounds and the ability to establish structure activity relationships (SAR). This multistep evaluation and optimization of ADME–PK properties continues until a defined drug profile is achieved.

2.2.1 Drug Discovery Strategies

The trend in the pharmaceutical industry has been to introduce ADME–PK screens early in the drug discovery process—before a molecule enters into the drug development phases (i.e., preclinical development, clinical development). In this way, the specific criteria of the molecule can be improved while timeline and financial costs associated with progressing an unsuitable drug candidate through the drug development pipeline can be avoided. Additionally, since less than 1 in 10 compounds that are recommended for development successfully proceed through development to become an approved new chemical entity (NCE) (Figure 2.4), information that provides even a small increase in overall success rates can significantly reduce the number of lead compounds needed to reach NCE goals.

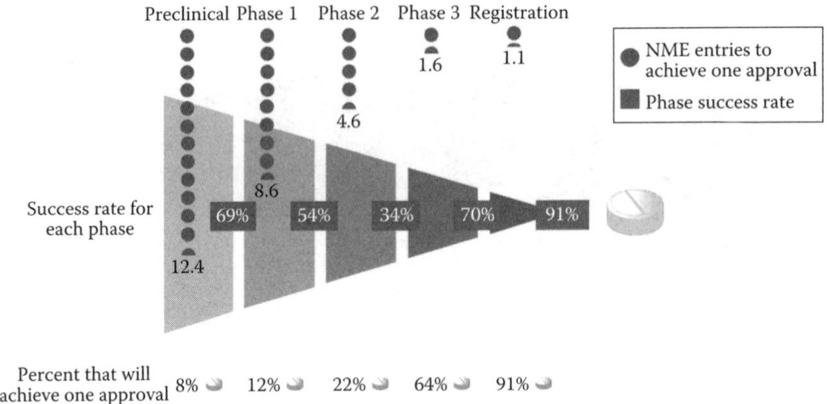

FIGURE 2.4 The NCE success rates from 2003 to 2007 for each phase of drug development. Less than 10% of lead compounds are approved. (Courtesy of KMR Group, Chicago, IL and the Pharmaceutical Benchmarking Forum. With permission.)

2.2.2 ADME–PK Screening

Traditionally, the focus of screening activities for lead optimization had been on SAR to modify or fine-tune the pharmaceutical properties associated with compound potency and specificity. As a consequence, ADME–PK screening took place at a later stage of drug discovery. Thus, compounds optimized for activity and potency were often found to lack properties that were important for safety and/or developability [16]. Drug discovery strategies now actively feature ADME–PK screens within a secondary screening context to optimize compounds around a predetermined set of pharmaceutical properties. The resulting ADME–PK data are typically generated in a format that can be used to directly compare the profiles of a related series of lead compounds. The PK profiles are increasingly important as potential for once-a-day dosing increases patient compliance and often provides a marketing advantage for an approved drug. Thus, standardized analytical methods are prevalent within a drug discovery environment. Furthermore, ADME–PK screening is no longer done in a sequential mode but in a parallel mode during lead optimization.

2.2.3 ADME–PK Screening Benchmarks

In 2004, Kola and Landis provided an updated perspective on the reasons for attrition in drug discovery [17]. In their report, PK/bioavailability was cited as the leading cause of attrition for lead compounds in drug discovery in 1991 (Figure 2.5). At this time, lead compounds were failing at a rate of 40% due to poor bioavailability. Consequently, the pharmaceutical industry quickly realigned resources in drug discovery to screen drug candidates for desirable bioavailability properties earlier in the process. By the year 2000, the attrition due to bioavailability decreased to below 10%! However, toxicology and commercial issues increased significantly to 20%.

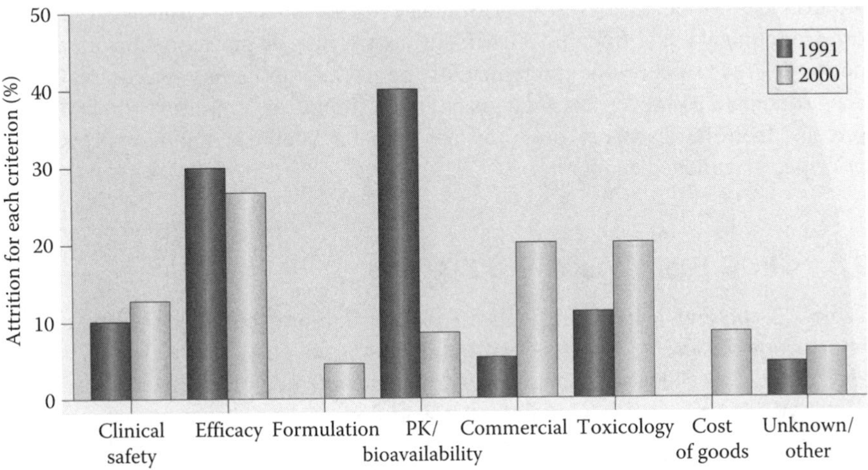

FIGURE 2.5 Reasons for attrition for lead compounds in drug discovery in 1991 and 2000. In 1991, PK/bioavailability was the leading cause of attrition in drug discovery. By 2000 the attrition due to bioavailability decreased to below 10% because drug candidates were routinely screened for desirable bioavailability properties earlier in the process. (Courtesy of Nature Publishing Group, 2004. With permission.)

And as a consequence, a more proactive study of toxicology and drug safety has recently emerged within drug discovery [18–20]. Interestingly, efficacy remained a critical issue and was essentially unchanged during this timeframe with an attrition rate of 28%–30%.

2.2.4 ADME–PK Screening Rationale

Organizations within a drug discovery setting tend to establish a series of compulsory screens to provide a comprehensive and prospective outlook on clinical response of the test drug. Screens for a specific drug discovery program can be modified to aggressively test properties that are deemed to be critical for clinical success. Drug discovery strategies can also be significantly altered by either clinical success or failure. For example, if a test drug fails due to a drug–drug interaction, then future drug discovery screens in this therapeutic area would likely feature the corresponding assays earlier in the process.

Conceptually, the strategy for ADME–PK screening in drug discovery can be compared to decathlon event as stated by Kerns and Di [21]. A specific ADME–PK screen is analogous to a specific athletic event. The overall "winner" is not necessarily the lead compound that exhibits the best performance in any single ADME–PK screen, but simply has adequate characteristics in all ADME–PK screens.

So it can be generalized that a standard series of ADME–PK screening assays is assembled during the initiation of a drug discovery program. A drug candidate must attain at least a moderate level of performance in a screen to be considered as a viable drug for clinical trials. Specific ADME–PK assays may be given high priority based on the desired target profile or predicted features (i.e., strengths, weaknesses) of the drug candidate. However, once data is obtained from these preliminary ADME–PK screens, then the discovery program may take on a more dynamic approach and priorities may indeed change.

2.3 DRUG DISCOVERY WORKFLOWS

Table 2.2 contains a list of ADME–PK assays that are routinely used to provide information via *in vitro* assays, lead optimization screens, and *in vivo* PK studies [22]. Each assay is developed and applied with the intent of providing a quick survey of ADME–PK properties. Certainly, the workflow of drug discovery activities can often be depicted in a linear, stepwise fashion according to the ADME–PK screens highlighted in Table 2.2. However, such a workflow would not take into consideration the dynamics of a drug discovery program nor would such a workflow depict the differences that exist between different programs (i.e., diabetes, cancer, cardiovascular). The fact is that the discovery process is often initially supported with a standard set of ADME–PK assays that can be viewed as somewhat compulsory and highly iterative.

TABLE 2.2
The Representative ADME/Tox Assays That Are Routinely Performed in Drug Discovery

***In Vitro* Assays**	**Lead Optimization Toxicology Screens**	**Pharmacokinetic Studies**
CYP450 enzyme inhibition and induction	Rodent safety screens	Bioavailability
Cytochrome P450 inhibition (LC–MS-based)	Multiple routes of administration	Bioequivalence
HERG channel inhibition	Single or repeat dosing	Dose ranging
Microsomal stability	Daily clinical observation	Linearity
Metabolite identification	Histopathology	Proportionality
Aqueous solubility		Toxicokinetics
Enzyme activity		Biliary excretion
Permeability		Mass balance
Cytotoxicity		Tissue distribution
Metabolic drug–drug interactions		Placental transfer
Protein binding		Metabolite profiling
		Surgical models
		Cerebral spinal fluid, lymph, urine
		Cardiovascular telemetry

2.3.1 Drug Discovery Objectives

To better understand drug discovery workflows, it is helpful to recognize a primary objective of ADME–PK screening programs: to determine dosing regimens (Figure 2.6). Insights into "how much?" and "how long?" with respect to dosing regimen are gained from PK screens [23,24]. The bioavailability and half-life is determined from the corresponding PK profiles obtained for each drug candidate. Furthermore, bioavailability and half-life are affected by absorption, clearance, and volume of distribution. ADME properties provide unique insight into these important parameters and are used to optimize PK properties. The dynamic relationship between ADME properties and PK provides the basis for the systematic optimization of lead compounds in drug discovery.

Despite the extensive use of ADME–PK screens in drug discovery, there is no one "correct" or standardized workflow for drug discovery. Often, ADME screening strategies are dictated by specific needs of the discovery program (i.e., therapeutic area, competitive landscape), recent precedent (i.e., success, failure), and/or the chemistry and disposition of a series of lead compounds.

As the discovery program evolves and matures with new findings and results, the initial process becomes modified to reflect strategic needs based on the strengths and/or weaknesses of the drug candidate and advances in competing laboratories. Thus, a highly dynamical model for drug discovery support can emerge.

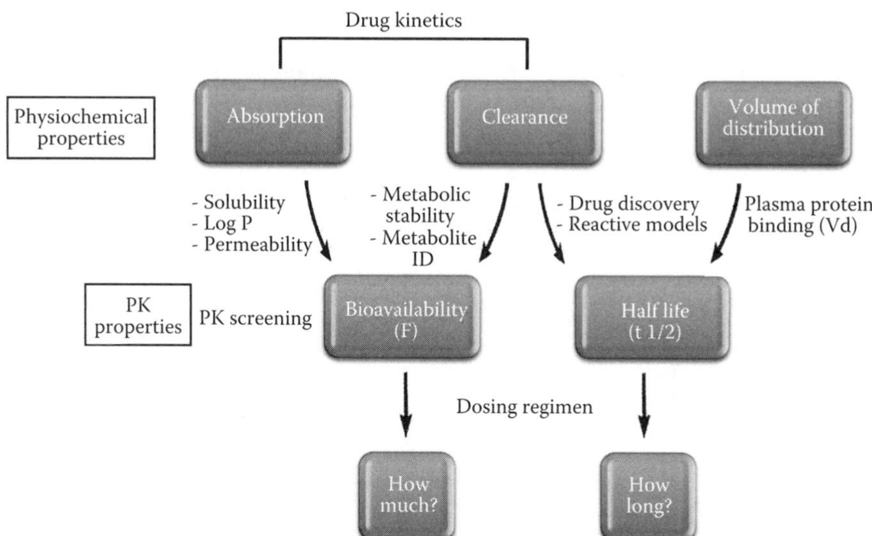

FIGURE 2.6 The ADME–PK screening data that are obtained in drug discovery and used to determine dosing regimen (i.e., "how much?" and "how long?"). A combination of assays associated with physicochemical properties (absorption, clearance, and volume of distribution) and PK properties (bioavailability and half life) are used to evaluate each drug candidate. (Courtesy of Milestone Development Services, Newtown, PA. With permission.)

2.4 ANALYTICAL METHODS

The analytical considerations within drug discovery are challenging. Drug metabolism-based studies require sensitive, selective, and robust analytical methods [15]. Industry preferences for pharmaceutical analysis are generally based on the sample type: nontrace/pure; nontrace/mixture; trace/pure; and trace/mixture (Figure 2.7). The sample type dictates the analytical tool or platform that will be required for analysis. The trace/mixture sample type has emerged as the preferred sample for accelerated drug discovery and development [13]. Hyphenated techniques such as high-performance liquid chromatography with ultraviolet detection (HPLC/UV) and LC–MS are well suited for drug discovery applications whereby samples obtained from an *in vitro* or *in vivo* study are subjected to sample preparation, chromatographic separation, and analytical detection.

The fast pace and highly interdisciplinary nature of drug discovery research further requires that these trace/mixture methods accommodate high-throughput analysis formats. Thus, there exists a balance between simplicity and complexity to provide methods that generate data and inspire confidence while maintaining speed.

As a general rule, the demands on drug discovery support dictate that there be (1) no repeat analyses and (2) no carryover. Both of these issues pose serious threats to fast analysis schemes and a high-throughput environment [25]. The recent literature contains more detailed descriptions of analytical methodologies that relate to LC–MS for drug discovery support. A review of LC–MS in the pharmaceutical industry [13] as well as specific analytical methodologies such as sample preparation [26], chromatography [27], electrospray ionization (ESI) [28], bioanalysis [29], and data analysis [30] provide comprehensive descriptions and references.

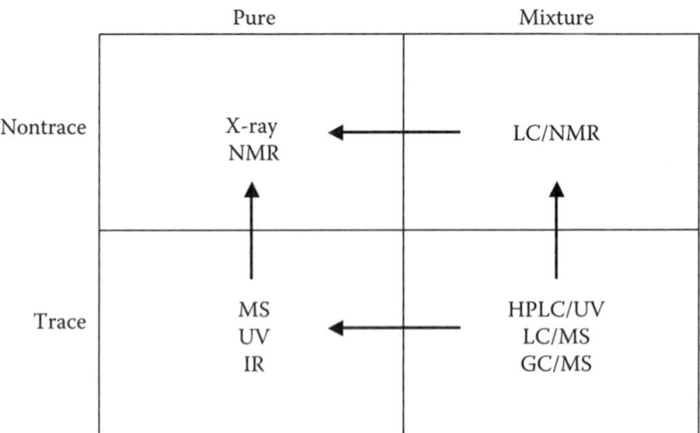

FIGURE 2.7 A structure analysis matrix that illustrates analysis preferences in the pharmaceutical industry based on four specific sample types: nontrace/pure; nontrace/mixture; trace/pure; and trace/mixture. (Courtesy of Milestone Development Services, Newtown, PA. With permission.)

2.4.1 Sample Preparation

Automated sample preparation is commonplace in the drug discovery laboratory with biological fluid samples such as plasma, serum, bile, or urine [26,31]. The primary goal for sample preparation with LC–MS-based methods in drug discovery is to separate the drug and/or metabolites from the endogenous materials in the sample matrix (e.g., proteins, salts, and metabolic by-products).

Sample preparation is often a critical step in the analytical method since most LC–MS instruments cannot accept the sample matrix directly. The removal of endogenous materials from the biological sample results in a concentrated analyte and sets the stage for a high-quality analysis. The analyte is essentially moved from a totally aqueous environment into a chosen percentage of organic/aqueous solvent that is appropriate for injection into a chromatographic system [31].

Many different sample preparation techniques are available to the drug discovery scientist. Off-line sample preparation procedures include protein precipitation, filtration, dilution followed by injection, liquid–liquid extraction (LLE), and solid-phase extraction (SPE). Typically, these procedures are performed in an automated, high-throughput mode that features a 96-well plate format. Online sample preparation procedures include SPE and turbulent flow chromatography (TFC) with conventional chromatographic media or restricted access media (RAM). These online approaches are often simple and easy to automate.

2.4.2 Chromatography

HPLC-based techniques have been a traditional mainstay of the pharmaceutical industry [13]. Analytical chromatography allows complex mixtures to be transformed into separated components. The analytical format and corresponding output is well understood.

Typical HPLC/UV methods require that each analyte component separate during the chromatographic run. Analytes are separated (i.e., baseline separation) based on chemical interaction with the stationary phase and mobile phase. A unique retention time is generally required for each analyte and provides the basis for identification and comparative results.

In combination with mass spectrometry, the purpose of the HPLC significantly changes when compared with UV detection-based methods. This change is due to the fact that the mass spectrometer is also used as a separation device. Therefore, chromatographic methods that are developed for LC–MS applications tend to have significantly less stringent requirements and baseline separation is not necessary. The result is a faster chromatographic run. Coeluting peaks are not a significant issue as with HPLC/UV. These peaks can sometimes be resolved with the mass spectrometer. Both retention time and molecular weight is used for identification. This advantageous combination is further realized with ultrahigh-performance liquid chromatography (UHPLC) separation formats (see Chapter 8) [32,33].

Fortunately, nearly all pharmaceutically relevant compounds are amenable to HPLC. Aside from the ability to separate a specific analyte(s), the HPLC also provides a unique measure of control with regard to the LC–MS analysis.

A reproducible percentage of organic/aqueous solvent can be attained to provide for highly sensitive and highly efficient chemistry environment for ionization. In most cases, this environment would promote the formation of an abundant protonated molecule $[M + H]^+$.

2.4.3 Ionization

The ionization process allows for the study of ions in the gas phase. Therefore, ionization for LC–MS analysis involves (1) conversion of a liquid into a gas and (2) the deposit of a charge onto the molecule(s) of interest. Once an analyte is in the gas phase and contains a charge, then the mass spectrometer is able to separate the resulting ions based on some property of mass (i.e., mass-to-charge, arrival time).

Significant advances with the ionization source have been made. The pioneering work of many outstanding scientists with moving belt [34–36], direct liquid introduction (DLI) [37–39], thermospray ionization (TSI) [40,41], and ESI [42–44] set the stage for the development of modern ionization sources for high-performance LC–MS analysis.

An important consideration for LC–MS method development is the extent to which the analyte signal will be affected by the sample matrix [29]. This situation is referred to as ion suppression. Ion suppression is the result of reduced ion signal from the sample matrix [45]. Detailed studies performed by King et al. determined that ion suppression originates in solution [46,47]. An increased chromatographic resolution combined with a more selective sample cleanup is perhaps the best way to reduce the impact of ion suppression.

2.4.4 Mass Spectrometry

If the science of mass spectrometry can be viewed as the study of ions in the gas phase, then a mass spectrometer can be thought of as a molecular weighing machine. The ability to assign molecular weight to a specific molecule(s) is a powerful capability for structure elucidation and quantitative analysis. The physical measurement of mass or molecular weight provides a universal descriptor of a drug compound throughout its lifetime in drug discovery and throughout the drug development continuum.

Mass spectrometers that are routinely used in drug discovery include the single quadrupole [48], triple quadrupole [49], quadrupole ion trap [50], time-of-flight [51], quadrupole time-of-flight [52], and Orbitrap [53] mass spectrometers. Mass spectrometers are primarily used to perform two functions: (1) discriminate against matrix interference and (2) maximize the signal for the ion(s) of interest [29].

The various scan modes of the mass spectrometer provide powerful methods of analysis and unique capabilities for information gathering [54,55]. For example, the full scan mode is used to survey the ions that are generated in the source to confirm or identify structure based on molecular weight assignment. The resulting full scan mass spectrum, therefore, contains an ion(s) that is indicative of molecular weight. Two dimensions of mass analysis or tandem mass spectrometry (MS/MS) provide powerful capabilities for qualitative and quantitative analysis. For example, the product ion mass spectrum provides information that is useful for structure identification. Strategies that feature substructure templates are routinely used for the identification of metabolites (see Chapter 5) [56,57].

The two mass spectrometry scan modes that are used for quantitative analysis are selected ion monitoring (SIM) and selected reaction monitoring (SRM). The SIM experiments involve the use of a single mass analyzer and are performed when the mass spectrometer is set to detect a selected mass that corresponds to the drug molecule. Sensitivity is enhanced since only the ion of interest is sampled [58]. The SRM mode is the most widely used quantitative scan mode and involves two dimensions of mass analysis. The SRM mode provides greater selectivity and enhanced limits of detection (LOD) as a specific precursor-product ion relationship is monitored [59]. The first mass analyzer is set to select the molecular ion of the analyte. The molecular ion enters a collision cell and undergoes collisionally induced dissociation (CID) to produce fragment ions diagnostic of structure. The most abundant fragment ion is selected and monitored by the second mass analyzer. Quantitative analysis using SRM is typically performed on a triple quadrupole mass spectrometer. Many drug discovery applications use a series of SRM experiments for the simultaneous quantitation of drug compounds during a single run (see Chapter 1).

2.4.5 Data Analysis and Information Management

The actual currency that is used to make decisions in drug discovery is information. Thus, LC–MS approaches have developed a unique partnership with information-intensive tools responsible for sample tracking, method development, data interpretation, and data storage.

Traditional, and perhaps, manual approaches for method development would be too time-consuming to perform within a drug discovery environment. As a result, automated method development routines, such as the procedure to generate optimized SRM tables for PK screening, are an essential requirement for drug discovery support [60].

Real-time and data-dependent software [61] has also become an essential component of qualitative analysis applications such as metabolite identification [62] and quantitative analysis such as PK screening [63]. Specialized software packages are able to perform tasks such as molecular weight confirmation, metabolite identification, peak integration, and calibration regression.

Since large quantities of data are generated quickly, data must be efficiently interpreted and reported in a facile manner that will support rapid decisions [64]. Data is formatted, reformatted, sorted, and filtered with a laboratory information management system (LIMS) or a process that involves the assembly and export of a text-based file (i.e., Microsoft Excel). The database is queried to provide the exact data that is needed for a specific report format.

Finally, software tools that can secure electronic records and enable safe storage of data have received considerable attention [65]. These technologies allow for more streamlined approaches (i.e., electronic notebooks, paperless laboratories) and provide for the legal defensibility of patents and NCEs.

2.5 APPLICATIONS

Accurate and timely answers to the iterative questions "what do I have?" and "how much is there?" are paramount to the success of a drug discovery program. The bottom-line is that the pharmaceutical industry prefers analysis formats for ADME–PK screening that can generate information quickly to make fast decisions.

The LC–MS/MS methods used to determine PK properties, bioavailability, and half-life are briefly described in this section followed by an ensemble of ADME screens (physicochemical properties, metabolite identification, metabolic stability, reactive metabolites, drug–drug interaction, and plasma protein binding) that are used to determine the effect of absorption, clearance, and volume of distribution on PK properties. The goal of this phase of drug discovery is to use the information obtained from these screens and optimize the ADME–PK parameters of a drug candidate to define a dosing regimen (i.e., "how much?" and "how long?") for clinical studies. A detailed description of these specific ADME–PK screening methodologies will follow in subsequent chapters.

2.5.1 *In Vivo* Pharmacokinetic Screening

The determination of the concentration of a drug in plasma is an accepted surrogate marker for drug exposure. The use of LC–MS-based approaches for the quantitative analysis of a drug or its metabolite in a physiological sample such as plasma or serum is perhaps the hallmark of PK studies in drug discovery. The PK profiles derived from LC–MS data provide bioavailability and half-life information for drug candidates. These PK profiles are used to directly compare lead compounds and select specific compounds for further study.

Several reviews provide a comprehensive overview of the application of LC–MS-based methods for PK screening in drug discovery [13,14,29]. These reviews highlight the use of LC–MS methods to provide high-throughput PK screening data in a variety of experimental formats such as "*n*-in-one dosing" or cassette dosing and plasma pooling methods as well as other novel screening approaches such as the cassette-accelerated rapid rat screen (CARRS) developed by Korfmacher et al. [63]. Most of these methods are performed on a triple quadrupole mass spectrometer platform. These methods are used to screen lead compounds at a lower limit of quantitation (LLOQ) of 1–10 ng/mL.

The screening of drug discovery candidate compounds based on PK properties using LC–MS approaches was introduced by scientists at Merck [66], Glaxo Wellcome [67,68], and Schering-Plough [69]. These methods were developed using animal models that involve the use of rats and/or dogs.

2.5.1.1 *n*-in-One and Cassette Dosing

The simultaneous PK assessment of multiple drug candidates in one animal is known as *n*-in-one dosing or cassette dosing [67,70]. Typically, the cassette is limited to five or fewer compounds. This parallel approach is used to increase productivity by increasing the number of compounds that can be analyzed and decreasing the number of animals required for screening. Precautions must be weighed heavily during experimental setup and method development to avoid interferences from analytes as well as corresponding metabolites. Risk is associated with these studies as the potential for drug–drug interaction must be taken into account [12,71].

2.5.1.2 Pooling by Time/Pooling by Compound

Pooling strategies for PK assessment can also be performed postdose. These strategies are referred to as pooling by time or pooling by compound and involve single compound administration. The pooling-by-time approaches involve the pooling of plasma obtained from identical time points for different compounds. The overall analysis time is decreased; however, sensitivity is decreased due to sample dilution. The pooling by compound strategy was developed by Hop et al. [72] and involved the pooling plasma from multiple time points from a given compound resulting in a single sample. Thus, a "time average" of exposure is attained and allows for comparisons of net exposure to be made.

2.5.1.3 Cassette-Accelerated Rapid Rat Screen

Another approach that has been developed to streamline and standardize the PK screening activities is known as CARRS [63]. This approach was developed by Korfmacher and coworkers and uses a 96-well plate that accommodates a total of six compounds dosed via single compound administration. One compound is dosed per rat and two rats are used per drug candidate. Plasma samples are collected at six time points resulting in a set of 12 rat plasma samples. The plasma samples are then pooled across time points for each rat resulting in a total of six pooled samples (i.e., one sample per time point). An abbreviated three-point calibration curve is used for each assay.

2.5.2 Physicochemical Properties

The early definition of a drug candidate's physicochemical properties provides great value to medicinal chemists and drug discovery project teams [73–75]. Solubility, acid dissociation constant (pK_a), lipophilicity, and permeability are the physicochemical properties that affect passive transport mechanisms of absorption. High-throughput LC–MS-based methods have been used to provide a quick and timely assessment of these critical properties in drug discovery. In this way, the so-called vital signs of absorption can be determined and the need for more detailed investigation of physicochemical properties can be assessed [25].

2.5.2.1 Solubility

The effective delivery of a drug to the biological site of action (e.g., oral or intravenous bioavailability) is influenced by the solubility of the drug. For example, a drug must be soluble for intestinal absorption to occur. The solubility of drug candidates can be determined by traditional "shake-flask" method [76,77], HPLC/UV-absorbance analysis [78], or with LC–MS-based methods [75].

2.5.2.1.1 Shake-Flask Method

Although shake-flask methods are manually intensive, new methods are always validated against this approach. The drug of interest is added to standard buffer solution contained in the flask until undissolved excess drug appears. The saturated solution is shaken to establish equilibrium and the sample is prepared for either microfiltration or centrifugation. The resulting supernatant solution is analyzed by HPLC/UV.

2.5.2.1.2 HPLC/UV and LC–MS Methods

Higher levels of throughput for solubility measurement can be achieved using HPLC/UV or LC–MS methods. Samples are prepared as 10 mM DMSO solutions. The typical upper limit for saturated solutions prepared from DMSO stock is 100 μL (w/1% DMSO content). The range is based on known ADME requirements and other early drug discovery screens. In general, the kinetic solubility determined from the DMSO solution is more related to early *in vitro* biological assays while the thermodynamic solubility determinations from well-defined solid material is more meaningful to late development programs [78]. Parallel sample preparation with 96-well plate formats and fast HPLC cycle times (2 min gradient analysis) are typically used with LC–MS methods to address high-throughput and increased workloads.

As with many activities associated with physicochemical property analysis, solubility measurements can follow a tiered workflow. For example, all lead optimization compounds can be assayed at pH 7.4, 6.5, 4.0, and 1.5 using DMSO solutions. Lead compounds that progress onward through the discovery process can be examined in solid form at pH 7.4 and 4.0 to facilitate downstream formulation activities [25].

2.5.2.2 Acid Dissociation Constant (pK_a) and Lipophilicity

The pK_a value provides a measure of the acid or basic properties of a compound. Specifically, the pK_a value indicates the tendency of a molecule or ion to keep a proton (H^+) at its ionization center(s). In terms of absorption, pK_a is an important criterion to help determine whether a molecule will be taken up by aqueous tissue components or lipid membranes (lipophilicity).

Lipophilicity is determined by measuring the equilibrium solubility of a compound in a lipophilic phase such as octanol to its solubility in an aqueous phase such as water. Analytical measurements relate pK_a to log P (the partition coefficient). Log P is a physicochemical parameter of aqueous and lipid solubility that influences the delivery of drugs to their target when the drug is in the neutral state.

Lipophilicity can be reported as log P, which is the intrinsic partitioning of the drug between octanol and water when the drug is in the neutral state. However, lipophilicity is often reported as log D (distribution coefficient), which represents the distribution of the drug at a specific pH, typically pH 7.4.

High-throughput formats use HPLC-based methods via correlation of log D with retention time [79,80]. The experimental setup of a lipophilicity screening assay involves the equilibration of the drug compound in water (or buffer) and octanol. The phases are diluted in water/methanol and analyzed by LC–MS. The values for log P are determined by correlation to log k [81]. A calibration curve is established from known log P of each standard versus the log k value calculated from the retention time (t_r). Hexachlorobenzene (log P = 6.11) can be used as a standard for log P calibration (log P values >5).

Many drug discovery researchers prefer to use software that can predict aqueous pK_a values. For example, the ACD/pK_a software program contains a vast library that contains experimental values for over 16,000 compounds in aqueous solutions and over 2000 compounds in nonaqueous solvents [82].

2.5.2.3 Permeability

The prediction of the important structural features that affect intestinal permeability is useful information to obtain early in the drug discovery process. The two most common models used to obtain fast, high-throughput measurements are the parallel artificial membrane permeation assay (PAMPA) and the cell line assays that feature cultured human colon adenocarcinoma cells (Caco-2). Each method uses a surrogate model to mimic intestinal absorption followed by LC–MS analysis.

2.5.2.3.1 Parallel Artificial Membrane Permeation Assay (PAMPA)

The PAMPA method is based on a 96-well plate platform that features a phospholipid membrane-soaked pad that separates an aqueous donor and receiver compartment. The membrane does not contain active transporters. The strategic use of PAMPA assays in drug discovery is based on the ability to screen large libraries and quickly determine trends based on the ability of compounds to permeate membranes by passive diffusion [83]. In this way, quantitative structure activity relationships (QSAR) can be initiated based on this mechanism of absorption [84].

2.5.2.3.2 Cultured Human Colon Adenocarcinoma Cells (Caco-2)

The use of the Caco-2 cell line assays provide more physiologically relevant data than the PAMPA assays since they express transporters so that both active and passive transport can be determined. Caco-2 cell monolayers are contained in a 96-well plate format and test compounds can be incubated to model and test absorption [85].

2.5.3 Metabolite Identification

The rapid structure identification of metabolites provides an early perspective on the metabolically labile sites or “soft spots” of a drug candidate [86]. This information is useful during lead optimization and can serve to initiate research efforts that deal with metabolism-guided structural modification and toxicity.

Metabolite identification using LC–MS techniques is based on the fact that metabolites generally retain most of the core structure of the parent drug [56,87,88]. Therefore, the parent drug and the corresponding metabolites would be expected to undergo similar fragmentations and produce mass spectra that indicate major substructures.

The success of LC–MS approaches relies on the performance of the ESI or atmospheric pressure chemical ionization (APCI) interface and the ability to generate abundant ions that correspond to the molecular weight of the drug and drug metabolites. The production of abundant molecular ions is an ideal situation for molecular weight confirmation via the adduct of the molecular ion (i.e., $[M+H]^+$, $[M+NH_4]^+$). The confident assessment of molecular weight of the respective drug metabolites allows for direct comparison with the parent drug. In this way, differences in molecular weight can be assessed and correlated with known metabolic pathways such as oxidation, reduction, and conjugation.

Typically, the LC–MS full scan mass spectra of a drug molecule contain abundant $[M + H]^+$ ions with little detectable fragmentation. The product ion mass spectrum contains product ions associated with diagnostic substructures of the drug molecule. There is no need, and more importantly, there is no time in a drug discovery setting to identify all fragment ions observed in the product ion mass spectra. Instead, streamlined approaches based on standard methods and structural template motifs are used [89].

This LC–MS-based methodology can be automated to fit specialized needs within drug discovery based on throughput [62]. The recent application of mass defect filtering [90] and high-resolution accurate mass analysis [33] provides further automated protocols for metabolite identification (see Chapters 5 and 6 for more on this topic).

2.5.4 Metabolic Stability

In vitro methods for the determination of metabolic stability provide a cost-effective solution to analyze the wealth of samples encountered in a drug discovery environment. Detail is sacrificed so that approximate and/or relative results can be obtained with significant savings in resources (i.e., instrumentation, animals) and time.

The use of fast gradient elution LC–MS techniques for metabolic screening was first described by Ackermann and coworkers in 1998 [91] and Korfmacher and coworkers in 1999 [69]. In the method developed by Ackermann et al., a HPLC column-switching apparatus is used to desalt and analyze lead candidates incubated with human liver microsomes. The resulting data can be quickly resolved into specific categories of metabolic stability: high (≥60%); moderate (≥30%–59%); low (≥10%–29%); and very low (<10%).

The methodology for metabolic stability screening introduced by Korfmacher et al. featured liver microsomal incubation systems with a highly automated LC–MS method. The method was developed to work in a high-throughput environment in support of lead optimization. Aside from providing for the capability of analyzing a large number of *in vitro* samples, the LC–MS/MS-based method featured assay routines with multiple compounds in one batch. A set of six samples of each compound was used where three samples correspond to zero time incubation and three samples correspond to a 20 min incubation time. Data processing was performed in an automated fashion as the largest peak in the chromatogram was assumed to correspond to the $[M+H]^+$ of the parent compound. Peak areas of the expected $[M+H]^+$ was obtained for each sample and the results were distributed using a standard report format.

The LC–MS methods introduced by Ackermann et al. and Korfmacher et al. paved the way for more powerful platforms for high-throughput metabolic stability screening that featured the use of LC–MS/MS [80,92]. These powerful methods relied on robotic sample preparation systems that were integrated with an automated LC–MS/MS system.

2.5.5 Reactive Metabolites

In many cases, drug metabolites are less toxic than the corresponding parent drug. In some instances, drugs can undergo metabolic reactions that lead to the formation of reactive species [93]. These reactive metabolites can bind to cell proteins and DNA to cause drug-induced toxicity and cell damage [94].

Most reactive metabolites are electrophiles that react with compounds that contain nitrogen or sulfur atoms. These reactive species are generally short-lived and highly unstable. Therefore, analysis and detection is challenging. The approach that is currently used to characterize reactive metabolites in drug discovery involves the use of trapping agents that form stable adducts with the reactive intermediates [95].

An isotopically labeled trapping agent such as glutathione (GSH) is a useful tool to conduct high-throughput screening for reactive metabolites (see Chapter 6) [96]. In this experiment, a mixture of GSH (γ-glutamylcysteinylglycine) and a corresponding stable-isotope labeled compound (GSX, γ-glutamylcysteinylglycine-$^{13}C_2$–^{15}N) are used to facilitate reactive metabolite screening and provide a unique mass spectrum signature that contains a doublet that differs in mass by 3 Da. Drug discovery lead compounds are prepared in human microsomal incubations and supplemented with GSH and GSX at an equal molar ratio.

Both positive and negative controls can be used to provide quality control and provide confidence in the method (i.e., elimination of false positives). This approach is essential for the detection of low abundant reactive metabolites.

2.5.6 Drug–Drug Interactions

The simultaneous administration of two different drugs can lead to unwanted and harmful drug–drug interactions. This situation can be caused by drug-induced inhibition or induction of cytochrome P450 (CYP) enzymes. Thus, CYP enzymes can have a significant effect on safety profile of coadministered drugs. The early determination of the inhibitory potency of a new drug provides the critical information necessary to assess the potential of the drug to induce biologically significant drug–drug interactions.

One of the first laboratories to proactively use LC–MS-based methods to screen drug discovery candidates for drug–drug interactions was GlaxoSmithKline in 1998 [97]. These assays can be performed either *in vitro* as described by Ayrton et al. [97] and Bu et al. [98] or *in vivo* [99,100]. The methods feature the use of CYP enzymes that are involved in oxidative biotransformation pathways via hydroxylation, *N*-demethylation, or *O*-desethylation.

The LC–MS-based methodology for drug–drug interaction screening involves the use of specific probe substrates that are known for their ability to be converted to a corresponding metabolite by a CYP enzyme in the presence of a potential inhibitor. The LC–MS-based screens are typically set up for high-throughput analysis with 96- or 384-well plate formats. Lead compounds are tested at either a single concentration (10 μM) or at multiple concentrations that span the anticipated maximum steady-state plasma levels (e.g., 0.5, 5, and 50 μM) to determine IC_{50} values.

Sample preparation usually involves a protein precipitation cleanup step. Fast chromatography methods that provide run times of less than 2 min are generally preferred and attained. The LC–MS/MS method is optimized for each CYP substrate's metabolite SRM transition. The selectivity of MS/MS reduces the need for complete chromatographic resolution of individual components. Therefore, analytical run times are significantly reduced. Data analysis must be highly streamlined using automated data analysis and reporting.

The results from these studies provide information on whether a drug may inhibit the biotransformation of another drug when the two are coadministered. LC–MS approaches continue to be used routinely to identify and select drug candidates that have a lower potential for drug–drug interactions.

The use of the appropriate probe substrate and corresponding experimental condition for the high-throughput *in vitro* drug interaction studies is critical when extrapolating the results to *in vivo* situations. Most of the *in vitro* findings that are obtained with one probe substrate are usually extrapolated to the compound's potential to affect all substrates of the same enzyme. Due to this practice, it is important to use the right probe substrate and to conduct the experiment under optimal conditions [101]. For example, it may be necessary to evaluate two or more probe substrates for CYP3A4-based drug interactions. Thus, the search for better probe substrates for some enzymes continues to be investigated.

2.5.7 Plasma Protein Binding

Plasma protein binding assays are performed in drug discovery with human plasma and with plasma obtained from species that contain proteins that nonspecifically bind to the drug [102]. The percentage of bound drug provides insight on the relationship between total *in vivo* plasma drug concentration and the pharmacologically available unbound drug.

As a useful reference point, the binding of drugs to plasma proteins results in a relatively low volume of distribution. Conversely, drugs that remain unbound in plasma are available for distribution to other organs as well as tissues and results in a large volume of distribution.

So why is the assessment of plasma protein binding important? First, the binding of a drug to plasma proteins results in a decrease in drug clearance and a prolonged half-life. Second, only the unbound drug (i.e., free concentration of drug) is available for elimination via excretion and/or metabolism. Thus, the determination of protein binding of a drug can allow for an early prediction of drug efficacy, disposition, and potential for drug–drug interactions. Some believe that the proactive use of these assays can be incorporated into models to estimate and predict metabolic stability, drug inhibition, drug transport, and intrinsic clearance [103].

The three most widely used *in vitro* protein-binding techniques are equilibrium dialysis, ultrafiltration, and blood cell partitioning [104]. These assays are based on the fact that most drugs bind reversibly to plasma proteins such as albumin, lipoproteins, and glycoproteins. *In vitro* plasma protein-binding assays feature a chromatographic separation (i.e., human serum albumin-immobilized column, 96-well plate) followed by MS/MS to determine the percentage of bound drug. The LC–MS-based assay, typically either equilibrium dialysis or ultrafiltration, allows for the relative ranking of lead compounds by percent binding.

2.6 CONCLUSIONS

The impact of early ADME and PK information in the drug discovery process has been significant. Studies that were difficult to envision 10 years ago are now compulsory. However, analytical scientists in the pharmaceutical industry are not able to rest on these accomplishments. Pharmaceutical companies are now faced with the potential loss in sales of $135 billion due to patent expirations between 2008 through 2012 [3]. The pressure on discovery and development scientists has never been greater. These new challenges will likely create new opportunities.

Pharma in the twenty-first century will be lean and nimble, concurrently focusing on reducing cycle time and increasing speed to market—with fewer resources. There will be a relentless focus on gaining knowledge that enables rapid decisions in discovery and early development to reduce late-stage attrition (Phase III and regulatory approval). New drugs will no longer be evaluated solely on clinical benefit. Criteria associated with economic impact will assume a more dominant role in decision-making. The absence of blockbuster drugs may increase attention on specialty markets and result in a greater number of drug candidates in discovery and development. Highly potent drugs will require reduced quantities of active pharmaceutical ingredient, but demand analytical methods with much greater sensitivity for metabolic studies.

The information presented in this chapter demonstrates that the widespread adoption of LC–MS in drug discovery has changed both how and when drug metabolism studies are performed. The combination of sensitivity, selectivity, and productivity that is provided on the LC–MS platform is ideally suited for industrial endeavors and has become essential figures of merit for the drug discovery process. The technical advances in LC–MS and innovative approaches to ADME–PK studies described in the following chapters are vital to meet the challenges that face drug discovery and development.

REFERENCES

1. Garnier, J.P., Rebuilding the R&D engine in big pharma, *Harv. Bus. Rev.*, 86, 68, 2008.
2. Ainsworth, S.J., Pharma adapts: Firms invent new strategies to deflect generic competition and stem growing safety concerns, *Chem. Eng. News*, 85, 13, Dec. 3, 2007.
3. IMS Health, 21st century pharma: Managing current challenges to ensure future growth, Executive Summary of June 12, 2008 IMS Pharma Strategy Series Webinar.
4. Ainsworth, S.J., Facing the giants: Drug companies struggle to deflect generics competition, appease cautious regulators, *Chem. Eng. News*, 86, 20, Dec. 1, 2008.
5. McCaughan, M., A new era in drug safety regulation, *Progressions*, (Ernst & Young), 24, 2007.
6. DiMasi, J.A. and Grabowski, H.G., The cost of biopharmaceutical R&D: Is biotech different? *Manage. Decis. Econ.*, 28, 469, 2007.
7. DiMasi, J.A., Hansen, R.W., and Grabowski, H.G., The price of innovation: New estimates of drug development costs, *J. Health Econ.*, 22, 151, 2003.

8. Corr, P.B., News and comment: Interview, *Drug Discov. Technol.*, 10, 1017, 2005.
9. Grabowski, H.G., Vernon, J., and DiMasi, J.A., Returns on research and development for 1990s new drug introductions, *Pharmacoeconomics*, 20, Suppl. 3, 11–12, 2002.
10. Lin, J.H. and Lu, A.Y.H., Role of pharmacokinetics and metabolism in drug discovery and development, *Pharmacol. Rev.*, 49, 403, 1997.
11. Thompson, T.N., Early ADME in support of drug discovery: The role of metabolic stability studies, *Curr. Drug Metab.*, 1, 215, 2000.
12. White, R.E., High-throughput screening in drug metabolism and pharmacokinetic support of drug discovery, *Ann. Rev. Pharmacol. Toxicol.*, 40, 133, 2000.
13. Lee, M.S., *LC/MS Applications in Drug Development*, John Wiley & Sons, Inc.: New York, 2002.
14. Korfmacher, W.A., Ed., *Using Mass Spectrometry for Drug Metabolism Studies*, CRC Press: Boca Raton, FL, 2005.
15. Lee, M.S., Ed., *Integrated Strategies for Drug Discovery Using Mass Spectrometry*, John Wiley & Sons, Inc.: New York, 2005.
16. Venkatesh, S. and Lipper, R.A., Role of the development scientist in compound lead selection and optimization, *J. Pharm. Sci.*, 89, 145, 2000.
17. Kola, I. and Landis, J., Can the pharmaceutical industry reduce attrition rates? *Nat. Rev. Drug Discov.*, 3, 711, 2004.
18. Baillie T.A. et al., Drug metabolites in safety testing, *Toxicol. Appl. Pharmacol.*, 182, 188, 2002.
19. Prueksaritanont, T., Lin, J.H., and Baillie, T.A., Complicating factors in safety testing of drug metabolites: Kinetic differences between generated and preformed metabolites, *Toxicol. Appl. Pharm.*, 217, 143, 2006.
20. US Food and Drug Administration Guidance for Industry Safety Testing of Drug Metabolites, 2008, available at: http://www.fda.gov/downloads/Drugs/GuidanceCompliance Regulatoryinformation/Guidances/ucm079266.pdf
21. Kerns, E.H. and Di, L., *Drug-Like Properties: Concepts, Structure Design and Methods: From ADME to Toxicity Optimization*, Elsevier/Academic Press: Oxford, U.K., 2008, p. 13.
22. Rourick, R.A., More throughput does not equal less flexible or less believable data, *Presented at the Eighth Annual Symposium on Chemical and Pharmaceutical Structure Analysis*, Princeton, NJ, October 17–20, 2005.
23. van de Waterbeemd, H. and Gifford, E., ADMET in silico modelling: Towards prediction paradise? *Nat. Rev. Drug Discov.*, 2, 192, 2003.
24. Thompson, T.N., Drug metabolism in vitro and in vivo results: How do these data support drug discovery? in *Using Mass Spectrometry for Drug Metabolism Studies*, Korfmacher, W.A., Ed., CRC Press: Boca Raton, FL, 2005, p. 35.
25. Hayward, M.J. and Lee, M.S., High throughput drug discovery support: Column chromatography/MS^n strategies from the beginning to a development candidate short course, *Presented at the 56th American Society for Mass Spectrometry Conference*, Denver, CO, May 31–June 6, 2008.
26. Wells, D.A., Sample preparation for drug discovery bioanalysis, in *Integrated Strategies for Drug Discovery Using Mass Spectrometry*, Lee, M.S., Ed., John Wiley & Sons, Inc.: New York, 2005, p. 477.
27. Kazakevich, Y. and Lobrutto, R., *HPLC for Pharmaceutical Scientists*, John Wiley & Sons, Inc.: Hoboken, NJ, 2007.
28. Cody, R.B., Electrospray ionization mass spectrometry, in *Applied Electrospray Mass Spectrometry*, Pramanik, B.N., Ganguly, A.K., and Gross, M.L., Eds., Marcel Dekker, New York, 2002, p. 1.

29. Ackermann, B.A., Berna, M.J., and Murphy, A.T., Advances in high throughput quantitative drug discovery bioanalysis, in *Integrated Strategies for Drug Discovery Using Mass Spectrometry*, Lee, M.S., Ed., John Wiley & Sons, Inc.: New York, 2005, p. 315.
30. Laycock, J.D. and Hartmann, T., Automation, in *Integrated Strategies for Drug Discovery Using Mass Spectrometry*, Lee, M.S., Ed., John Wiley & Sons, Inc.: New York, 2005, p. 511.
31. Wells, D.A., *High Throughput Bioanalytical Sample Preparation Methods and Automation Strategies*, Elsevier Science: Amsterdam, the Netherlands, 2003.
32. Churchwell, M.I. et al., Improving LC–MS sensitivity through increases in chromatographic performance: Comparisons of UPLC–ES/MS/MS to HPLC–ES/MS/MS, *J. Chromatogr. B*, 825, 134, 2005.
33. Bateman, K.P. et al., MSE with mass defect filtering for in vitro and in vivo metabolite identification, *Rapid Commun. Mass Spectrom.*, 21, 1485, 2007.
34. Smith, R.D. and Johnson, A.L., Deposition method for moving ribbon liquid chromatograph-mass spectrometer interfaces, *Anal. Chem.*, 53, 739, 1981.
35. Hayes, M.J. et al., Moving belt interface with spray deposition for liquid chromatography/mass spectrometry, *Anal. Chem.*, 55, 1745, 1983.
36. Games, D.E. et al., A comparison of moving belt interfaces for liquid chromatography mass spectrometry, *Biomed. Mass Spectrom.*, 11, 87, 1984.
37. Yinon, J. and Hwang, D.G., Metabolic studies of explosives. II. High-performance liquid chromatography-mass spectrometry of metabolites of 2,4,6-trinitrotoluene, *J. Chromatogr.*, 339, 127, 1985.
38. Lee, E.D. and Henion, J.D., Micro-liquid chromatography/mass spectrometry with direct liquid introduction, *J. Chromatogr. Sci.*, 23, 253, 1985.
39. Lant, M.S., Martin, L.E., and Oxford, J., Qualitative and quantitative analysis of ranitidine and its metabolites by high-performance liquid chromatography-mass spectrometry, *J. Chromatogr.*, 323, 143, 1985.
40. Blakely, C.R. and Vestal, M.L., Thermospray interface for liquid chromatography/mass spectrometry, *Anal. Chem.*, 55, 750, 1983.
41. Irabarne, J.V., Dziedzic, P.J., and Thomson, B.A., Atmospheric pressure ion evaporation-mass spectrometry, *Int. J. Mass Spectrom. Ion Phys.*, 50, 331, 1983.
42. Whitehouse, C.M. et al., Electrospray interface for liquid chromatographs and mass spectrometers, *Anal. Chem.*, 57, 675, 1985.
43. Bruins. A.P., Covey, T.R., and Henion, J.D., Ion spray interface for combined liquid chromatography/atmospheric pressure ionization mass spectrometry, *Anal. Chem.*, 59, 2642, 1987.
44. Fenn, J.B. et al., Electrospray ionization for mass spectrometry of large biomolecules, *Science*, 246, 64, 1989.
45. Buhrman, D.L., Price, P.I., and Rudewicz, P.J., Quantitation of SR 27417 in human plasma using electrospray liquid chromatography-tandem mass spectrometry: A study of ion suppression, *J. Am. Soc. Mass Spectrom.*, 7, 1099, 1996.
46. King, R. et al., Mechanistic investigation of ionization suppression in electrospray ionization, *J. Am. Soc. Mass Spectrom.*, 11, 942, 2000.
47. King, R., Ion formation from complex solutions: Understanding matrix effects and ionization suppression, *Presented at the Ninth Annual Symposium on Chemical and Pharmaceutical Structure Analysis*, Princeton, NJ, October 17–19, 2006.
48. Doerge, D.R. et al., Analysis of methylphenidate and its metabolite ritalinic acid in monkey plasma by liquid chromatography/electrospray ionization mass spectrometry, *Rapid Commun. Mass Spectrom.*, 14, 610, 2000.

49. Murphy, A.T. et al., Determination of xanomeline and active metabolite, *N*-desmethylxanomeline in human plasma by liquid chromatography-atmospheric pressure chemical ionization mass spectrometry, *J. Chromatogr. B. Biomed Appl.*, 668, 273, 1995.
50. Tiller, P.R. et al., Drug quantitation on a benchtop liquid chromatography-tandem mass spectrometry system, *J. Chromatogr. A.*, 771, 119, 1997.
51. Clauwaert, K.M. et al., Investigation of the quantitative properties of the quadrupole orthogonal acceleration time-of-flight mass spectrometer with electrospray ionization using 3,4-methylendioxymethamphetamine, *Rapid Commun. Mass Spectrom.*, 13, 1540, 1999.
52. Yang, L., Wu, N., and Rudewicz, P.J., Applications of new liquid chromatography-tandem mass spectrometry technologies for drug development support, *J. Chromatogr. A*, 926, 43, 2001.
53. Bateman, K. et al., Full scan data acquisition for rapid quantitative and qualitative analysis using a bench-top non-hybrid ESI-Orbitrap mass spectrometer, *Presented at the 56th American Society for Mass Spectrometry Conference*, Denver, CO, May 31–June 6, 2008.
54. McLafferty, F.W., *Tandem Mass Spectrometry*, John Wiley & Sons Inc.: New York, 1983.
55. Yost, R.A. and Boyd, R.K., Tandem mass spectrometry: Quadrupole and hybrid instruments, *Methods Enzymol.*, 193, 154, 1990.
56. Lee, M.S. and Yost, R.A., Rapid identification of drug metabolites with tandem mass spectrometry, *Biomed. Environ. Mass Spectrom.*, 15, 193, 1988.
57. Clarke, N.J. et al., Systematic LC/MS metabolite identification in drug discovery, *Anal. Chem.*, 73, 430A, 2001.
58. Fouda, H. et al., Quantitative analysis by high-performance liquid chromatography atmospheric pressure chemical ionization mass spectrometry: The determination of the renin inhibitor CP-80, 794 in human serum, *J. Am. Soc. Mass Spectrom.*, 2, 164, 1991.
59. Covey, T.R., Lee, E.D., and Henion, J.D., High-speed liquid chromatography/tandem mass spectrometry for the determination of drugs in biological samples, *Anal. Chem.*, 58, 2453, 1986.
60. Hiller, D.L. et al., Rapid screening technique for the determination of optimal tandem mass spectrometric conditions for quantitative analysis, *Rapid Commun. Mass Spectrom.*, 11, 593, 1997.
61. Decaestecker, T.N. et al., Evaluation of automated single mass spectrometry to tandem mass spectrometry function switching for comprehensive drug profiling analysis using a quadrupole time-of-flight mass spectrometer, *Rapid Commun. Mass Spectrom.*, 14, 1787, 2000.
62. Lopez, L.L. et al., Identification of drug metabolites in biological matrices by intelligent automated liquid chromatography/tandem mass spectrometry, *Rapid Commun. Mass Spectrom.*, 12, 1756, 1998.
63. Korfmacher, W.A. et al., Cassette-accelerated rapid rat screen: A systematic procedure for the dosing and liquid chromatography/atmospheric pressure ionization tandem mass spectrometric analysis of new chemical entities as part of new drug discovery, *Rapid Commun. Mass Spectrom.*, 15, 335, 2001.
64. Lewis, K., Samples to answers: An optimized process for discovery, *Presented at the 10th Annual Symposium on Chemical and Pharmaceutical Structure Analysis*, Langhorne, PA, October 22–25, 2007.
65. Julian, R., Long-term data access using open standards, *Presented at the Ninth Annual Symposium on Chemical and Pharmaceutical Structure Analysis*, Princeton, NJ, October 17–19, 2006.

66. Olah, T.V., McLoughlin, D.A., and Gilbert, J.D., The simultaneous determination of mixtures of drug candidates by liquid chromatography/atmospheric pressure chemical ionization mass spectrometry as an *in vivo* drug screening procedure, *Rapid Commun. Mass Spectrom.*, 11, 17, 1997.
67. Berman, J. et al., Simultaneous pharmacokinetic screening of a mixture of compounds in the dog using API LC/MS/MS analysis for increased throughput, *J. Med. Chem.*, 40, 827, 1997.
68. Shaffer, J.E. et al., Use of "n-in one" dosing to create an *in vivo* pharmacokinetics database for use in developing structure-pharmacokinetic relationships, *J. Pharm. Sci.*, 88, 313, 1999.
69. Korfmacher, W.A. et al., Development of an automated mass spectrometry system for the quantitative analysis of liver microsomal incubation samples: A tool for rapid screening of new compounds for metabolic stability, *Rapid Commun. Mass Spectrom.*, 13, 901, 1999.
70. Wu, J.T. et al., Direct plasma sample injection in multiple-component LC-MS-MS assays for high-throughput pharmacokinetic screening, *Anal. Chem.*, 72, 61, 2000.
71. White, R.E. and Manitpisitkul, P., Pharmacokinetic theory of cassette dosing in drug discovery screening, *Drug Metab. Dispos.*, 29, 957, 2001.
72. Hop, C.E.C.A. et al., Plasma-pooling methods to increase throughput for *in vivo* pharmacokinetic screening, *J. Pharm. Sci.*, 87, 901, 1998.
73. Lipinski, C.A. et al., Experimental and computational approaches to estimate solubility and permeability in drug discovery and development settings, *Adv. Drug Del. Rev.*, 23, 3, 1997.
74. Kerns, E.H., High throughput physicochemical profiling for drug discovery, *J. Pharm. Sci.*, 90, 1838, 2001.
75. Guo, Y. and Shen, H., pKa, solubility, and lipophilicity, in *Optimization in Drug Discovery*, Yan, Z. and Caldwell, G.W., Eds., Humana Press: Totowa, NJ, 2004, p. 1.
76. Grant, D.J.W. and Higuchi, T., *Solubility Behavior of Organic Compounds*, John Wiley & Sons, Inc.: New York, 1990.
77. Yalkowsky, S.H. and Banerjee, S., *Aqueous Solubility: Methods of Estimation for Organic Compounds*, Marcel Dekker: New York, 1992.
78. Avdeef, A., Physicochemical profiling (solubility, permeability and charge state), *Curr. Top. Med. Chem.*, 1, 277, 2001.
79. Lombardo, F. et al., ElogPoct: A tool for lipophilicity determination in drug discovery, *J. Med. Chem.*, 43, 2922, 2000.
80. Kerns, E.H. and Di, L., High throughput metabolic and physicochemical property analysis: Strategies and tactics for impact on drug discovery success, in *Integrated Strategies for Drug Discovery Using Mass Spectrometry*, Lee, M.S., Ed., John Wiley & Sons, Inc.: Hoboken, NJ, 2005.
81. Zhao, Y. et al., High-throughput logP measurement using parallel liquid chromatography/ultraviolet mass spectrometry and sample-pooling, *Rapid Commun. Mass Spectrom.*, 16, 1548, 2002.
82. ACD/PhysChem Suite, ACD/pK_a DB, Advanced Chemistry Development, Toronto, Canada (see http://www.acdlabs.com/products/phys_chem_lab/pka/).
83. Chaturvedi, P.R., Decker, C.J., and Odinecs, A., Prediction of pharmacokinetic properties using experimental approaches during early drug discovery, *Curr. Opin. Chem. Biol.*, 5, 452, 2001.
84. Riley, R.J., Martin, I.J., and Cooper, A.E., The influence of DMPK as an integrated partner in modern drug discovery, *Curr. Drug Metab.*, 3, 527, 2002.
85. Caldwell, G.W. et al., In vitro permeability of eight beta-blockers through caco-2 monolayers utilizing liquid chromatography/electrospray ionization mass spectrometry, *J. Mass Spectrom.*, 33, 607, 1998.

86. Fernandez-Metzler, C.L., Subrumanian, R., and King, R.C., Application of technological advances in biotransformation studies, in *Integrated Strategies for Drug Discovery Using Mass Spectrometry*, Lee, M.S., Ed., John Wiley & Sons, Inc.: New York, 2005.
87. Perchalski, R.J., Wilder, B.J., and Yost, R.A., Structural elucidation of drug metabolites by triple quadrupole mass spectrometry, *Anal. Chem.*, 54, 1466, 1982.
88. Lee, M.S., Yost, R.A., and Perchalski, R.J., Tandem mass spectrometry for the identification of drug metabolites, *Annu. Rep. Med. Chem.*, 21, 313, 1986.
89. Kerns, E.H. et al., Buspirone metabolite structure profile using a standard liquid chromatographic-mass spectrometric protocol, *J. Chromatogr. B*, 698, 133, 1997.
90. Zhu, M. et al., Detection and characterization of metabolites in biological matrices using mass defect filtering of liquid chromatography/high resolution mass spectrometry data, *Drug Metab. Dispos.*, 34, 1722, 2006.
91. Ackermann, B.L. et al., Increasing the throughput of microsomal stability screening using fast gradient elution LC/MS, in *Proceedings of the 46th ASMS Conference on Mass Spectrometry and Allied Topics*, Orlando, FL, 16, 1998.
92. Jenkins, K.M. et al., Automated high throughput adme assays for metabolic stability and cytochrome p450 inhibition profiling of combinatorial libraries, *J. Pharm. Biomed. Anal.*, 34, 989, 2004.
93. Kalgutkar, A.S. et al., On the diversity of oxidative bioactivation reactions on nitrogen-containing xenobiotics, *Curr. Drug Metab.*, 3, 379, 2002.
94. Zhou, S. et al., Drug bioactivation, covalent binding to target proteins and toxicity relevance, *Drug Metab. Rev.*, 27, 41, 2005.
95. Ma, S. and Subrumanian, R.J., Detecting and characterizing reactive metabolites by liquid chromatography/tandem mass spectrometry, *J. Mass Spectrom.*, 41, 1121, 2006.
96. Yan, Z. and Caldwell, G.W., Stable-isotope trapping and high-throughput screenings of reactive metabolites using the isotope ms signature, *Anal. Chem.*, 76, 6835, 2004.
97. Ayrton, J. et al., Application of a generic fast gradient liquid chromatography tandem mass spectrometry method for the analysis of cytochrome p450 probe substrates, *Rapid Commun. Mass Spectrom.*, 12, 217, 1998.
98. Bu, H.-Z. et al., High-throughput cytochrome p450 (CYP) inhibition screening via a cassette probe-dosing strategy. VI. Simultaneous evaluation of inhibition potential of drugs on human hepatic isozymes CYP 2A6, 3A4, 2C9, 2D6 and 2E1, *Rapid Commun. Mass Spectrom.*, 15, 741, 2001.
99. Frye, R.F. et al., Validation of the five-drug "pittsburgh cocktail" approach for assessment of selective regulation of drug-metabolizing enzymes, *Clin. Pharmacol. Ther.*, 62, 365, 1997.
100. Scott, R.J. et al., Determination of a "gw cocktail" of cytochrome p450 probe substrates and their metabolites in plasma and urine using automated solid phase extraction and fast gradient liquid chromatography tandem mass spectrometry, *Rapid Commun. Mass Spectrom.*, 13, 2305, 1999.
101. Yuan, R. et al., Evaluation of cytochrome p450 probe substrates commonly used by the pharmaceutical industry to study *in vitro* drug interactions, *Drug Metab. Disp.*, 30, 1311, 2002.
102. Tang, C. et al., Effect of albumin on phenytoin and tolbutamide metabolism in human liver microsomes: An impact more than protein binding, *Drug Metab. Dispos.*, 30, 648, 2002.
103. Masimirembwa, C.M., Bredberg, U., and Andersson, T.B., Metabolic stability for drug discovery and development: pharmacokinetic and biochemical challenges, *Clin. Pharmacokinet.*, 42, 515, 2003.
104. Cohen, L.H., Plasma protein-binding methods in drug discovery, in *Optimization in Drug Discovery*, Yan, Z. and Caldwell, G.W., Eds., Humana Press: Totowa, NJ, 2004, p. 111.

3 PK Principles and PK/PD Applications

Hong Mei and Richard A. Morrison

CONTENTS

3.1 INTRODUCTION

Both quantitative and qualitative data generated by liquid chromatography/mass spectrometry (LC/MS) are important for pharmacokinetic (PK) and pharmacodynamic (PD) studies during the drug discovery process. PK parameters are used to describe how the "body reacts to the drug," while PD parameters are used to describe how the "drug affects the body." Efforts in improving PK properties in drug discovery have contributed to the reduction of attrition rate due to poor PK/bioavailability [1] in the clinic. Attrition rate due to lack of efficacy might be reduced with a better understanding of the relationship between PK properties (e.g., drug exposure) and PD properties (e.g., efficacy, duration). Historically, PK and PD were assessed in separate experiments precluding robust PK/PD analyses. A combined PK/PD study links the concentration–time profile to the intensity and duration of pharmacological response and reveals the temporal relationship between drug exposure and response. A thorough understanding of PK and PK/PD during the drug discovery phase should result in a reduction of the attrition rate of drug candidates at the proof-of-concept phase.

During lead optimization, PK studies are designed to address exposure-related issues including low and variable absorption, high and variable first-pass effect, inappropriate half-life (too short or too long), effect of food, gender, formulation, and dose. PK/PD studies are designed to address the relationship between exposure and efficacy. One should keep in mind that the desired PK properties also depend on the PK/PD relationship. For example, a short half-life ($t_{1/2}$) compound might be acceptable for a relatively slower PD effect or for an agonist mechanism where tolerance could occur.

This chapter is written for nonpharmacokinetic scientists, with the focus on understanding principles of PK and PK/PD and their application in drug discovery.

3.2 KEY PK PARAMETERS FOR DRUG DISCOVERY APPLICATIONS

PK parameters can be characterized as "descriptive" or "mechanistic" parameters. The descriptive parameters are those used to describe exposure, while the mechanistic ones are the intrinsic properties of compounds controlling the extent and duration of exposure. Understanding the true meaning and relationship between these two categories is important for improving the chemical and physicochemical properties of drug candidates.

Major descriptive PK parameters of exposure are peak concentration (C_{max}), trough concentration (C_{min}), area under the plasma concentration versus time curve (AUC), and bioavailability (F) as illustrated in the time–plasma concentration profile in Figure 3.1. Primary mechanistic PK parameters contributing to the extent of drug exposure are clearance (CL) and volume of distribution (V_{dss}). The ratio of V_{dss}-to-CL is the mean residence time (MRT), an intrinsic parameter that characterizes the residence time of drug molecules in the body. CL, V_{dss}, and MRT can be called dispositional PK parameters as well. Fraction absorbed (F_a), gut bioavailability (F_{gut}), and hepatic bioavailability (F_h) are the three major mechanistic parameters that control the total bioavailability (F).

3.2.1 C_{MAX}, C_{MIN}, AND AUC

C_{max}, C_{min}, and AUC are used to describe the extent of exposure. The magnitude of exposure reflects the integrated effect of the dispositional properties of a compound, such as bioavailability, absorption time, central volume of distribution, and

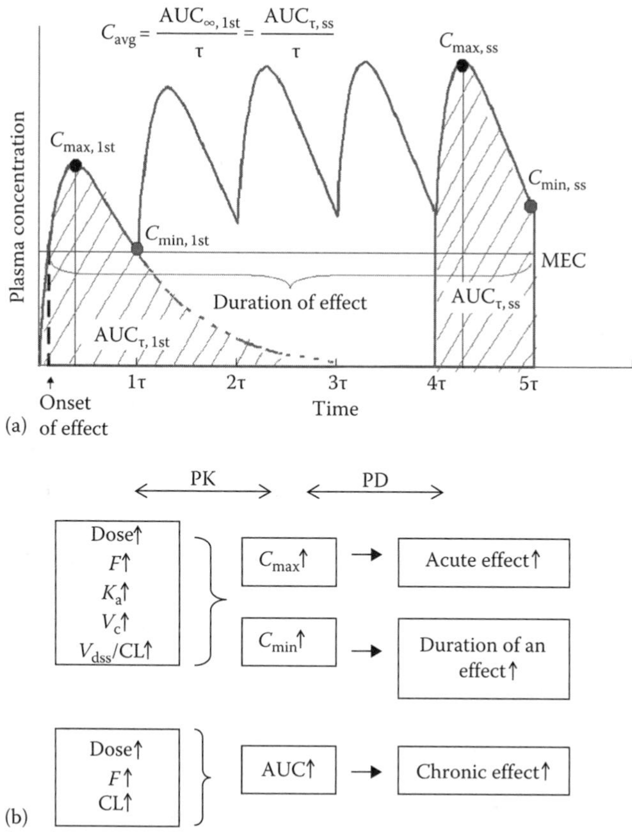

FIGURE 3.1 Schematic description of (a) exposure parameters and (b) their correlation with pharmacological or toxicological effects. MEC, minimum effective concentration.

elimination rate. In addition, these exposure parameters are also used to correlate with various types of pharmacological or toxicological effects.

C_{max} and C_{min} are the peak concentration and trough concentration at the end of the dosing interval τ, respectively. C_{max} and C_{min} are time-dependent exposure parameters reflecting the range of exposure in the dosing interval. The ratio of C_{max}/C_{min} is also a useful parameter reflecting the fluctuation of exposure. C_{max}, C_{min}, and C_{max}/C_{min} ratio reflect the integrated effect of bioavailability, volume of distribution, mean absorption time (MAT), and elimination half-life. C_{max} and C_{min} increase with increasing dose and bioavailability as long as solubility and permeability are not limiting and elimination is not saturated. C_{max} and C_{min} also increase with decreasing central volume of distribution. C_{max} and C_{min} increase with shorter MAT and longer elimination $t_{1/2}$, respectively, when MAT is shorter than elimination $t_{1/2}$. There are situations when we would like to manipulate C_{max} and C_{min} for different purposes. For example, to determine the safety margin versus cardiac effects, a C_{max} target of 10–30-fold greater than the therapeutic C_{max} is desired. To obtain such a high C_{max}

can be a challenge for low solubility compounds. The following approaches can be used to increase C_{max}:

1. Formulation with better solubility
2. Reduction of particle size
3. Changing crystalline form to amorphous form
4. Double dosing with a few hours between doses

C_{max} is often used to correlate with acute pharmacological/toxicological effects, such as blood pressure, heart rate, nausea, and seizures. A low C_{max}/C_{min} ratio profile is often preferred clinically to avoid possible C_{max}-related acute toxicity. C_{max}/C_{min} ratio can be improved with a more frequent dosing interval. However, once-a-day dosing is the most compliant and commercially viable dose regimen and is typically targeted during discovery. A desired effective $t_{1/2}$ would be ≥10 h for once-a-day dosing for a compound to maintain a $C_{max}/C_{min} \leq 4$, when MAT is around 1 h. Unlike C_{max} and C_{min}, AUC is a time-averaged parameter reflecting overall exposure (Figure 3.1a). Factors contributing to the magnitude of AUC are dose, bioavailability, and clearance. AUC increases with increased dose as long as solubility and permeability are not limiting and elimination is not saturated (Figure 3.1b). Time factors such as absorption time and elimination $t_{1/2}$ have no effect on AUC. Since AUC is a time-averaged exposure it is often used to correlate with chronic efficacy or chronic toxicity.

"Exposure multiple or exposure ratio" is a common term used in the pharmaceutical industry to describe the fold of exposure of C_{max} or AUC observed at doses used in toxicity studies, compared to the C_{max} or AUC at the highest marketed dose in humans or projected therapeutic dose based on preclinical data. This number provides an estimate for the safety margin of a compound and is helpful for guiding the dose selection of drug safety studies.

The accumulation ratio (R_A) can be obtained by comparing exposure parameters at steady state after multiple dosing at a fixed dosing interval (τ) to the corresponding value at the first dose:

$$R_{A,AUC} = \frac{AUC_{\tau,ss}}{AUC_{\tau,1st}} \tag{3.1}$$

$$R_{A,C_{max}} = \frac{C_{max,ss}}{C_{max,1st}} \tag{3.2}$$

$$R_{A,C_{min}} = \frac{C_{min,ss}}{C_{min,1st}} \tag{3.3}$$

where

$R_{A,AUC}$, $R_{A,C_{max}}$, and $R_{A,C_{min}}$ are accumulation ratio of AUC, C_{max}, and C_{min}, respectively

$AUC_{\tau,ss}$ and $AUC_{\tau,1st}$ are AUCs during a dosing interval at steady state and the first dose, respectively

$C_{max,ss}$ and $C_{max,1st}$ are the peak concentration at steady state and first dose, respectively
$C_{min,ss}$ and $C_{min,1st}$ are the trough concentration at the steady state and first dose, respectively

Accumulation ratio based on C_{max}, C_{min}, or AUC can be different. Accumulation estimated from C_{min} comparison tends to overestimate, while AUC comparison gives the accumulation of overall exposure, and is not that sensitive to the variability of C_{min} values [2]. Estimating possible accumulation is also important for designing multiple dosing studies in drug safety studies. If a plasma concentration profile is available for a certain dose, superposition [3] should be used to estimate the plasma profile at steady state. R_A can be estimated for the different exposure parameters.

In summary, descriptive exposure parameters are useful and reflect the overall dispositional properties of compounds. They can be used for screening and ranking compounds, comparing exposure changes after changing test conditions, and investigating the relationship between exposure and efficacy/toxicity.

3.2.2 Clearance

Clearance is an intrinsic PK parameter that controls the extent of drug exposure. Clearance by definition is a proportionality factor between the rate of elimination and its concentration at the site of measurement. Based on this definition, the following equation can be derived:

$$CL_s = \frac{F \cdot \text{Dose}}{\text{AUC}} \tag{3.4}$$

where F is the bioavailability. As shown, systemic clearance is the ratio of bioavailable dose to AUC, a parameter reflecting elimination efficiency. Clearance can be understood as a flow; it is a time-averaged value corresponding to a volume of plasma completely cleared per unit time. As a compound flows through different elimination organs, it can be extracted by each organ at a different extent, and systemic CL is a sum of all organ clearances, as expressed by:

$$CL_s = CL_{liver} + CL_{kidney} + CL_{lung} + CL_{other} \tag{3.5}$$

where CL_{liver}, CL_{kidney}, CL_{lung}, and CL_{other} are clearance of liver, kidney, lung, and other organs, respectively.

CL can be easily converted to a more meaningful number called blood-flow-normalized CL that reflects the elimination efficiency. This is related to the notion that CL is a flow that takes a compound out of the systemic circulation. For example, the maximum liver clearance should not exceed the liver blood flow, since the clearing rate cannot exceed the delivery rate. The value of liver clearance is therefore more meaningful when compared to its blood flow as expressed by the extraction ratio $E = CL_h/Q_h$, where CL_h is hepatic clearance and Q_h is hepatic blood flow. Typically, the liver is the major organ for elimination. Liver has a relatively high blood flow. Compounds can be arbitrarily categorized as low, moderate, and high clearance using the CL/Q_h less than 0.3, between 0.3 and 0.7, and greater than 0.7, respectively. Comparing systemic CL with hepatic blood flow in each species is very informative when one needs to compare CL across species in terms of elimination efficiency. Table 3.1 compares different clearance values in rat, monkey, dog, and human using the CL/Q_h value. For example, a CL of 50 mL/min/kg in rat equates to 26 mL/min/kg in monkey, 19 mL/min/kg in dog, and 13 mL/min/kg in human, in terms of elimination efficiency.

In vivo systemic CL is usually estimated by the AUC obtained after intravenously (IV) dosing a compound using Equation 3.4 with F assumed to be unity (i.e., 100% bioavailability). The three frequently ignored factors that cause errors for CL estimation in a drug discovery environment are low solubility, compound instability in plasma, and blood cell partitioning. Many drug discovery compounds have very low solubility, and can precipitate in the dosing vial or the circulation in small animals, causing an overestimation of both CL and central volume of distribution. Using lower dosing concentration or slower IV infusion rates can prevent such errors. Significant partitioning into blood cells can cause an apparent high CL for a compound when using plasma concentrations for calculations. Blood CL (CL_b) value should be adjusted for compounds with significant blood cell penetration using this equation:

$$CL_b = CL_p \cdot \frac{C_p}{C_b} \tag{3.6}$$

where

CL_p is compound clearance in plasma

C_p and C_b are compound concentration in plasma and whole blood, respectively

TABLE 3.1
Liver Blood Flow and Liver Blood Flow Normalized Clearance in Common Species

	CL (mL/min/kg)			
CL/Qh	**Rat**	**Monkey**	**Dog**	**Human**
1.0	84	44	31	21
0.9	76	40	28	19
0.8	67	35	25	17
0.7	59	31	22	15
0.6	50	26	19	13
0.5	42	22	16	11
0.4	34	18	12	8
0.3	25	13	9	6
0.2	17	9	6	4
0.1	8	4	3	2

Some early PK studies in drug discovery are carried out without a full understanding of the stability in plasma. Poor plasma stability can cause an overestimation of systemic CL. Measures for stabilizing labile compounds need to be taken to obtain an accurate CL when a compound proceeds to a lead stage.

Renal CL (CL_r) can be easily estimated by comparing the amount of compound excreted in urine to plasma AUC using the following equation:

$$CL_r = \frac{A_{ex,t}}{AUC_{0-t}} \tag{3.7}$$

or

$$CL_r = \frac{A_{ex,\infty}}{Dose} \cdot CL_s \tag{3.8}$$

The total amount excreted in urine up to time t ($A_{ex,t}$) can be estimated by measuring the total volume of urine and the compound concentration in urine, while the entire excreted amount ($A_{ex,\infty}$) can be estimated using a similar method with the confirmation that no more compound can be detected in the last portion of urine collection. Early PK studies usually do not collect urine that requires use of metabolic cages. Renal CL usually is estimated when a compound proceeds to a lead stage. In a routine metabolite identification or mass balance study, urine is collected and can be used for renal CL estimation.

The estimation of *in vivo* hepatic CL requires cannulation of the portal vein. By comparing AUC of portal vein dosing and systemic dosing, CL_h can be estimated with the following equations:

$$F_h = \frac{AUC_{pv}^{sys}/Dose_{pv}}{AUC_{iv}^{sys}/Dose_{iv}} \tag{3.9}$$

and

$$CL_h = (1 - F_h) \cdot Q_h \tag{3.10}$$

where

AUC_{pv}^{sys} and AUC_{iv}^{sys} are AUCs of systemic sampling from portal vein dosing and systemic IV dosing, respectively

Q_h is the standard liver blood flow listed in Table 3.2

Hepatic clearance can also be estimated from *in vitro* intrinsic metabolic CL obtained by incubation of a compound with hepatocytes or liver microsomes. This method requires fewer resources than the *in vivo* approach and is more suitable for screening a large number of compounds; however, there are documented cases when *in vitro* clearance does not accurately predict *in vivo*. Intrinsic metabolic CL (CL_{int}) is a parameter that only reflects the intrinsic ability of liver to metabolize

TABLE 3.2
Scaling Factors and Liver Blood Flow for Estimating *In Vivo* Hepatic Clearance from *In Vitro* Intrinsic CL

	Rat	Monkey	Dog	Man
Body weight (kg)	0.25	5	10	70
Liver weight (g/kg bodyweight)	40	18	27	24
Microsomal yield (mg/g liver)	45	45	45	32
Microsomal yield (mg/kg BW)	1800	810	121	768
Scaling factor: μL/min/mg protein to mL/min/kg	1.8	0.81	1.21	0.77
Hepatocellularity (10E6 cells/g liver)	128	99	188	99
Hepatocellularity (10E6 cells/kg BW)	5120	1782	5076	2376
Scaling factor: μL/min/10E6 cells to mL/min/kg	5.1	1.8	5.1	2.4
Hepatic blood flow (mL/min/kg)	84	44	31	21

drug molecules without any limitation. However, there are biological factors that can limit the accessibility of drug molecules to these enzymes, such as protein binding and liver blood flow. There are several models that incorporate these limitations for extrapolating *in vitro* intrinsic CL to *in vivo* hepatic CL. The following well-stirred model is the most popular:

$$\mathrm{CL_h} = \frac{\mathrm{CL_{int}} \cdot \mathrm{SF} \cdot \dfrac{f_{\mathrm{u,blood}}}{f_{\mathrm{u,incubation}}} \cdot Q_\mathrm{h}}{\mathrm{CL_{int}} \cdot \mathrm{SF} \cdot \dfrac{f_{\mathrm{u,blood}}}{f_{\mathrm{u,incubation}}} + Q_\mathrm{h}} \tag{3.11}$$

where

SF is a scaling factor for converting the unit of $\mathrm{CL_{int}}$ from μL/mL/million cells (hepatocytes) or μL/mL/mg (microsomes) to mL/min/kg (body weight)

$f_{\mathrm{u,blood}}$ and $f_{\mathrm{u,incubation}}$ are free fraction in blood and free fraction in incubation, respectively

Q_h is liver blood flow

The scaling factors and liver blood flow rates of the common species are listed in Table 3.2. By comparing renal CL and hepatic CL to total systemic CL, the major route of elimination usually can be identified.

The following relationship can be easily obtained by rearranging Equation 3.4

$$\frac{\mathrm{AUC}}{\mathrm{Dose}} = \frac{F}{\mathrm{CL}} \tag{3.12}$$

Thus dose-normalized AUC reflects overall exposure efficiency, and it actually equals the delivery efficiency (F) divided by the elimination efficiency CL as

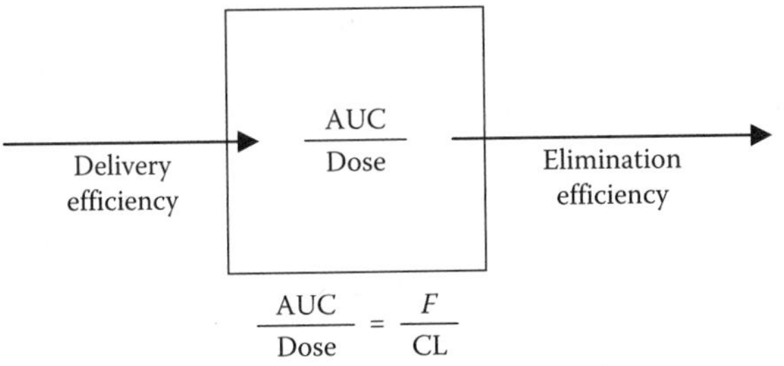

FIGURE 3.2 Illustration of exposure efficiency (AUC/dose) controlled by F and CL. (From Kwon, Y., *Handbook of Essential Pharmacokinetics, Pharmacodynamics and Drug Metabolism for Industrial Scientists*, Kluwer Academic/Plenum Publishers, New York, 2001. With permission.)

shown in Figure 3.2. Therefore, dose-normalized AUC reflects the combined effects of bioavailability and clearance and can be used for screening or ranking overall exposure efficiency of drug discovery compounds, which will be discussed later. Comparing dose-normalized AUC is comparing the exposure efficiency. During rising dose studies, in addition to comparing the increase of absolute AUC values at increasing doses, a comparison of dose-normalized AUC provides valuable information on the changes of exposure efficiency at increased dose levels. When the exposure efficiency remains the same at different dose levels, it is called linear PK or dose-proportional PK. Figure 3.3 shows how to use dose-normalized AUC to evaluate the PK linearity or dose proportionality or the exposure efficiency. The mechanistic cause for changes in exposure efficiency with increasing dose is mostly due to the saturation of an absorption or elimination mechanism, such as saturated solubility/dissolution or saturation of transporters, protein binding or enzymes. For example, the possible mechanisms for decreased exposure efficiency with increasing doses could be reduced absorption due to saturated concentration at higher dose and/or increased CL due to increased free fraction as a result of saturation of plasma protein binding [4]. Similarly, the possible causes for increased exposure efficiency

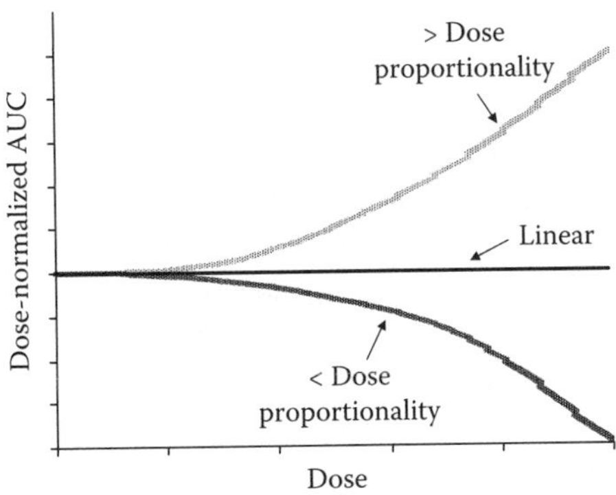

FIGURE 3.3 Dose-normalized AUC for evaluation of exposure linearity.

could be enhanced absorption from saturation of efflux in gut and/or decreased CL resulting from saturation of metabolic enzymes [4]. There are also other mechanisms for nonlinear PK. The autoinduction of metabolic enzymes is a common cause for increased CL after multiple dosing.

Oral clearance is a common term used in the pharmaceutical industry to describe the apparent CL of an orally dosed compound, expressed as CL/F = Dose/AUC. Oral clearance is the reciprocal of dose-normalized AUC. Oral clearance reflects the overall exposure efficiency, without differentiating the efficiency of compound delivery or elimination, while IV clearance only reflects the exposure efficiency caused by elimination. Oral clearance is equal or higher than IV clearance. When bioavailability is high, oral clearance is very close to IV clearance. If the bioavailability is low, either due to low absorption or high first-pass effect, oral clearance can be much higher than IV clearance. High oral clearance indicates very low exposure efficiency. When absorption is complete ($F_a\%$ = 100), oral CL is actually the intrinsic hepatic clearance for a compound when hepatic metabolism is the major elimination route. Neither oral clearance nor intrinsic CL is limited by the blood flow.

3.2.3 Bioavailability

Bioavailability is defined as the percentage of administered dose that reaches the systemic circulation unchanged. It measures the delivery efficiency of a compound into the circulation. Total bioavailability is an important PK parameter for an orally dosed drug. Compounds with low bioavailability not only require higher doses but also generate more variable exposures. Total bioavailability (F) is estimated by comparing dose-normalized AUC from a non-IV route to that from an IV route. When oral bioavailability is too low, the cause needs to be investigated, in order to synthesize new analogs that address the issue. Therefore understanding the factors that contribute to oral bioavailability and methods for evaluating these factors would be helpful. Oral bioavailability can be dissected into three major components: fraction absorbed (F_a), gut bioavailability (F_{gut}), and hepatic bioavailability (F_h). Each component relates to different properties of a compound as shown in Figure 3.4.

F_a is the fraction of dose absorbed into enterocytes from the intestinal lumen after oral administration. The two major processes involved are (1) the dissolution of solid particles into gastrointestinal (GI) fluid and (2) the permeation of molecules across intestinal membranes.

The solid form of a compound must disintegrate and dissolve in the GI fluid before absorption can occur. In general, disintegration occurs much faster than dissolution. Compounds with high lipophilicity and/or crystallinity can have slow dissolution that can be the rate-limiting process during absorption. The reduction of particle size or other formulation strategies are effective approaches to improve the absorption for such compounds. On the other hand, interstudy variability in exposure and efficacy can be very large when formulation and particle size are not very well controlled, especially in early drug discovery.

After molecules dissolve into GI fluids, they need to permeate through intestinal membrane to reach the systemic circulation. There are several potential pathways for

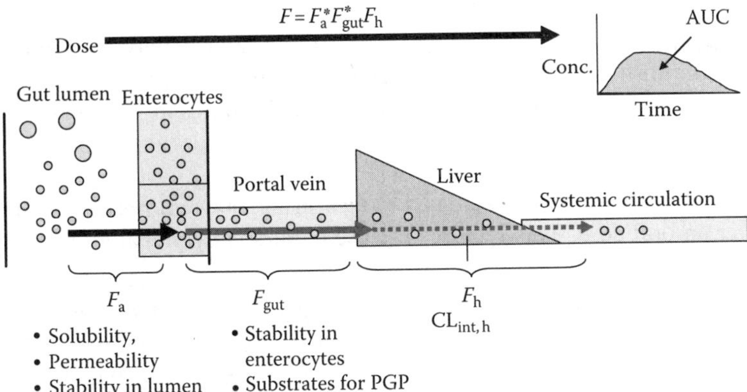

FIGURE 3.4 Schematic illustration of the role of fraction absorbed (F_a), gut bioavailability (F_{gut}), and liver bioavailability (F_h) in total oral bioavailability (F).

intestinal permeation: (1) transcellular passive diffusion through the epithelial cells for unionized molecules; (2) paracellular diffusion through the tight conjunctions for very small hydrophilic molecules (molecular mass <350 Da); and (3) transporter-mediated uptake [5]. Due to comparatively large capacity, passive transcellular diffusion is the major pathway for most small molecules (molecular mass <500–700 Da). Only the unionized molecules within the right range of lipophilicity can partition into the intestinal membrane to complete the passive permeation process. For highly water-soluble compounds, permeation can be a rate-limiting step for overall absorption. If there are no uptake transporters to facilitate the absorption, and the molecular size is too big to pass through the tight conjunctions, the absorption of this class of compounds is very poor. *In vitro* passive permeability can be assessed using Caco-2 cells or a parallel artificial membrane permeability assay (PAMPA) system. Studies have established the correlation between F_a with Caco-2 permeability [6] as well as PAMPA permeability [7] with effective permeability (P_{eff}) measured using human small intestine. Due to the reported interlaboratory differences between *in vitro* permeability and *in vivo* F_a, each laboratory needs to "calibrate" with compounds of known F_a [8].

Absorption is dependent on both dissolution and permeability; therefore it depends on pK_a and lipophilicity. In general log $D_{7.4}$ within the range of −0.5 to 2 is considered to be optimal for oral absorption [9].

The fraction absorbed (F_a) can be evaluated by direct and indirect methods. Direct estimation requires radiolabeled compounds.

IV/PO studies of radiolabeled compounds: F_a can be estimated by comparing dose-normalized AUC of radioactivity after IV and oral dosing of a radiolabeled compound with the following equation assuming linear PK:

$$F_a = \frac{\text{AUC}_{\text{radioacitivity,PO}} / \text{Dose}_{\text{radioacitvity,PO}}}{\text{AUC}_{\text{radioactivity,IV}} / \text{Dose}_{\text{radioactivity,IV}}} \quad (3.13)$$

where

$AUC_{radioactivity,PO}$ and $AUC_{radioactivity,IV}$ are AUCs of radioactivity after oral and IV administration of a radioactive compound, respectively

$Dose_{radioactivity,PO}$ and $Dose_{radioactivity,IV}$ are radioactive doses for oral and IV administration, respectively.

Mass balance: F_a can also be estimated by measuring total radioactivity in urine and bile after an oral dose. This requires the use of surgically prepared bile duct cannulated animals.

Indirect method: F_a can be estimated indirectly from IV/PO PK of cold compounds with the following assumptions: (1) liver is the major organ for elimination, (2) linear PK, (3) plasma concentration is the same as blood concentration, and (4) there is no gut first-pass loss.

$$F_a = \frac{F}{F_h} = \frac{F}{1 - \dfrac{CL}{Q_h}} \tag{3.14}$$

where F is obtained by the comparison of dose-normalized AUC after IV and oral dosing, and CL is the systemic CL. Mathematically, this method is overly sensitive to variability under the following two conditions: (1) when F_h is very close to F, one can get $F_a >> 1$ (from a practical point of view, we can ignore this sensitivity issue since it can be concluded that F_a is high) and (2) when CL/Q_h is close to 1, the denominator of this equation can be very small. This equation can generate a wide range of F_a with a small change in CL/Q_h at this condition. Caution should be taken to estimate F_a when CL/Q_h is high [10].

The first-pass effect refers to presystemic elimination by the gut or liver. After being absorbed into enterocytes, compounds can be metabolized by gut and liver metabolic enzymes or effluxed by transporters before they reach the systemic circulation.

F_{gut} is the fraction of drug absorbed into the enterocytes that escapes presystemic intestinal elimination. It is the fraction delivered into the portal vein after gut first pass. Compounds with significant gut first pass tend to have high exposure variability since the enzymes or transporters responsible for the first pass could be induced, inhibited, or saturated at different extent among individuals [11]. Therefore, this undesired property should be screened out in the lead optimization stage.

It is difficult to estimate F_{gut} in animals, since it requires difficult surgery and portal vein sampling [12]. Without a special port into the portal vein, F_{gut} can be estimated from IV/PO PK studies with radiolabeled compounds, with an assumption that liver is the major organ for elimination:

$$F_{gut} = \frac{F}{F_a \cdot F_h} = \frac{F}{F_a \cdot \left(1 - \dfrac{CL}{Q_h}\right)} \tag{3.15}$$

where

F is obtained by comparing the dose-normalized AUC of compound after IV and oral dosing

F_a is obtained by comparing the dose-normalized AUC of radioactivity after IV and oral dosing

CL is the systemic CL obtained after IV dosing

Q_h is the liver blood flow

Like the indirect method for F_a estimation, this method is also sensitive to experimental variability when CL/Q_h is close to 1.

CYP3A is currently believed to be the major enzyme in humans responsible for gut first pass. The prediction of F_{gut} in humans at the preclinical stage remains one of the most challenging tasks due to the heterogeneous distribution of intestinal enzymes and transporters in the whole length of small intestine and the possible interplay of CYP3A with efflux transporters [13]. Various models have been proposed to quantitatively estimate F_{gut} in humans [11], however, no method is currently well accepted. Extrapolation of F_{gut} of CYP3A substrates from preclinical species to humans requires careful consideration, since the gut first pass of CYP3A substrates in dog can be negligible [14], while in monkeys it can be higher than that in humans [15,16]. There are examples where monkey tends to show lower total bioavailability than humans for substrates of CYP3A as well as other CYPs [17]. Qualitatively the following two approaches can provide some gut bioavailability estimation in humans due to metabolism via intestinal CYP3A:

1. CYP3A contribution and activity. It was shown that major CYP enzymes in the human small intestine is CYP3A (80%) and CYP2C9 (15%), and CYP3A content in the small intestine is less than 1% of liver [18]. Intestinal CYP3A is the same isoform as the hepatic CYP3A since they have the same cDNA [19]. Therefore the metabolism extent in gut due to CYP3A should be reflected by its turnover rate in human liver microsomes. Based on a retrospective approach, it was suggested that for a compound to show significant gut first pass, it has to be both a major CYP3A substrate (>80% is metabolized through CYP3A) and have a hepatic intrinsic CL value exceeding 100 mL/min/kg [20].
2. Grapefruit juice effect. Oral grapefruit juice is a specific inhibitor of intestinal CYP3A, without affecting hepatic CYP3A [21]. Therefore, AUC and C_{max} changes after taking grapefruit juice can be used as an index that reflects the extent of gut first pass [11].

A quantitative estimation of F_{gut} for CYP3A substrates in human was established recently with the consideration of the interplay of CYP3A and efflux transporters [22]. Through the simulation, it was suggested that of P-glycoprotein efflux increased the level of CYP3A metabolism by up to 12-fold in the distal intestine. The model also suggested that F_{gut} is highly sensitive to changes in intrinsic metabolic clearance but less sensitive to changes in intestinal permeability.

Hepatic bioavailbility (F_h) is the fraction of absorbed drug entering the liver that escapes presystemic hepatic elimination. Compounds with high hepatic first pass also tend to show high variability in exposure.

Compared to gut first pass, liver first pass is easier to estimate using CL_{int} measured *in vitro*, due to the homogenous physiology of the liver. F_h can be estimated with CL_{int} obtained by *in vitro* system with this equation:

$$F_h = 1 - \frac{CL_{int} \cdot SF \cdot \dfrac{f_{u,blood}}{f_{u,incubation}}}{CL_{int} \cdot SF \cdot \dfrac{f_{u,blood}}{f_{u,incubation}} + Q_h} \tag{3.16}$$

where

SF and Q_h is the scaling factor and liver blood flow listed in Table 3.2

$f_{u,blood}$ and $f_{u,incubation}$ is the free fraction in blood and incubation, respectively.

In vivo F_h can be estimated from IV PK studies with an assumption that liver is the major elimination organ:

$$F_h = 1 - \frac{CL}{Q_h} \tag{3.17}$$

As discussed previously, the estimation is sensitive to experiment variability when CL approaches the Q_h value.

With portal vein dosing, *in vivo* F_h can be determined using Equation 3.9 by comparing dose-normalized AUC after portal vein dosing and systemic IV dosing. Liver metabolism plays an important role not only in systemic drug elimination but also in bioavailability. Assuming that liver metabolism is the major route of elimination, the following relationship of AUC and CL can be established:

$$AUC = \frac{F_a \cdot F_{gut} \cdot F_h \cdot Dose}{CL_h} = \frac{F_a \cdot F_{gut} \cdot \left(1 - \dfrac{CL_h}{Q_h}\right)}{CL_h} \tag{3.18}$$

As reflected in the above equation, increasing liver metabolism not only reduces the delivery efficiency by reducing the bioavailability due to first-pass metabolism but also increases elimination efficiency by increasing CL. This is the "double" impact of liver metabolism on overall exposure. For compounds with kidney as the major elimination organ, high CL only impacts the elimination efficiency and does not affect delivery efficiency, since the kidney is not a first-pass organ.

As discussed above, dose-normalized AUC reflects the exposure efficiency contributed by CL and *F*. *F* is an important parameter for orally administered compounds and is more tangible than an oral AUC value. It is generally accepted that $F \geq 20\%$

is desired for orally administered drugs to minimize interindividual variability and to avoid large clinical doses. Oral AUC can be a valuable screening parameter that filters out compounds with $F < 20\%$. A retrospective study was performed in order to address how frequently a high oral AUC correlates with good bioavailability, what is the maximum F for a given AUC, and what is an acceptable oral AUC threshold for 10 mg/kg screening dose in rats [23]. In this study, the relationship of AUC and F was analyzed with 100 compounds as well as with simulations for a variety of scenarios. It was demonstrated that the maximum F for a given AUC or minimum AUC for a given F occurs when absorption is complete. Based on the empirical and theoretical relationships described above, a lower limit for acceptable rat oral AUC (500 ng h/mL) was proposed since this is the lowest possible AUC for a drug dosed at 10 mg/kg with 20% oral bioavailability. Analogously, the lowest possible AUC for a drug with 50% oral bioavailability can be estimated at 2000 ng h/mL which then defines a "moderate" AUC range between 500 and 2000 ng h/mL; AUC values higher than 2000 ng h/mL are thus considered "high." With a database of about 5000 compounds, it was shown that more than half (54%) of screening compounds had low AUC values (<500 ng h/mL), 25% had moderate AUC (500–2000 ng h/mL), and 21% had high AUC (>2000 ng h/mL). Therefore, this high-throughput rat oral PK screen is a very effective primary filter such that the majority of compounds can be removed from further consideration with a single experiment requiring less than 16 mg of drug.

Similar concepts can be applied to oral monkey or oral dog screening. Table 3.3 shows the threshold values of 20% and 50% F for rat, monkey, and dog at the oral screening dose of 10 mg/kg, respectively.

It should be noted that this approach can give a false positive prediction for compounds with low F_a but high F_h, such as acidic compounds. Acidic compounds can have high plasma AUC due their extremely high binding to plasma albumin, even with a very small percentage of the dose entering the circulation. In addition, the extensive plasma protein binding and ionization at physiological pH limits distribution of acids into other tissues. Typically, acids tend to have very low CL, such that high AUC values can be observed even with low oral bioavailability. Such false positive compounds can be identified and filtered out with a full IV/PO study.

TABLE 3.3
AUC Threshold Values for $F \geq 20\%$ and $F \geq 50\%$ at an Oral Dose of 10 mg/kg in Rat, Monkey, and Dog

	AUC ng/mL h		
	Rat	Monkey	Dog
$F \geq 20\%$	≥500	≥1000	≥1350
$F \geq 50\%$	≥2000	≥3800	≥5400

3.2.4 Volume of Distribution

The distribution into different tissues is a kinetic process governed by the organ blood flows, binding to tissues and plasma proteins, regional pH gradients, and permeability of cell membranes. The extent of distribution depends on how extensively a compound can go into tissues and how tightly it is bound to by tissues, as shown by the following equation:

$$V_{dss} = V_p + V_t \cdot \frac{f_{u,p}}{f_{u,t}} \tag{3.19}$$

where

V_p and V_t are physiological volumes of plasma and tissues, respectively
$f_{u,p}$ and $f_{u,t}$ are fraction unbound in plasma and tissues, respectively

The compound-related factors that contribute to the extent of distribution are the macromolecule binding in plasma and tissue, lipophilicity, and ionization state. The basic principles are (1) only unbound compound (ionized or unionized) can distribute from capillaries to interstitial fluid; (2) only unionized unbound compound with proper lipophilicity can permeate through cell membranes (transcellular passive diffusion) and enter cells; and (3) some ionized free compounds might distribute through tight junctions between cells or by absorptive transporters or receptor-mediated endocytosis. Different degrees of hydrophobic and electrostatic binding to macromolecules in plasma and tissues account for the distribution differences among neutrals, acids, and bases. Log *D* governs the hydrophobic interaction between compounds and macromolecules, and its impact on plasma protein binding and tissue macromolecules binding are similar. In other words, changes in log *D* will change the hydrophobic interaction of a compound to protein in plasma and to macromolecules in tissue in the same direction with a similar extent ($f_{u,p}/f_{u,t} \approx 1$). Log *P* is the only factor that determines the distribution for neutral compounds, thus their volume of distribution is typically around 0.4–1 L/kg [4]. In addition to the hydrophobic binding, the electrostatic binding plays an important role for the distribution of acids and bases. As shown in Figure 3.5, the distribution equilibrium of bases is driven toward tissues ($f_{u,p}/f_{u,t} >> 1$) due to higher binding affinity of the positively charged basic groups to the negatively charged phospholipids in the tissue. Therefore, basic compounds tend to have higher volume of distribution. On the other hand, acids are limited to extracellular space ($f_{u,p}/f_{u,t} << 1$) due to the following factors: (1) negatively charged acids have low affinity for negatively charged tissue phospholipids, which are the major macromolecules in tissues and (2) negatively charged acidic groups have higher affinity for the positively charged lysine groups in plasma albumin, which retain the acidic compounds in plasma. Most acids have volume of distribution less than 0.4 L/kg, close to extracellular volume [4].

The volume of distribution is a proportionality factor that relates the amount of drug inside the body to the concentration in plasma. Compounds that are highly tissue-bound (e.g., lipophilic bases) tend to have a small percentage of the dose remaining in the circulation. Thus, plasma concentration is low and volume of

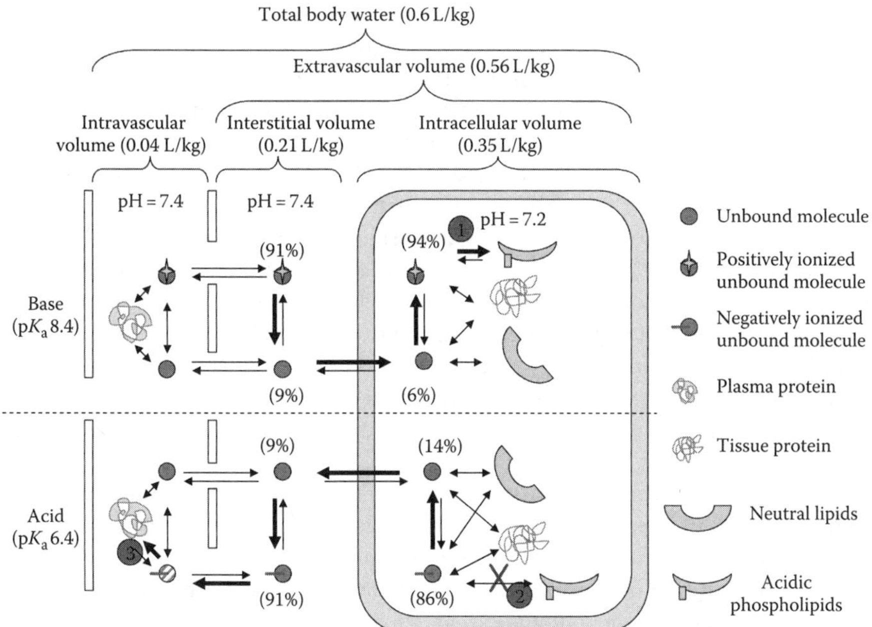

FIGURE 3.5 Schematic diagram of compound distribution and volume values for physiological spaces. Only unbound molecules can leave vascular space. Only unionized and unbound molecules can partition into cell membrane. The interaction of unbound molecules with intracellular macromolecules (protein, neutral lipids, and acidic lipids) is dependent on its lipophilicity and ionization states. ❶ The positively charged basic groups have high affinity to negatively charged phospholipids due to charge attraction. ❷ Negatively charged acidic groups have low affinity to the negatively charged phospholipids due to charge repulsion. ❸ Negatively charged acidic groups have high affinity to the positively charged lysine groups in albumin.

distribution is high. Compounds that distribute poorly into tissues tend to have high plasma concentrations and low volume of distribution. Volume of distribution is an apparent volume or an imaginary space, not a physical volume. However, by comparing this apparent volume with the physical volume of plasma, extravascular fluid, and interstitial fluid, the extent of extravascular distribution can be assessed. Volume of distribution that is lower than total body water (0.6 L/kg) is considered as low and limited distribution; volume of distribution in the range of 0.6–2 L/kg is considered as moderate distribution; and volume of distribution greater than 2 L/kg is considered as high distribution [4]. In general, the larger the volume of distribution, the longer time it takes to eliminate the compound, since the drug is sequestered from the rapidly perfused elimination organs (liver and kidney).

There are several terms of volume distribution, central volume of distribution (V_c or V_1), peripheral tissue volume of distribution (V_2), volume of distribution at steady state (V_{dss}), and volume of distribution of beta phase (V_β). Central volume of distribution corresponds to the rapidly equilibrating space consisting of blood and highly perfused tissues, such as liver, spleen, and kidney. Peripheral tissue volume of distribution corresponds to the more slowly equilibrating space that is not highly perfused, such as fat tissue and bone. V_{dss} equals the sum of V_1 and V_2. For a compound that only distributes in the central volume, its plasma concentration versus time profile after IV bolus administration on a semilog plot contains only one phase and declines monoexponentially. For such compounds, $V_c = V_{dss} = V_\beta$. For example, the volume of distribution for most acidic compounds can be described by the central volume of distribution. For a compound that takes time to distribute into different types of tissues, its plasma concentration versus time profile after IV bolus administration on a semilog plot contains more than one phase. This type of distribution is described by a simplified two-compartment model representing the whole body as shown in Figure 3.6. After IV bolus administration, the compound quickly distributes into the central volume, V_c. The magnitude of V_c can be estimated by the ratio of the dose amount to the concentration at zero time (C_0), and C_0 can be estimated by extrapolation of the initial concentration values back to time zero. The highest C_0 is limited by the plasma volume, which is around 0.05 mL/kg [24]. When a compound takes time to fully distribute into peripheral tissues, the volume of distribution increases with time. When this distribution reaches equilibrium at a certain time point called t_{ss}, the average compound concentration in tissues and extracellular fluids is the highest. The volume of distribution at this time point is called the volume of distribution at steady state (V_{dss}). After the time of distributional steady state, volume of distribution continues to increase to a stage of pseudodistribution equilibrium, where the amount of a compound between the plasma (central compartment) and the tissue (peripheral compartment) remains constant. The volume of distribution at and after this time point is called V_β. V_β is influenced by the relative rates of distribution and elimination, and therefore, it is not a pure distribution term. V_c (V_1) and V_{dss} are more frequently used in drug discovery to reflect instant and steady-state distribution, respectively. For this type of distribution, one would observe $V_\beta > V_{dss} > V_c$. These different terms of volume of distribution reflected the kinetic process of distribution. The ratio of V_{dss}-to-V_c is also an indicator of distribution pattern. When this ratio is around 1, the compound has a consistent distribution, and it sees the body as one well-mixed compartment. When the V_{dss}/V_c is much greater than 1, the distribution is different in different tissues, and the compound sees the body as multiple compartments. With the same effective $t_{1/2}$, the multicompartment compounds have higher C_{max}/C_{min} ratio than that of one-compartment compounds. For human PK predictions, multicompartment model provides a more detailed time-dependent exposure information such as C_{max} than the one-compartment model for compounds with $V_{dss}/V_c >> 1$. In practice, the two-compartment model is sufficient to describe the distribution pattern for a majority of compounds with a once-a-day or more frequent dosing regimen.

Actually, it is more meaningful to understand V_{dss} as a time-averaged parameter for distribution, just like CL is a time-averaged parameter for elimination. Thus, V_{dss} can be used to compare the overall extent of volume of distribution of different compounds, despite their differences in the pattern of distribution.

There are three common approaches to determining the volume of distribution as follows:

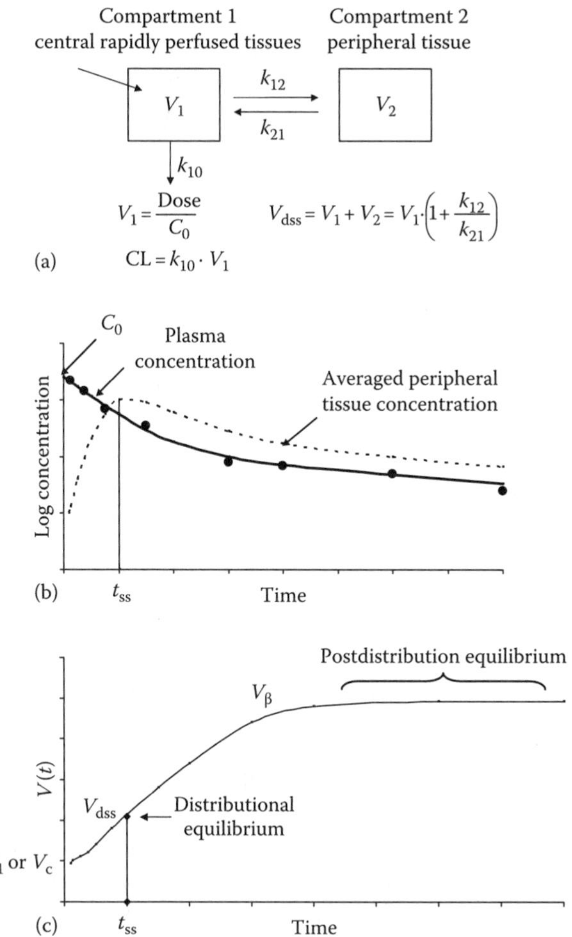

FIGURE 3.6 Compartmental analysis for different terms of volume of distribution. (Adapted from Kwon, Y., *Handbook of Essential Pharmacokinetics, Pharmacodynamics and Drug Metabolism for Industrial Scientists*, Kluwer Academic/Plenum Publishers, New York, 2001. With permission.) (a) Schematic diagram of two-compartment model for compound disposition. Compound is administrated and eliminated from central compartment (compartment 1) and distributes between central compartment and peripheral compartment (compartment 2). V_1 and V_2 are the apparent volumes of the central and peripheral compartments, respectively. k_{10} is the elimination rate constant, and k_{12} and k_{21} are the intercompartmental distribution rate constants. (b) Concentration versus time profiles of plasma (—) and peripheral tissue (---) for two-compartmental disposition after IV bolus injection. C_0 is the extrapolated concentration at time zero, used for estimation of V_1. The time of distributional equilibrium is t_{ss}. V_{dss} is a volume distribution value at t_{ss} only. V_β is the volume of distribution value at and after postdistribution equilibrium, which is influenced by relative rates of distribution and elimination. (c) Time-dependent volume of distribution for the corresponding two-compartmental disposition. V_1 is the starting distribution space and has the smallest value. Volume of distribution increases to V_{dss} at t_{ss}. Volume of distribution further increases with time to V_β at and after postdistribution equilibrium. V_β is influenced by relative rates of distribution and elimination and is not a pure term for volume of distribution.

1. *In vivo* IV plasma concentration profile
   Due to the kinetic feature of distribution, V_c, V_{dss}, and V_β can only be determined with plasma concentration profile obtained after an IV administration. The equations for calculating different terms of volume of distribution are

$$V_c = \frac{\text{Dose}_{\text{iv}}}{C_{p(0)}} \tag{3.20}$$

   where $C_{p(0)}$ is plasma concentration at zero time, which can be back-extrapolated using the initial concentrations after IV bolus injection (Figure 3.6).

$$V_{dss} = \frac{\text{AUMC}_{\infty,\text{iv}}}{\text{AUC}_{\infty,\text{iv}}} \cdot \frac{\text{Dose}_{\text{iv}}}{\text{AUC}_{\infty,\text{iv}}} = \text{MRT} \cdot \text{CL} \tag{3.21}$$

   where AUMC is the area under the first moment curve (plasma concentration × time versus time curve) and the ratio of AUMC-to-AUC is the MRT.

$$V_\beta = \frac{\text{CL}}{\beta} \tag{3.22}$$

   where β is the slope of the terminal phase of the plasma concentration profile.
2. *In vitro* physiological approach
   Volume of distribution at steady state (V_{dss}) can also be predicted with *in vitro* measured compound affinity to different tissues. This affinity is quantitated by tissue to plasma partition coefficients (K_p or K_{pu}) obtained by *in vitro* incubation of different tissue homogenates with compound of interest [25,26]. Coupled with the known physiological information in preclinical and clinical species, such as lipid, protein, and water content in each tissue, this mechanistic approach should provide good prediction in humans. However, this approach is very labor intensive; therefore it is not typically used in most drug discovery applications.
3. *In silico* approach
   Extensive research has established the relationship between the extent of distribution of compounds and their physicochemical properties. With this information, V_{dss} can be quite successfully predicted using *in silico* models [27–30]. *In silico* prediction of distribution is based on physicochemical properties that relates to passive transmembrane diffusion and tissue binding, and it only predicts V_{dss}. The other factors that contribute to distribution, such as transporter-mediated distribution, were not taken into account. These algorithms are based on the assumption that all compounds will dissolve in intra- and extracellular tissue water, and the unionized portion will partition into the neutral lipids and neutral phospholipids located within tissue cells. For compounds categorized as a strong base (at least one basic group (pK_a >7), an additional mechanism of electrostatic interaction with tissue acidic phospholipids is incorporated. Acids and weak bases are assumed

to bind primarily with albumin in plasma, while neutral compounds bind to lipoproteins [30]. The input values for these algorithms are log *P*/*D*, and plasma protein binding, which can be predicted from chemical structures or measured from experiments. Recently, a pure *in silico* prediction using 31 computed descriptors from the chemical structure provided a good prediction for V_{dss} [31]. Software such as ADMET predictor in GastroPlus also provides good predictions for V_{dss}.

In summary, the extent of passive distribution depends on macromolecule binding in plasma and tissue, lipophilicity, and pK_a. The extent of distribution can be evaluated by comparing the apparent volume to the physical volume of body water. Distribution is a kinetic process as reflected by the different terms of volume of distribution V_c, V_{dss}, and V_β and their ratios. V_{dss} is a time-averaged parameter that should be used to compare the extent of distribution among lead optimization candidates. It takes a longer time for a compound with large V_{dss} to be cleared from the body.

3.2.5 Mean Residence Time and Different Terms for Half-Life

MRT is the average time that molecules of a drug reside in the body after drug administration. It is defined as the time needed for the initial dose to decrease to 1/*e* (36.8%) of the starting value, where *e* is the base of the natural logarithm, with a value of 2.718. When MRT is obtained by IV bolus injection (MRT_{iv}), it is a pure intrinsic dispositional property that reflects the loss of the compound from the body. Half-life ($t_{1/2}$) is the time needed for 50% reduction in concentration or amount. Both parameters characterize the elimination rate of compounds; however, MRT_{iv} is not dependent on the distributional pattern and is a time-averaged parameter that applies throughout the entire concentration versus time profile. Half-life, on the other hand, typically describes the rate of decline for a particular phase of the concentration versus time profile (e.g., the apparent terminal elimination phase). For a compound with one-compartment distribution character:

$$\mathrm{MRT_{iv}} = 1.44t_{1/2} \text{ or } t_{1/2} = 0.693\mathrm{MRT_{iv}} \tag{3.23}$$

However, the relationship between MRT_{iv} and $t_{1/2}$ becomes more complicated for multicompartment compounds.

Like different terms for volume of distribution, different terms of half-life are used to reflect different distributional or elimination phases. If a compound shows multicompartmental distribution, it has a different $t_{1/2}$ in each phase. As shown in Figure 3.7, after IV bolus administration, Compound A achieved an instant distribution with an MRT_{iv} of 9 h and $t_{1/2}$ of 6 h. In contrast, Compound B demonstrated a two-phase distribution. The half-life of the earlier α phase ($t_{1/2,\alpha}$) is 1.3 h and the half-life of the latter β phase ($t_{1/2,\beta}$) is 13 h. Even though Compound B has a different terminal $t_{1/2}$ compared to Compound A, they have the same MRT_{iv}, indicating that the average time of these two compounds staying in the body is the same. Noncompartmental PK calculations are typically done during drug discovery, and the $t_{1/2}$ reported is the terminal $t_{1/2}$. Often, this terminal $t_{1/2}$ represents only a small percentage of the dosed

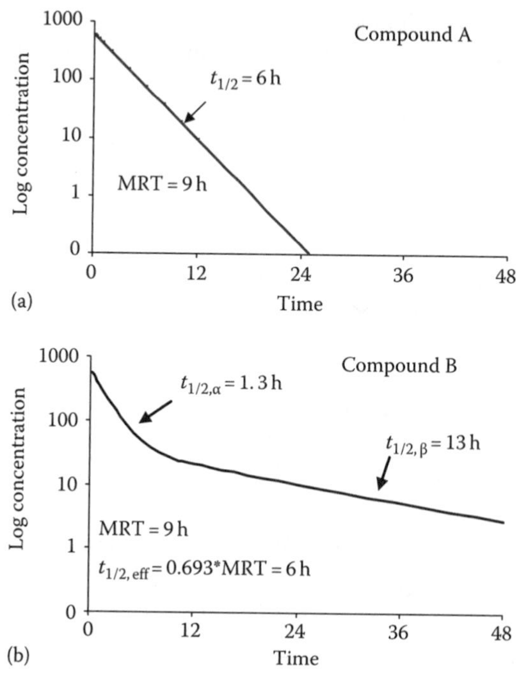

FIGURE 3.7 Different terms of half-life for one- and two-compartment models.

molecules and can be much longer than the MRT; therefore, terminal $t_{1/2}$ can be misleading since it can correspond to a tiny fraction of the dose that is slowly eliminated. MRT_{iv} is proportional to the ratio of V_{dss}/CL. Compounds with the same V_{dss}/CL ratio have the same MRT_{iv}, despite their differences in extent and pattern of distribution and clearance. Whether a compound fits a one-compartment or a multicompartment model, MRT_{iv} is a general term representing the average time that molecules reside in the body. Due to the familiarity of $t_{1/2}$ (as opposed to MRT_{iv}) to most people, and a demand for an elimination half-life that represents the major fraction of the dose, an "effective $t_{1/2}$" that equals to 0.693*MRT_{iv} was introduced [3] as the general term to represent the average elimination rate. This parameter provides a way for comparing effective $t_{1/2}$ for compounds with different distributional features and works well for compounds where the terminal phase only reflects a small fraction of the dose.

In a drug discovery environment, the elimination rate is used to estimate accumulation after multiple dosing. Many terms of half-lives were introduced with the attempt to simplify multicompartment kinetics for the estimation of accumulation. A recent article by Sahin and Benet compared and commented on various terms of half-life [32]. The accumulation after multiple dosing is not only a function of elimination rate but also a function of dosing interval for multicompartmental distribution compounds. In addition, the accumulation of C_{max} is a function of absorption rate [32]. Furthermore, the accumulation for C_{max}, C_{min}, and AUC can be different with the same compound and same dosing interval. Therefore, the half-life calculated based on accumulation ratios from different exposure parameters and with different dosing intervals for the same compound can be different. It is not practical to use

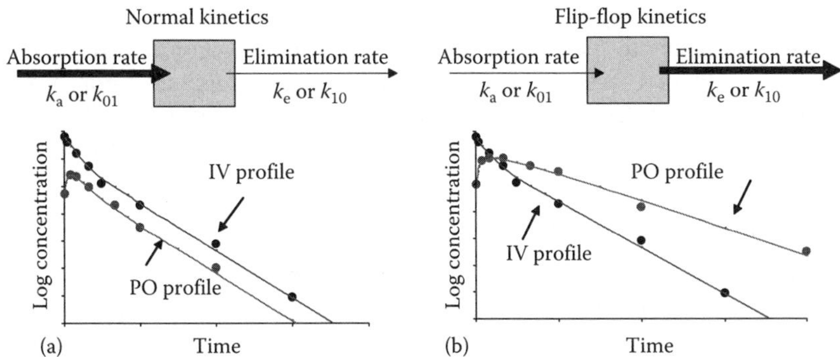

FIGURE 3.8 Schematic description of normal kinetics and flip-flop kinetics.

so many terms of half-life in drug discovery representing accumulation for various conditions. Accumulation ratios for C_{max}, C_{min}, and AUC are best estimated using the superposition of the observed plasma profile from a single dose with selected dosing intervals, without the need to use all the confusing terms of half-life.

When elimination rate is faster than the absorption rate, one observes a so called flip-flop profile that is an apparent switch of the slope of absorption and elimination. The much slower absorption rate becomes rate limiting and now describes the terminal decay phase. Flip-flop kinetics can be detected by comparing the plasma concentration profiles obtained by IV and PO route, where the elimination phase after IV dosing is more rapid than the one after PO dosing (Figure 3.8). It is very common to see flip-flop kinetics when $MRT_{iv} < 2$ h. The application of flip-flop kinetics is very important in the development of sustained-release and controlled-release formulation in pharmaceutical manufacturing. It can be used to overcome a short intrinsic MRT_{iv} for a high-efficacy compound and provide an apparent longer $t_{1/2}$. However, a good absorption in colon should be confirmed when an apparent $t_{1/2}$ longer than 6 h is desired. Flip-flop kinetics can also be achieved with a prodrug strategy where formation of the active drug is rate limiting. This might be a better approach for prolonging $t_{1/2}$ in some cases, since it is does not depend on the physiology of the small intestine or a good colonic absorption. The apparent $t_{1/2}$ in this case is dependent on the formation rate of drug from the prodrug and the volume of distribution of prodrug. Slower formation rate and larger volume of distribution of the prodrug will prolong the apparent $t_{1/2}$ of the active drug.

In summary, in our experience, most compounds in drug discovery display multicompartmental distribution. For these compounds, MRT_{iv} (not the terminal half-life) should be used for comparing the elimination rate and superposition of plasma profile (not the single term half-life) should be used for estimating the accumulation after multiple dosing.

3.3 PK/PD IN DRUG DISCOVERY

Currently, PK/PD studies are an important tool to facilitate the drug discovery process by (1) validating the new target involvement with *in vitro* to *in vivo* PK/PD correlation [33]; (2) ranking compounds based on both PK and PD properties;

(3) understanding the causes for apparent PK/PD disconnects; (4) evaluating the key exposure parameters for efficacy [34], thereby guiding selection of dose and sampling time for PK/TK studies; and (5) forecasting human dose and regimen, which can inform a go/no go decision. This section will introduce some basic PD and PK/PD principles that are important for understanding the value of PK/PD studies.

3.3.1 Basic Concepts in Pharmacodynamics

Pharmacodynamics studies the time course of drug effect. It provides information such as the onset, intensity, and duration of the drug effect. It is designed to reveal the relationship between compound exposure at the effect site and the magnitude of pharmacological response. Due to the difficulties of measuring the drug exposure at the effect site, the equilibrated plasma concentrations at steady state are frequently used as a surrogate. Plasma concentration could be maintained at steady state with constant infusion administration. However, this approach is not widely used in drug discovery due to many factors. For efficiency, most pharmacologic effects are point estimates during the screening stage, without knowing the status of PK equilibrium or the temporal PD response. Whether this approach is appropriate depends on the characteristics of the PD response and distributional characteristics of the compound. When a full-time course PK study is combined with PD measurements, one can estimate steady-state PD parameters with proper modeling without the need to achieve and maintain steady-state plasma concentrations. Full-time course PK/PD studies are important to fully understand the PK/PD relationship and for predicting efficacious plasma concentrations in humans. Understanding the differences of pharmacological responses and PD models provides the basis for designing and interpreting PK/PD studies.

Pharmacological responses can be described with the following features:

1. Reversible versus irreversible response
2. Direct versus indirect response
3. Continuous versus categorical response

3.3.1.1 Reversible versus Irreversible Response

The effect of a drug on a biological system is usually evaluated by comparing the response change relative to the baseline level prior to treatment. If a baseline response is restored when drug is discontinued, it is a reversible response; otherwise it is an irreversible response. Most drug effects are reversible, such as beta blockers for lowering heart rate. Pharmacological effect for some antibiotics and anticancer agents are irreversible, due to irreversible binding to the target. For example, the effect of taxol is irreversible due to its irreversible binding to microtubules [35]. For irreversible responses, C_{max} is the key exposure parameter that correlates to efficacy.

3.3.1.2 Direct versus Indirect Response

If an observed pharmacological response is generated immediately after the initial interaction with a target, and the intensity of the response is directly correlated with the drug concentration at the effect site, it is a direct response. For example, the heart rate reduction by a beta blocker is a direct effect. If an observed pharmacological effect is generated after a series of biological events following the initial interaction, and with a significant time delay, it is an indirect response. For example, the effect of recombinant human erythropoietin on the red blood cell production is an indirect response, since the production time and life span of the red blood cells are 4 and 120 days, respectively. In other words, it will take at least four days to see the initial effect. Both direct and indirect responses are reversible responses.

Mechanistically, direct and indirect responses can be differentiated by the turnover rate of the involved biological processes. The turnover rates of electrical and chemical signals are less than 1 min, turnover rates for most enzymes and mRNAs are in the range of 6–12 h, and turnover rates for proteins, cells, and tissues are in the range of days to 2–3 weeks [36]. If a response has a turnover rate less than a few minutes, it can be considered a direct response; any response with a turnover rate longer than several hours can be considered an indirect response. Experimentally, direct and indirect responses can be differentiated by comparing the time course of plasma concentration and time course of effect or by plotting the concentration and effect using the data obtained at various times as shown in Figure 3.9. If it is an indirect response, a hysteresis would be observed—that is, the intensity of the effect is lower at earlier time points than at the later ones with the same plasma drug concentration levels. If it is a direct response, a clear correlation curve between concentration and effect would be observed. For direct responses, the response–concentration curve established by one time point should be the same as the one established by different time points. In other words, direct pharmacological responses only depend on

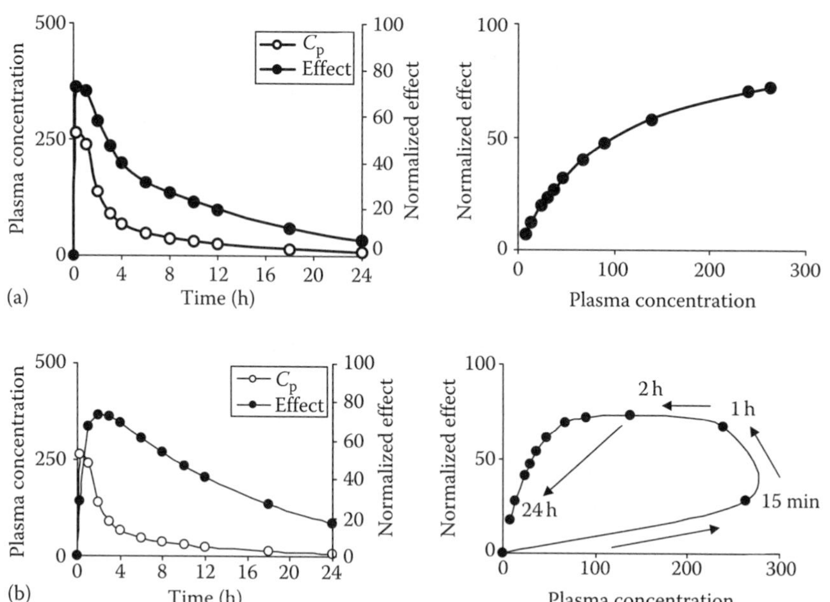

FIGURE 3.9 Schematic description of hysteresis loop. (a) Direct response curves and (b) indirect response curves.

concentration, not on time, and the PK/PD relationship can be built with data from one time point at different dose levels.

Not all hysteresis profiles (delayed responses) are due to an indirect response. Distributional delay and slow dissociation of drug–target complex are also common causes for hysteresis. Distributional delay is common for compounds with targets located in the central nervous system due to the blood brain barrier. There are other factors that can cause a hysteresis and prosteresis (clockwise hysteresis) curves: sensitization of targets or development of tolerance; formation of active metabolites or antagonistic metabolites; and upregulation or downregulation of receptors. For this kind of PK/PD situation, using one time point, especially selecting the peak time of plasma concentration, can be misleading (see details in the section of PK/PD for acute pharmacology models).

3.3.1.3 Direct E_{max} Model and Target Binding

A direct response is directly connected with the interaction of a biological target. Most biological targets are proteins, such as receptors, enzymes, ion channels, and transporters. The following uses receptor binding as an example to illustrate the E_{max} model for direct responses derived from the initial target binding.

$$[\mathrm{C}]+[\mathrm{R}] \underset{k_{\mathrm{off}}}{\overset{k_{\mathrm{on}}}{\rightleftarrows}} [\mathrm{RC}] \xrightarrow{\varepsilon} \text{Effect} \tag{3.24}$$

where

[C] is the drug concentration at receptor site
[R] is the concentration of unbound or inactivated receptor
[RC] is the concentration of bound or activated receptor complex
ε is the proportionality factor between effect and activated receptor complex

The following equation can be derived when the interaction is at equilibrium:

$$[\mathrm{RC}] = \frac{R_{\max} \cdot [\mathrm{C}]}{K_{\mathrm{d}} + [\mathrm{C}]} \tag{3.25}$$

where

R_{max} is the maximum receptor binding
K_d is the dissociation constant of drug–receptor complex (k_{off}/k_{on})
k_{on} is a second-order formation rate constant with a unit of concentration^{-1} time^{-1}
k_{off} is a first-order dissociation rate constant, with a unit of time^{-1}

Thus, K_d has a concentration unit. In fact, K_d corresponds to the drug concentration that generates 50% of the maximum receptor binding, and it is an EC_{50} in concept. The percentage of receptor occupancy can be obtained by rearranging the above equation:

$$\frac{[\mathrm{RC}]}{R_{\max}} = \frac{[\mathrm{C}]}{K_{\mathrm{d}} + [\mathrm{C}]} \tag{3.26}$$

The direct effect (E) generated by the receptor interaction is proportional to the concentration of receptor complex [RC], therefore the above equation can be converted to

$$\frac{E}{E_{\text{max}}} = \frac{\varepsilon \cdot [\text{RC}]}{\varepsilon \cdot R_{\text{max}}} = \frac{[\text{C}]}{K_{\text{d}} + [\text{C}]} \tag{3.27}$$

The direct E_{max} model is obtained by rearranging the above equation:

$$E = \frac{E_{\text{max}} \cdot [C]}{K_{\text{d}} + [C]} = \frac{E_{\text{max}} \cdot [C]}{EC_{50} + [C]} \tag{3.28}$$

where EC_{50} is the drug concentration that generates 50% of maximum effect. If plasma concentration is equilibrated with the concentration at the receptor site, then [C] can be replaced by free plasma concentration.

The E_{max} model reflects a capacity limited nonlinear relationship between drug concentration and drug effect. The maximum drug effect is limited by the amount of the biological target, and this is a hallmark feature of quantitative pharmacology [36]. The general structure for such capacity-limited response is

$$Y = \frac{\text{Capacity} \cdot x}{\text{Sensitivity} + x} \tag{3.29}$$

as seen in the derived receptor-binding equation, Michaelis–Menton equation for enzyme kinetics and Hill equation for hemoglobin binding. Two PD parameters in the E_{max} model are E_{max} and EC_{50}, reflecting the capacity (maximum efficacy) and sensitivity (potency), respectively.

If the mechanism is antagonism, the following inhibitory E_{max} model should be used:

$$E = (E_{\text{max}} - E_0) \cdot \left(1 - \frac{C}{C + \text{IC}_{50}}\right) \tag{3.30}$$

where

E_{max} is the effect generated by an agonist or the baseline response
E_0 is the maximum inhibitory effect
IC_{50} is the concentration for 50% inhibition between E_{max} and E_0

The response of antagonists usually is expressed by a relative value, such as percentage of baseline. For the cases where $E_0 = 0$, the maximum inhibitory effect is 100% and IC_{50} is the only PD parameter obtained. IC_{50} is a valuable parameter for comparing the potency for antagonists among compounds and among preclinical species and human.

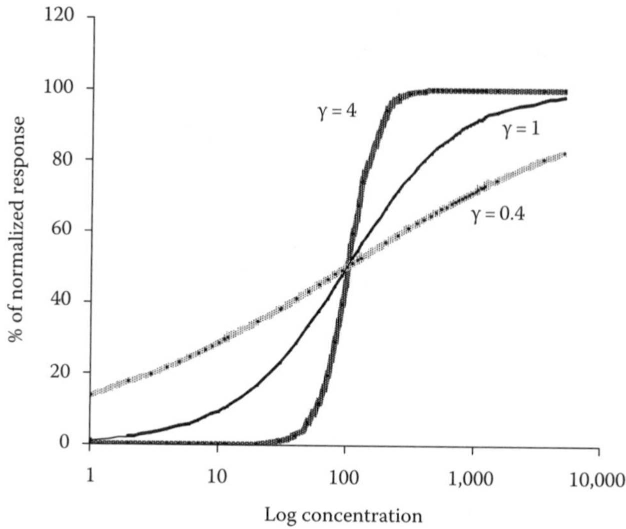

FIGURE 3.10 γ and steepness of response curve in the sigmoid E_{max} model.

The most general direct PD model is the following sigmoidal E_{max} model:

$$E = \frac{E_{max} \cdot C^{\gamma}}{EC_{50}^{\gamma} + C^{\gamma}} \tag{3.31}$$

Compared to the E_{max} model, this model has a sigmoid coefficient γ that reflects the steepness of the response curve as shown in Figure 3.10. When γ is >4, the response curve shifts from a gradual response to "all or none" categorical response; and a γ less than unity could be an indication of active metabolites and/or multiple receptor sites [4].

3.3.2 Methodology

The methodology of PD modeling can be divided into three categories: empirical, semimechanistic, and mechanism based, with the confidence in prediction ranging from low to high. Empirical methodology uses descriptive models that rely on the sigmoidal E_{max} model. It is relatively easy to build. Most acute responses in drug discovery can be described by the simple E_{max} model. However assumptions need to be made when extending the results to the downstream PD responses. These assumptions should be verified as the drug progresses into drug development. The application of an empirical model for prediction should at least be verified by a reference compound with the same or similar mechanism and known clinical exposure–response relationship. If data from an animal screening model cannot be fitted with a direct response model, the possible causes and mechanisms need to be explored, and semimechanistic methods containing both mechanism and empirical factors can be developed. This approach can also be very effective when the key PD factor is mechanism based while others are empirically assumed or approximated. The empirical factors and assumptions can be replaced by mechanistic factors in the "learn-confirm" paradigm along the drug development processes to formulate a complete mechanism-based model [37]. Mechanism-based modeling contains specific mathematical expressions to quantify processes on the causal path between drug administration and effect. When developed with translational biomarkers these models are much better for extrapolation and prediction. There are many processes between drug administration and the final pharmacologic effect. The rate-limiting step in the following major steps need to be addressed: (1) distribution to target site, (2) target binding and activation, (3) PD interactions, (4) transduction, (5) homeostatic feedback processes, and (6) the effects of the drug on disease progression. A thorough translational PK/PD model should include predictions in the following stages: (1) from *in vitro* test system to the *in vivo* situation, (2) from *in vivo* animal studies to humans, (3) from healthy volunteers to patients, and (4) from intra- to interindividual variability in drug effects [38]. Even though this requires a thorough and in-depth multidisciplinary approach that usually takes a long time and large resources due to many "learn-confirm" cycles, there is an increasing trend to develop mechanism-based models starting from drug discovery to maximize the likelihood for a successful proof-of-concept study in patients. A true mechanism-based model should have distinctive parameters separating drug-specific properties from biological or system properties.

3.3.3 Variety of PD Models and Principles of Scaling PD Parameters

The general form of pharmacological responses can be expressed as a function of drug-specific parameters, drug concentration, and drug-independent biological system parameters. Drug-specific parameters reflect target affinity, such as receptor affinity and intrinsic potency, which can be predicted on the basis of *in vitro* bioassays [38]. System parameters are those rate constants that reflect the production rate, turnover rate, transduction process, homeostatic feedback, or disease progression of involved biological targets. Mager et al. provided the details of a variety of PD models in their review paper [39]. For a direct response, the PK/PD model is quite simple. The effect is simply a hyperbolic Hill function of drug concentration at the effect site. If there is no distributional delay, the direct response model can be established with single time point PK/PD data. For other responses that involve rate constants of a biological system, such as production rate or turnover rate of the involved biological entities, PK/PD models become more complicated. For these models, the drug effect is not only a function of drug concentration but also a function of time as reflected by the rate constants of a biological system. Such models can only be established with full-time course data at several dose levels.

The basic principles for scaling PD parameters from preclinical species to human are (1) EC_{50} and E_{max} are weight-independent parameters. However, differences in free fraction and intrinsic potency should be considered when extrapolating EC_{50} from preclinical species to humans and (2) turnover rate or other system rate constants in PD models should be scaled using the principle of allometry with exponent b around −0.25 [36].

There are many applications of PK/PD studies in drug discovery. The prediction of human dose and regimen are important to make sure that the dose amount is developable and the dosing regimen meets the target candidate profile. In order to use preclinical PK/PD studies to predict the human dose and dose regimen, the following two issues need to be addressed: (1) the exposure parameter that correlates with efficacy and (2) the human relevance of the preclinical model.

3.3.4 Evaluating the Key PK Exposure Parameter that Correlates with Efficacy

In the absence of interspecies difference in target affinity or plasma protein binding, the exposure that drives the effect in preclinical species should be very close to the exposure for the same extent of effect in humans [40,41], thus determination of the key exposure parameter that drives the efficacy is critical for the prediction of clinical efficacy. In order to determine which exposure parameter correlates to efficacy, and the relationship between compound half-life and duration of the effect, the first thing that needs to be addressed is identifying the rate-limiting step in the PK process and the PD process. Similar to the MRT of a compound, the activated (or inhibited) biological entity that produces a pharmacological response also has a MRT or a turnover rate. There are many kinetic and dynamic steps from the time of dosing a drug to the time of the final effect, and the rate-limiting step is the one that controls the whole process.

For a direct drug response, where the involved biological process has a very short turnover rate in the scale of less than few minutes, the drug concentration has to be kept at or greater than the effective concentration to maintain the duration of desired effect. Drug effect will be lost immediately when the drug concentration is lower than this level. For this type of response, the duration of effect is controlled by the time above the effective concentration; therefore, compound $t_{1/2}$ is an important dispositional PK parameter for such responses. For example, the decongestion effect of pseudoephedrine is through direct interaction with α-adrenergic receptors in the mucosa of the respiratory tract producing vasoconstriction that results in shrinkage of swollen nasal mucous membranes, reduction of tissue hyperemia, edema, and nasal congestion, and an increase in nasal airway patency. The plasma concentration of pseudoephedrine needs to be maintained $\geq$230 ng/mL for this decongestion effect (internal data). In fact the duration of a direct effect is not only a function of compound half-life but also a function of dosing interval. For example, if the maximum absorbable dose generates a plasma concentration profile where effective concentration occurs in the middle of the plasma concentration profile as shown in Figure 3.11a, the effective $t_{1/2}$ that controls the duration is not the terminal $t_{1/2}$ of 22 h. With an 8 h dosing interval to maintain the plasma concentration profile above the effective concentration, the calculated effective $t_{1/2}$ from AUC accumulation ratio is 1 h, same as the $t_{1/2}$ in α phase. Therefore, the half-life for drug effect in this case is $t_{1/2,\alpha}$. If the plasma concentration profile was not generated at the maximum absorbable dose, the dose could be raised to the extent where plasma concentration at 24 h (C_{24h}) is equal to the target effective concentration as shown in Figure 3.11b, a once-a-day

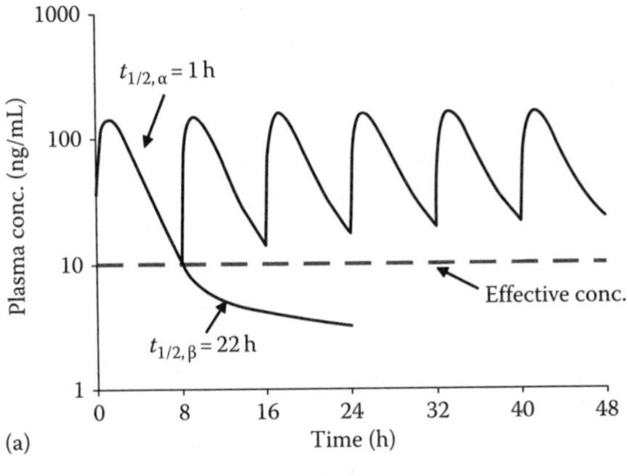

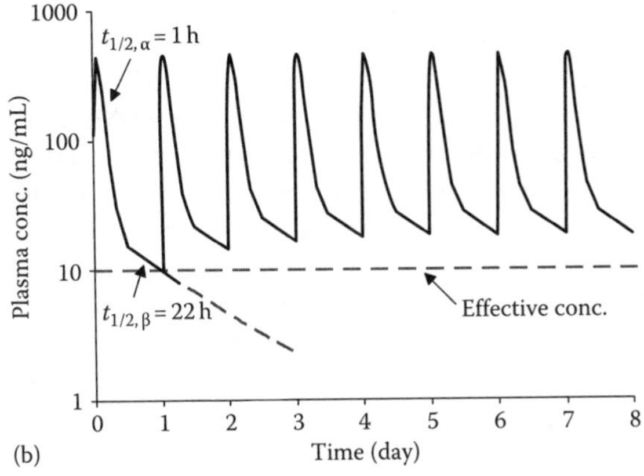

FIGURE 3.11 The relationship of effective $t_{1/2}$, dosing interval, and duration of effect. (a) Effective $t_{1/2} = 1$ h for TID and (b) effective $t_{1/2} = 4$ h for QD.

(QD) dosing regimen could be employed. The effective $t_{1/2}$ calculated using the AUC accumulation ratio from a 24-h dosing interval is 4 h, between the $t_{1/2,\alpha}$ and $t_{1/2,\beta}$ values, and not the same as the effective $t_{1/2}$ of 7.5 h calculated from MRT_{iv}. Therefore, which $t_{1/2}$ controls the duration of a direct response depends on many factors, such as maximum absorbable dose, the timing of the effective concentration, and the frequency of dosing.

For other drug responses where the turnover rate of the involved biological process is much slower than the elimination rate of the drug, the duration of effect is dominated by the turnover rate of the PD response. The onset for such a drug response is usually slower and the drug effect lasts longer than expected based on the drug concentration. The magnitude of such chronic responses is better correlated with daily averaged exposure (AUC or C_{ave}). CL is the important dispositional PK parameter in these cases. For example, typically the effect of tumor growth inhibition (TGI) is correlated with AUC, not trough concentration, due to the relatively slow tumor growth rate (see the section of PK/PD for TGI).

Most drug responses measured in early drug discovery correspond to acute or subchronic responses. There are potentially more downstream biological steps or cascades to the final drug effect. The PD response observed in acute or subchronic studies could be a direct response, while the final clinical response might involve a biological process that has slower turnover or homeostatic compensation. Identifying the rate-limiting step between drug administration and the final clinical effect is very important in determining which exposure parameter correlates with efficacy and should be used in the prediction of human dose and regimen, and which dispositional PK parameter is the key parameter to improve during lead optimization. This requires a thorough understanding of the mechanism involved in any downstream pathways. If a biological target is a novel one, early PK/PD models can only be built with an assumption-rich approach, and these assumptions should be verified or changed along the whole development process when more rigorous PK/PD studies are designed and performed on the final clinical candidate in the clinic. A thoroughly built mechanism-based PK/PD model can be applied for compounds with the same target or a new target that involves similar downstream pathways.

In summary, understanding the rate-limiting step in the whole PK and PD process is very important for focusing the efforts in optimizing ADME properties and determining the exposure parameter that correlates to final clinical effect. If the duration of the effect is controlled by the PD response, efforts should not be wasted on prolonging PK $t_{1/2}$. In this case, the efficacy is better correlated with a time-averaged exposure parameter (AUC-to-dosing interval ratio), thus CL and F are important parameters to focus. Also, the difference between $t_{1/2}$ in PK and $t_{1/2}$ in PD should be taken into consideration for dose regimen selection. Even when PK $t_{1/2}$ is the rate-limiting half-life, the effective PD $t_{1/2}$ for multicompartment compounds depends not only on elimination half-life but also on the dosing interval and timing of the effective concentration in the plasma concentration profiles.

3.3.5 Establishing the Translatability of Preclinical Models

Various approaches can be used to evaluate the translatability of preclinical models. However, the translatability is preferably built through biomarkers that can be quantified in preclinical models as well as in the clinic. The development of sitagliptin, a dipeptidyl pepetidase-4 (DPP-4) inhibitor, for diabetes control is a good example. The inhibition of target DPP-4 was quantitatively linked with glucose tolerance in mice and humans through PK/PD studies in preclinical models and in the clinic. It was demonstrated that a near maximal glucose-lowering effect after a single oral dose was associated with inhibition of plasma DPP-4 activity of 80% or greater, corresponding to a plasma sitagliptin concentration of 100 nM or greater in both preclinical species [42] and human [43,44]. This relationship can be now used for developing compounds with the same mechanism.

When no quantitative biomarker is available, the translatability can be built through a reference compound that shares the same or similar mechanism. The translatability of a preclinical pain model for pregabalin was established with the reference compound—gabapentin [45,46]. The two- to threefold higher potency of pregabalin relative to gabapentin was observed in preclinical species with an effect compartment model [47]. This relative potency comparing to the reference compound of gabapentin was then "translated" to the established E_{max} model of gabapentin where plasma AUC is correlated with the pain score. Using a reference compound to build the translatability is very important for situations lacking quantitative biomarkers.

3.3.6 PK/PD for Acute Pharmacology Models

Most PD screening models are acute in nature, such as the extent of receptor binding [48], threshold values for neuropathic pain [49], signal intensity in electroencephalogram (EEG) [48], and concentration change of amyloid beta monomers in cerebrospinal fluid [50]. The concept of receptor occupancy is increasingly applied in mechanism-based PK/PD modeling to correlate initial target engagement to the effect, and human PET studies can provide occupancy directly in humans, which assists in the "learn-confirm" process. Such a model can also link the *in vitro* intrinsic efficacy to a direct clinical response [38]. The possible response delay caused by distribution delay or slow target dissociation in these acute models are commonly ignored in the early drug discovery stage. The measurement of responses at peak plasma concentration is often used. Potency established with data from one time point might be an overestimation or underestimation, depending on the time of PD measurement. Therefore, single time point measurements might provide misleading information for ranking and comparing the potency to a reference compound. With a full-time course PK/PD study at a higher dose level and proper PK/PD models, such as an effect compartment model or/and a receptor-binding kinetics model [48], potency (EC_{50}) can be accurately estimated and translated to clinical studies.

The hysteresis observed in these acute efficacy models can be modeled by invoking a hypothetical effect compartment [51]. The distribution rate constant from the central compartment to the effect compartment is termed as k_{eo}, and it reflects the rate of equilibration between these two compartments. The magnitude of k_{eo} also reflects the distribution delay. A smaller k_{eo} indicates a slower equilibration and longer distribution delay. The equilibration half-life ($t_{1/2,k_{eo}}$) is equal to $0.693/k_{eo}$. The effect compartment concentration will be 50%, 75%, 87.5%, 93.75%, and 97% of the plasma concentration after 1, 2, 3, 4, and 5 equilibration half-lives, respectively. As illustrated in Figure 3.12, if EC_{50} was evaluated only at a time point prior to equilibrium with plasma concentration C_p, the EC_{50} would be higher than the true value, since C_p is higher than the concentration in the effect compartment (C_e) before equilibrium. If EC_{50} was evaluated at a time point at or after equilibrium, the estimated EC_{50} would be closer or higher than the true EC_{50} depending on the $t_{1/2,k_{eo}}$. If $t_{1/2,k_{eo}}$ is smaller than 0.5 h, EC_{50} estimated with a single time point at or after 2 h is very close to the true value. However, when $t_{1/2,k_{eo}}$ is longer, for example at 1.5 h, EC_{50} estimated at or after the equilibrium is an underestimation. For this situation, the correct EC_{50} can be estimated with full-time course data, with the effect compartment model that has a rate constant k_{eo} that reflects the distribution delay.

Another model can be used to account for the hysteresis if the initial target interaction is the receptor binding, where the response delay is due to the slow dissociation constant (k_{off}) [48]. Both affinity rate constant (k_{on}) and dissociation rate

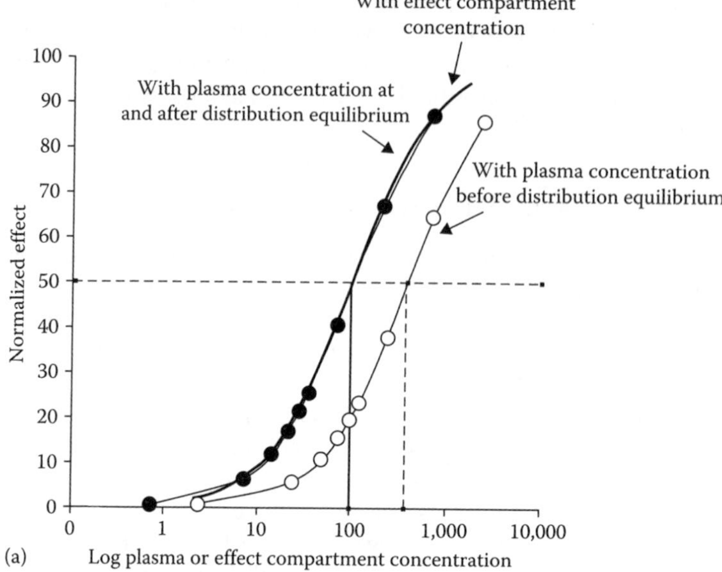

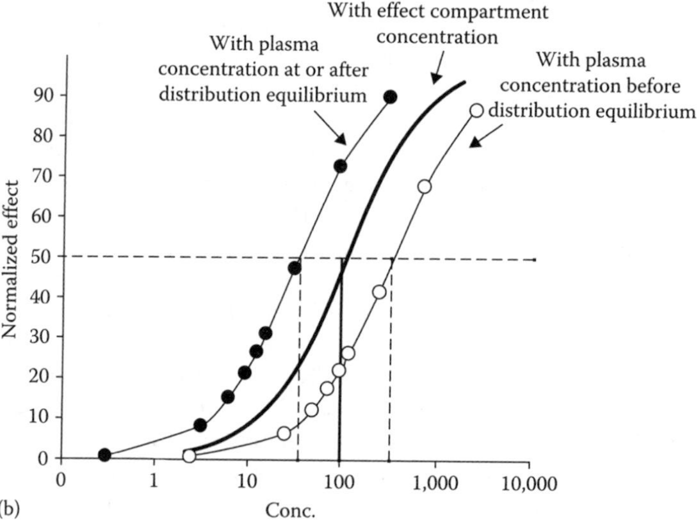

FIGURE 3.12 Left shift and right shift of response curves caused by distributional delay when the response curve was established with plasma concentration at one time point. (a) Distributional delay with $t_{1/2,k_{eo}}$ of 0.5 h. (b). Distribution delay with $t_{1/2,k_{eo}}$ of 1.5 h. Response curves established at 0.5 h (void circle line) is a right shift, with an overestimation of EC_{50}. Response curves established at or after equilibrium is the same as the true curve if $t_{1/2,k_{eo}}$ is 0.5 h or shorter; but a left shift, if $t_{1/2,k_{eo}}$ is longer, such as 1.5 h, with an underestimation of EC_{50}.

constant (k_{off}) can be estimated from the time course of receptor binding and plasma concentration. EC_{50} equals k_{off}/k_{on}, and the dissociation half-life for activated receptor equals $0.693/k_{off}$. EC_{50} established at the time before equilibrium would be an overestimation. For this model, EC_{50} estimated at or after the time of equilibrium with single time point data is closer to the true EC_{50}.

Both distributional delay and slow dissociation of the drug–target complex are compound-dependent properties, not system-dependent properties. One cannot extrapolate the observation from one compound to another even if they have the same biological target. Without full-time course data, one cannot tell if there is a hysteresis and whether this hysteresis could significantly affect the accuracy of estimated EC_{50} established with point estimates.

As a matter of fact, the target dissociation constant (k_{off}) is an important drug PD parameter that contributes to the duration and extent of the effect. The dissociation half-life correlates better with the extent of the effect than the dissociation constant K_d. For the same target and same ratio of k_{off}/k_{on} (same K_d or EC_{50}), the compound with a smaller k_{off} has a higher magnitude of effect [52], and longer duration [53]. When the dissociation half-life of activated target is longer than the elimination half-life of compound, the duration of effect is controlled by the dissociation half-life. Therefore, improving k_{off} in lead optimization could be more important than improving compound half-life. In addition, improving k_{off} could also reduce possible off-target toxicity, since the residence time of compound in other tissues is shorter than that in the target tissue.

3.3.7 PK/PD for Subchronic Pharmacology Models

In the later stage of lead optimization, the lead compound will be tested in subchronic pharmacology models in many cases. A recently published PK/PD model of TGI is used as an example to illustrate the advantages of a mechanism-based model [54].

Human tumor xenografts in nude mice are a well-established preclinical model for evaluating the efficacy of anticancer agents. The traditional method measures the TGI for evaluating efficacy of anticancer agents. TGI is obtained by comparing the tumor weight of the treated group to the control group. TGI is dependent on many factors such as dose regimen, inoculation time and amount, and variability of tumor growth. Thus it is hard to rank compounds when they are evaluated at different times. The translatability of TGI to humans is poor and has to be based on many assumptions. Instead of using TGI, this PK/PD model separated drug-specific properties from tumor growth properties. As shown in Figure 3.13, k_1 and k_2 are drug-specific parameters and λ_0 and λ_1 are tumor-specific parameters. The parameter k_1 is a first-order rate constant, reflecting transitional process for cell death caused by the drug; k_2 is a second-order rate constant, reflecting antitumor potency of the drug; λ_0 is a first-order rate constant, reflecting tumor cell aggressiveness during the exponential tumor growth phase; λ_1 is a zero-order rate constant, reflecting tumor cell aggressiveness during the linear growth phase.

The parameter k_2 is a drug-specific parameter that reflects the potency of TGI. It is not dependent on the inoculation size, dose regimen, or time of evaluation. Therefore, it can be used for ranking potency for compounds being evaluated at different times

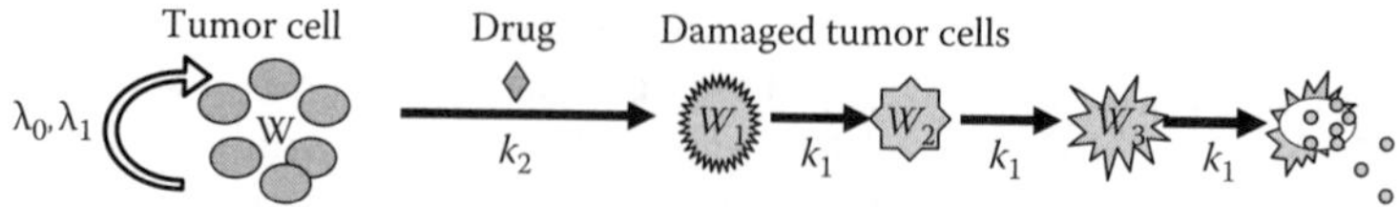

FIGURE 3.13 Schematic description of mechanism-based model for TGI. (Adapted from Rocchetti, M. et al., *Eur. J. Cancer*, 43, 1862, 2007. With permission.)

with different protocols. It can also be used for comparing the potency to a reference compound that has the same or similar mechanisms.

It can be derived from this model that the reciprocal of k_2 ($1/k_2$) equals a threshold exposure (AUC_T) required for complete suppression of tumor growth. This AUC_T equals a threshold concentration (C_T) multiplied by the mean tumor growth time, thus it represents a total AUC for the length of mean tumor growth time. For example, $1/k_2$ (AUC_T) represents a total AUC for four days, when $\lambda_0 = 0.25$ day^{-1}.

$$1/k_2 = \underset{\substack{\text{Threshold} \\ \text{concentration for} \\ \text{tumor growth rate=0}}}{\underset{\downarrow}{C_T}} \times \underset{\substack{\text{Tumor mean} \\ \text{growth time}}}{\underset{\downarrow}{1/\lambda_0}} = \underset{\substack{\text{Threshold total AUC for} \\ \text{tumor grow rate=0, in the} \\ \text{mean tumor growth time.}}}{\underset{\downarrow}{AUC_T}} \tag{3.32}$$

Unlike the E_{max} model where potency is a concentration term (EC_{50}), the potency in this chronic model is an AUC that equals a threshold concentration multiplied by the mean production (growth) time of the involved biological entity (tumor). This tumor growth rate is the rate-limiting step for all the processes involved starting from drug absorption to the final inhibition of the tumor growth.

As mentioned above, drug potency is a body weight independent parameter, and it should be the same across species. Therefore, $1/k_2$ obtained from a mouse model can be used to predict the total AUC during the mean tumor growth time in humans. It was demonstrated that $1/k_2$ obtained from human tumor xenografts in nude mice correlates to total AUC in a three-week cycle in human for 10 anticancer agents at their average therapeutic dose [55].

The daily AUC_T required for static tumor growth can be calculated from the following formula:

$$AUC_{T,daily} = \frac{1}{k_2} \cdot \lambda_0 \tag{3.33}$$

Therefore, the daily AUC for TGI not only depends on the potency $1/k_2$ but also depends on the tumor growth rate. Since tumor growth rate is slower in man (lower λ_0) than that in mice, the $AUC_{T,daily}$ in man should be lower than that in mouse. Similarly, the threshold concentration C_T in man should be lower than that in mouse, due to slower tumor growth rate, as derived in the following equations:

$$\frac{1}{k_2} = C_{T,animal} \cdot \frac{1}{\lambda_{0,animal}} = C_{T,human} \cdot \frac{1}{\lambda_{0,human}} \tag{3.34}$$

$$\frac{C_{\mathrm{T,animal}}}{C_{\mathrm{T,human}}} = \frac{\lambda_{0,\mathrm{amimal}}}{\lambda_{0,\mathrm{human}}} \tag{3.35}$$

Another advantage of this mechanism-based approach is the minimization of the impact of the variability of individual tumor growth rate on the estimation of potency $1/k_2$. The tumor growth rate in the exponential phase (λ_0) and linear phase (λ_1) can be obtained from the same mouse using the tumor growth data obtained before and after treatment, respectively, rather than comparing to a vehicle-dosed control group. These individual growth rates serve as the individual control and provides better estimation on the drug-specific parameter k_2 for each animal.

The above is a good example demonstrating the following advantages for mechanism-based modeling:

1. Separating drug-specific and biological-specific properties
2. Reducing the impact of variability of biological properties on the drug-specific properties
3. Better predictivity (translatability)

3.3.8 Challenges in PK/PD Modeling

The challenges of PK/PD modeling exist at both the technical and nontechnical levels. PK/PD modeling during lead optimization is still a relatively new concept for the pharmaceutical industry. Most pharmaceutical companies are using a model-aided rather than a model-based process. The acceptability of model-based PK/PD has not been fully embraced by drug discovery organizations. The implementation of model-based drug development requires pharmaceutical companies to foster innovation, make a change of mindset, align adaptive processes, and create a close collaboration among different functional groups [56]. Obtaining reliable, reproducible, and quantifiable pharmacological responses, identifying translatable biomarkers, distinguishing PD variability and PK variability are still technically difficult during the drug discovery stage; however, gaining a better appreciation of these issues preclinically can reduce attrition rates at the proof-of-concept stage.

REFERENCES

1. Kola, I. and Landis, J., Can the pharmaceutical industry reduce attrition rates? *Nat. Rev. Drug Discov.*, 3(8), 711, 2004.
2. Robbie, G. J. and Marroum, P. J., Dependence of effective half-life on estimation method, *Abstract #1047 AAPS Annual meetings*, Orlando, FL, 1999. www.aapsj.org/abstracts/AM_1999/725.htm
3. Gibaldi, M., *Pharmacokinetics*, 2nd edn. New York and Basel: Marcel Dekker, Inc., 1984.
4. Kwon, Y., *Handbook of Essential Pharmacokinetics, Pharmacodynamics and Drug Metabolism for Industrial Scientists*. New York: Kluwer Academic/Plenum Publishers, 2001.
5. Goldberg, M. and Gomez-Orellana, I., Challenges for the oral delivery of macromolecules, *Nat. Rev. Drug Discov.*, 2(4), 289, 2003.

6. Sun, D. et al., Comparison of human duodenum and Caco-2 gene expression profiles for 12,000 gene sequences tags and correlation with permeability of 26 drugs, *Pharm. Res.*, 19(10), 1400, 2002.
7. Avdeef, A. et al., PAMPA—critical factors for better predictions of absorption, *J. Pharm. Sci.*, 96(11), 2893, 2007.
8. Artursson, P. and Borchardt, R. T., Intestinal drug absorption and metabolism in cell cultures: Caco-2 and beyond, *Pharm. Res.*, 14(12), 1655, 1997.
9. Smith, D. A., Humphrey, M. J., and Charuel, C., Design of toxicokinetic studies, *Xenobiotica*, 20(11), 1187, 1990.
10. Nomeir, A. A. et al., Estimation of the extent of oral absorption in animals from oral and intravenous pharmacokinetic data in drug discovery, *J. Pharm. Sci.*, Feb. 18, 2009. [Epub ahead of print.] Published online in Wiley Interscience (www.interscience.wiley.com). DOI 10.1002/jps. 21705.
11. Kato, M., Intestinal first-pass metabolism of CYP3A4 substrates. *Drug Metab. Pharmaco.*, 23(2), 87, 2008.
12. Kunta, J. R. et al., Differentiation of gut and hepatic first-pass loss of verapamil in intestinal and vascular access-ported (IVAP) rabbits, *Drug Metab. Dispos.*, 32(11), 1293, 2004.
13. Pang, K. S., Komura, H., and Iwaki, M., Modeling of intestinal drug absorption: roles of transporters and metabolic enzymes (for the Gillette Review Series), Species differences in in vitro and in vivo small intestinal metabolism of CYP3A substrates, *Drug Metab. Dispos.*, 31(12), 1507, 2003.
14. Lee, Y. H. et al., Differentiation of gut and hepatic first-pass effect of drugs: 1. Studies of verapamil in ported dogs, *Pharm. Res.*, 18(12), 1721, 2001.
15. Nishimura, T. et al., Asymmetric intestinal first-pass metabolism causes minimal oral bioavailability of midazolam in cynomolgus monkey, *Drug Metab. Dispos.*, 35(8), 1275, 2007.
16. Komura, H. and Iwaki, M., Species differences in in vitro and in vivo small intestinal metabolism of CYP3A substrates, *J. Pharm. Sci.*, 97(5), 1775, 2008.
17. Chiou, W. L. and Buehler, P. W., Comparison of oral absorption and bioavailablity of drugs between monkey and human, *Pharm. Res.*, 19(6), 868, 2002.
18. Paine, M. F. et al., The human intestinal cytochrome P450 "pie", *Drug Metab. Dispos.*, 34(5), 880, 2006.
19. Lown, K. S., Ghosh, M., and Watkins, P. B., Sequences of intestinal and hepatic cytochrome P450 3A4 cDNAs are identical, *Drug Metab. Dispos.*, 26(2), 185, 1998.
20. Kato, M. et al., The intestinal first-pass metabolism of substrates of CYP3A4 and P-glycoprotein-quantitative analysis based on information from the literature. *Drug Metab. Pharmaco.*, 18(6), 365, 2003.
21. Uno, T. and Yasui-Furukori, N., Effect of grapefruit juice in relation to human pharmacokinetic study, *Curr. Clin. Pharmacol.*, 1(2), 157, 2006.
22. Badhan, R. et al., Methodology for development of a physiological model incorporating CYP3A and P-glycoprotein for the prediction of intestinal drug absorption, *J. Pharm. Sci.*, 98(6), 2180, 2009.
23. Mei, H., Korfmacher, W., and Morrison, R., Rapid in vivo oral screening in rats: reliability, acceptance criteria, and filtering efficiency, *AAPS J.*, 2006, 8(3), E493, 2006.
24. Davies, B. and Morris, T., Physiological parameters in laboratory animals and humans, *Pharm. Res.*, 10(7), 1093, 1993.
25. Ballard, P., Leahy, D. E., and Rowland, M., Prediction of in vivo tissue distribution from in vitro data 1. Experiments with markers of aqueous spaces, *Pharm. Res.*, 17(6), 660, 2000.
26. Ballard, P. et al., Prediction of in vivo tissue distribution from in vitro data. 2. Influence of albumin diffusion from tissue pieces during an in vitro incubation on estimated tissue-to-unbound plasma partition coefficients (Kpu), *Pharm. Res.*, 20(6), 857, 2003.

27. Poulin, P. and Theil, F. P., A priori prediction of tissue:plasma partition coefficients of drugs to facilitate the use of physiologically-based pharmacokinetic models in drug discovery, *J. Pharm. Sci.*, 89(1), 16, 2000.
28. Poulin, P. and Theil, F. P., Prediction of pharmacokinetics prior to in vivo studies. I. Mechanism-based prediction of volume of distribution, *J. Pharm. Sci.*, 91(1), 129, 2002.
29. Poulin, P. and Theil, F. P., Prediction of pharmacokinetics prior to in vivo studies. II. Generic physiologically based pharmacokinetic models of drug disposition, *J. Pharm. Sci.*, 91(5), 1358, 2002.
30. Rodgers, T. and Rowland, M., Mechanistic approaches to volume of distribution predictions: understanding the processes, *Pharm. Res.*, 24(5), 918, 2007.
31. Lombardo, F. et al., A hybrid mixture discriminant analysis-random forest computational model for the prediction of volume of distribution of drugs in human, *J. Med. Chem.*, 49(7), 2262, 2006.
32. Sahin, S. and Benet, L. Z., The operational multiple dosing half-life: A key to define drug accumulation in patients and to designing extended release dosage forms, *Pharm. Res.*, 25(12), 2869, 2008.
33. Lindstrom, E. et al., Neurokinin 1 receptor antagonists: Correlation between in vitro receptor interaction and in vivo efficacy, *J. Pharmacol. Exp. Ther.*, 322(3), 1286, 2007.
34. Drusano, G. L. et al., Use of preclinical data for selection of a phase II/III dose for evernimicin and identification of a preclinical MIC breakpoint, *Antimicrob. Agents Chemother.*, 45(1), 13, 2001.
35. Makrides, V. et al., Evidence for two distinct binding sites for tau on microtubules, *Proc. Natl. Acad. Sci.*, 101(17), 6746, 2004.
36. Mager, D. E. and Jusko, W. J., Development of translational pharmacokinetic-pharmacodynamic models, *Clin. Pharmacol. Ther.*, 83(6), 909, 2008.
37. Sheiner, L. B., Learning versus confirming in clinical drug development, *Clin. Pharmacol. Ther.*, 61(3), 275, 1997.
38. Danhof, M. et al., Mechanism-based pharmacokinetic-pharmacodynamic (PK-PD) modeling in translational drug research, *Trends Pharmacol. Sci.*, 29(4), 186, 2008.
39. Mager, D. E., Wyska, E., and Jusko, W. J., Diversity of mechanism-based pharmacodynamic models, *Drug Metab. Dispos.*, 31(5), 510, 2003.
40. Leipist, E. I. and Jusko, W. J., Modeling and allometric scaling of s(+)-ketoprofen pharmacokinetics and pharmacodynamics: A retrospective analysis, *J. Vet. Pharmcolol. Ther.*, 27, 211, 2004.
41. Yassen, A. et al., Animal-to-human extrapolation of the pharmacokinetic and pharmacodynamic properties of buprenorphine, *Clin. Pharmacokinet.*, 46(5), 433, 2007.
42. Kim, D. et al., Discovery of potent and selective dipeptidyl peptidase IV inhibitors derived from beta-aminoamides bearing substituted triazolopiperazines, *J. Med. Chem.*, 51(3), 589, 2008.
43. Herman, G. A. et al., Effect of single oral doses of sitagliptin, a dipeptidyl peptidase-4 inhibitor, on incretin and plasma glucose levels after an oral glucose tolerance test in patients with type 2 diabetes, *J. Clin. Endocrinol. Metab.*, 91(11), 4612, 2006.
44. Bergman, A. J. et al., Pharmacokinetic and pharmacodynamic properties of multiple oral doses of sitagliptin, a dipeptidyl peptidase-IV inhibitor: A double-blind, randomized, placebo-controlled study in healthy male volunteers, *Clin. Ther.*, 28(1), 55, 2006.
45. Lockwood, P. A. et al., The use of clinical trial simulation to support dose selection: application to development of a new treatment for chronic neuropathic pain, *Pharm. Res.*, 20(11), 1752, 2003.
46. Miller, R. et al., How modeling and simulation have enhanced decision making in new drug development, *J. Pharmacokinet. Pharmacodyn.*, 32(2), 185, 2005.
47. Feng, M. R. et al., Brain microdialysis and PK/PD correlation of pregabalin in rats, *Eur. J. Drug Metab. Pharmacokinet.*, 26(1–2), 123, 2001.

48. Danhof, M. et al., Mechanism-based pharmacokinetic-pharmacodynamic modeling: Biophase distribution, receptor theory, and dynamical systems analysis. *Annu. Rev. Pharmacol. Toxicol.*, 47, 357, 2007.
49. Field, M. J. et al., Gabapentin and pregabalin, but not morphine and amitriptyline, block both static and dynamic components of mechanical allodynia induced by streptozocin in the rat, *Pain*, 80(1–2), 391, 1999.
50. Best, J. D. et al., In vivo characterization of Abeta(40) changes in brain and cerebrospinal fluid using the novel gamma-secretase inhibitor N-[cis-4-[(4-chlorophenyl)sulfonyl]-4-(2,5-difluorophenyl)cyclohexyl]-1,1,1-trifluoromethanesulfonamide (MRK-560) in the rat, *J. Pharmacol. Exp. Ther.*, 317(2), 786, 2006.
51. Sheiner, L. B. et al., Simultaneous modeling of pharmacokinetics and pharmacodynamics: application to d-tubocurarine, *Clin. Pharmacol. Ther.*, 25(3), 358, 1979.
52. Copeland, R. A., Pompliano, D. L., and Meek, T. D., Drug-target residence time and its implications for lead optimization, *Nat. Rev. Drug Discov.*, 5(9), 730, 2006.
53. Gabrielssson, J. et al., Early integration of pharmacokinetic and dynamic reasoning is essential for optimal development of lead compounds: Strategic considerations, *Drug Discov. Today*, 14(7–8), 358, 2009.
54. Simeoni, M. et al., Predictive pharmacokinetic-pharmacodynamic modeling of tumor growth kinetics in xenograft models after administration of anticancer agents, *Cancer Res.*, 64(3), 1094, 2004.
55. Rocchetti, M. et al., Predicting the active doses in humans from animal studies: A novel approach in oncology, *Eur. J. Cancer*, 43(12), 1862, 2007.
56. Zhang, L., Pfister, M., and Meibohm, B., Concepts and challenges in quantitative pharmacology and model-based drug development, *AAPS J.*, 12, 12, 2008.

4 Mass Spectrometry for *In Vitro* ADME Screening

Inhou Chu

CONTENTS

4.1 INTRODUCTION

A recent survey showed that the attrition rate related to absorption, metabolism, distribution, and excretion (ADME) and pharmacokinetic (PK) failures has dropped from 40% to 50% in 1991 [1–3] to approximately 10% in 2000 [2,4]. Such a dramatic improvement within approximately one decade did not happen by chance. It was driven by the recognition of the impact of ADME and PK attributes on the safety, efficacy, and developability of new drug candidates and allocating necessary resources to address them early in drug discovery. However, recent advances in synthetic processes, which have generated enormous numbers of new chemical entities (NCEs) over the past few years, have created the need to design new high-throughput solutions to allow rapid assessment of the drug metabolism and pharmacokinetic (DMPK) attributes of potential drug candidates without compromising the quality of the data [5]. As a result, innovative *in vitro* ADME and PK approaches [6–15] are now routinely used in the pharmaceutical industry to evaluate DMPK attributes. Therefore, compounds with undesirable DMPK characteristics can be screened out as early, as quickly and as cheaply as possible, so that available resources can be focused on the promising drug candidates that survive this screening process.

Since the unambiguous determination of either the parent compound and/or possible metabolites is essential for successful DMPK screens, these higher-throughput, mostly cell-based, DMPK screening assays usually take advantage of the speed, sensitivity, and specificity of high-performance liquid chromatography–tandem mass spectrometry (HPLC–MS/MS) [16–22]. Examples include projecting absorption, metabolism, PKs, drug–drug interactions (DDIs), and protein binding. In this chapter, several higher-throughput *in vitro* screening assays that are performed as part of new drug discovery are discussed. The methodologies for those screens and how HPLC–MS/MS is used to meet the current and future demands are presented.

4.2 CYP INHIBITION

The majority of marketed drugs are metabolized oxidatively by a family of heme-containing enzymes, namely, cytochrome P450s (CYPs) [23]. Inhibition or induction of one or more of these drug-metabolizing enzymes by coadministered drugs is one of the major sources of DDI, which can lead to adverse drug effects including fatalities [24]. This is particularly important since many patients, especially the elderly population, are on multiple medications.

Since human liver CYP450 inhibition can be performed in a high-throughput format and *in vivo* DDI can be predicted by *in vitro* methods, it comes as no surprise that enzyme inhibition is one of the earliest and most important DMPK screening assays for NCEs in drug discovery [25–40]. Screening potential CYP inhibitory ability of NCEs is based on either liquid chromatography–mass spectrometry (LC–MS) or fluorometric detection [26]. The fluorometric assay, which uses recombinant CYPs and fluorogenic probe substrates, is low cost, provides rapid turnaround, and is higher throughput than LC–MS approaches. However, this assay suffers from lack of probe specificity and interference (intrinsic fluorescence and/or fluorescence quenching from compound and metabolites) therefore limiting the use of this technique [30,41]. LC–MS/MS and liver microsomes, on the other hand, has become the most common platform for CYP inhibition assays due to its high sensitivity and specificity [42] and is accepted by FDA [43].

While innovative procedures [27,44] and instrumentation [28] have shown promise in increasing the throughput of the CYP inhibition assay, the approach that we took for screening NCEs is to combine multiple CYP inhibition assays into one incubation using human liver microsomes. This approach has been developed and validated in-house and was similar to the cassette probe-dosing principal that adapted by other groups [45–48]. Realizing that the majority of drugs are metabolized by CYPs 3A4, 2D6, and 2C9 [49], and taking into account the cost of reagents, we selected the evaluation of the inhibition of these three isoforms (three-in-one assay) as the first CYP inhibition screen. If an NCE is advanced to become a potential recommendation candidate, more rigorous inhibition assays are carried out in which these three isoforms as well as CYP 2C19 and 1A2 are evaluated in individual assays. Because of the well-documented substrate-dependent inhibition of CYP 3A4 [50], a second substrate (midazolam) is used to evaluate the inhibition of CYP 3A4 for those NCEs that are being recommended for development.

The CYP inhibition assay utilizes the 96-well plate format with a robotic system, where both incubation and analysis are performed in the same plates. The setup of the sample plates is shown in Figure 4.1. For each compound, both direct inhibition and metabolism/mechanism-based inhibition, which is caused by a metabolite of the NCE that is either a more potent direct reversible inhibitor (metabolism-based) or a time-dependent irreversible inhibitor (mechanism-based), are evaluated. Both direct and mechanism-based inhibitors could result in inhibitory DDIs [51,52].

The three-in-one assay is conducted using pools of human liver microsomes from 10 to 20 donors. The microsomes are frozen in small aliquots and stored at −80°C until use. In this assay, hydroxytolbutamide, dextrorphan, and 6-β-hydroxytestosterone that are formed from tolbutamide hydroxylase (CYP 2C9 reaction) [53], dextromethorphan *O*-demethylase (CYP 2D6 reaction) [53,54], and testosterone 6-β-hydroxylase (CYP 3A4 reaction) [55], respectively, are quantified by LC–MS/MS in an approximately 1 min run time.

In the coincubation experiment, the NCE is incubated at 0.3, 3, and 30 μM in wells containing a mixture of microsomal protein (final concentration 0.2–0.4 mg/mL) and probe substrates at concentrations that approximate the K_m for each reaction (tolbutamide [200 μM], dextromethorphan [16 μM], and testosterone [100 μM] for CYPs 2C9, 2D6, and 3A4, respectively) in a 50 mM Tris-acetate buffer, pH 7.4, containing 150 mM KCl. The plates are prewarmed at 37°C for 5 min. The reactions are initiated by the addition of 20 μL of a 10 mM solution of NADPH prepared in buffer (final concentration, 1 mM) followed by a brief shaking. The total volume of the reaction mixture is 200 μL. Following an incubation period of approximately 15 min, the reactions are terminated by the addition of 35% perchloric acid followed

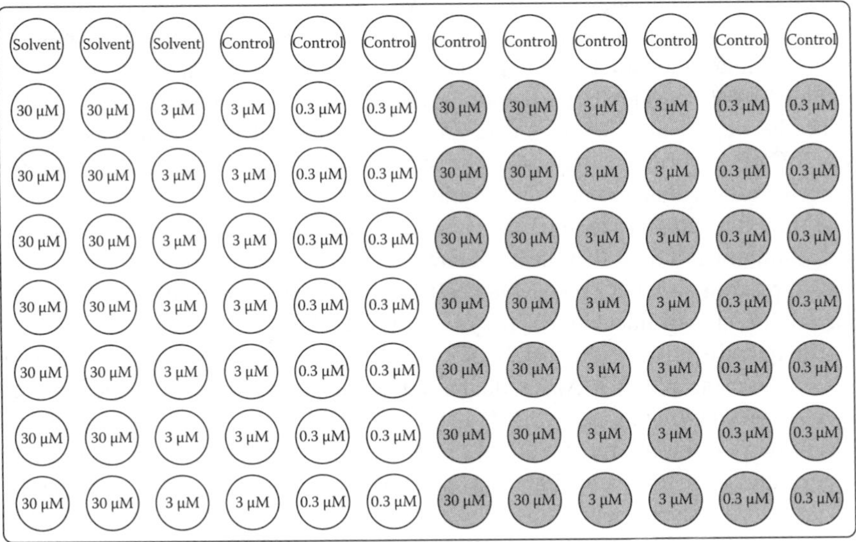

FIGURE 4.1 A template for the three-in-one CYP inhibition assay. Wells 1–12 of the first row contain solvent, substrates, or prototype inhibitors of each CYP, which serve as quality controls. Rows 2–8 contain NCEs at the concentrations indicated.

by shaking and centrifugation for protein precipitation. In the preincubation experiment, the NCE is first incubated with liver microsomes and NADPH for 30 min, then the probe substrates are added and the reactions are allowed to proceed as with the coincubation experiment.

The potential of DDI of an NCE is expressed as IC_{50} values, which is the NCE concentration that reduces the enzyme activity by 50%, and is calculated as follows:

$$\% \text{ inhibition} = \left[1 - \left\{\frac{[\text{metabolite}]_{\text{NCE}}}{[\text{metabolite}]_{\text{control}}}\right\}\right] \times 100\%$$

where the % inhibition is determined based on metabolite concentrations observed in the incubations with ($[\text{metabolite}]_{\text{NCE}}$) and without the NCE ($[\text{metabolite}]_{\text{control}}$). The IC_{50} values are determined by semilog plotting of NCE concentrations on the *x*-axis versus percentage of remaining activity on the *y*-axis, and the concentration that results in the 50% inhibition of enzyme activity is calculated from the graph.

NCEs are generally classified as direct inhibitors if the IC_{50} values obtained from both pre- and coincubation experiments are similar (within threefold), or higher under the preincubation conditions. If the IC_{50} under the preincubation conditions is much lower compared to that under coincubation (greater than threefold, i.e., greater inhibition), the NCE is likely to be a metabolism/mechanism-based inhibitor. Both pre- and coincubation IC_{50} values are required to determine whether or not the NCE is a direct and/or metabolism/mechanism-based inhibitor.

For direct inhibition, NCEs with IC_{50} values >10 μM are considered to be less likely to cause inhibitory DDIs. NCEs with IC_{50} values <1 μM are considered potent inhibitors and are likely to produce DDIs. These potent inhibitors are usually excluded from consideration except in rare occasions. The fate of NCEs with IC_{50} values between 1 and 10 μM is determined by additional factors such as potency, which CYP isoform, therapy area, the stage of the discovery program, PKs, and projected efficacious plasma concentrations.

It should be emphasized that the organic cosolvent concentration should be kept to a minimum as it may alter the basal enzyme activity. Also, a compound could be a weak direct inhibitor but a potent metabolism/mechanism-based inhibitor. Therefore, it is important to evaluate both types of inhibition [52].

The quantification of the metabolites of CYPs 2D6, 3A4, and 2C19 reactions is carried out using a SCIEX API 3000 mass spectrometer, running in the positive ion, selected reaction monitoring (SRM) mode. The HPLC gradient system used consists of two solvent mixtures. Solvent mixture A (SMA) consists of 94.9% H_2O, 5% MeOH, and 0.1% formic acid and solvent mixture B consists of 94.9% MeOH, 5% H_2O, and 0.1% formic acid. The analytical column used is a Develosil Combi-RP5, 5 μm, 3.5 × 20 mm (Phenomenex, Inc., Torrance, CA), with a mobile phase flow rate of 1.1 mL/min. A short run time (1 min) HPLC method is used because of the specificity of the mass spectrometer and the lack of matrix effects (this was thoroughly

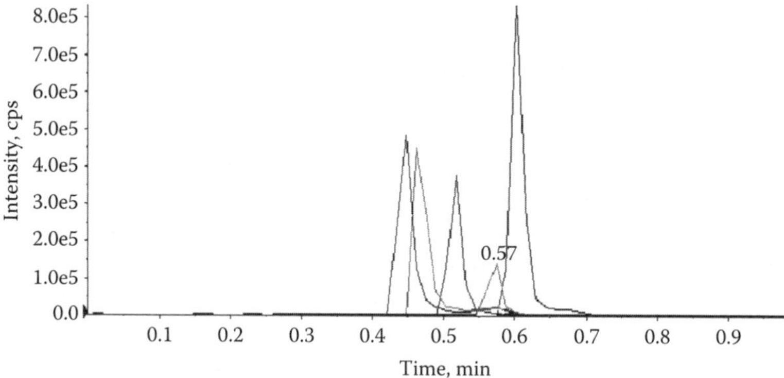

FIGURE 4.2 A representative chromatogram of the three-in-one CYP inhibition assay.

examined during the validation of the assay). A representative ion chromatogram is shown in Figure 4.2.

A potential issue with the short run time of the three-in-one assay is the possibility of matrix effects caused by the coelution of the analytes of interest with other components in the reaction mixture [45]. Also, MS signal suppression usually occurs at high concentrations (in our case, at NCE concentration of 30 μM), which if occurred would generate high IC_{50} value that will not result in compound rejection. On the other hand, if signal enhancement occurs, unusual results would be apparent (concentration of probe substrate metabolite(s) increases with the increase in concentration of NCE), which could be quickly identified. In these rare cases, the assay would be repeated with a more rigorous LC–MS/MS analysis. Furthermore, to minimize the potential for signal suppression, atmospheric pressure chemical ionization (APCI) is used as the ionization mode instead of electrospray ionization (ESI). It has been well documented that the potential for signal suppression in the APCI mode is much less than that of the ESI [56–58]. The throughput for the three-in-one assay is eight 96-well plates/day/instrument. For 30 NCEs, four 96-well plates are used (two for preincubation and two for coincubation); therefore, eight plates would generate 360 IC_{50}s daily.

If a compound is identified as a potential drug candidate, CYP inhibition evaluation is repeated for the selected compounds as well as for any major or active metabolites that are identified. In these more robust evaluations, individual CYP assays with extended concentration ranges are used. For CYP 2D6, CYP 2C9, CYP 2C19 (using *S*-mephenytoin as a substrate), and CYP 3A4 individual assay, the LC–MS/MS conditions are exactly the same as those used in the three-in-one screening assay. For the CYP 1A2 assay, phenacetin is used as the specific substrate and the same solvent mixtures used above. As for CYP 3A4 assay using midazolam as a substrate, because of the presence of an isobaric metabolite (1′-hydroxymidazolam) of the targeted metabolite (4-hydroxymidazolam), a Kromasil C4, 3.5 μm, 50×4.6 mm column (Phenomenex, Inc., Torrance, CA) instead of the Combi-RP5 column was used.

4.3 CACO-2 PERMEABILITY

Oral administration of therapeutics is the preferred route for drug delivery. Drugs are primarily absorbed through the intestinal lumen by four major pathways: transcellular, paracellular, facilitated diffusion, and active transporter [59,60]. Lipophilic compounds, which partition into the lipid bilayer of the cell membrane, are absorbed through transcellular passive diffusion driven by a concentration gradient. On the other hand, low molecular weight, hydrophilic compounds may be absorbed by passing through cell–cell junctions (paracellular route). Since the junctions are tight and represent a small surface area compared with the enterocytes, this mode of transport is slow and is primarily for water-soluble low molecular weight compounds (<350 Da). Compounds can also be absorbed by a transmembrane transporter without requiring energy, that is, facilitated diffusion that is driven by concentration gradient, or with the expenditure of energy, that is, active transport, which could proceed against concentration gradient. Because a reversible drug–carrier complex is formed with the last two modes of transport, this process could be fairly selective. Also, as the number of carriers is limited, the process is saturable either by the compound itself at higher concentrations or by other compounds that interact with the same carrier. It is worth noting that a compound could be absorbed via one or more modes of transport with varying degrees.

The majority of NCEs are evaluated for potential oral absorption as they are mainly targeted for oral delivery. Therefore, *in vitro* models to assess the potential for oral absorption have become major components of drug discovery programs [59,60]. The most widely used model to assess intestinal permeability is the human colon adenocarcinoma cell line, that is, Caco-2 cells [61], which is a cell line that exhibits morphological and functional characteristics of small intestinal cells. Caco-2 cells form tight junctions, develop microvilli, express brush border enzymes and some major drug-metabolizing enzymes [62], and an array of transporters found *in vivo* [63]. Several studies have shown a good correlation between Caco-2 permeability and oral absorption in humans [64,65]. With this Caco-2 assay, NCEs can be ranked according to their ability to cross the Caco-2 monolayer in a relatively short period of time. Specific structural features that facilitate transport of a chemical series through the Caco-2 monolayer can be delineated and incorporated into the design of new drug candidates. Two *in vitro* assays are preferred in the projection of human absorption using Caco-2 monolayers (permeability assay and the efflux transport [bidirectional] assay), which will be discussed in the following sections.

4.3.1 PERMEABILITY

Caco-2 monolayers are incubated at 37°C for 30 min in a CO_2 incubator with transport media (TM), prior to initiating a transport experiment. The TM for the apical side is made of Hank's balance salt solution (HBSS) containing 10 mM HEPES buffer (pH 7.4) and 25 mM D-glucose (a recent study suggested that simulated intestinal fluid might be a better media to use in the apical compartment especially for highly lipophilic compounds [66,67], and this new media will be investigated in the future).

This buffer is used in both the unidirectional (permeability) and the bidirectional (Efflux, Section 4.3.2). In the permeability assay, NCEs are placed into the apical compartment of the Transwell. Duplicate samples are taken immediately after compound addition from the apical compartment (zero time) and then after 2 h from both the apical and basolateral compartments for LC–MS/MS analysis. The integrity of the monolayer is confirmed by the measurement of the transepithelial electrical resistance (TEER), which must be above a certain limit to be used for transport experiments. In addition, with each experiment a transcellular and a paracellular marker are included for quality control.

Analytical samples are collected into 96-well plates and mixed with three volumes of acetonitrile containing an internal standard. Sample–acetonitrile mixtures are vortexed, centrifuged, and the supernatant is subjected to LC–MS/MS analysis in the same 96-well plate. Caco-2 screening data are presented as apparent permeability coefficient (Papp), which is calculated as follows:

$$\text{Papp} = (\text{CB}_t/\text{CA}_0) \times V_b/(T * S)$$

where

CB_t is the concentration of the compound in the basolateral side at time T

CA_0 is the concentration of the compound in the apical side at the beginning of the experiment (zero time)

V_b is the volume of the basolateral compartment of the transwell

T is the sampling time (120 min)

S is the Caco-2 membrane surface area [68]

4.3.2 Efflux Transport

During the process of absorption from the lumen side of the intestine, efflux transporters (such as P-glycoprotein) that are located on the apical membrane of the enterocytes could pump certain compounds back into the intestinal lumen after they partition into the cell membrane [69]. These efflux transporters could hinder absorption and may influence the overall disposition of a compound. A bidirectional transport system has been set up to determine if the NCE is a substrate for efflux transporters, primarily P-glycoprotein (P-GP). In this assay, the permeability of an NCE is determined for both the apical-to-basolateral (A-B) and the basolateral-to-apical (B-A) directions. If a compound possesses a higher permeability in the B-A compared to the A-B direction (larger than twofold), this would suggest that it is a substrate for an efflux transporter. Also, the role of efflux transporters in limiting the permeability of a drug candidate can be further delineated by using a specific inhibitor [20]. If a compound is a substrate of the efflux system, the apparent apical to basolateral permeability would likely to be enhanced by the inhibitor. The assay conditions are similar to that of the permeability experiment except that in the basolateral compartment 4% bovine serum albumin (BSA) is added to the TM. The addition of BSA has been shown to reduce nonspecific binding that is often encountered with highly lipophilic compounds [70]. It was reported that the addition of BSA has

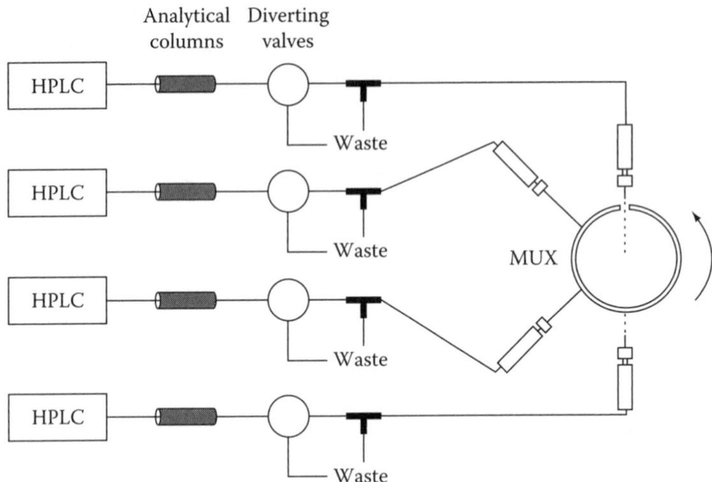

FIGURE 4.3 A schematic representation of the 4-sprayer MUX interface.

posted analytical issue with other laboratory [66]; however, no interference from BSA was observed in our analyses.

Unlike CYP inhibition assays, where common analytes are quantified regardless of the compound tested, for Caco-2 screening, the NCE itself is analyzed, requiring method development. Consequently, a short run time assay such as that used in the enzyme inhibition is not feasible because of the potential for matrix effects, which is difficult to estimate for each compound. For this reason, we took a different strategy to increase the LC–MS/MS throughput for the Caco-2 assay. The multiplexed inlet system (MUX), which allows for four data acquisitions simultaneously [71], has been adapted for the Caco-2 LC–MS/MS assays. A schematic representation of the four-way MUX interface (Waters Corp.) is shown in Figure 4.3. The MUX ion source is equipped with four electrospray needles around a rotating disk that contains an orifice allowing independent sampling from each sprayer. Four samples are simultaneously introduced into the mass spectrometer. Thus, both method development and sample analysis times/compound are greatly reduced [72]. The reproducibility of the Caco-2 assay was evaluated by the determination of apparent permeability of propranolol in six separate runs. The coefficient of variation (%CV) was 11.2% [37]. Over 100 compounds/week (combined permeability and efflux samples) are routinely evaluated on a single instrument with the MUX system.

4.4 HEPATOCYTE CLEARANCE

The availability of human tissues has allowed several human DMPK parameters to be evaluated in drug discovery before the compound enters into clinical studies. One such assay is the determination of human hepatocyte clearance. Since liver is the major site of metabolism for most drugs, screening compounds for metabolic stability in liver preparations is usually one of the primary screenings employed in drug discovery. This parameter allows an early look at the potential metabolic stability of a drug candidate in animals and humans. In an early screening paradigm, microsomal stability assay is usually used as a primary DMPK screen by many laboratories [73–75] because of its lower cost and ease of preparation compared to hepatocytes (Figure 4.4). However, microsomes lack some of phase II enzymes and require additional phase II cofactors to be added during incubation. For this reason, we decided to carry out the *in vitro* clearance studies in human (from individual subjects) or animals (pooled from several animals) cryopreserved hepatocytes [21,76–78]. The hepatocytes are thawed in a water bath at 37°C before use. Hepatocytes suspension is centrifuged at room temperature for 5 min and the supernatant is discarded. The hepatocyte pellets are gently resuspended in medium to a final density of approximately 1 million cells/mL. The viability of the cell suspension is determined by trypan blue staining, immediately prior to and post the experiment. Test compounds are prepared in methanol and added to hepatocyte suspension in a Waymouth medium at a final concentration of 1 μM. Incubations are carried out in 24-well plates in a shaker at 37°C in a CO_2 incubator. Aliquots of 100 μL solution are taken from the incubation mixture at preset time points (up to 2 h) and mixed with two volumes of

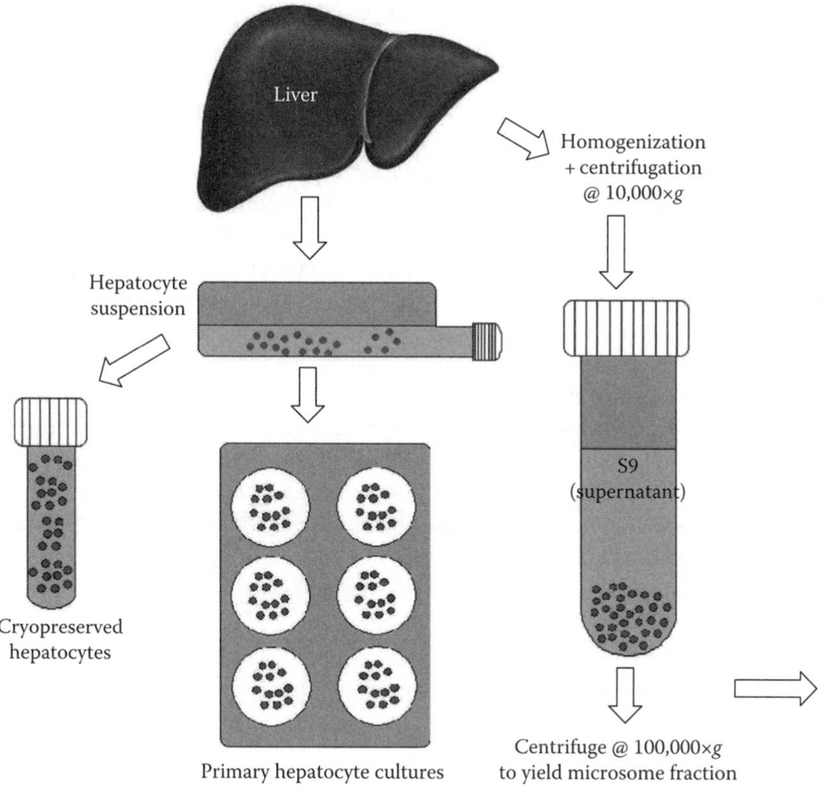

FIGURE 4.4 *In vitro* system for collecting hepatocytes and microsomes for clearance study.

acetonitrile in a 96-well plate. The plates are sonicated, vortexed, and centrifuged prior to LC–MS/MS analysis. The intrinsic hepatocyte clearance is calculated as described earlier [78] using the following equation:

$$\mathrm{CL_{in}} = \frac{C_0 - C_{120\,\mathrm{min}}}{\mathrm{AUC}_{0-120\,\mathrm{min}}} \times \frac{V}{N} (\mu\mathrm{L/min/M\ cells})$$

where

C_0 and $C_{120\mathrm{min}}$ are the concentrations of the compound in μM at 0 and 120 min, respectively

AUC is the area under the concentration versus time curve (μM/min)

V is the volume of incubation mixture (1000 μL)

N is the number of hepatocytes in millions

The predictability of this *in vitro* model was evaluated by the determination of the intrinsic clearances of 64 compounds and comparing the values to the *in vivo* clearance [38], which indicated a good relationship between the *in vivo* and *in vitro* results. Some fundamental assumptions need to be kept in mind when using this model: (1) *in vitro* enzyme kinetics are applicable to *in vivo* kinetic properties, (2) intrinsic clearance follows first-order kinetics, and (3) liver is the major organ responsible for the clearance of test compounds.

Samples analysis was carried out using a generic LC gradient elution and a Waters Aquity UPLC (ultra-performance LC) system [77,79,80] with either 0.1% formic acid or 10 mM ammonium acetate as mobile phases. Separation was carried out on a BEH C18 column (50×2.1 mm; 1.7 μm particle size) with mobile phase flow rate of 1 mL/min flow rate. The total turnaround time (run time plus injection time) was 2 min/sample. This combination of a sub-2 μm particles column packing stationary phase and a high LC flow rate increased the speed, resolution, and sensitivity of LC system (Figure 4.5). A generic LC method was used and that dramatically reduced the LC method development time. For compounds that showed interference either from the formation of the metabolites or from the matrix, a longer LC gradient is used (turnaround time 3.5 min/sample), which should provide sufficient resolution to eliminate interference from potential metabolites (Figure 4.6). In the initial screening mode, no calibration curve is included in the sample set and the throughput is 200 compounds/instrument/week. For the potential recommendation candidates, the assay is repeated (N=3) with calibration standards.

4.5 PLASMA PROTEIN BINDING

It has been generally believed that the pharmacological activity, safety, and tissue distribution of a drug is related to the unbound fraction of a drug in plasma rather than the total concentration, due to its ability to cross the membranes and interact with receptors [81–88]. For the majority of drugs, the free fraction in human plasma is usually constant among individuals; hence the therapeutic effect is most often correlated with the total concentration. However, this approach may not be appropriate

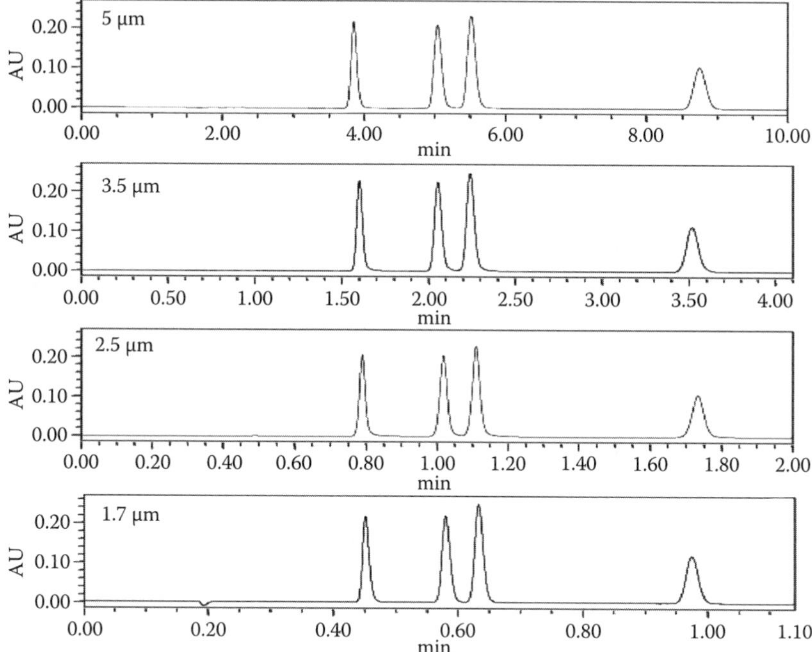

FIGURE 4.5 Improved throughput from a 5 μm particle size column (total run time 10 min) to 1.7 μm particle size UPLC column (total run time 1.1 min; http://www.waters.com/webassets/cms/library/docs/720001507en.pdf).

in some cases [85]. For example, it is known that a variety of disease states and conditions can lead to a significant decrease in protein binding, hence increasing the free fraction that may produce some clinical consequences, particularly for drug with a narrow therapeutic window [85–87]. Therefore, knowing the free fraction (f_u) is helpful in understanding the PKs as well as the PK/pharmacodynamic (PD) relationship of a drug candidate.

Typically in the lead characterization stage, protein binding is determined for human and animals (mouse, rat, dog, monkey as well as the pharmacology species, if different). Knowledge of plasma protein binding in animals is essential in scaling-up PK parameters (mainly volume of distribution and clearance) from animal models to humans [85,89,90]. Four assays have been evaluated, which include ultrafiltration [92–95], equilibrium dialysis [91], rapid equilibrium dialysis (RED) [99], and immobilized human serum albumin (HSA) LC–MS [96–98]. The ultrafiltration method utilizes a pressure gradient to force the aqueous component (containing the free drug) through a membrane with a molecular weight cut-off limit of 10,000 Da. This method is more efficient than the traditional equilibrium dialysis assay but suffers the limitation of nonspecific binding of the drug to the membrane, which has limited its use in early drug discovery. Equilibrium dialysis is usually conducted in a 96-well dialysis plate with top and bottom chambers separated by a semipermeable membrane, with a molecular weight cut-off of 10,000 Da. Nonspecific binding is not a major issue in equilibrium dialysis but

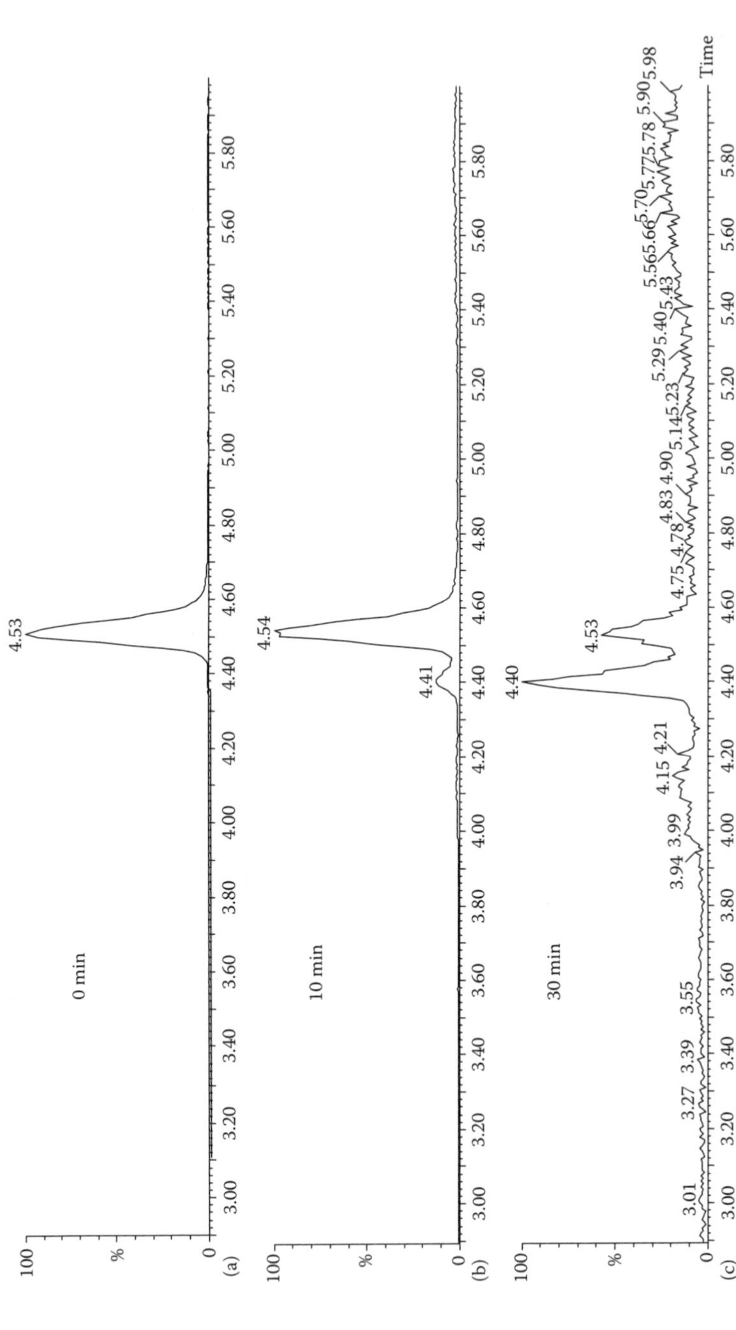

FIGURE 4.6 Mass chromatograms showing a metabolite generated in hepatocyte incubations as indicated by comparing the 0, 10, and 30 min incubations.

this assay suffers major drawbacks in long equilibration time (~18h), long assay preparation time, and complexity of automation [93,99]. The RED device reduces the equilibrium time by increasing the contact area between the plasma and buffer [99]. However, evaluation in our laboratory showed that the equilibrium between the buffer and plasma side of many proprietary compounds tested was not reached even after 8h of dialysis. This means same-day turn around time cannot be achieved with this device, which provides little advantage over the transitional equilibrium dialysis device. Immobilized HSA LC–MS is by far the fastest technique among the four methods [100]. Up to 20 compounds can be assayed in little more than an hour [88]. However, HSA LC–MS may not be suitable for compounds bound mainly to other plasma proteins (e.g., α-1-acid glycoprotein [AGP]) and compounds with high HSA affinity might have difficulty eluting out of the column (% protein bound >99%). This technique is more suitable for testing large chemical libraries to establish the structure–activity relationship (SAR) within a chemical series. For the above mentioned reasons, we decided that equilibrium dialysis, which is less susceptible to experimental artifacts [93], should be the primary method to assess protein binding.

The screening assay is conducted in a 96-well equilibrium dialysis device (Harvard Apparatus, Holliston, MA). A volume of 200μL of plasma containing 10μM of the NCE is placed into the top chamber (plasma side) followed by the addition of an equal volume of phosphate-buffered saline (PBS), pH 7.5, into the bottom chamber (buffer side). The dialysis device is kept in a rotator for 20h at 37°C using an incubator with 10% CO_2 as the atmosphere environment. Twenty microliters from the plasma side and 100μL from the buffer side are aliquotted into a sampling plate. In order to make the analytical matrix the same for both plasma and buffer sides, 100μL of PBS buffer and 20μL of blank plasma are added to plasma and buffer samples, respectively. Samples are then mixed with two volumes of acetonitrile containing internal standard, vortexed, centrifuged, and analyzed by LC–MS/MS. The percentage of plasma protein binding is calculated as

$$\% \text{ bound} = 100 \times (C_p - C_b)/C_p$$

where C_p and C_b are the compound concentrations in plasma and buffer chambers, respectively.

The accuracy of the assay for compounds with percent protein-bound exceeding 93% is excellent with %CV<0.5 (internal results). This is very important as the accuracy of measurement becomes increasingly important when the free concentration decreases. We also have excellent interday reproducibility with the %diff, in most of cases, <1% for majority of the compounds tested (data not shown). The plate format for the equilibrium dialysis is shown in Figure 4.7. The recovery, in this case, was calculated by comparing the combined concentrations of plasma and buffer sides samples to the time zero samples taken in the beginning of the assay. With the new format, the throughput of the assay was increased to 48 sets/analyst/week. For the few promising drug candidates, the assay is repeated with calibration curves and four replicates.

A	Time 0	Blank	B1	B2	P1	P2	Time 0	Blank	B3	B4	P3	P4
B	Time 0	Blank	B1	B2	P1	P2	Time 0	Blank	B3	B4	P3	P4
C	Time 0	Blank	B1	B2	P1	P2	Time 0	Blank	B3	B4	P3	P4
D	Time 0	Blank	B1	B2	P1	P2	Time 0	Blank	B3	B4	P3	P4
E	Time 0	Blank	B1	B2	P1	P2	Time 0	Blank	B3	B4	P3	P4
F	Time 0	Blank	B1	B2	P1	P2	Time 0	Blank	B3	B4	P3	P4
G	Time 0	Blank	B1	B2	P1	P2	Time 0	Blank	B3	B4	P3	P4
H	Time 0	Blank	B1	B2	P1	P2	Time 0	Blank	B3	B4	P3	P4

FIGURE 4.7 Plasma protein binding plate format.

4.5.1 Impact of pH and Temperature on the Plasma Protein Binding Assay

Many factors may affect the degree of plasma protein binding *in vivo* such as age, disease state, pregnancy, etc. [101]. However, under well-controlled *in vitro* settings, only few factors might impact the outcome of the assay, such as pH and temperature. In order to fully understand the importance of these factors, we have conducted in-house studies that indicated that careful control of both factors is vital for accurate plasma protein binding results.

pH: There are two major drug-binding proteins in plasma, albumin and AGP. It is widely known that albumin, which accounts for 50%–60% of the measured serum protein, has several pH-dependent conformations [102], and this alternation in the conformation will lead to changes in drug binding to the protein. In addition, recent findings [103–107] suggested that albumin binding is highly dependent on the ionization state, which is highly influenced by the pH, of the compound. Those two factors are the main reason that most of the drugs show a pH-dependent binding to plasma protein [105,106], which strongly suggests that a pH-controlled environment is crucial for accurate measurement of plasma protein binding.

To minimize the pH shift during the equilibrium dialysis incubation, a buffer system was employed. However, our in-house study, as well as the observations from other laboratories, indicated that pH of the fresh plasma after one freeze-and-thaw cycle is elevated to 7.80 (unpublished observation). After 20h incubation inside the incubator without CO_2, the pH of the plasma increased from 7.8 to 8.20, which is a direct result of CO_2 evaporation from the carbonic acid–bicarbonate buffer that regulates the physiological pH of plasma [108]. This pH shift can be compensated by using 10% CO_2 in the atmosphere in a CO_2 incubator [108]. This procedure restores the plasma pH back to the vicinity of 7.4 indicating that adjusting the plasma pH prior to the experiment is not necessary [109]. In general, nonpolar, cationic compounds had a higher likelihood of displaying pH-dependent binding [104,105]. Also, the change in plasma protein binding due to pH change is not species-dependent but is drug-specific.

Temperature: Temperature is another factor that could affect the plasma protein binding. Phenytoin and aztreonam are two compounds previously evaluated [86,110–112]. For both compounds, the extent of protein binding significantly decreased as temperature increased. Paradoxically, more recent studies have shown a different trend, where the degree of propofol binding to HSA increased at elevated temperatures [86]. These observations reenforces the importance of temperature control in the determination of the unbound fraction for a drug candidate.

4.6 CYP ISOZYME PROFILING

Most drugs are substrates for one or more cytochromes P450 that are responsible for catalyzing drug biotransformation [113–115]. Since this biotransformation process directly impacts the systemic drug exposure, identifying the major CYP isozymes responsible for the metabolism of the drug candidate in humans provides valuable information regarding bioavailability, potential interindividual variability, potential DDI, and the polymorphic metabolism of a drug candidate [45,47,48]. For example, if a compound is metabolized solely by a polymorphic CYP such as CYP 2D6, the decision to advance such a candidate may need to be reconsidered.

Human liver microsomes (usually used at 1 mg/mL, as an initial concentration) are incubated with 1 μM of the drug candidate in the presence of 1 mM NADPH with and without individual selective human CYP inhibitors (ketoconazole, quinidine, sulfaphenazole, tranylcypromine, and furafylline). In addition, microsomes from insect cells expressing individual recombinant CYPs (r-CYP) are incubated with the drug candidate. If the major human metabolites of the drug candidate are known in advance (from human hepatocyte metabolism studies), the disappearance of the substrate and the appearance of metabolites are monitored allowing the assignment of each metabolite to the CYPs responsible for its formation, which is an added value. Following incubation, protein is precipitated with perchloric acid, the samples are centrifuged and the supernatant analyzed by LC–MS/MS. The appearance of individual metabolites and the disappearance of the drug candidate from incubations with liver microsomes and individual r-CYPs would provide preliminary information regarding which are the major isozymes responsible for the metabolism of the drug candidate. This would allow the identification of potential liabilities of drug candidate and the formulation of a plan to address them in phase I trials.

4.7 hERG MOCK RUN

The decision by Merck to recall the blockbuster drug Vioxx, a COX-2 inhibitor, as a result of an increased risk of heart attack in patients highlighted the importance of understanding the cardiac liabilities of a drug candidate before it reaches the market. In fact, of the 38 drugs withdrawn from the market in the past 15 years, nearly half were withdrawn because of cardiotoxicity [116,117]. For example, terfenadine, an antihistamine, was withdrawn from the market because it was shown to cause QT interval prolongation and cardiac arrhythmias. This effect has been traced to the blockade of the cardiac delayed rectifier potassium current known as I(Kr). This current is conducted by tetrameric pores with the individual subunit encoded by the human ether-a-go-go-related gene (hERG) [118], which is important for the normal repolarization of the heart muscle. Blockade of this hERG potassium channel is believed to be the predominant cause of drug-induced QT prolongation. Thus, elimination of NCEs that are potent inhibitors of the hERG channel (near the anticipated therapeutic concentration) as early as possible in the discovery stage is important in order to eliminate this potential risk.

The most widely used technique for the evaluation of hERG channel interaction is the voltage clamp. A detailed description of the experimental setup has been described elsewhere [119]. The hERG interaction is measured and reported as the % inhibition of the hERG current compared to the vehicle control at various concentrations of the NCE. The concentration that inhibits 50% (IC_{50}) is calculated, whenever possible. The concentrations of NCE used in the assay are carefully selected based on the expected maximum plasma concentration (C_{max}) at the pharmacologically active dose (usually based on studies in animal models) and the human plasma protein binding for the NCE.

To ensure the accuracy of the result, the concentrations of the compound before and after the hERG voltage clamp study are determined by LC–MS/MS. Compound loss during the study could be due to low solubility in the working buffer, adsorption to the surface of the vessel used to prepare the final dilution, and adsorption to the reservoir on the voltage clamp rig or to the tubing used to deliver the NCE solution from the reservoir to the recording chamber. Sampling the buffer for the determination of compound concentration during the voltage clamp experiment is not feasible as this may disturb the voltage measurement. Therefore, a mock experiment is subsequently carried out as with the actual run but without the cells. The concentrations of test article are measured in samples that have flowed through the drug reservoir and tubing of a voltage clamp rig (posttubing samples) and also in samples that did not flow through the tubing (pretubing samples). By measuring pre- and posttubing concentrations, the source of any loss could be narrowed down for further investigations, if needed. Sample analysis is carried out by LC–MS/MS. Usually, the solubility of the compound in the hERG buffer is determined in advance of the voltage clamp experiment to make sure that the compound is soluble at the intended concentrations.

4.8 EMERGING NEW "HIGH-THROUGHPUT" TECHNIQUES

The intrinsic properties of the current LC–MS/MS technique, particularly the HPLC separation speed and the autosampler cycle-time, hindered the ability of LC–MS/MS to become a true "high-throughput" analytical technique. To address this issue, several new techniques have emerged in recent years, which include BioTrove's RapidFireTM XC-MS system [120], Phytronix technologies' laser diode thermal desorption (LDTD)-MS system, and Applied Biosystems/MDS SCIEX's FlashQuant workstation [121,122]. The major advantage of these techniques is the removal of HPLC which dramatically reduced the run-to-run cycle time. In addition, the sample introduction process was redesigned to further reduce the sample introduction rate. In the case of RapidFire system (BioTrove, Woburn, MA), which is a "stand-alone" platform that can be easily connected to any mass spectrometer, a computer-controlled fluidic robot system replaced the conventional HPLC and autosampler for sample injection (Figure 4.8). Since no HPLC separation is available, sample cleanup might be needed in some assays to remove undesired reaction components (salts, detergents, etc.) that might compromise the ionization efficiency of the analytes. This system has been successfully validated for a CYP inhibition

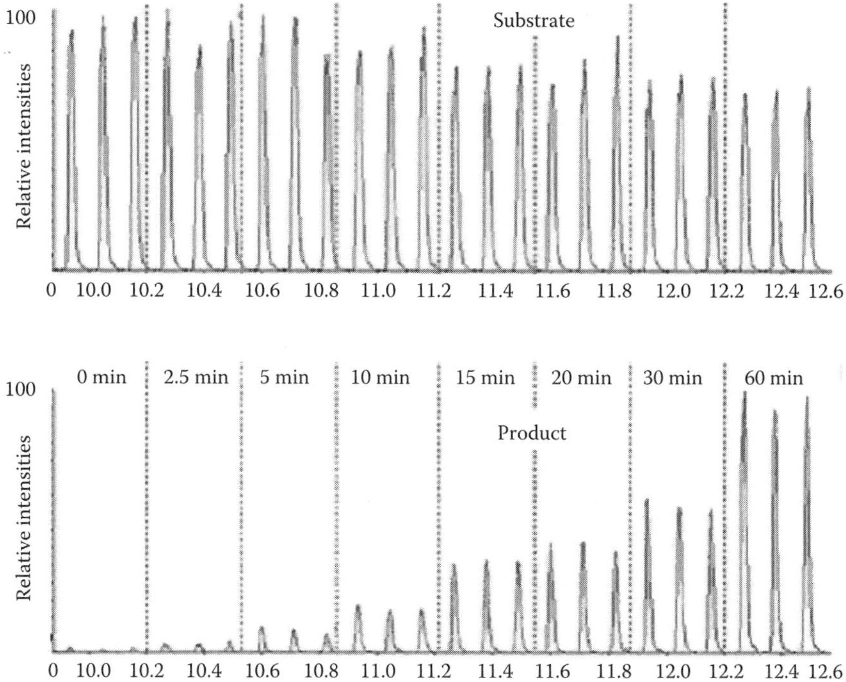

FIGURE 4.8 A demonstration of the speed of RapidFire system. The 24 samples (in triplicates) for the 0–60 min time course experiment were analyzed under 2.5 min. (Courtesy of BioTrove, Woburn, MA, http://www.biotrove.com. With permission.)

assay with the throughput approaching the fluorescence-based methods (3500 data points per 8 h shift on a single instrument [120].

The other two techniques, FlashQuant and LDTD_MS, both utilize laser desorption to facilitate samples introduction into mass spectrometer. In FlashQuant workstation, samples were spotted on a stainless steel plate after desalting (Figure 4.9). The desalting step is critical in this technique to ensure the formation of uniform crystals on each sample spot and the reproducibility of the results. Samples are then introduced into the mass spectrometer through MALDI. The matrix interference associated with using MALDI is resolved by using a noninterfering matrix as well as the specificity of tandem mass spectrometry (MS/MS) [121,122].

In LDTD_MS, each sample (1–10 μL) is spotted in a stainless steel insert on a 96-well polypropylene plate (LazWell) without applying any matrix, and allowed to dry (either with or without applying N_2). Upon operation (Figure 4.10) the laser thermally desorbs the dried samples by heating the bottom of the stainless steel insert. The desorbed molecules are carried by a preheated carrier gas through the transfer tube into the APCI source and ionized by the corona discharge needle [30]. The vaporization condition (i.e., laser power and duration) need to be carefully optimized so that the temperature is high enough to desorb the samples but not too high to cause thermal degradation of the analyte.

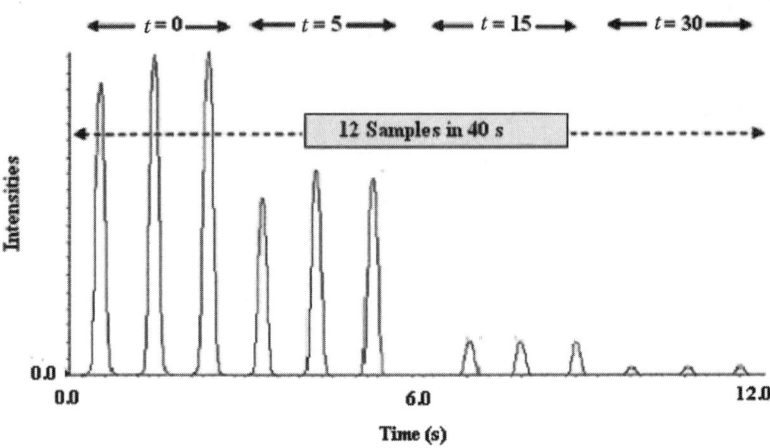

FIGURE 4.9 A representative chromatogram of microsomal stability profiling of buspirone analyzed by FlashQuant workstation. Samples (~2 µL) are spotted and air-dried on a MALDI plate. Solid samples are irradiated by a focused laser light, and the resulting ions are analyzed under the MRM mode by triple quadrupole analyzer. Ninety-six samples were analyzed in 5 min and 384 samples were analyzed in 20 min. (Courtesy of Applied Biosystem and MDS SCIEX, Foster City, CA.)

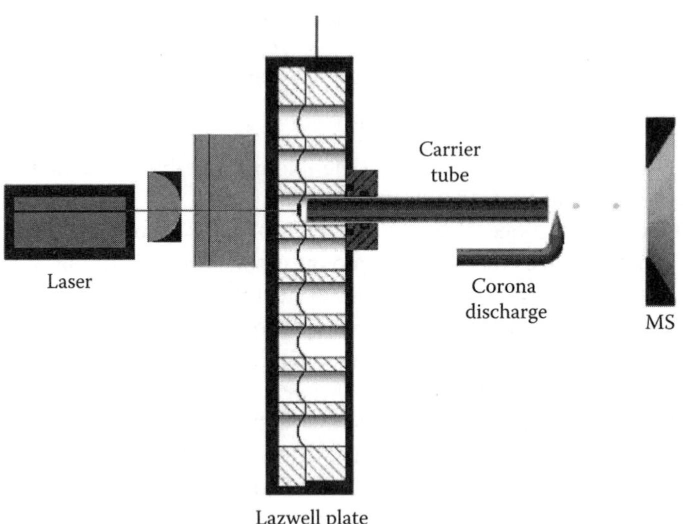

FIGURE 4.10 A Schematic representation of LDTD-MS instrumentation. In this system, samples (~2 µL) are deposited into the LazWell plate and thermally desorbed into gas-phase using an IR laser. The desorbed molecules are carried by the carrier gas into the corona discharge region and undergo APCI. Ions generated are transferred into mass spectrometer for analysis. (Courtesy of Phytronix technologies, Inc., Quebec, Canada, http://www.ldtd-ionsource.com/eng/ldtd/the-ldtd-technology.asp)

With these innovative sample introduction techniques, 96 samples can be analyzed in less than 20 min, which is compared to >2.5 h for traditional LC–MS/MS. However, as in the case for the BioTrove's RapidFire system where no LC separation is involved, both techniques rely solely on the specificity of MS/MS to differentiate different compounds. To take full advantage of FlashQuant and LDTD_MS on complex system (e.g., hepatocyte incubation assay or *in vivo* studies), matrix effects, isobaric interferences, metabolic interferences, and MS cross-talk need to be carefully investigated prior to applying those techniques in samples analysis.

4.9 PRECAUTIONS FOR USING MASS SPECTROMETRY IN HIGH-THROUGHPUT SETTINGS

It is commonly believed that the matrix effect in the *in vitro* assays is low due to the relatively "clean" matrix compared to plasma or serum. Therefore, a common practice for increasing the throughput of *in vitro* assays is using single ion monitoring (SIM), to eliminate an optimization step of the mass spectrometer, or by using a shorter LC run time. While this is true in most cases, in many occasions adequate LC–MS/MS separation is still needed for accurate measurements [107]. These observations have been reported by other investigators [123–125] and are illustrated in the following three examples:

The first example emphasizes the need to use SRM mode over SIM mode. This is illustrated in Figure 4.11 (in a Caco-2 experiment), where two peaks with identical *m/z* were detected in the LC–MS mode, while in the LC–MS/MS mode they produced different product ions allowing the recognition of the analyte of interest.

An example of matrix effect is shown in Figure 4.12. In this case, a drop in the internal standard area counts for the buffer-side samples (a protein-binding study), analyzed using a fast HPLC gradient, indicated possible signal suppression. A modified HPLC gradient, which extended the internal standard retention time from 1.5 to 3 min eliminated the signal suppression. In this case, only the internal standard signal is affected and no suppression was observed for the NCE tested. However, if the situation is reversed, we would observe a consistent internal standard signal and mistakenly think the results are acceptable.

A second example emphasizing the importance of good chromatography in the Caco-2 assay is presented in Figure 4.13. In this case, an additional peak was observed under SRM after the 2 h transport experiment, presumably a rearrangement product or a metabolite with the same *m/z*. If the chromatography did not provide adequate separation of the two components, an inaccurate permeability estimate would have resulted.

As these three examples emphasize, caution must be taken when developing bioanalytical assays, not only for *in vivo* studies but also for *in vitro* screening assays. It is especially important for higher-throughput LC–MS/MS assays so that the short turn around time is not achieved at the expense of data quality.

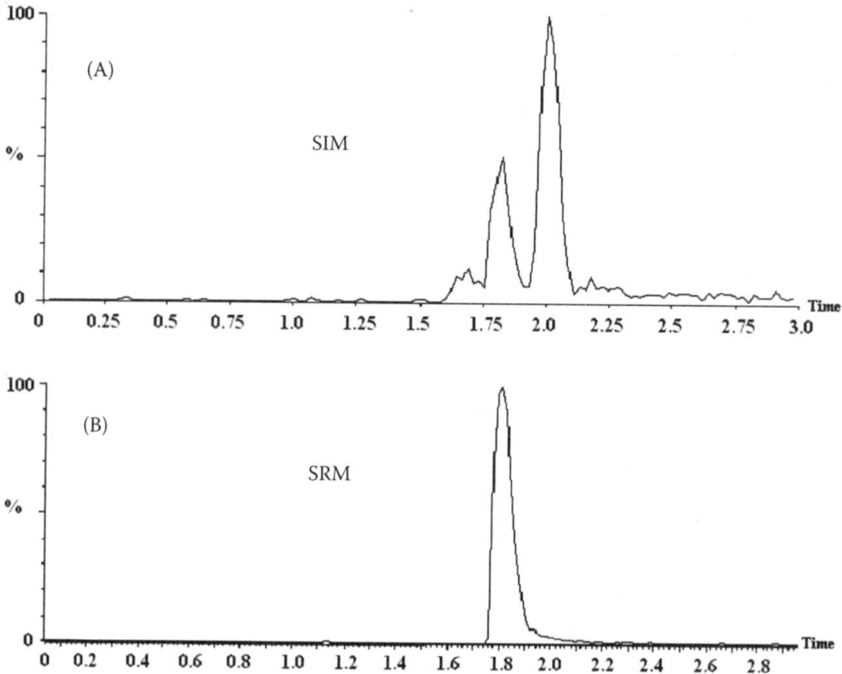

FIGURE 4.11 SIM and SRM chromatograms for the same sample. In the SIM mode (A), several interference peaks are observed while in the SRM mode (B), only the analyte is apparent. (Adapted from Fung, E.N. et al., *Rapid Commun. Mass Spectrom.*, 17, 2147, 2003. With permission.)

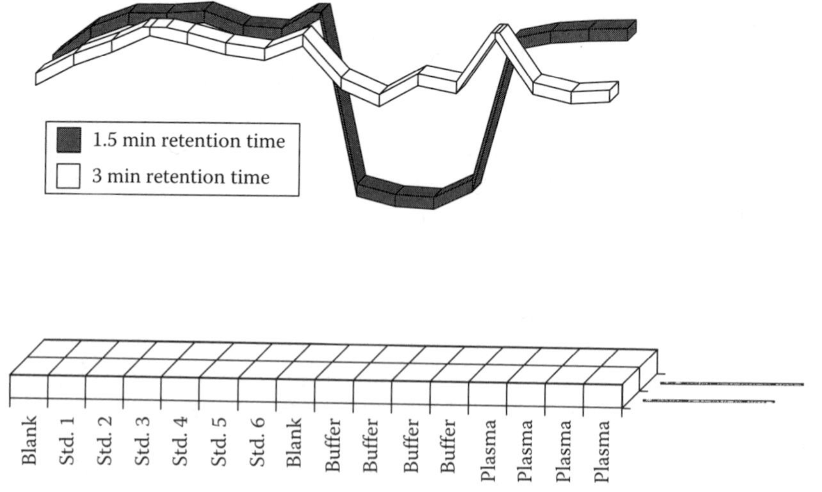

FIGURE 4.12 Internal standard ion intensity in a protein-binding run. *Note* the drop of ion intensity in the buffer side samples when a shorter run time is used.

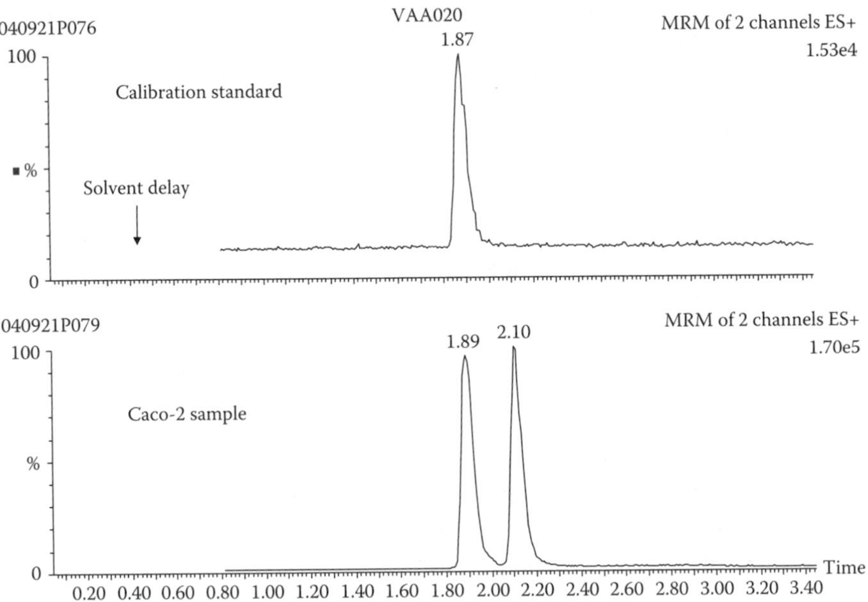

FIGURE 4.13 Mass chromatogram of a Caco-2 experiment. *Note*: A single peak is observed for the calibration standards and 0h samples (not shown), but an additional peak is observed in the 120min sample.

4.10 CONCLUSIONS

Drug discovery is a complex, multistage optimization process in which target-class chemicals are continuously modified to improve their drug-like properties, which must possess most, if not all, of the following properties [126]:

- Shows desired pharmacological properties
- Reaches the target organ
- Resides in the body and target organ long enough to have clinically meaningfully effect
- Shows sufficient specificity toward target receptor to minimize the off-target activities
- Has the proper physical–chemical properties for development

The likelihood of a new chemical entity to become a drug candidate is based on a balance between all of the above properties.

A recent study showed, despite the 147% R&D budget increase from 1993 to 2003, the number of FDA-approved new drugs has increased by only 7% [16]. This relatively lack of return on investment has forced the pharmaceutical industry to rethink their approaches and investments. Cost-effective and less labor-intense *in vitro* models are incorporated into drug discovery to weed out compounds with undesirable ADME and PK attributes. To quickly identify the potential drug candidates among all the possible leads, many new techniques and approaches have evolved but none has made more impact in all phases of drug discovery and development than LC–MS/MS [127], and turnaround time as short as 10 min per 96-well plate has been achieved by some newly developed technologies. Today, the LC–MS/MS based technologies have been implemented from early target identification to late-stage clinical assessment. Several *in vitro* screening assays described in this chapter, which include CYP inhibition, Caco-2 permeability, efflux transport evaluation, hepatocyte clearance, protein binding, isozyme profiling, and hERG [128], utilize LC–MS/MS assays to extract decision-making information for an NCE. Compounds that pass through these assays will move forward to more lengthy and costly PK studies in animals.

While the specificity and sensitivity provided by the LC–MS/MS paved the way for not only the increased throughput but also improved data quality for pharmaceutical research and development, we have to emphasize that understanding the relevance, precision, and limitation of each high-throughput assay is essential, particularly if the assay is involved in go/no go decisions. Another important consideration is that when setting-up a sequence of screens in a decision tree, it should be made to eliminate compounds with undesirable attributes as early as possible, which would focus the resources on those candidates with a better chance for success in development.

REFERENCES

1. Li, A.P., Screening for human ADME/Tox drug properties in drug discovery, *Drug Discov. Today*, 6, 357, 2001.
2. Kola, I. and Landis, J., Can the pharmaceutical industry reduce attrition rates? J. *Nature Rev.*, 3, 711, 2004.
3. Kennedy, T., Managing the drug discovery/development interface, *Drug Discov. Today*, 2, 436, 1997.
4. Frank, B. and Hargreaves, R., Clinical biomarkers in drug discovery and development, *Nature Rev.*, 2, 566, 2003.
5. Edward, P., Application of technologies and parallel chemistry for the generation of actives against biological targets, *Drug Discov. Today*, 13, 464, 2008.
6. Caldwell, G.W. et al., In vitro permeability of eight-blockers through Caco-2 monolayers utilizing liquid chromatography/electrospray ionization mass spectrometry, *J. Mass Spectrom.*, 33, 607, 1998.
7. Wang, Z. et al., Determination of in vitro permeability of drug candidates through a Caco-2 cell monolayer by liquid chromatography/tandem mass spectrometry, *J. Mass Spectrom.*, 35, 71, 2000.
8. Bu, H.Z. et al., High-throughput Caco-2 cell permeability screening by cassette dosing and sample pooling approaches using direct injection/on-line guard cartridge extraction/tandem mass spectrometry, *Rapid Commun. Mass Spectrom.*, 14, 523, 2000.
9. Markowska, M. et al., Optimizing Caco-2 cell monolayers to increase throughput in drug intestinal absorption analysis, *J. Pharmacol. Toxicol. Methods*, 46, 54, 2001.
10. Larger, P. et al., Utility of mass spectrometry for in vitro ADME assays, *Anal. Chem.*, 74, 5273, 2002.
11. Hakala, K. et al., Development of LC/MS/MS methods for cocktail dosed Caco-2 samples using atmospheric pressure photoionization and electrospray ionization, *Anal. Chem.*, 75, 5969, 2003.

12. Laitinen, L. et al., N-in-one permeability studies of heterogeneous sets of compounds across Caco-2 cell monolayers, *J. Pharm. Res.*, 20, 187, 2003.
13. Jenkins, K.M. et al., Automated high throughput ADME assays for metabolic stability and cytochrome P450 inhibition profiling of combinatorial libraries, *J. Pharm. Biomed. Anal.*, 34, 989, 2004.
14. Sunberg, S.A., High-throughput and ultra-high-throughput screening: Solution-and cell-based approaches, *Curr. Opin. Biotech.,* 11, 47, 2000.
15. Davies, A.M., Volkov, Y., and Spiers, J.P., Drug discovery automation, *Screening*, 3, 21, 2008.
16. Espada, A. et al., Application of LC/MS and related techniques to high-throughput drug discovery, *Drug Discov. Today*, 13, 417, 2008.
17. Korfmacher, W.A., Foundation review: Principles and applications of LC-MS in new drug discovery, *Drug Discov. Today*, 10, 1357, 2005.
18. Chu, I. and Nomeir, A.A., Utility of mass spectrometry for in vitro ADME assays, *Curr. Drug Metab.*, 7, 467, 2006.
19. Chu, I. and Nomeir, A.A., In vitro DMPK screening in drug discovery, role of LC-MS/MS, in *Progress in Pharmaceutical and Biomedical Analysis*, Vol. 6, Chowdhury, S., Ed., Elsevier, the Netherlands, p. 105, 2005.
20. Smalley, J. et al., Development of a quantitative LC–MS/MS analytical method coupled with turbulent flow chromatography for digoxin for the in vitro P-gp inhibition assay, *J. Chrom.* B., 854, 260, 2007.
21. Lau, Y.Y. et al., Evaluation of a novel in vitro Caco-2 hepatocyte hybrid system for predicting in vivo oral bioavailability, *Drug Metab. Dispos.*, 32, 937, 2004.
22. Li, C. et al., Development of in vitro pharmacokinetic screens using Caco-2, human hepatocyte, and Caco-2/human hepatocyte hybrid systems for the prediction of oral bioavailability in humans, *J. Biomol. Screen.*, 12, 1084, 2007.
23. Smith, D.A., Ackland, M.J., and Jones, B.C., Properties of cytochrome P450 isoenzymes and their substrates. Part 1: Active site characteristics, *Drug Discov. Today*, 2, 479, 1997.
24. Honig, P.K. et al., Terfenadine–ketoconazole interaction: Pharmacokinetic and electrocardiographic consequences, *JAMA*, 269, 1513, 1993.
25. Tang, Z.-M. and Kang, J.-W., Enzyme inhibitor screening by capillary electrophoresis with an on-column immobilized enzyme, *Anal. Chem.*, 78, 2514, 2006.
26. Li, D. et al., Comparison of cytochrome P450 inhibition assays for drug discovery using human liver microsomes with LC–MS, rhCYP450 isozymes with fluorescence, and double cocktail with LC–MS, *Inter. J. Pharm.*, 335, 1, 2007.
27. Bhoopathy, S.B. et al., A novel incubation direct injection LC/MS/MS technique for in vitro drug metabolism screening studies involving the CYP 2D6 and the CYP 3A4 isozymes, *J. Pharm. Biomed. Anal.*, 37, 739, 2005.
28. Rainville, P.D. et al., Sub one minute inhibition assays for the major cytochrome P450 enzymes utilizing ultra-performance liquid chromatography/tandem mass spectrometry, *Rapid Commun. Mass Spectrom.*, 22(9), 1345, 2008.
29. Yao, M. et al., Development and full validation of six inhibition assays for five major cytochrome P450 enzymes in human liver microsomes using an automated 96-well microplate incubation format and LC–MS/MS analysis, *J. Pharm. Biomed. Anal.*, 44, 211, 2007.
30. Wu, J. et al., High-throughput cytochrome P450 inhibition assays using laser diode thermal desorption-atmospheric pressure chemical ionization-tandem mass spectrometry, *Anal. Chem.*, 79, 4657, 2007.
31. Rostami-Hodjegan, A. and Tucker, G., "In silico" simulations to assess the "in vivo" consequences of "in vitro" metabolic drug–drug interactions, *Drug Discov. Today: Technol.*, 1, 441, 2004.

32. Saraswat, I.D. et al., High-throughput method for enzyme kinetic studies, *J. Biomol. Screening*, 8, 544, 2003.
33. Di Marco, A. et al., Development and validation of a high-throughput radiometric CYP2C9 inhibition assay using tritiated diclofenac, *Drug Metab. Dispos.*, 33, 359, 2005.
34. Di Marco, A. et al., Development and validation of a high-throughput radiometric cyp3a4/5 inhibition assay using tritiated testosterone, *Drug Metab. Dispos.*, 33, 349, 2005.
35. Rodrigues, A. and Lin, J., Screening of drug candidates for their drug–drug interaction potential, *Curr. Opin. Chem. Biol.*, 5, 396, 2001.
36. Chu, I. et al., Validation of higher-throughput high-performance liquid chromatography/atmospheric pressure chemical ionization tandem mass spectrometry assays to conduct cytochrome P450s CYP2D6 and CYP3A4 enzyme inhibition studies in human liver microsomes, *Rapid Commun. Mass Spectrom.*, 14, 207, 2000.
37. Obach, R.S. et al., The utility of in vitro cytochrome p450 inhibition data in the prediction of drug-drug interactions, *J. Pharmacol. Exp. Ther.*, 316, 336, 2006.
38. Turpeinen, M. et al., Cytochrome P450 (CYP) inhibition screening: Comparison of three tests, *Eur. J. Pharm. Sci.*, 29, 130, 2006.
39. Pelkonen, O. et al., Prediction of drug metabolism and interactions on the basis of in vitro investigations, *Basic Clin. Pharmacol. Toxicol.*, 96, 167, 2005.
40. Zlokarnik, G. et al., High throughput P450 inhibition screens in early drug discovery, *Drug Discov. Today*, 10, 1443, 2005.
41. Cohen, L.H., Remley, M.J., and Vaz, A., In vitro drug interactions of cytochrome p450: An evaluation of fluorogenic to conventional substrates, *Drug Metab. Dispos.*, 31, 1005, 2003.
42. Wang, J., Urban, L., and Bojanic, D., Maximising use of in vitro ADMET tools to predict in vivo bioavailability and safety, *Expert Opin. Drug Metab. Toxicol.*, 3, 641, 2007.
43. Bell, L. et al., Evaluation of fluorescence- and mass spectrometry-based CYP inhibition assays for use in drug discovery, *J. Biomol. Screen.*, 13, 343, 2008.
44. Youdim, K.A. et al., An automated, high-throughput, 384 well cytochrome P450 cocktail IC_{50} assay using a rapid resolution LC-MS/MS end-point, *J. Pharm. Biomed. Anal.*, 48, 92, 2008.
45. Turpeinen, M. et al., Multiple P450 substrates in a single run: Rapid and comprehensive in vitro interaction assay, *Eur. J. Pharm. Sci.*, 24, 123, 2005.
46. Bu, H.Z. et al., High-throughput cytochrome P450 (CYP) inhibition screening via a cassette probe-dosing strategy. VI. Simultaneous evaluation of inhibition potential of drugs on human hepatic isozymes CYP2A6, 3A4, 2C9, 2D6 and 2E1, *Rapid Commun. Mass Spectrom.*, 15, 741, 2001.
47. Dierks, E.A. et al., A method for the simultaneous evaluation of the activities of seven major human drug-metabolizing cytochrome P450s using an in vitro cocktail of probe substrates and fast gradient liquid chromatography tandem mass spectrometry, *Drug Metab. Dispos.*, 29, 23, 2001.
48. Testino, S.A., Jr. and Patonay, G., High-throughput inhibition screening of major human cytochrome P450 enzymes using an in vitro cocktail and liquid chromatography–tandem mass spectrometry, *J. Pharm. Biomed. Anal.*, 30, 1459, 2003.
49. Mizutani, T., PM frequencies of major CYPs in Asians and Caucasians, *Drug Metab. Rev.*, 35, 99, 2003.
50. Wrighton, S.A. et al., The human CYP3A subfamily: practical considerations, *Drug Metab. Rev.*, 32, 339, 2000.
51. Newton, D.J., Wang, R.W., and Lu, A.Y.H., Cytochrome P450 inhibitors. Evaluation of specificities in the in vitro metabolism of therapeutic agents by human liver microsomes, *Drug Metab. Dispos.*, 23, 154, 1995.

52. Nomeir, A.A., Palamanda, J., and Favreau, L., in *Optimization in Drug Discovery: In Vitro Methods*, Yan, Z and Caldwell G, Ed., Totowa, NJ: Humana Press, pp. 245–262, 2004.
53. Veronese, M.E. et al., Tolbutamide and Phenytoin hydroxylation by cDNA expressed human liver cytochrome P4502C9, *J. Biochem. Biophys. Res. Commun.*, 175, 1112, 1991.
54. Chen, Z.R., Somogyi, A.A., and Bochner, F., Simultaneous determination of dextromethorphan and three metabolites in plasma and urine using high-performance liquid chromatography with application to their disposition in man, *Ther. Drug Monit.*, 12, 97, 1990.
55. Waxman, D.J. et al., Steroid hormone hydroxylase specificities of eleven cDNA-expressed human cytochrome P450s, *Archives Biochem. Biophys.*, 290, 160, 1991.
56. Sunner, J., Nicol, G., and Kebarle, P., Factors determining relative sensitivity of analytes in positive mode atmospheric pressure, *Anal. Chem.*, 60, 1300, 1988.
57. Kebarle, P. and Tnag, L. From ions in solution to ions in the gas phase, *Anal. Chem.*, 65, 972A, 1993.
58. King, R. et al., Mechanistic investigation of ionization suppression in electrospray ionization, *J. Am. Soc. Mass Spectrom.*, 11, 942, 2000.
59. Wilson, G., Cell culture techniques for the study of drug transport, *Eur. J. Drug Metab. Pharm.*, 15, 159, 1990.
60. Stevenson, C.L., Augustijns, P.F., and Hendren, R.W., Use of Caco-2 cells and LC/MS/MS to screen a peptide combinatorial library for permeable structures, *Int. J. Pharm.*, 177, 103, 1999.
61. Yamashita, S. et al., Optimized conditions for prediction of intestinal drug permeability using Caco-2 cells, *Eur. J. Pharm. Sci.*, 10, 195, 2000.
62. Borlak, J. and Zwadlo, C., Expression of drug-metabolizing enzymes, nuclear transcription factors and ABC transporters in Caco-2 cells, *Xenobiotica*, 33, 927, 2005.
63. Maubon, N. et al., Analysis of drug transporter expression in human intestinal Caco-2 cells by real-time PCR, *Fundam. Clin. Pharmacol.*, 21, 659, 2007.
64. Gres, M.C. et al., Correlation between oral drug absorption in humans, and apparent drug permeability in TC-7 cells, a human epithelial intestinal cell line: Comparison with the parental Caco-2 cell line, *Pharm. Res.*, 15, 726, 1998.
65. Polli, J.E. and Ginski, M., Human drug absorption kinetics and comparison to Caco-2 monolayer permeabilities, *J. Pharm. Res.*, 15, 47, 1998.
66. Fossati, L. et al., Use of simulated intestinal fluid for Caco-2 permeability assay of lipophilic drugs, *Int. J. Pharm.*, 360, 148, 2008.
67. Ingels, F. et al., Effect of simulated intestinal fluid on drug permeability estimation across Caco-2 monolayers, *Int. J. Pharm.*, 274, 221, 2004.
68. Eddy, E.P. et al., In vitro models to predict blood-brain barrier permeability, *Adv. Drug Deliv. Rev.*, 23, 185, 1977.
69. Hunter, J. et al., Functional expression of P-glycoprotein in apical membranes of human intestinal Caco-2 cells. Kinetics of vinblastine secretion and interaction with modulators, *J. Biol. Chem.*, 268, 14991, 1993.
70. Krishna, G. et al., Permeability of lipophilic compounds in drug discovery using in vitro human absorption model, Caco-2, *Int. J. Pharm.*, 222, 77, 2001.
71. Roddy, T.P. et al., Mass spectrometric techniques for label-free high-throughput screening in drug discovery, *Anal. Chem.*, 79, 8207, 2007.
72. Fung, E.N., Chu, I., and Nomeir, A.A., Higher-throughput screening for Caco-2 permeability utilizing a multiple sprayer liquid chromatography/tandem mass spectrometry system, *Rapid Commun. Mass Spectrom.*, 17, 2147, 2003.
73. Halladay, J.S. et al., Metabolic stability screen for drug discovery using cassette analysis and column switching, *Drug Metab. Lett.*, 1, 67, 2007.

74. Kantharaj, E. et al., Simultaneous measurement of metabolic stability and metabolite identification of 7-methoxymethylthiazolo[3,2-a]pyrimidin-5-one derivatives in human liver microsomes using liquid chromatography/ion-trap mass spectrometry, *Rapid Commun. Mass Spectrom.*, 19, 1069, 2005.
75. Jouin, D. et al., Cryopreserved human hepatocytes in suspension are a convenient high throughput tool for the prediction of metabolic clearance, *Eur. J. Pharm. Biopharm.*, 63, 347, 2006.
76. Jones, K., Slamon, D., and Gill, H., *Everything You Needed to Know About ADME, but Were too Afraid to Ask*! Cyprotex Discovery Ltd., http://www.cyprotex.com/news/165, 2006.
77. Plumb, R. et al., A rapid simple approach to screening pharmaceutical products using ultra-performance LC coupled to time-of-flight mass spectrometry and pattern recognition, *J. Chromatogr. Sci.*, 46, 193, 2008.
78. Lau, Y.Y. et al., Development of a novel in vitro model to predict hepatic clearance using fresh, cryopreserved, and sandwich-cultured hepatocytes, *Drug Metab. Dispos.*, 30, 1446, 2002.
79. Yamashita, T. et al., High-speed solubility screening assay using ultra-performance liquid chromatography/mass spectrometry in drug discovery, *J. Chromatogr. A.*, 1182, 72, 2008.
80. Swartz, M.E., Ultra performance liquid chromatography (UPLC): An introduction, *Sep. Sci. Redefined*, May, 8, 2005.
81. Fichtl, B., Nieciecki, A.V., and Walter, K., Tissue binding versus plasma binding of drugs: General principles and pharmacokinetic consequences, *Adv. Drug Res.*, 20, 117, 1991.
82. Kalvaas, J.C. and Maurer, T.S., Use of plasma and brain unbound fractions to assess the extent of brain distribution of 34 drugs: Comparison of unbound concentration ratios to in vivo P-Glycoprotein efflux ratios, *Biopharm. Drug Dispos.*, 23, 327, 2002.
83. Devane, C.L., Clinical significance of drug plasma protein binding and binding displacement drug interactions, *Psychopharmacol. Bull.*, 36, 5, 2002.
84. Kratochwil, N.A. et al., Predicting plasma protein binding of drugs: A new approach, *Biochem. Pharm.*, 64, 1355, 2002.
85. Benet, L.Z. and Hoener, B., Changes in plasma protein binding have little clinical relevance, *Clin. Pharmacol Ther.*, 71, 115, 2002.
86. Dawidowicz, A.J., Kobielski, M., and Pieniadz, J., Anomalous relationship between free drug fraction and its total concentration in drug–protein systems: I. Investigation of propofol binding in model HSA solution, *Eu. J. Pharm. Sci.*, 34, 30, 2008.
87. Dawidowicz, A.J. et al., Influence of propofol concentration in human plasma on free fraction of the drug, *Chem. Biol. Int.*, 159, 149, 2006.
88. Ebert, S.C., Application of pharmacokinetics and pharmacodynamics to antibiotic selection, *Pharm. Therapeut.*, 29, 244, 2004.
89. Banik, G.M., In silico ADME-Tox prediction: The more, the merrier, *Curr. Drug Discov.*, May, 31, 2004.
90. Segall, M.D., Measurement of drug–protein binding by immobilized human serum albumin-HPLC and comparison with ultrafiltration, *Future Drug Discov.*, 81, 2004.
91. Banker, M.J., Clark, T.H., and Williams, J.A., Development and validation of a 96-well equilibrium dialysis apparatus for measuring plasma protein binding, *J. Pharm. Sci.*, 92, 967, 2003.
92. Taylor, S. and Harker, A., Modification of the ultrafiltration technique to overcome solubility and non-specific binding challenges associated with the measurement of plasma protein binding of corticosteroids, *J. Pharm. Biomed. Anal.*, 41(1), 299, 2006.

93. Lee, K.-J. et al., Modulation of nonspecific binding in ultrafiltration protein binding studies, *Pharm. Res.*, 20, 1015, 2003.
94. Zhang, J. and Musson, D.G., Investigation of high-throughput ultrafiltration for the determination of an unbound compound in human plasma using liquid chromatography and tandem mass spectrometry with electrospray ionization, *J. Chromatogr. B*, 843, 47, 2006.
95. Musson D.G. et al., Assay methodology for the quantitation of unbound ertapenem, a new carbapenem antibiotic, in human plasma, *J. Chromatogr. B Analyt. Technol. Biomed. Life Sci.*, 1, 783, 2003.
96. Valko, K. et al., Fast gradient HPLC method to determine compounds binding to human serum albumin. Relationships with octanol/water and immobilized artificial membrane lipophilicity, *J. Pharm. Sci.*, 92, 2236, 2003.
97. Hage, D.S. and Austin, J., High-performance affinity chromatography and immobilized serum albumin as probes for drug- and hormone-protein binding, *J. Chromatogr. B*, 739, 39, 2000.
98. Singh, S.S. and Mehta, J., Measurement of drug–protein binding by immobilized human serum albumin-HPLC and comparison with ultrafiltration, *J. Chromatogr. B*, 834, 108, 2006.
99. Waters, N.J. et al., Validation of a rapid equilibrium dialysis approach for the measurement of plasma protein binding, *J. Pharm. Sci.*, 97, 4586, 2008.
100. Li, F. et al., Drug-protein binding screening using tandem column affinity chromatography-tandem mass spectrometry, in preparation.
101. Vita, M. et al., Stability, pKa and plasma protein binding of roscovitine, *J. Chromatogr. B*, 821, 75, 2005.
102. Leonard, W.J., Jr., Viaji, K.K., and Foster, J.F.A., A structural transformation in bovine and human plasma albumins in alkaline solution as revealed by rotatory dispersion study, *J. Biol. Chem.*, 238, 1984, 1963.
103. Ermondi, G., Lorenti, M., and Caron, G., Contribution of ionization and lipophilicity to drug binding to albumin: A preliminary step toward biodistribution prediction, *J. Med. Chem.*, 47, 3949, 2004.
104. Kochansky, C.J. et al., Impact of pH on plasma protein binding in equilibrium dialysis, *Mol. Pharm.*, 5, 438, 2008.
105. Wan, H. and Rehngren, M., High-throughput screening of protein binding by equilibrium dialysis combined with liquid chromatography and mass spectrometry, *J. Chromatog. A*, 1102, 125, 2006.
106. Hinderling, P.H. and Hartmann, D., The pH dependency of the binding of drugs to plasma proteins in man, *Ther. Drug Monit.*, 27, 71, 2005.
107. Chu, I. and Nomeir, A.A., In vitro DMPK screening in drug discovery, role of LC-MS/MS, identification and quantitation of drugs, metabolites and metabolizing enzymes by LC-MS, in *Progress in Pharmaceutical and Biomedical Analysis*, Vol. 6, Chowdhury, S.K. Ed., Elsevier, the Netherlands, Chapter 5, 2005.
108. Lui, C.Y. and Chiou, W.L., pH Shift during equilibrium dialysis, *Biopharm. Drug Disp.*, 7, 309, 1986.
109. Fura, A. et al., Shift in pH of biological fluids during storage and processing: Effect on bioanalysis, *J. Pharm. Biomed. Anal.,* 32, 513, 2003.
110. Kodama, H. et al., Effect of temperature on serum protein binding characteristics of phenytoin in monotherapy paediatric patients with epilepsy, *J. Clin. Pharm. Thera.,* 26, 175, 2001.
111. Kodama, H. et al., Temperature effect on serum protein binding kinetics of phenytoin in monotherapy patients with epilepsy, *Eu. J. Pharm. Biopharm.*, 47, 29, 1999.
112. Singh, B. and Bansal, S., In vitro protein binding studies on aztreonam: Temperature effects thermodynamics, *Indian J. Pharm. Sci.*, 60, 270, 1998.

113. Pelkonnen, O. and Breimer, D.D., Role of environmental factors in the pharmaceutical of drugs, *Pharmacokinetics of Drug*, Welling, P.G. and Balant, L.P. Eds., Springer: Berlin, p. 289, 1994.
114. Code, E.L. et al., Human Cytochrome P4502B6, *Drug Metab. Dispos.*, 25, 985, 1997.
115. Linget, J.-M. and du Vignaud, P.J., Automation of metabolic stability studies in microsomes, cytosol and plasma using a 215 Gilson liquid handler, *J. Pharm. Biomed. Anal.*, 19, 893, 1999.
116. Brown, A.M., Drugs, hERG and sudden death, *Cell Calcium*, 35, 543, 2004.
117. Dykens, J.A. and Will, Y., The significance of mitochondrial toxicity testing in drug development, *Drug Discov. Today*, 12, 777, 2007.
118. Aronov, A.M., Predictive in silico modeling for hERG channel blockers, *Drug Discov. Today*, 10, 149, 2005.
119. Wood, C., Williams, C., and Waldron, G.J., Patch clamping by numbers, *Drug Discov. Today*, 9, 434, 2004.
120. Frank, M., Ozbal, C., and Larkin, S., Ultra-high throughput determination of CYP450 enzyme inhibition through mass spectrometry, Agilent Application Note, January, 2008.
121. Zhong, F., Ghobarah, H., Scott, G., and Lebre D., *The Application Notebook*, Applied Biosystems/MDS SCIEX, September 24, 2007.
122. Zhong, F. et al., High throughput metabolic stability screening by MALDI triple quadrupole analysis, *The Application Note Book*, September, 24, 2007.
123. Jemal, M. and Ouyang, Z., The need for chromatographic and mass resolution in liquid chromatography/tandem mass spectrometric methods used for quantitation of lactones and corresponding hydroxy acids in biological samples, *Rapid Commun. Mass Spectrom.*, 14, 1757, 2000.
124. Matuszewski, B.K., Constanzer, M.L., and Chavez-Eng, C.M., Matrix effect in quantitative LC/MS/MS analyses of biological fluids: A method for determination of finasteride in human plasma at picogram per milliliter concentrations, *Anal. Chem.*, 70, 882, 1998.
125. Jemal, M. and Xia, Y.-Q., The need for adequate chromatographic separation in the quantitative determination of drugs in biological samples by high performance liquid chromatography with tandem mass spectrometry, *Rapid Commun. Mass Spectrom.*, 13, 97, 1999.
126. Fang, A.S. et al., Mass spectrometry analysis of new chemical entities for pharmaceutical discovery, *Mass Spectrom. Rev.*, 27, 20, 2008.
127. Jamieson, C. et al., Medicinal chemistry of hERG optimizations: Highlights and hang-ups, *J. Med. Chem.*, 49, 5029, 2006.
128. Kassev, D.B., Applications of high-throughput ADME in drug discovery, *Curr. Opin. Chem. Biol.*, 8, 339, 2004.

5 Metabolite Identification Strategies and Procedures

Ragu Ramanathan, S. Nilgün Çömezoğlu, and W. Griffith Humphreys

CONTENTS

5.1 INTRODUCTION

Discovering and developing safe, efficacious, and commercially successful drugs is a long, complex, risky, and costly process. The process has yielded many medicines that have had dramatic positive influences on public health. A typical drug discovery and development process leading up to registration involves a discovery phase, a preclinical development phase, and clinical Phases I, II, and III. According to current estimates, advancing a small molecule drug through all the phases can take between 10 and 15 years and cost over $800 million. Although the pharmaceutical industry's investment in research and development has been constantly increasing, the number of new molecular entities (NMEs) reaching the market has been steadily decreasing.

In a recent review article, Kola and Landis [1] elegantly captured the root causes for the high rate of attrition in drug development and proposed several methods to reduce attrition. Based on the data presented in the review, about 1 out of every 10–12 NMEs entering first-in-human (FIH) clinical trials becomes a drug and about 60% of the failures in the clinic were attributed to lack of efficacy and safety. As documented in the review by Kola and Landis [1], and several other industry surveys, the cost of terminating an NME in clinical Phases II, III, and beyond is much higher than terminating an NME in Phase I clinical trials or before [2]. Furthermore, the late clinical stage failure of torcetrapib [3] and the postmarketing withdrawals of fenfluramine-phentermine (Fen-Phen) [4] and Vioxx [5,6] were especially costly to the industry as a whole as well as the sponsors. According to a recent report, the United States holds the world's best drug safety record [7] and over the last 20 years, only about 3% of the approved medicines have been withdrawn from the U.S. market. Despite the safety record, pharmaceutical companies across the globe, in collaboration with regulatory agencies such as the U.S. Food and Drug Administration (US FDA) and the European medicines agency (EMEA), are constantly striving to improve the safety profiles of all drugs.

How can late-stage clinical attrition be reduced and yet still produce a safe and efficacious drug? This is the number one question that every member of the biopharmaceutical industry is trying to answer. Reducing early-stage clinical attrition is a difficult task because animal models are not perfect predictors of efficacy and safety in humans [8]. Many factors can confound the safety and efficacy predictions used to move a new chemical entity (NCE) into the clinic, among them absorption, distribution, metabolism, and excretion (ADME) properties and the mechanism of toxicity differences between the preclinical species and humans.

A historic perspective on the development of drug metabolism (DM) science has been elegantly reviewed by Murphy [9–11] in a three-part series published in the *Drug Metabolism and Disposition* journal. The reasons for DM studies are many but one of the key goals is helping to ensure human safety. During early discovery phase *in silico* methods are used to weed out compounds with undesirable ADME properties [12,13]. Higher throughput DM studies are conducted in the screening mode, using *in vitro* systems such as microsomes, hepatocytes, and/or S9 fractions, to select compounds with optimized clearance characteristics. As a next step in discovery, *in vitro* systems are used to establish metabolic pathways across species and to determine whether any metabolite would contribute to the pharmacological activity or have toxicological effect, and to see if major *in vitro* formed human metabolites are adequately exposed in nonclinical species selected for safety evaluation studies. Selected discovery compounds are also subjected to reactive metabolite and covalent-binding assays to weed out compounds that form reactive intermediates. Later, lead compounds that are likely to advance into further development are assessed *in vivo* using mice, rats, rabbits, dogs, and/or monkeys. Subsequently, during Phase I and/or II clinical trails, metabolites in humans are identified following drug administration to assure that the nonclinical species undergoing safety assessment are adequately exposed to human metabolites of the drug.

Liquid chromatography-mass spectrometry (LC–MS) is the most commonly used technique among the strategies and procedures available for evaluating and optimizing drug metabolism and pharmacokinetic (DMPK) properties at various stages of drug discovery and development [14–17]. Quantitative LC–MS–based bioanalytical techniques available for evaluation of pharmacokinetics (PK) properties such as oral bioavailability, half-life, and drug–drug interactions have been the subject of several reviews [18–23] as well as the focus of Chapters 1 through 4. Information on PK parameters in animal models are necessary for the start of the FIH studies, especially in the prediction of the efficacious exposure levels necessary to exert an NCE's pharmacological activity and predict safe doses. Another related area that is important for safe clinical trials is the metabolic properties of an NCE. Metabolism information helps to avoid developing an NCE that has the potential to exhibit large interpatient variability in PK parameters of an NCE through polymorphic enzyme-mediated metabolism and/or undergoing metabolic drug–drug interactions. In addition, an NCE dose may need adjustment due to formation of pharmacologically active metabolite(s) and/or reactive metabolites. Therefore, the guidance on *Safety Testing of Drug Metabolites* (STDM) [24] suggests assessing species differences in metabolism of a drug (*in vitro* studies and/or nonclinical animal studies) as early as possible.

The focus of this chapter is LC–MS–based strategies and procedures available for detection and characterization of metabolites at various stages of drug discovery and development. In addition, radioactivity and/or ultraviolet (UV) detection techniques are discussed with respect to their utility in quantifying metabolites.

5.2 METABOLITE IDENTIFICATION AT DIFFERENT STAGES OF DRUG DISCOVERY AND DEVELOPMENT

In the discovery phase, metabolite identification is usually performed with a combination of *in vitro* and *in vivo* experiments using samples from different species in order to compare metabolite exposures. The structural identification of major circulating metabolites formed in nonclinical animal models as well as the metabolites formed in human *in vitro* systems is needed for the metabolites to be synthesized and their pharmacological activities and/or toxicological implications to be determined [25]. In addition, metabolite identification can lead to the discovery of candidates with satisfactory clearance/PK properties and/or improved safety profile. Following are some examples of metabolites that were later developed as drugs: desloratadine from loratadine, acetaminophen from phenacetin, morphine from codeine, minoxidil sulfate from minoxidil, fexofenadine from terfenadine, and oxazepam from diazepam.

In the development phase, preliminary metabolite identification experiments can be conducted in selected samples from human single ascending dose (SAD) or multiple ascending dose (MAD) studies [26–37]. Definitive biotransformation profiles of drug candidates are determined in nonclinical species and in humans by administration of the radiolabeled form of the drug candidate. These studies also provide the quantitative assessment (mass balance) of the excretion pathways of the drug and are required before the drug candidate enters Phase III clinical trials. Isotopes used in metabolism and mass balance studies usually are ^{14}C or ^{3}H [38]. Profiling and identification of metabolites are accomplished with the use of various analytical methods and techniques that include high-performance liquid chromatography (HPLC) for metabolite profiling and full-scan MS (LC–MS), product ion scans (LC–MS/MS), and NMR for metabolite identification. Figure 5.1 summarizes some of the strategies and procedures used for metabolite identification at different stages of drug discovery and development.

5.2.1 Sample Preparation Techniques for Metabolite Identification

The presence of proteins, lipids, and other endogenous material, which potentially interferes with the detection of drug-derived material, makes it difficult for identification of metabolites in complex biological matrices. By some estimates, during the metabolite profiling/identification process, a significant amount of time is spent in developing the appropriate sample preparation/cleanup method. The degree and extent of sample preparation/cleanup varies depending on an assay's intended use. While discovery stage metabolite identification assays are developed in a manner that does not demand efficient extraction recovery of all the metabolites, assays developed to support definitive metabolite profiling and clinical studies have to possess quantitative/uniform recoveries of the metabolites and should be reproducible.

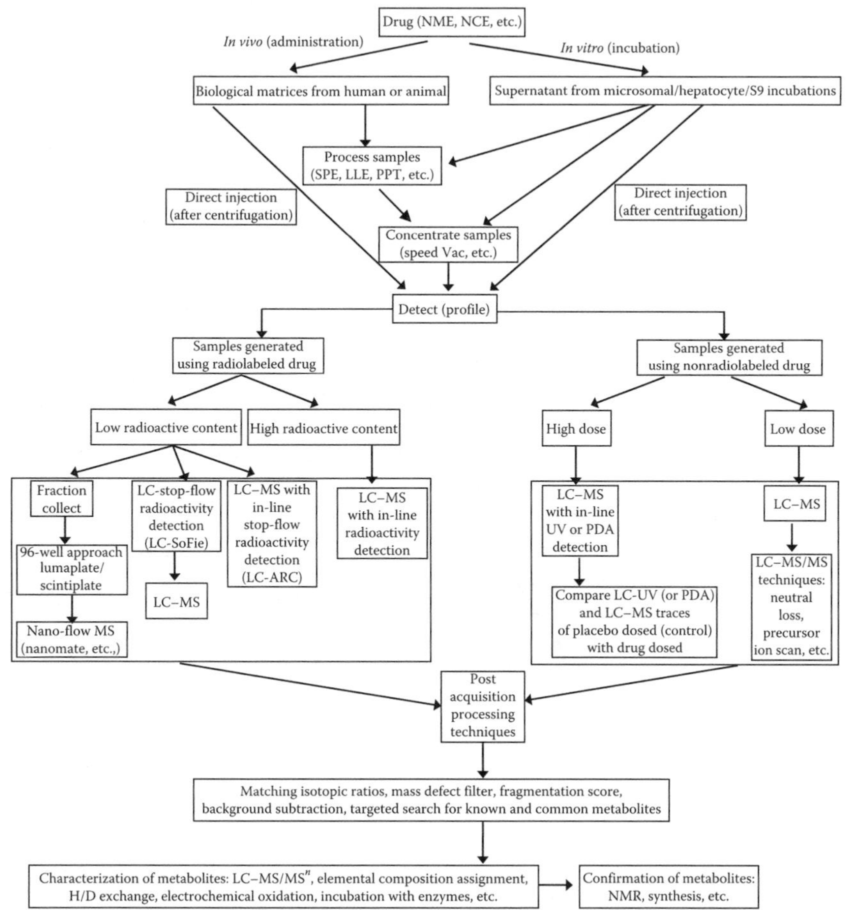

FIGURE 5.1 Some of the common metabolite identification strategies and procedures.

Irrespective of whether an assay is being developed to support a development or a discovery study, common goals of sample preparation/cleanup include the following: (1) obtaining a representative sample in solution; (2) removing co-eluting/interfering matrix components with minimum drug and metabolites losses; (3) obtaining sufficient concentration for MS detection; (4) limiting the number of steps; (5) obtaining quantitative extraction recovery of both metabolites and the drug; (6) not degrading or modifying the drug or its metabolites; and (7) maintaining ruggedness and reproducibility.

A generic bioanalytical sample preparation scheme is presented in Figure 5.2. The most preferred sample processing route involves LC–MS analysis after automated solid-phase extraction (SPE), liquid–liquid extraction (LLE), or protein precipitation (PPT) without the evaporation/concentration steps [39]. This is the case for most *in vitro* samples and the high-throughput quantitative PK samples, which are analyzed following PPT and centrifugation. In many of the high-throughput assays, to increase throughput, the LC–MS/MS analysis time of the first sample is used

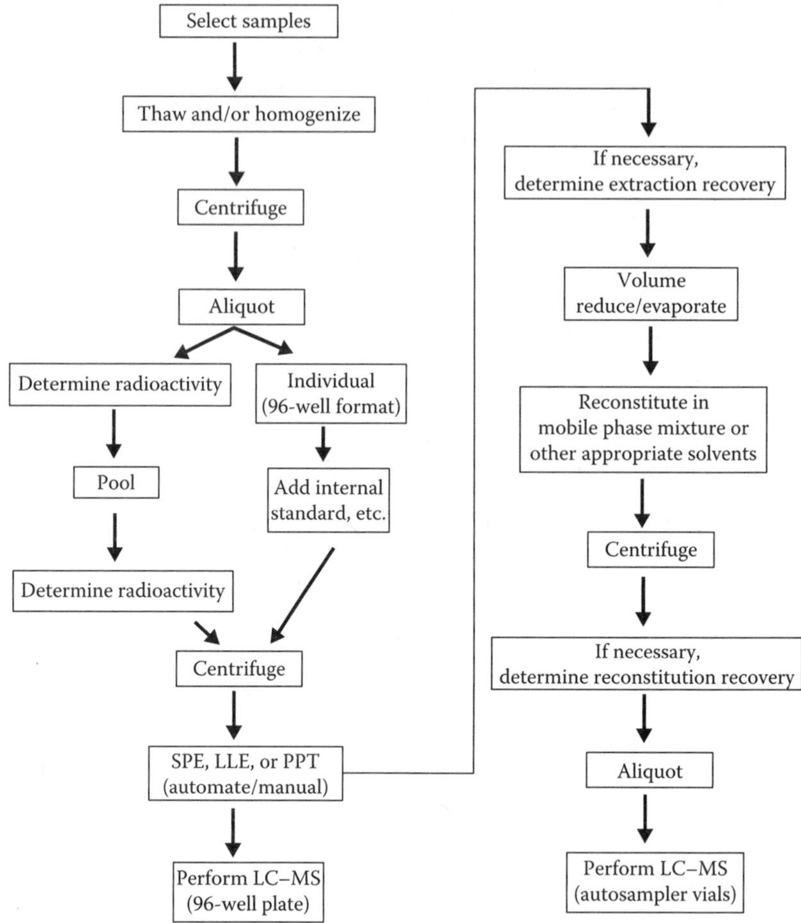

FIGURE 5.2 Bioanalytical sample processing strategies.

for online automated processing/extraction/cleanup of the second sample. In comparison to manual off-line sample processing, automated online sample processing methods are less prone to experimental errors and are more time efficient. The more complex *in vivo* samples containing endogenous material usually have metabolites at much lower concentrations than in the *in vitro* samples [40]. As a result, the protonated molecules for the minor metabolites in the full-scan total ion chromatograms are not readily discernable from the background ions and detection of the metabolites becomes difficult. While direct analysis of some samples such as bile is possible, most often plasma, urine, and feces have to be extracted using SPE, LLE, or PPT or a combination of methods before volume reduction/evaporation and LC–MS analysis. The volume reduction/evaporation steps, shown in Figure 5.2, allow for the generation of samples with sufficient concentration of metabolites for identification of unknown metabolites and to successfully apply lower duty cycle full-scan LC–MS as well as perform multiple MS/MS experiments.

Unlike in quantitative PK analysis, in which inter- and intrasubject/animal variability are important to understand clearance, absorption, drug–drug interactions, and polymorphism-related changes between subjects/animals, metabolite profiling/identification studies are performed using pooled representative samples from clinical and/or nonclinical studies. It is not practical however to pool the samples from clinical study subjects who are identified as either slow or extensive metabolizers of an NCE. Presence of such slow- and extensive metabolizers come to light through quantitative PK analysis of the NCE and if the study had involved a radiolabeled NCE, determination of radioactivity in plasma and excreta can also give some indication about the intersubject variabilities in the metabolism of the NCE. Therefore, it is highly desirable to pool samples (Figure 5.2) based on PK and/or radioactivity information.

Traditional sample pooling approaches used in metabolite profiling/identification studies involve pooling across animals/subjects per dose group and gender. For plasma metabolite profiling/identification, equal volumes of plasma from each animal/subject is added together to generate pooled plasma samples at three or four time points per dose group and gender. Plasma time points for metabolite profiling are selected based on estimated time of maximum drug concentration (t_{max}) or half-life ($t_{1/2}$) of the dosed compound. For example, for an NCE, if the t_{max} is around 1.5 h, then postdose blood for metabolite profiling is collected at 2, 6, 12, and 24 h. If blood collection is limited due to lack of drug material or number of animals, profiling 2, 8, and 24 h plasma samples is sufficient to understand the biotransformation pathways. Metabolite profiling/identification at different time points provides information about primary and secondary metabolites as well as indicating whether any metabolites are likely to accumulate. An alternative pooling approach involves pooling various volumes of plasma across all the time points in a time proportional manner to generate a 0–24 h pooled plasma for each subject and then pooling across subjects to generate one plasma sample. This pooling method is often referred to as the "time proportional pooling" or "area under the curve (AUC) pooling" scheme because determination of the parent NCE's AUC, using a quantitative bioanalysis method, allows one to determine the AUC concentrations of its major metabolites using the qualitative metabolite profiles obtained using the AUC pooling method. This approach, which has been investigated by several groups in support of the recent safety testing of metabolites guidance [41–43], is time and resource efficient and provides similar representative semiquantitative AUC estimates of the metabolites over a desired time interval (0–24, 0–48 h, etc.).

Bile, urine, and feces samples for metabolite profiling are collected in 0–8 and 8–24 h block collection period or in one 0–24 h block collection period. Development stage definitive metabolite profiling and mass balance studies generally involve urine and feces collection in 24 h block periods over 7, 10, or 15 days or until more than 85% of the radioactivity is excreted. Urine and fecal sample time points selected for metabolite profiling generally cover over 90% of the radioactivity excreted in the respective matrix.

5.2.2 Separation Techniques for Metabolite Identification

Chemical and physical properties of the majority of drugs and their metabolites make them well suited for chromatographic separation using reverse-phase HPLC. Reversed-phase HPLC is achieved using a nonpolar stationary phase and a combination of polar mobile phases. Chromatographic conditions for the separation of a drug and its metabolites are selected by optimizing the column composition, column particle size, column temperature, and mobile phase composition (solvent type, solvent composition, pH, and additives). Since MS ionization techniques require volatile buffers at low concentrations (usually less than 25 mM), varying the mobile phase additives may or may not optimize the separation of a drug and the metabolites. Several research groups have shown that increasing the column temperature above 60°C is helpful to achieve improved separation, routine bioanalysis at column temperatures above 60°C is not practical [44]. The column temperature for most metabolite identification experiments are maintained between 25°C and 40°C and changing the column temperature within the 15°C can only lead to small changes in the chromatographic selectivity/elution order. Therefore, the column chemistry and/or the particle size are considered to be important parameters in optimizing the separation of a drug and its metabolites from the components of the matrix.

The column chemistry can be altered by changing either the packing material or the bonded phase or both. The packing material, used in LC–MS experiments, usually is based on 3–5 μm silica that has been treated with RMe_2SiCl, where R most commonly is either $C_{18}H_{37}$ or C_8H_{17}. Other common bonded phases include C4, cyano, and phenyl. The pore-size of most of the packing material is between 80 and 100 Å. Although the selectivity and retention times are not expected to vary between C18 columns from different vendors, they do vary in some instances and most of the separation issues that cannot be achieved by changing the mobile phase composition can be optimized by changing the bonded phase. The availability of alternate selectivity from different bonded phases comes in handy when one is trying to separate co-eluting metabolites.

Alternatives to conventional reverse-phase columns include monolithic, hydrophilic interaction liquid chromatography (HILIC), and ultraperformance liquid chromatography (UPLC; columns with sub-2 μm particles) columns. Some of the attributes of monolithic columns, which was introduced to the analytical community during the 1990s, include high permeability, low pressure drop, and good separation efficiency [45–47]. Because of these advantages monolithic phases have also been used for online extraction before LC–MS analysis [48,49]. Although several laboratories have evaluated the applications of monolithic columns for the bioanalysis of NCEs and their metabolites [45–47,50–52], the main focus of these reports is increasing throughput in quantitative bioanalysis rather than for metabolite profiling and identification. Dear et al. [49] used debrisoquine and its hydroxylated metabolites, formed following incubation with human liver microsomes (HLM), to show that the resolution and selectivity gain of short monolithic columns allows the conventional column analysis time of 30 min to be reduced to 5 min. The same group also showed advantages associated with using monolithic columns, during quantitative bioanalysis, for separating 3′-azido-3′-deoxythymidine (AZT) from AZT-glucuronide and paracetamol from paracetamol-glucuronide [53].

The performance of monolithic and sub-2 μm particle UPLC columns was evaluated for the detection and characterization of metabolites in urine following administration of a single 500 mg acetaminophen dose to healthy volunteers [54]. MS detection following UPLC separation resulted in approximately three times the sensitivity and allowed the detection of additional glucuronide (Figure 5.3), glutathione, and sulfate metabolites in comparison to the monolithic column–based method. As shown in Figure 5.3, the chromatographic peaks from the monolithic column were wider than those detected with UPLC and resulted in higher limits of detection for the metabolites. Furthermore due to better chromatographic separation efficiency, ion suppression was minimal with the UPLC approach and resulted in improved sensitivity.

Castro-Perez et al. [55] compared the performance of a conventional HPLC with that of a UPLC and demonstrated improvements in chromatographic resolution and peak capacity. These improvements led to reduction in ion suppression and increased MS sensitivity. Comparison of the mass spectrum obtained using HPLC with that from UPLC revealed that the higher resolving power of the UPLC–MS system resulted in a much cleaner mass spectrum than that obtained using the HPLC–MS system. The sensitivity improvement directly resulted in a higher ion count in the UPLC mass spectrum (855 vs 176). The additional sensitivity attainable with the UPLC approach is again demonstrated in the extracted ion chromatogram (XIC) of desmethyl-dextromethrophan-glucuronide metabolite (*m/z* 434). While HPLC–MS resulted in a peak-to-peak signal-to-noise (S/N) ratio of 25:1 for the metabolite, UPLC–MS provided an improved S/N ratio of 115:1. The increased S/N ratio achieved using the UPLC was attributed to improved peak resolution and a reduction in ion suppression resulting from other co-eluting metabolites and endogenous compounds. In a similar comparison evaluation Wang et al. [56] used reversed-phase UPLC and 1.7 μm particle column and developed a fast, sensitive, and rugged method to detect testosterone and its four *in vitro* formed hydroxyl metabolites. In comparison to conventional HPLC–MS approach, the UPLC approach provided improved LC–MS run time, separation, and sensitivity (see Chapter 8).

It has been shown that HILIC can be used as an alternative to conventional reverse-phase HPLC. In HILIC, highly polar analytes are retained longer to elute away from the early eluting matrix components. To increase the retention of highly polar compounds, a polar stationary phase such as bare silica gel or polar-bonded phase is used in HILIC columns. Similar to the normal-phase chromatography, HILIC also requires a high percentage of a nonpolar mobile phase. However, unlike the traditional normal-phase chromatography, which involves nonpolar solvents such as hexane and methylene chloride and avoids water as one of the mobile phase components, hydrophilic interaction requires some water in the mobile phase. The retention of the polar analytes decreases with increasing water content. In conventional reverse-phase HPLC, very high water content is required to retain polar analytes. The high water content in turn hinders the ionization and desolvation process during LC–MS [57,58]. Therefore, HILIC allows one to elute highly polar analytes with small amounts of water and maintain good LC–MS sensitivity [58]. In a recent

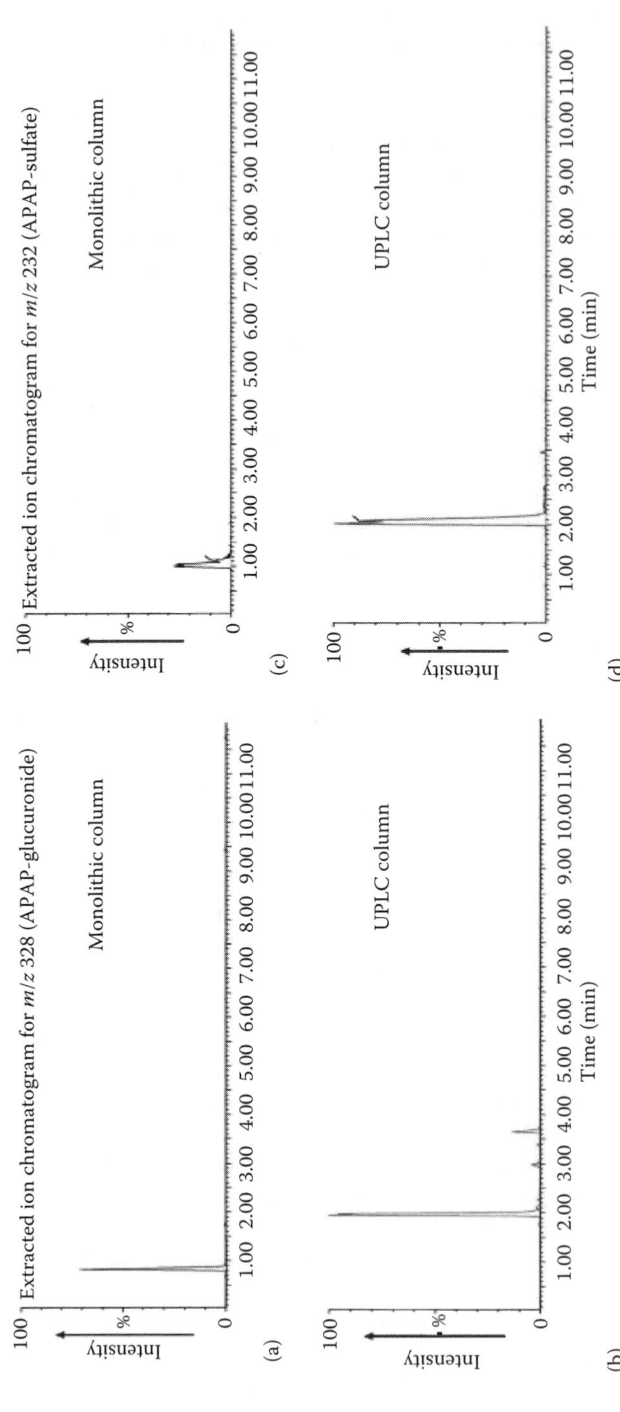

FIGURE 5.3 Comparison of XICs of acetaminophen-glucuronide ($m/z = 328$) (panels a and b) and acetaminophen-sulfate ($m/z = 232$) (panels c and d) obtained using either a monolithic column (Panels a and c) or an ACQUITY UPLC column (panels b and d). Positive ion mode ESI-MS following injection of urine following administration of acetaminophen (APAP). (Reprinted from Johnson, K.A. and Plumb, R., *J. Pharm. Biomed. Anal.*, 39, 805, 2005. With permission.)

study, Xue et al. [57] compared HPLC–MS/MS and HILIC–MS/MS for analysis of muraglitazar in human plasma and showed that the sensitivity of the LC–MS/MS assay can be improved by using HILIC.

5.2.3 Ionization Techniques for Metabolite Identification

Atmospheric pressure ionization (API) techniques are the most commonly used techniques in DM studies. Since the ionization occurs at the atmospheric pressure, API can be characterized as a "soft" ionization technique. There are three commonly used API sources [59] that can directly couple LC with MS: electrospray ionization (ESI) [60], atmosphere pressure chemical ionization (APCI) [61–63], and atmospheric pressure photoionization (APPI) [64,65]. The properties of the compound, such as its structure, polarity, and molecular weight, lead to the selection of one of these ionization techniques for sample analysis.

ESI, the softest ionization technique among the API techniques, is the most preferred method for metabolite identification studies. Due to its gentle ionization, it is a method of choice for Phase II conjugated metabolites and is also the commonly used technique for the analysis of peptides, proteins, carbohydrates, and oligonucleotides. Disadvantages of ESI technique include its susceptibility to ion suppression effects from high concentrations of buffer, salt, and other endogenous material present in sample solutions [57,58]. Other ionization techniques, APCI and APPI, offer some advantages such as better efficiency and sensitivity [58]. APCI and APPI have better tolerance to salts and matrix effects when compared to ESI [57]. Since the ionization in APCI occurs mainly in the gas phase in contrary to ESI, in which the analyte is ionized within the solution prior to its introduction into the source, APCI is subjected to less ion suppression and provides a wider dynamic detection range than ESI [66–69].

APCI is compatible with higher HPLC flow rates (1–2 mL/min) than ESI (0.1–0.5 mL/min). APCI has been widely utilized for the analysis of compounds within a selected range of polarity and volatility that excludes compounds with high molecular weight (>1000 Da). Both ESI and APCI can be used for compounds with low molecular weight (<600 Da) [70]. APPI, the newest ionization technique among the three [65,71], is a source similar to the APCI source, in which vaporization of the HPLC effluent occurs using a heated nebulizer. The APPI source employs a krypton discharge lamp to ionize the analytes via an ionizable dopant (toluene or acetone), instead of a corona discharge as in APCI. The photons generated by the lamp ionize the analyte. The high sensitivity of APPI might be due to the energy generated that is lower than the ionization potential (IP) of the most commonly used HPLC solvents, but within the IP range of many organic compounds. APPI might have an advantage for the analysis of nonpolar compounds [72]. APPI however is still a developing technique that has a great potential in the analysis of metabolites, and it needs to be evaluated further to be widely utilized in metabolite identification.

Another ionization technique that has made some inroads into the metabolite identification arena is the nanospray ionization (NSI). NSI is a low-flow (10–500 nL/min) ESI technique with many advantages over conventional-flow ESI (~200 μL/min) for drugs and metabolites. Advantages of using NSI include decreased sample consumption and increased sensitivity. Decreased sample consumption allows multiple MS experiments to be performed on isolated metabolites. NSI can be used for LC–MS or direct infusion-MS analysis of drugs and metabolites [73]. Several groups have shown that NSI is less susceptible to ion suppression effects and can be used to achieve normalized MS responses from drugs and their metabolites [74–78]. In a most recent study, Erve et al. [73] used an NSI coupled to a LTQ-Orbitrap and demonstrated the detection and characterization of prazosin rat metabolites present in neat bile and urine and extracts of feces, plasma, and brain homogenate. Limited susceptibility of the NSI source allowed the researchers to analyze bile and urine following a cleanup step using ZipTips packed with C4 or C18 reversed-phase material. In addition, the group also subjected the extracts of plasma, feces, and brain homogenate to a ZipTip cleanup step to avoid clogging of the NSI source due to presence of matrix materials and particulates. Although the NSI approach increased the metabolite identification throughput, elimination of the chromatographic separation step leads to difficulty in identification of isomeric metabolites.

Most recently introduced ionization techniques for pharmaceutical analysis include the direct analysis in real time (DART) [79–81] and the desorption electrospray ionization (DESI) [82–84]. Both DESI and DART allow gas-phase ions to be formed at atmospheric pressure directly from an NCE's native environment without sample extraction or preparation and/or chromatographic separation. A solvent is electrosprayed at the surface of a condensed phase target substance. Volatilized ions containing the electrosprayed droplets and the surface composition of the target are formed from the surface and introduced into the mass analyzer [76–78]. An analogous technique to DESI that does not require the electrospray solvent is DART [85]. The possibility of increasing the sample throughput by the elimination of sample preparation and HPLC separation has prompted several DMPK researchers to investigate DESI and DART for quantitative and/or qualitative determination of drugs and metabolites present in plasma, whole blood, tissues, and urine [81]. However, for both of these ionization techniques to make an impact in the DMPK arena, the following limitations have to be addressed: interference from salts and endogenous material present in the samples, instability/in-source fragmentation of metabolites, variability in detection limits with sample placement, reproducibility, and sensitivity [84,86,87]. For more on DART, see Chapter 13.

5.2.4 Mass Analyzers for Metabolite Identification

A mass analyzer separates ions according to their mass-to-charge (*m*/*z*) ratios. Several mass analyzers have been used in metabolism studies, including quadrupole mass filters (QMFs), linear (two-dimensional [2D] or linear ion trapping [LIT]), and three-dimensional (3D or quadrupole ion trap [QIT]) ion traps, hybrid quadrupole orthogonal time-of-flights (Q-TOFs), and Fourier transform mass spectrometers (Orbitrap and Fourier transform ion cyclotron resonance [FTICR]). Ion traps and QMFs are low-resolution mass analyzers with a resolution in the low-thousand range, whereas TOF and FTMS (Orbitrap and FTICR) are high-resolution analyzers that generate a minimum resolution of 15,000 (TOF) or over 30,000 (FTMS).

In the ion trap instruments, API formed ions are transferred with the aid of radiofrequency (RF) only octapole and/or hexapole ion guides into the trap. Upon introduction of ions from an API source, traps (2D, 3D, and FTICR) are capable of manipulating and performing multiple MS experiments, using various timed events, with the same ion population. Therefore, traps are classified into tandem-in-time mass analyzers. The QITs utilize a cylindrical ring and two end-cap electrodes to create a 3D quadrupolar field for ion storage, manipulation, and mass analysis. To alleviate some of the limitations associated with QIT, two new ion trap designs (LTQ and Q-Trap), based on the 2D or LIT technology, were introduced. The absence of quadrupole field in the center line allows ions to be trapped in an axial mode for mass analysis or transmitted to a second mass analyzer for high-resolution mode detection (Orbitrap or FTICR) [88–90]. Some of the functional improvements in 2D traps over 3D traps include increased ion storage capacity, faster scanning capabilities, and improvements in detection and trapping efficiencies.

The FTICR utilizes a combination of strong magnetic field (x- and y-directions) and electric field (z-direction) to store ions of various m/z in a cylindrical or cubic cell [90]. An electric field applied to a set of plates is used to excite the trapped ions for detection and/or MS/MS fragmentation. Upon on-resonance or off-resonance excitation, ions are detected using their respective image current imposed on another set of opposite plates. The frequency with which ions of a particular m/z registers an image current on the detection plates is recorded, amplified, and then the time domain signal is fast Fourier transformed (FT) into frequency domain signal before being converted to mass spectral data. Several research groups have evaluated the utility of FTICR-MS for metabolite detection, characterization, and identification [90–92]. However, today, application of FTICR-MS for metabolite profiling studies are considered an "over kill" because most of the metabolite profiling tasks can be achieved using less expensive and more user-friendly TOF or Orbitrap-based hybrid tandem mass spectrometers. Some of the other limitations, such as the "TOF effect," associated with the application of FTICR mass spectrometer–based systems for metabolite profiling have also been discussed [90].

Another mass analyzer type that uses FT to generate mass spectral data is the Orbitrap. Unlike in FTICR, where a combination of strong magnetic field and electric fields is required for trapping, manipulation, and high-resolution mass measurement, the Orbitrap uses a combination of electric fields applied between sections of a roughly egg-shaped outer electrode and an inner spindle electrode for high-resolution mass measurement [90]. Externally formed ions are transferred into a "C-trap" and then ion packets are injected into the Orbitrap to orbit around the spindle and induce image current on the egg-shaped outer electrode. Using amplification and fast FT, the image current is transformed into mass spectral data. Since MS/MS using a stand-alone Orbitrap is not yet a possibility, the Orbitrap is used in combination with either LIT or transmission quadrupoles for performing the MS/MS or MS^n event and then the fragments are mass measured using the Orbitrap. The combination of two mass analyzers allows parallel experiments to be performed with the LIT and the Orbitrap. For example, while the Orbitrap is measuring a full-scan accurate mass LC–MS spectrum of an analyte, which usually requires longer acquisition time

(~400 ms for mass measurement with 30,000 mass resolving power), MS/MS or MS^n experiments on the co-eluting metabolites can be performed in the LIT (<100 ms per low-resolution MS/MS in the LIT). These fast scanning features from two mass analyzers along with high mass resolution mode mass measuring capability makes the LTQ-Orbitrap an ideal platform for metabolite structural elucidation [88,93].

As opposed to ion traps, in TOF and QMF instruments, the ions generated in the ion source travel in a beam and pass through the analyzer to the detector. Therefore, TOFs, QMFs, and any permutation of these two mass analyzers are categorized into tandem-in-space instruments. The QMF utilizes four parallel conducting rods placed such that a combination of RF and DC voltages permits the passage or filtering of only a single *m/z* value. A mass spectrum is generated by varying the frequencies and amplitudes of the RF and DC fields. Low operating RF and DC voltages make QMFs tolerant to high vacuum conditions (10^{-6} torr range) and make these instruments most ideal for coupling with LC systems. Among the mass analyzer types, TOF is conceptually the simplest of all. Ions formed from either an API or a matrix-assisted laser desorption ionization (MALDI) source are briefly stored and thermalized and then released from the source region by an electrical field pulse and accelerated down the TOF flight tube. High *m/z* ions travel at a slower velocity and reach the detector later than the faster ions with low *m/z*. Calibration of the accelerating field and resulting flight times permits mass analysis for unknowns. Hybrid instruments combining QMF and TOF mass analyzers (Q-TOF) have become common in many metabolite identification laboratories across the globe. Another emerging hybrid mass analyzer type involves the combination of ion trap and TOF mass analyzers (IT-TOF). IT-TOFs provide MS^n capability for structural elucidation of metabolites.

In the process of metabolite identification by MS, detection of a metabolite molecular ion is the start of the process. Next, the product ion scan is the major experiment that obtains structural elucidation information on the detected metabolite. The first step of a product ion scan is referred to as the MS/MS or MS^2. Subsequent generation of product ion spectra from the fragment ion obtained from the previous step is defined as MS^3 fragmentation. The MS^n data is used to assign the fragmentation mechanisms that help the elucidation of drug metabolites. Generation of MS^n data was shown to be helpful in determining the sites of biotransformation [94,95]. Other MS techniques such as precursor ion and constant neutral loss (CNL) scan and accurate mass measurements are powerful tools for detection of drug metabolites in complex biological mixtures. Selected reaction monitoring (SRM) experiments may be used for the verification of proposed biotransformation pathways. These scan functions are available on various types of mass analyzers.

In general, the DM process of biotransformation involves Phase I and/or Phase II metabolic reactions. As shown in Table 5.1, the Phase I biotransformations include reactions such as oxidation, reduction, and hydrolysis. Glucuronidation, sulfation, and glutathione conjugations are some of the common Phase II biotransformations. Some of the common atmospheric ionization (API)-based LC–MS techniques for detecting Phase I and Phase II reactions include (1) targeted searches for the parent NCE's precursor ion mass offset by a net transformation mass (Table 5.1); (2) SRM transitions involving monitoring for a metabolite ion (M+16, M+32, M−14, etc.)

TABLE 5.1
Examples of Common Oxidation (Phase I) and Conjugation (Phase II) Biotransformations

Type of Biotransformation		Net Transformation	*m/z* Shift (Nominal)	*m/z* Shift (Exact)
Oxidation	Hydroxylation/ N-oxidation/S-oxidation	+O	+16	+15.9949
	Dihydroxylation	+2O	+32	+31.9898
	Dehydrogenation or reduction	$-H_2$	−2	−2.0156
	Demethylation	$-CH_2$	−14	−14.0234
	Deethylation	$-C_2H_4$	−28	−28.0468
	Depropylation	$-C_3H_6$	−42	−42.0468
	Oxidative deamination	$-NH_3$, +O	+1	+0.9843
	Oxidative dechlorination	−Cl, +OH	−18	−17.9662
	Oxidative defluorination	−F, +OH	−2	−1.9957
	Hydration	$+H_2O$	+18	+18.0105
	Methyl to an acid	$-H_2$, $+O_2$	+30	+29.9742
Conjugation	Glucuronidation	$+C_6H_8O_6$	+176	+176.0321
	Sulfation	$+SO_3$	+80	+79.9568
	Glutathione conjugation	$+C_{10}H_{17}N_3O_6S$	+307	+307.0837
		$+C_{10}H_{15}N_3O_6S$	+305	+305.0681
	Cysteine–glycine conjugation	$+C_5H_{10}N_2O_3S$	+178	+178.0410
	Cysteine conjugation	$+C_3H_7NO_2S$	+121	+121.0196
	N-acetyl-cysteine conjugation	$+C_5H_9NO_3S$	+163	+163.0301
	Glycine	$+C_2H_3NO$	+57	+57.0213
	Glutamine	$+C_5H_{10}N_2O_3$	+145	+145.0609
	Taurine	$+C_2H_5NO_2S$	+107	+107.0038

and a fragment ion mass offset by a net transformation mass (Table 5.1); (3) precursor ion scans based on a major parent NCE fragment ion and/or a major parent NCE fragment ion mass offset by a net transformation mass (Table 5.1); (4) neutral loss scans based on the parent NCE's fragmentation; (5) neutral loss scans based on common Phase II transformations; and (6) data-dependent experiments targeting peaks above a certain ion chromatographic threshold.

5.2.5 Oxidative Biotransformations

Oxidative biotransformations, which constitutes the major portion of Phase I reactions, can be catalyzed by either cytochrome P450s (CYP450) or nonmicrosomal enzymes such as flavin-containing monooxygenases (FMOs), monoamine oxidase (MAOs), alcohol dehydrogenase, and aldehyde dehydrogenase. As listed in Table 5.1, microsomal oxidations include aromatic and aliphatic hydroxylation, N-oxidation, S-oxidation (sulfoxidation and sulfonation), N-hydroxylation, N-, O-, S-dealkylation, deamination, dehalogenation, desulfation, epoxidation, alcohol oxidation, dehydrogenation, and oxidation of cyclic amines to lactams. Although MAOs, FMOs, alcohol dehydrogenase, and aldehyde dehydrogenase have been associated with several types of Phase I biotransformations, their involvement is of lesser importance. Among the 7703 different CYP450s identified across living organisms, 55 different CYP450s belonging to 16 families are present in humans [96–98]. Among these enzymes, more than 85% of the DM reactions are mediated by CYP3A4, CYP1A2, CYP2D6, CYP2C8, CYP2C9, and CYP2C19 [99,100]. In addition to being present in the liver, CYP450s are also present in the small intestine, lungs, and kidneys [101].

The guidance [24] on STDM states the following with respect to oxidative biotransformations, "metabolites formed from Phase I reactions are more likely to be chemically reactive or pharmacologically active and, therefore, more likely to need safety evaluation." Several recent reviews summarize atmospheric ionization (API)-based LC–MS approaches available for detecting and characterizing oxidative (Phase I) reaction products. Some of the special techniques and strategies such as APCI source fragmentation [102,103] and hydrogen–deuterium exchange (HDX) [104,105] available to distinguish between N- and C-oxidations have been explored by various research groups.

5.2.6 Conjugation Biotransformations

Conjugation biotransformations involve modification of a functional group such as –OH, $-NH_2$, –SH, or –COOH by bulky and polar groups such as glucuronides, sulfates, amino acids, and/or glutathiones. Among these reactions, glucuronidation is the most common conjugative reaction in mammals, and it is catalyzed by uridine diphosphoglucuronosyltransferase (UGT) enzymes [106]. Out of the 117 mammalian UGTs identified to date, about 22 enzymes have been found in humans. UGT1A4, UGT1A1, UGT1A8, UGT1A9/1A10, and UGT2B7 are collectively responsible for more than 80% of the UGT-mediated metabolism. Similar to the CYP450s, UGTs are mainly expressed in the liver but present to a lesser extent in the intestine, kidneys, and lungs. Most often, conjugation reactions terminate the pharmacological activity of a drug. During the course of metabolite identification, it is very common to incubate the biological samples with β-glucuronidase and sulfatase to elucidate the oxidative (Phase I) metabolite giving rise to glucuronide and sulfate metabolites, respectively [107–109]. Enzymatic treatments are usually conducted following incubations or administration of radiolabeled NCE rather than nonradiolabeled NCE so that comparison of metabolic profiles obtained with and without enzymatic treatments readily provides qualitative and quantitative information about oxidative and conjugative metabolites [109–111]. The guidance on STDM states the following with respect to conjugative biotransformations, "Phase II conjugation reactions generally render a compound more water soluble and pharmacologically inactive, thereby eliminating the need for further evaluation. However, if the conjugate forms a toxic compound such as acylglucuronide [112], additional safety assessment may be needed" [24]. Although limited, few conjugative metabolites have been found to be pharmacologically active and these include the phenolic glucuronide conjugate of ezetimibe [113] and morphine-6-glucuronide [104,114].

The reasons for identification of metabolites are manyfold but all boil down to human safety of drugs under clinical investigation. Initially metabolites of a drug are evaluated using *in silico* methods and then characterized with *in vitro* systems (microsomes, hepatocytes, S9 fractions, etc.) before assessing using mouse, rat, rabbit, dog, and/or monkey models. Subsequently, metabolites in humans are identified following drug administration to assure that the nonclinical species undergoing safety assessment are adequately exposed to human metabolites of the drug [25].

5.3 DISCOVERY STAGE METABOLITE IDENTIFICATION

Following oral administration, unlike following intravenous administration (or other nonoral routes such intramuscular and sublingual) where some of the metabolism pathways are avoided, the liver serves as the primary site of metabolism. Therefore, it is not surprising that during the drug discovery process emphasis is placed on enzymes present in the liver, especially on cytochrome P450s and UGTs. Since many compounds have to be screened using *in vitro* methods at the discovery stage, most often *in silico* methods are used as first-tier approaches to weed out NCEs with undesirable ADME properties.

5.3.1 *In Silico* Methods

Very early in the discovery phase, in the absence of *in vitro* or *in vivo* data, *in silico* strategies and procedures have been used to aid in drug design, especially to optimize metabolic clearance properties and to predict human PK properties [115]. *In silico* packages such as Admet Predictor, Pre-ADME, ChemSilico, KnowItAll, and Pharma Algorithms are capable of predicting properties such as permeability, absorption, elimination half-life, plasma-protein binding, blood–brain penetration, and oral bioavailability [116,117]. However, due to the complexity of biotransformation, species differences in metabolism, and limited data sets, *in silico* methods to predict metabolites have been very slow to develop. As shown in Figure 5.4, *in silico* methods available for metabolite prediction are categorized into either local or global techniques [118–121]. The global techniques are knowledge based and function to provide information about metabolites based on known metabolism data and/or rules built on known metabolism of substructures of NCEs. Examples of the global techniques include MetabolExpert [122], Meta [123,124], and Meteor [125]. The local techniques are based on quantitative structure metabolism relationships (QSMR) derived from structural, physicochemical, and/or quantum mechanical properties of an NCE [126,127]. Table 5.2 lists some of the *in silico* software packages available for assisting metabolite prediction.

In a recent study, *in silico* docking predictions were used to evaluate interactions of 20 commonly used cancer drugs with CYP2D6. The results obtained led to the identification of a new oxidative metabolite of metoclopramide [119,128]. The formation of this *in silico* predicted metabolite was later confirmed by LC–MS following incubation of metoclopramide with HLM and recombinant CYP2D6 [119].

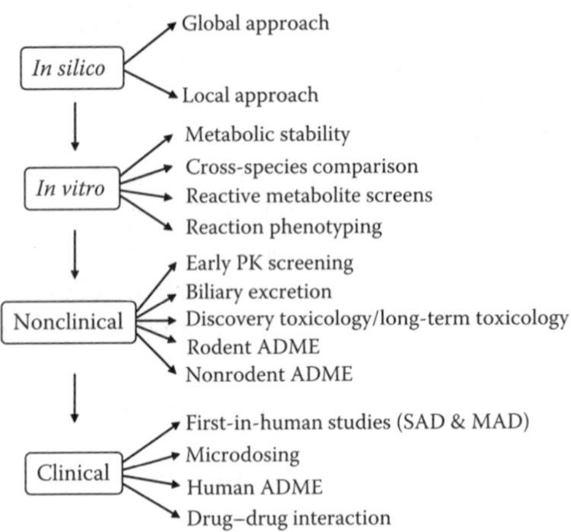

FIGURE 5.4 Metabolite identification work flow and the various ADME studies from which metabolite identification work processes are likely to originate from.

TABLE 5.2
Commonly Used *In Silico* Metabolite Prediction Software

Prediction Software	Web Site	Capabilities
MEXAlert	http://www.compudrug.com/	Biotransformation products
MetabolExpert	http://www.compudrug.com/	Biotransformation products
MetaDrug	http://www.genego.com/metadrug.php	Phase I and II biotransformation products; off-target effect prediction
Meteor	http://www.lhasalimited.org/index.php	Biotransformation products
MetaSite	http://www.moldiscovery.com/soft_metasite.php	Biotransformation products and the CYPs involved (crystal structures are available for prediction of involvement of the following major enzymes: CYP1A2, CYP2C9, CYP2C19, CYP2D6, CYP3A4)
Drug Bank	http://www.drugbank.ca./	DM prediction and drug interaction prediction

Source: Adapted from Wishart, D.S., *Drugs R. D.*, 8, 349, 2007. With permission.

Although there are several examples where *in silico* methods have been claimed to detect possible metabolism sites and lead to altered structures of NCEs, the details of these examples are not routinely published due to the proprietary nature of the drug discovery and development process [126]. Increasing computing power combined with development of sophisticated databases with commitment from DM scientists to submit entries can only increase the utility of the software packages/databases by medicinal chemists to design away or design in metabolic soft or hot

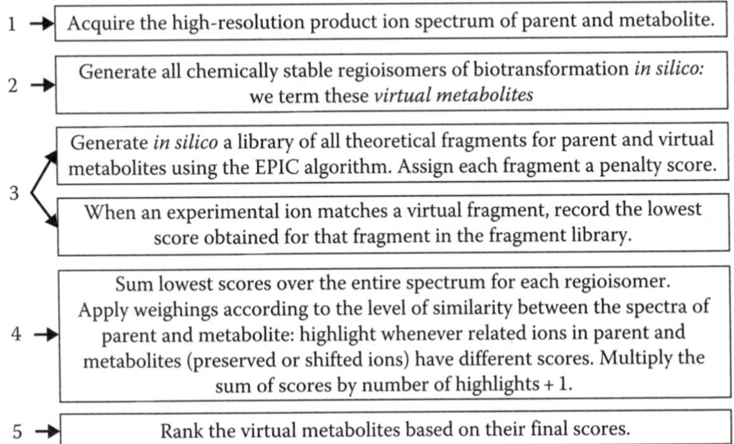

FIGURE 5.5 Steps involved in application of IsoScore for the automated localization of biotransformations. (Reprinted from Leclercq, L. et al., *Rapid Commun. Mass Spectrom.*, 23, 39, 2008. With permission.)

spots. The capabilities and predictions of the *in silico* techniques are questionable when used as a stand-alone tool. However, when *in silico* techniques are used in conjunction with *in vitro*, *in vivo*, and/or LC–MS experimental data, the combined complimentary approaches can help to streamline the metabolite identification process [128–137]. One such method, IsoScore (Figure 5.5), involves accurate mass measurements of NCEs, their metabolites, and their MS/MS fragmentation and comparison of the data with information generated using *in silico* methods [138]. Mathspec, another software approach, utilizes fragmentation data from high-resolution MS/MS spectra to systematically assemble possible parts of the molecule into rational molecules [132].

As a next step, LC–MS-based high-throughput assays are used to weed out NCEs exhibiting poor metabolic stability [139–143], protein binding [144–146], and Caco2-permeability [147–149]. These *in vitro* studies are used to select NCEs with viable DMPK properties such as the capability to enter circulation rapidly or slowly, remain in circulation for appropriate duration for producing efficacy and the ability to get eliminated without accumulation or causing toxic effects. For example, while it is highly desirable for an anesthetic NCE to be cleared rapidly, an anti-inflammatory NCE should be cleared slower and possess high protein-binding properties. Some of the *in vitro* systems listed in Table 5.3 are used in the first-tier assays to select NCEs for further optimization and characterization in, slightly more expensive and time-consuming, nonclinical animal studies.

5.3.2 Metabolic Stability Studies

Among the high-throughput assays, metabolic stability studies have evolved as a possible starting point for metabolite identification. Traditional metabolic stability

TABLE 5.3
Commonly Used Cell Systems for Generation of *In Vitro* Metabolite Identification Samples

System	Origin	Comments
Liver microsomes	Liver tissue	Phase I metabolism
Intestinal microsomes	Intestinal tissue	Phase I metabolism
Cytosolic fractions	Liver tissue	Phase II metabolism
Supersomes	Liver tissue (Baculovirus-insect-cell expressed)	Specific CYP/UGT-mediated metabolism
S9 fractions	Liver tissue	Phase I and II metabolism/ detecting DNA damage
Isolated perfused liver	Liver	Hepatotoxicity/metabolism
Liver slices	Liver	Hepatotoxicity/metabolism
Hepatocytes	Liver	Hepatotoxicity/metabolism
HepG2	Human hepatoma cell line	Hepatotoxicity/metabolism
HLE	Human lens epithelial cell line	Hepatotoxicity/metabolism
BC2	Human hepatoma cell line	Hepatotoxicity/metabolism
DNA microarray	Rat liver	Phase I and II metabolism
Genotoxicity	COMET assay Ames test	Detecting DNA damage
Renal toxicity Hepatotoxicity	DNA microarray	Changes in gene expression
Cytotoxicity	MTT assay SRB assay Clonogenic assay	Cell proliferation
hERG potassium channels	I_{kr} assay	Electrophysiology study

Source: Adapted from Ruiz-Garcia, A. et al., Pharmacokinetics in drug discovery, *J. Pharm. Sci.*, 97(2), 654, 2008. With permission.

studies, conducted to select an NCE with the most appropriate *in vitro* stability, involve incubation of an NCE or series of NCEs with human or rat liver microsomes or hepatocytes and LC–MS/MS monitoring for the disappearance of an NCE with respect to the incubation time. Most laboratories prefer using hepatocytes due to availability of oxidative and conjugative enzymes as well as all the cofactors. In addition, *in vitro–in vivo* correlation using hepatocytes have been shown to be more successful than liver microsomes [136,137]. Since the introduction of API-based LC–MS technology, metabolic stability studies have been performed in the high-throughput mode with fully automated sample preparation, LC–MS/MS analysis, data integration, and reporting [150–154]. Most often metabolically unstable NCE may not be an ideal candidate to advance further into development rather an ideal candidate for identifying metabolic hot spots within an NCE series or molecules with similar scaffold. Since most metabolic stability studies are conducted at physiologically relevant drug concentration (1–10 μM), identification of metabolites is a challenging task and requires sensitive detection techniques. Although metabolite identification is not routine and not always required, availability of sensitive mass spectrometers (QIT, LIT, TOF, Q-TOF, and LTQ-Orbitrap), which can operate in the full-scan LC–MS mode or targeted MS/MS or data-dependent MS/MS mode at high speed and selectivity, has allowed DMPK scientists to monitor both the disappearance of an NCE and the formation of its metabolites [90,155–157].

5.3.3 Early PK Screening

Initial *in vivo* evaluation to rank order the NCEs with suitable PK properties, following *in silico* and *in vitro* optimization, is usually performed using rats or mice. The requirement of smaller quantities of dosing material, the ability to maintain large number of animals from the same strain, and limited interanimal variability are some of the reasons for selecting rodents for the first set of *in vivo* evaluations. Several potential methods to accelerate the high-throughput *in vivo* PK analysis have been explored. In addition to automated sample preparation methods (PPT, SPE, and LLE), cassette dosing [158,159], cassette analysis [160–162], and cassette-accelerated rapid rat screening (CARRS) [163] have been evaluated as options for further accelerating the PK screening step. Most of these PK screening assays involve oral dosing of the NCE to two to three rats or mice and determining mainly the PK parameters. Cai et al. [160,161] and King et al. [164] explored the options to obtain metabolite information at selected PK time points while quantifying NCEs present in plasma. Metabolite information at this stage not only helped to understand the *in vivo* stability of an NCE but also helped to troubleshoot the quantitative assays with respect to co-eluting metabolites and their possible contribution to signal suppression or enhancements. The recent introduction of fast scanning hybrid mass spectrometers such as the Q-Trap 5500 and high-resolution mass spectrometers such as the Exactive and the LTQ-Orbitrap have revived the interest in obtaining metabolite information during high-throughput rodent PK studies [157,165,166].

5.3.4 Cross-Species Comparison

At the characterization phase of late discovery, before an NCE is recommended for full development, preliminary information about the NCE's biotransformation pathways is very useful for selecting appropriate rodent (rat or mouse) and nonrodent (dog or monkey) species for short- and long-term toxicological and/or safety evaluations. The initial interspecies comparison of an NCE is normally achieved using *in vitro* preparations involving microsomes or hepatocytes. The goal at this stage is to detect and characterize major human *in vitro* metabolites and confirm formation of the human metabolites in the toxicology species. At this stage, *in vitro* incubation parameters such as linearity with respect to metabolite production, number of cells, and incubation time are usually not optimized. Low concentrations are often selected to be therapeutically relevant and the high concentration incubations are used for aid in metabolite identification and if necessary to get quantitative estimates of the metabolites using UV detection online with LC–MS [157].

An alternate approach is to get quantitative estimates of metabolites using radiolabeled version of the NCE and simultaneously profile, detect, and characterize metabolites using an LC–MS system connected online with a radioactivity detector. At the discovery stage, *in vitro* cross-species comparison studies can be conducted using a tritium (^{3}H)-labeled form of the NCE rather than a ^{14}C-labeled form of the NCE due mainly to ease of incorporation of tritium into an NCE, low cost, and quick turnaround time [167]. Most often discovery stage cross-species comparison studies are performed for internal decision-making purposes and are an invaluable tool for selecting an NCE from a series for further advancement into development, selecting toxicological species that have exposure to human *in vitro* metabolites, and excellent concentrated source of metabolites useful for identification of active and/or reactive metabolites.

5.3.5 Reactive Metabolite Screens

Adverse drug reactions (ADRs) have led to withdrawals of troglitazone, trovafloxacin, bromfenac, tolcapone, pemoline, nefazodone, and ximelagatran and "black box" warnings for several more recently approved drugs [168–172]. One of the main problems with detecting ADRs is the low incidences of these types of reactions and the difficulty in being able to observe these types of reactions in a preclinical species. ADRs, which resulted in withdrawals or black box warning of drugs, were only detected either during the large-scale clinical trials (Phase IIb or Phase III) or the postmarketing period. Although the mechanisms responsible for ADRs are not fully understood, the formation of reactive metabolites and their subsequent binding to cellular macromolecules have been shown to be involved in these types of reactions [172,173]. Although some reactive metabolite intermediates such as epoxides that have a relatively long half-lives can be detected in *in vivo* samples, many reactive metabolites such as quinones have short half-lives and direct detection and identification of such reactive metabolites in biological samples is very challenging. It is difficult, but it is not impossible, to predict the propensity of an NCE to form reactive metabolites based on functional groups and/or *in vitro*–based trapping and screening assays, predictive software, and structural alert publications.

A variety of *in vitro*–based assays have been developed to screen for reactive metabolites and among them GSH-trapping experiments are the most commonly used across the pharmaceutical companies for minimizing risks associated with reactive metabolite formation [174–176]. GSH, a tripeptide (γ-L-glutamyl-L-cisteinylglycine), is known to form conjugates with quinones, quinoneimines, arene oxides, nitrenium ions, imine methides, and Michael acceptors via either displacement reactions or nucleophilic additions [176,177]. Under positive ionization mode MS/MS conditions, GSH conjugates undergo fragmentation to lose a diagnostic neutral molecule corresponding to the pyroglutamic acid (129 Da) moiety, allowing the use of CNL scan to screen for GSH adducts. Dieckhaus et al. [178] showed that monitoring of the γ-glutamyl-dehydroanalyl-glycine ion (*m/z* 272) using the precursor ion scan mode in electrospray negative ion mode is a more sensitive and selective screening technique, in comparison to CNL of 129, for detecting GSH adducts [178].

More recently, using a mixture of GSH and stable isotope–labeled GSH allowed Yan et al. [179] to use the mass difference of 3 Da observed between the nonlabeled and the labeled GSH adducts for unbiased data-dependent–based confirmation of GSH adducts [180]. In addition to GSH trapping, incubations with potassium cyanide and/or sodium cyanide followed by monitoring for CNL of 27 (cyano moiety) has been demonstrated as an alternative screening method for detecting reactive metabolite formation [174]. These LC–MS/MS-based screening assays have been problematic because they fail to provide quantitative estimates of the GSH adducts. To overcome this limitation UV has been used online with MS and trapping experiments with radiolabeled forms of GSH, potassium cyanide, and sodium cyanide have been explored [181–183]. Once reactive metabolite screening is complete the next step involves characterization of the metabolite using additional MS/MS experiments and/or NMR experiments and if necessary, working with the medicinal chemists to design out the structural components that are liable to form reactive metabolites. Fluorescence-based assays also have been explored to detect GSH adducts [184]. Several review articles focusing on the detection and characterization of reactive metabolites have been published (see also Chapter 6) [185–188].

5.3.6 Reaction Phenotyping

Identification of the enzymes responsible for the metabolism of a drug can be accomplished and their relative contribution to the metabolism of an NCE can be determined by reaction phenotyping studies, which are extremely important in drug discovery process [189]. Reaction phenotyping studies result in determination of human CYP450 enzymes involved in a particular biotransformation pathway. Reaction phenotyping studies are essential for determining the potential for drug interactions and polymorphic impact on drug disposition. The discovery stage reaction phenotyping approach involves incubation of NCEs in HLM and monitoring the biotransformation with and without a known specific chemical inhibitor of the CYP450 enzyme. Since CYP1A2, 2C9, 2C19, 2D6, and 3A4 are involved in biotransformation of majority of the NCEs, these five enzymes and their involvement in metabolism are assessed using automated high-throughput LC–MS/MS assays [190,191]. Since the complete biotransformation pathways of an NCE are not known at this stage of drug discovery, assay assessment mainly involves monitoring disappearance of the parent NCE in presence and absence of known CYP inhibitors. Although not routine, qualitative LC–MS-based metabolite identification experiments at this stage are not ruled out [192].

As a second-tier definitive approach, cDNA expressed enzyme systems are used to characterize the isoforms responsible for formation of primary metabolites of an NCE, to determine the kinetics of formation of primary metabolites and to predict drug–drug interaction liabilities [193,194]. At this stage of drug discovery, involvement of CYPs as well as FMOs and UGTs are also assessed [195,196]. In general, an NCE is incubated in cDNA expressed enzyme systems and the rate of disappearance of the parent or the rate of appearance of the metabolite(s) is monitored by an analytical method. LC–MS/MS is the method of preference to determine the concentration of the parent drug or the metabolites as a function of time. If the metabolite(s) standard is available, measuring the metabolite concentration is preferred; in the absence of a metabolite standard, the metabolite concentrations have been measured semiquantitatively with respect to the LC–MS or the UV absorbance of the parent drug to rank order the NCEs for further development.

Usually in the later stages of the drug discovery or during early development, upon availability of radiolabeled form of the NCE, definitive reaction phenotyping studies are conducted. In these studies the metabolite concentration is determined by using HPLC-radiometric detection and all major metabolites formed by various isozymes are identified using LC–MS techniques. For example, Ghosal et al. [197] used radiometric detection online with LC–MS for identification of human liver CYP450 enzymes involved in the biotransformation of vicriviroc (VCV), a CCR5 receptor antagonist. Incubation of VCV with HLM and cDNA expressed enzymes provided evidence for involvement of CYP3A4, CYP3A5, and CYP2C9 in the metabolism of VCV and formation of four of the six metabolites were via multiple enzymes and two of the six metabolites were mainly via CYP3A4 (Figure 5.6). The involvement of multiple enzymes and formation of the major metabolites via

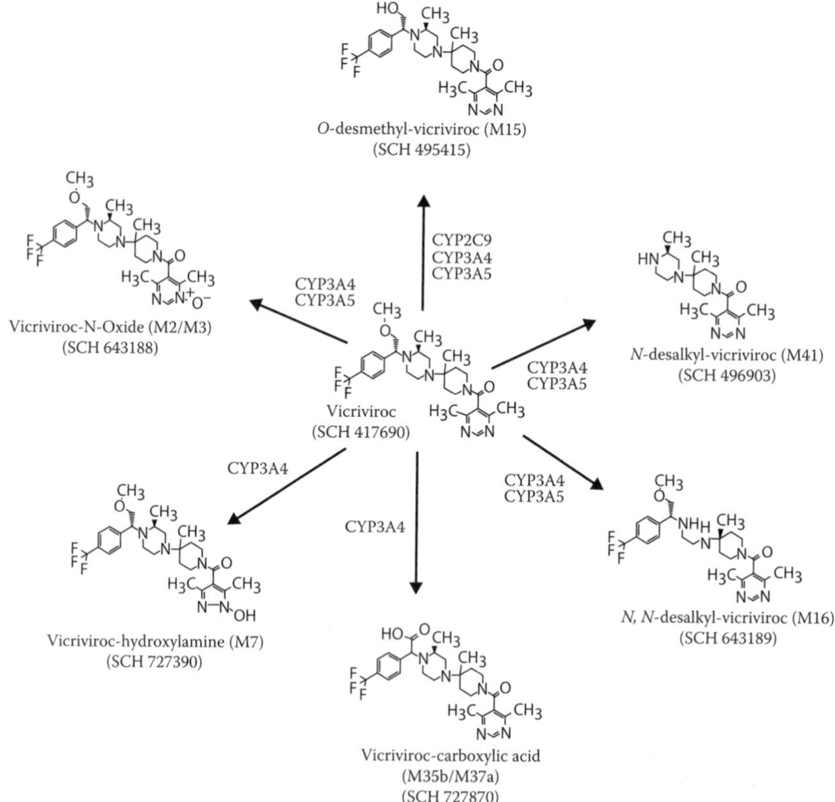

FIGURE 5.6 Biotransformation of VCV in HLM and cDNA-expressed human P450 enzymes. (Reprinted from Ghosal, A. et al., *Drug Metab. Dispos.*, 35, 2186, 2007. With permission.)

CYP3A4 are considered a plus for an NCE. If the major metabolic pathway of an NCE exclusively involves CYP2C9, 2C19, and 2D6, which exhibit genetic polymorphism, then drug–drug interactions and NCE/metabolite accumulations are some of the possible liabilities. If the same major metabolite is also being formed by other nonpolymorphic CYPs then safety associated with accumulation of NCE/metabolites and/or drug–drug interactions are less problematic. The importance of reaction phenotyping and the analytical methods available to study NCE liabilities have been published in several review articles [196,198–201].

5.3.7 Rodent and Nonrodent Biliary Excretion

Early in the discovery, in the absence of the radiolabeled drug, the bile duct cannulated (BDC) studies are mostly conducted in rats, but BDC dogs and monkeys are also used depending on the toxicology species used in the long-term safety evaluation studies. Following a single oral administration of an NCE, bile is collected for ≤24 h in rats (usually in 0–8 and 8–24 h blocks) and up to 72 h in dogs and monkeys. LC–MS analysis of bile samples provides preliminary information about conjugated metabolites and GSH adducts. In the later stages of the drug development, BDC animal studies conducted with the radiolabeled drug provides more definitive biliary excretion profiles of the NCE. During the excretion of biliary glucuronide and sulfate conjugates into the gastrointestinal (GI) tract, conjugated metabolites can be hydrolyzed during the passage through the intestine [202]. Once hydrolyzed, the aglycone regardless of whether its the NCE or its primary metabolite (usually a hydroxyl metabolite of the NCE) can reenter the systemic circulation, this process is usually referred to as enterohepatic recirculation [113,203]. Enterohepatic recirculation results in altered PK profiles, prolonged exposure to the NCE and delayed excretion of the drug-derived material. Therefore, to understand the role of conjugative metabolism in the overall metabolic clearance of an NCE, biliary profiles should be obtained. Since BDC studies do not generate full mass balance data, a separate study using non-BDC animals is necessary to understand the complete mass balance (percent urinary and fecal excretions) and individual contribution of each metabolite. If the data shows that the majority of the NCE is excreted in the feces as unchanged NCE, following cleavage of the glucuronide conjugates in the gut, collection of bile during the human ADME study should be considered to determine the unabsorbed portion of the NCE being excreted in the feces [204–206].

Overall, rodent and nonrodent biliary excretion studies with BDC animals are considered for the following reasons: (1) to determine the amount of NCE absorbed; (2) for development of metabolite profiling LC–MS method (bile, a complex concentrated matrix with plenty of endogenous matrix components, is an ideal source for optimizing the chromatographic separations of the metabolites); (3) for optimizing MS conditions for detecting and characterizing conjugated metabolites such as glucuronides and sulfates; and (4) to confirm the propensity of an NCE to form reactive metabolites under *in vivo* conditions (usually, LC–MS detection and characterization of GSH conjugates in bile or derivatives/degradants of GSH (mercapturic acid) in urine indicates the presence of reactive metabolites).

5.4 DEVELOPMENT STAGE METABOLITE IDENTIFICATION

In the development stage, it is a common practice to investigate the metabolism using a radiolabeled form of the NCE in appropriate *in vitro* systems, animal species, and humans. ^{14}C and ^{3}H are the choice of isotopes in most of the radiolabeled studies [38]. The *in vitro* studies with radiolabeled drug candidate usually includes species comparison studies that involve incubations with liver microsomes and S9 fraction, hepatocytes, and tissue slices from various species, and human c-DNA expressed CYP enzymes. *In vivo* metabolism studies include the nonclinical animal ADME studies and human ADME studies. In comparison to discovery stage metabolite profiling and characterization studies, development stage studies are typically conducted using longer HPLC methods with the aim of separating, detecting, and quantifying (using radioactivity detection [RAD]) a majority of the metabolites. Most often data from definitive metabolism studies are included in new drug application (NDA) submissions as well as in an approved drug's packaging inserts to help doctors and patients understand more about the drug being prescribed and its possible implications.

5.4.1 Definitive *In Vitro* Studies

In vitro metabolite profiles using a radiolabeled drug are determined to compare DM across different toxicology species and humans and to quantify (using RAD) the metabolites generated. As opposed to discovery stage *in vitro* studies, in development, the *in vitro* incubation parameters such as linearity with respect to metabolite production, number of cells, and incubation time are optimized, and the drug concentrations used for incubations are clinically relevant. The *in vitro* incubation systems include liver microsomes and S9 fraction, hepatocytes, and tissue slices from various species, and human c-DNA expressed CYP enzymes (see Table 5.3 for a complete list). The incubation samples are analyzed by HPLC-radiodetector or simultaneously with LC–MS and radiodetector using the flow split method. Other *in vitro* studies conducted in drug development include detailed reaction phenotyping studies to identify the enzymes involved in the metabolism of the drug candidate, inhibition studies to investigate the potential of the drug as an inhibitor of CYP enzymes, and induction studies to see if the drug induces CYP enzymes.

The data obtained from definitive *in vitro* studies provide information to understand the potential of the drug candidate for drug–drug interaction and help to make decisions about the clinical studies that should be conducted to assess potential interactions [207]. The relationship between an NCE's metabolism, clearance, enzyme induction, and inhibition is schematically shown in Figure 5.7. Detailed information about drug–drug interaction studies can be found in PhRMA position papers by Bjornsson et al. [200,201] and in book chapters by Zhang et al. [208] and Iyer and Zhang [209].

5.4.2 Definitive Nonclinical ADME Studies

Unlike the discovery stage nonclinical ADME studies where radiolabel NCE may not always be available, development stage definitive ADME studies are always

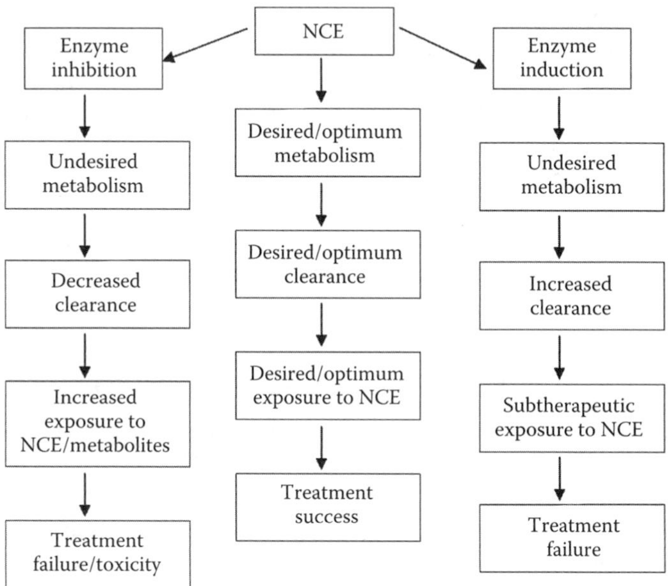

FIGURE 5.7 Relationship between an NCE's metabolism, clearance, enzyme induction, and inhibition.

conducted using radiolabeled NCE. At this stage of development, pharmacologically relevant doses are used in the ADME studies and the stability of the radiolabel in biological matrices as well as purity of the radiolabeled NCE are determined. Most often, the position for radiolabel addition is carefully designed so that during biotransformation, the radiolabel is not lost. Loss of the radiolabel or degradation associated with radioactivity label addition can lead to incomplete mass balance. In addition to determining the mass balance, drug-related material elimination routes, biotransformation pathways, gender differences in metabolism are also evaluated in definitive nonclinical ADME studies. Although not practical, accounting for the entire administered radioactivity in the excreta (urine, feces, etc.) allows for achieving 100% mass balance. In practice, mass balance above 90%, 85%, and 80% respectively in rat, dog, and human, over reasonable sample collection intervals (7–14 days), is considered complete [210]. Metabolite profiling at this stage often involves separation of co-eluting metabolites, radiochromatographic quantitation, and identification of all major circulating as well as urinary and fecal metabolites. Usually samples from across time intervals (0–24, 24–48, 48–96 h, etc.) and across animals ($n = 3$ per gender) are pooled so that a representative pooled sample is used for profiling as well as for LC–MS metabolite detection and characterization. Sample pooling schemes are designed so that the representative metabolite profiles contain a major portion (>90%) of the radioactivity excreted in each matrix. In addition, if necessary, sample extraction schemes are fine tuned to achieve uniform extraction across metabolites.

As shown in Figure 5.1 strategies for radioactivity-based metabolite quantification and metabolite detection and characterization are selected based on the radioactivity content of a sample and the dose. If the administered dose is low and the plasma samples contain very low radioactivity, some of the sensitive techniques discussed in Section 5.5 may have to be selected for profiling the plasma samples. Finally all definitive nonclinical ADME study data are used in support of nonclinical drug safety evaluations and for comparison with human ADME metabolic profiles to assure that the animal models being used for long-term safety testing are adequately exposed to all the human metabolites. Therefore most of the nonclinical ADME metabolite profiles used for comparison with human ADME metabolite profiles and for NDA submission-related activities and publications originate from definitive nonclinical ADME studies.

5.4.3 Microdose of Radioactivity for Metabolite Identification

One of the major hurdles faced by the biopharmaceutical industry is the late-stage clinical attrition. Reducing the clinical attrition is a difficult task because animal models are not perfect predictors of efficacy and safety in humans, especially when it comes to anticancer NCEs. As one of the new tools to make the drug development process faster and safer and to eliminate less promising NCEs earlier in the clinical testing with limited investments and resources, microdosing strategies are being supported by the US FDA [211] and the EMEA [212]. Since microdosing studies precede traditional Phase I studies (e.g., SAD, MAD), these studies are also referred to as "Phase 0" or "Exploratory IND" studies.

A microdose is defined as the smaller of either 1/100 of the expected pharmacologically effective dose determined using *in vitro* and preclinical animal models or a maximum of 100 μg. Although the informational requirements for microdosing studies are more flexible than for traditional IND studies, the FDA guidelines suggest performing a single-dose animal toxicity study, with a week of observation period, to be completed using doses above the human subtherapeutic doses to show no risk of toxicity before starting microdosing studies. A safety margin of about 1000-fold is recommended between animals and humans.

In a nutshell, microdosing studies involve the administration of a single subtherapeutic dose of a radiolabeled NCE to limited number healthy adult volunteers for no more than 7 days to (1) assess whether the mechanism of action observed in preclinical models translates to humans, (2) select the most promising lead candidate from multiple NCEs, (3) assess human PK and/or metabolism, and (4) to explore an NCE's distribution [211]. If necessary, assuming linear PK, the PK and distribution data from microdosing can be used to estimate the PK at therapeutically relevant doses. A recent evaluation of five compounds showed that microdose studies are most reliable for predicting the therapeutic dose level PK of small molecule NCEs that possesse linear PK, short half-lives, and moderate metabolism [213–215].

Since the doses are very small, conventional LC–MS techniques are sometimes not sensitive enough to assay samples from microdosing studies. Often accelerator mass spectrometry (AMS) and positron emission tomography (PET) are required for obtaining PK and distribution information, respectively. As described in Section 5.4, although AMS is capable of quantifying ^{14}C-labeled compounds with attomole (10^{-18} M) sensitivity, the technique is not useful for distinguishing between an NCE and its metabolites unless additional chromatographic separation and fractionation is undertaken before the AMS step. To overcome this limitation, several laboratories have demonstrated the applications of conventional LC–MS techniques to detect NCEs and/or metabolites from microdosing studies [216–219]. Recently, Seto et al. [220] demonstrated the utility of fast scanning Q-Trap 5500 hybrid mass spectrometer for detecting metabolites following an oral microdose of atorvastatin, ofloxacin, methimazole, omeprazole, or tamoxifen to rats. SRM-IDA-based methods in combination with a list of possible metabolites for each drug were used to quantify and identify the circulating as well as urinary metabolites from the microdose study. For more on AMS and microdosing see Chapter 14.

5.4.4 First-in-Human Studies (SAD and MAD)

During the transition of an NCE from discovery to development, following application of scaling factors (allometric scaling) to account for differences between nonclinical animals and humans, the efficacious human dose can be predicted using PK and pharmacodynamics (PD) data from a variety of *in vitro* and *in vivo* studies [221,222]. Adverse dose ranges and the therapeutic safety margins can be coarsely determined using discovery stage toxicology studies and adjusted using discovery stage ADME data. As a result of the recent TGN1412-related tragic events, doses proposed for the FIH clinical studies are being carefully scrutinized by the pharmaceutical industry scientists as well as the regulatory authorities [223–225]. Around the time of discovery to development transition of an NCE, the predicted efficacious human dose can be used to estimate the NCE needs for IND-enabling toxicological and FIH studies. The certified form of the NCE or active pharmaceutical ingredient (API) for IND-enabling toxicological studies usually becomes available once all the salt and physical forms of the NCE are finalized and large-scale manufacturing issues have been worked out. Upon completion of the IND-enabling rodent and nonrodent toxicological studies, more refined safe human doses to be used in the FIH study usually come to light and the pharmaceutical company is ready to file for a traditional IND (rather than the exploratory IND) to start Phase I clinical trials.

FIH studies, which are part of Phase I clinical trials, start with SAD and advance into MAD administration. Goals of SAD and MAD studies are to determine the potential toxicity of an NCE and safe dosage ranges on a small number (20–100) of healthy adult volunteers. Therefore, traditionally, these studies have included determination of parent exposure and PK parameters over the dose range but have not focussed on identification of metabolites. However, the recent publication of the FDA guidance on the safety testing of metabolites [24] has encouraged sponsors to include some aspects of metabolite identification at this stage [29–36].

A historical account, leading up to the publication of the FDA's guidance on safety testing on metabolites of small molecule drug products, has been discussed [26–36]. The FDA guidance on metabolites states the following "We encourage the identification of differences in DM between animals used in nonclinical safety assessments and humans as early as possible during the drug development process [24]. The discovery of disproportionate drug metabolites late in drug development can potentially cause development and marketing delays" [24]. Furthermore, the guidance places emphasis on major human circulating metabolites by stating "Generally, metabolites identified only in human plasma or metabolites present at disproportionately higher levels in humans than in any of the animal test species should be considered for safety assessment. Human metabolites that can raise a safety concern are those formed at greater than 10 percent of parent drug systemic exposure at steady state" [24]. A flow diagram (Figure 5.8) showing some of the studies needed to determine safety of a human drug metabolite was also included in the guidance. The STDM guidance triggered a paradigm shift across the pharmaceutical industry and placed emphasis on obtaining *in vivo* human metabolism information earlier rather than waiting for human ADME studies (see Section 5.4.5), which usually takes place late in the Phase

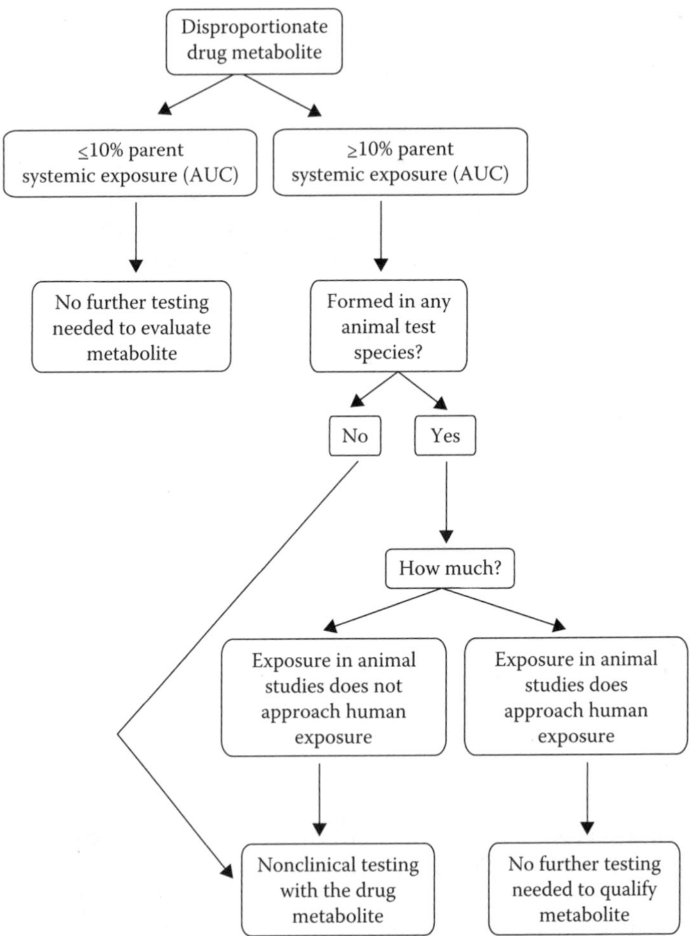

FIGURE 5.8 Decision tree flow diagram showing the steps necessary for assuring safety of human metabolites. (Reprinted from Food and Drug Administration (US FDA), *Guidance for Industry: Safety Testing of Drug Metabolites*, 1, 2008. http://www.fda.gov/downloads/Drugs/GuidanceComplianceRegulatoryInformation/Guidances/UCM079266.pdf. With permission.)

I or Phase II. Furthermore, metabolite exposures at steady state are never available following human ^{14}C-ADME studies because these studies are always conducted using a single dose administration.

Requirements for obtaining *in vivo* human metabolism information early in the development of an NCE and the opportunity to determine metabolite concentrations at steady state has persuaded several of the pharmaceutical companies to take advantage of the SAD and MAD studies to get a glimpse of the metabolites present in human plasma and urine [24]. In SAD studies, urine (0–24 h) and blood samples (3–4 time points) can be collected from placebo- and NCE-dosed healthy volunteers from the top two or three dose groups. Since the volume of plasma samples from FIH studies are limited, one option is to use urine samples (pooled across subjects) to optimize/develop extraction, chromatographic, and MS conditions for metabolite detection activities. Once plasma extraction/reconstitution methods are optimized, pooled plasma from NCE- and placebo-dosed subjects are analyzed. LC–MS profiles of placebo-dosed subjects are used to eliminate matrix ions, dose formulation–related ions, and other background ions so that drug-related ions can be readily identified in the NCE-dosed samples.

In a typical drug development program, the MAD study takes place several months after the conclusion of the SAD study dosing. Therefore, SAD study samples provide the opportunity to get the metabolite identification activities started earlier rather than waiting until the MAD or ^{14}C-human ADME study. For example, if any of the metabolites are human specific or pharmacologically active and separate toxicological evaluations have to be performed (Figure 5.8), complete identification, synthesis, and *in vitro* and *in vivo* testing activities can get started 1–2 years in advance and avoid any potential delays in the drug approval process. Additionally, metabolite profiling of SAD study plasma samples provides information about some of the possible co-eluting metabolites that may have to be separated in the regulated quantitative bioanalytical method so that metabolites are not interfering with the quantitative determination of the NCE and the PK calculations. Following a MAD study, Day 1 plasma and urine LC–MS metabolic profiles are compared with those from either Day 7 or Day 14 to determine if any of the metabolites have the chance to accumulate and potentially lead to toxicity. A comparison of plasma metabolites detected in humans with those from laboratory animals provides assurances that the animals used in short- and long-term safety testing are exposed to metabolites formed in humans. One point to note is that the vast majority of definitive nonclinical ADME studies are conducted following a single dose administration and short- and long-term toxicological studies are conducted following single and multiple doses. If a disproportionate metabolite is observed following multiple dose administration to humans (MAD study), if feasible (study protocol, sample integrity, etc.), plasma samples collected following multiple dose administration to nonclinical species (toxicological studies) can be used to evaluate for the disproportionate metabolite before embarking on separate experiments to evaluate the metabolites in toxicological studies.

Although there are several benefits in detecting and characterizing metabolites in human SAD and MAD study samples, there are also several analytical shortcomings and caveats that need to be taken into considerations before using the FIH metabolism data to derive decision regarding disproportionate circulating metabolites. The major challenges are (1) inability to provide assurances that none of the major metabolites are being overlooked due to difficulties in LC–MS ionization, (2) lack of metabolite quantification information, and (3) the possibility of selective extraction (during sample processing) of certain metabolites. Technological advances in MS, column, and separation techniques coupled with knowing the chemical and physical properties of a parent NCE somewhat help to overcome the limitations associated with (1) and (3) and these limitations are completely addressed during the course of ^{14}C-human ADME studies. However, limitation (2) is important to satisfy the STDM guidance and if necessary to trigger monitoring of disproportionate metabolites in the SAD and MAD studies. The widely different LC–MS response observed from many structurally related compounds limits the use of LC–MS in the full-scan detection mode (the initial mode used for FIH metabolite detection) for relative quantitative determination of drugs and/or metabolites in biological matrices. The most logical and conventional approach involves fully characterizing a metabolite, synthesizing and developing an LC–MS/MS–based assay for quantitative estimation of the metabolite. However, LC–MS/MS–based quantitative bioanalytical monitoring can become resource- and cost-intensive daunting task if an NCE is metabolized to multiple metabolites and several of them need to be quantitatively monitored. To be resource- and cost savvy and yet develop and bring safer drugs to market, several groups have evaluated UV- [28,157], NMR- [34], radioactivity- [226], and nano-ESI–based [78,227] approaches to get better estimates of the metabolites in FIH study samples, some of these techniques are discussed in Section 5.6 and a diagram (Figure 5.9) outlines some of the possible strategies that can be used with SAD and MAD studies.

5.4.5 Human ADME

A human ADME study conducted using radiolabeled form of the NCE (usually referred to as ^{14}C-human ADME) is a study that in most cases is required prior to start of the Phase III clinical trials and the data from the human ADME study are important components of the nonclinical sections of the NDA submission and drug label. The objectives of ^{14}C-human ADME studies are to determine the total disposition of the NCE, encompassing mass balance, routes of NCE/metabolite elimination, and biotransformation pathways.

The majority of the ^{14}C-human ADME studies are conducted with a small number of healthy adult subjects (often between 6–8) and if bile collection is needed, a small group of additional subjects are included [228]. Traditionally, due to ethical reasons, male subjects are selected for the ^{14}C-ADME studies. Before the start of the ^{14}C-ADME studies, study sponsors have the responsibility to determine stability of the radiolabel, purity of the radiolabel (distinguishing degradants from metabolites is very important), and conduct tissue distribution studies in nonclinical species preferably using quantitative whole-body autoradiography (QWBA) to detect radioactivity in tissues, organs, and excreta to determine the safe radioactivity dose. Nonclinical tissue distribution study data are extrapolated and used to show that radioactivity exposure of a specific tissue/organ will be well below the allowable limits to humans [229,230]. Most of the ^{14}C-human ADME studies consider a total radioactivity dose of 100 μCi or less to be safe [231].

Although regulatory guidance does not specify what is an acceptable mass balance recovery, a retrospective study evaluating mass balance achieved across various species and humans for 27 NCEs suggests that on average a mass balance of 80% or greater is achieved with most of the ^{14}C-human ADME studies [210,232,233]. The level of recovery ensures that no pathway accounting for >20% of the fractional clearance is unaccounted for in clinical pharmacology investigations. A drug's clearance pathways are used to design and prioritize drug–drug interaction, renal impairment, and hepatic impairment studies before large-scale Phase III clinical studies [210,234]. Clearance amounts over time combined with the NCE bioanalytical quantitative data can also allow the determination of slow- and extensive-metabolizers of an NCE.

Metabolite profiling and identification using samples from the ^{14}C-human ADME studies allow determination of disproportionate metabolites or confirmation of metabolites already identified in SAD and MAD studies. Additionally, extraction recoveries and chromatographic separations can be optimized as well as the quantities of each metabolite brought to light with the radiochromatographic profiles. Radiochromatographic profiles from ^{14}C-human ADME matrices are compared with those from nonclinical species to provide assurances that the nonclinical species undergoing short- and long-term toxicological studies are exposed to human metabolites. Finally, thorough identification of metabolites in plasma, urine, and feces from ^{14}C-human ADME studies are imperative to understand if the parent NCE and/or any of the metabolites have the potential to accumulate due to renal or hepatic impairment.

5.5 QUANTIFICATION OF METABOLITES: METABOLITE PROFILES

The quantification of metabolites becomes crucial during the development and clinical stages of drug discovery and development process. The widely different LC–MS response observed from many structurally related compounds, however, limits the use of LC–MS in the full-scan detection mode for quantitative determination of drugs and/or metabolites in biological matrices. LC–MS/MS–based assays involving SRM techniques are the preferred method to obtain quantitative information on an NCE and its metabolite in biological matrices. These LC–MS/MS techniques involve selection of a precursor ion, fragmentation using collision-induced dissociation (CID), and detection of a fragment ion [235–239]. To obtain quantitative information using such LC–MS/MS techniques, synthesis of reference standards is required so that LC–MS/MS responses from unknown quantities of drugs and/or metabolites can be related back to LC–MS/MS responses obtained using known quantities of drugs and/or metabolites (generating standard curves). Usually a standard curve must be generated for each and every molecule that needs to be quantified [19]. The standard curve of a drug cannot be used to quantify its metabolite(s) and a standard curve of one metabolite cannot be used to quantify another metabolite from the same drug because structural alterations from metabolism or biotransformation may lead to alterations in ionization efficiencies, extraction efficiencies (SPE, liquid–liquid, etc.), and/or interferences from endogenous matrix ions. In the absence of a reference standard, for obtaining quantitative information on a metabolite or when it is necessary to obtain information on multiple metabolites, radiolabeled (^{14}C or ^{3}H)

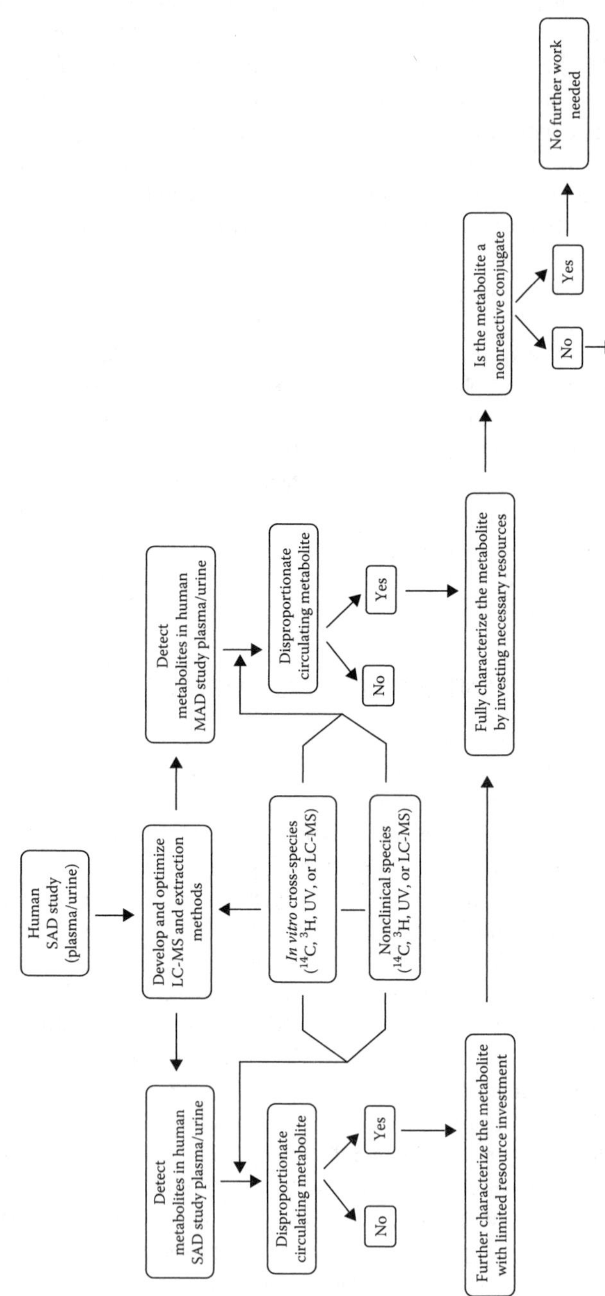
Human SAD study (plasma/urine)
Develop and optimize LC-MS and extraction methods
In vitro cross-species (^{14}C, ^{3}H, UV, or LC-MS)
Nonclinical species (^{14}C, ^{3}H, UV, or LC-MS)
Detect metabolites in human SAD study plasma/urine
Detect metabolites in human MAD study plasma/urine
Disproportionate circulating metabolite
No
Yes
Disproportionate circulating metabolite
No
Yes
Further characterize the metabolite with limited resource investment
Fully characterize the metabolite by investing necessary resources
Is the metabolite a nonreactive conjugate
No
Yes
No further work needed

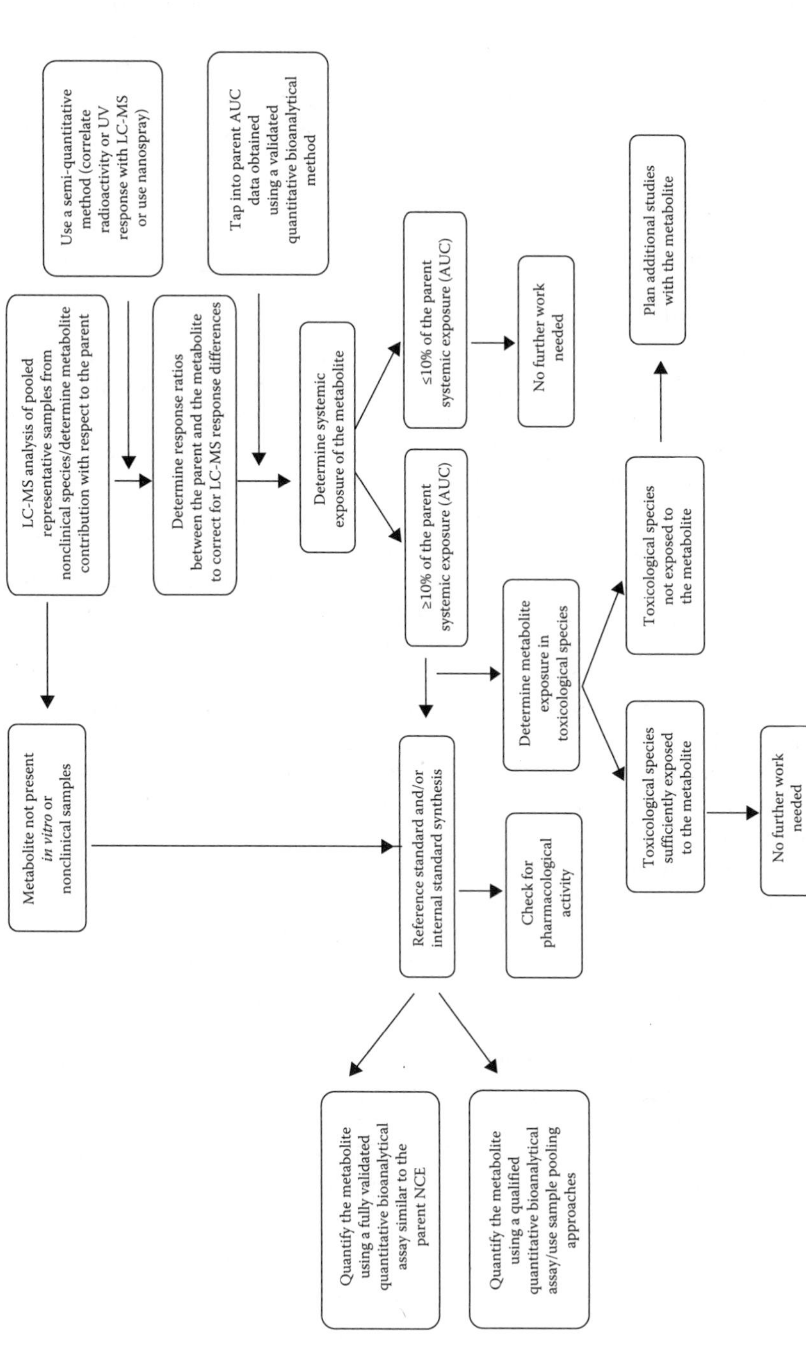

FIGURE 5.9 A proposed work flow for addressing metabolite in safety testing guidance and detecting and characterizing human metabolites in SAD and MAD studies.

drugs are administered or incubated and the samples are chromatographically separated and metabolites are quantified using one of several available RAD systems. This process is generally referred to as "metabolite profiling." RAD detection can be achieved by one of the following five approaches: (1) flow-through radioactivity detectors, (2) liquid chromatography-accurate radioisotope counting (LC-ARC), (3) microplate scintillation counting (MCS), (4) microplate imager (MI), and (5) AMS.

5.5.1 Flow-Through Radioactivity Detectors

Metabolites of drug candidates in various *in vivo* and *in vitro* systems are generally profiled and simultaneously metabolites are identified by coupling a mass spectrometer with a radioactivity flow detector and splitting the HPLC flow postcolumn [240]. Either solid (heterogeneous detection) or liquid (homogeneous detection) cells can be used with these radioactivity detectors [219,240]. For example, HPLC flow from a Waters Alliance model 2690 pump was split to deliver 80% to a Packard Model 500TR flow scintillation analyzer (FSA) and the rest to a QIT mass spectrometer for profiling and characterizing metabolites in plasma, urine, and feces following administration of ^{14}C-loratadine [232] or ^{14}C-desloratadine [233] to healthy volunteers. The same split flow method was used to confirm presence of all the human metabolites in preclinical species and to characterize additional metabolites detected in the preclinical species [109]. Most recently, Cuyckens et al. [241] demonstrated the application of high-pressure liquid chromatography online RAD with simultaneous LC–MS analysis and showed that an NCE and its metabolites can be separated with increased efficiency and enhanced sensitivity compared to that of conventional HPLC-RAD-MS.

Flow-through radioactivity detectors have a limit of detection (LOD) of approximately 300 cpm for ^{14}C when using liquid cells and even higher LOD when solid cells are employed [241]. This kind of poor sensitivity of the radioactivity flow detectors limits their application to the samples with low levels of radioactivity. For analysis of drug-related compounds with low levels of radioactivity (e.g., plasma metabolites), sensitivity can be improved by fraction collection during HPLC analysis and off-line liquid scintillation counting (LSC) of the fractions with addition of liquid scintillant [242–244]. However, this process is very time-consuming and labor intensive and also the isolated metabolites cannot be recovered. Also volatile compounds in the fractions could be lost during the concentration of fractions before LSC, which results in low recovery for a HPLC run. Recently, development of new RAD techniques offers improved methods as alternatives for detection of low radioactivity. These techniques include MCS, LC-ARC—stop flow or dynamic flow, and AMS. If necessary, MCS and LC-ARC can be coupled with a mass spectrometer for detection and characterization of metabolites.

5.5.2 Liquid Chromatography-Accurate Radioisotope Counting

LC-ARC is a novel online RAD method that is 10–20 times more sensitive than the conventional flow-through radioactivity detectors [219] and manufactured by Aim Research Company. It is a powerful technique to detect compounds with low levels of radioactivity during metabolite identification. The LC-ARC system utilizes an advanced stop-flow (StopFlow) counting technology that decreases the LOD to 5–20 dpm for ^{14}C and 10–40 dpm for ^{3}H and increases the sensitivity [219,242,245]. The successful application of LC-ARC in metabolism studies has been reported for both ^{14}C- and ^{3}H-labeled compounds [219,242,245]. In these studies LC-ARC was coupled with a mass spectrometer that allowed acquisition of mass spectroscopic data online. The DynamicFlow™ radioisotope detector is another instrument developed by Aim Research Co. for HPLC and ultra-high-pressure liquid chromatography (UPLC). DynamicFlow is a novel method that was developed based on the StopFlow technique. In the DynamicFlow method there is no need to stop the HPLC flow like in StopFlow to increase the sensitivity. Detection of low levels of radioactivity, down to 15 dpm, is possible with a DynamicFlow detector without stopping the HPLC flow [242,244]. A slow down of the HPLC flow rate can increase the sensitivity further when needed [243]. Like StopFlow radioisotope detector, the DynamicFlow radioisotope detector can also be interfaced with an MS for online metabolite identification.

5.5.3 Microplate Scintillation Counting

Since its introduction in 2000, MSC has quickly become a method of choice for profiling of low-level radioactive metabolites in many metabolism studies [246]. In this method, fractions are collected into 96-well microplates during HPLC analysis and the radioactivity in the collected fractions is determined by off-line scintillation counting. The sensitivity is increased by increasing the counting time to 2–5 min per well from 2 to 12 s resident time used in the radioactivity flow-through detectors. In addition to increased sensitivity other advantages of this method are that it generates less radioactive waste and it is less time-consuming and labor intensive compared to conventional LSC. Currently, there are two types of MSC instruments, TopCount and MicroBeta, that are both manufactured by Perkin Elmer and use two kinds of plates, Lumaplates® and Scintiplates®, respectively. Lumaplate is a microplate coated with a solid scintillator (yttrium silicate) at the bottom, in which addition of LSC cocktail prior to counting is not needed, and used with TopCount instrument [247–252]. Scintiplate is a 96-well polystyrene-based microplate, which scintillates on exposure to beta radiation and does not need the addition of scintillation cocktail. Scintiplate is used with MicroBeta scintillation counter. Zhu et al. [248] demonstrated that TopCount was 50–100-fold more sensitive than online radioactivity flow detector and approximately twofold more sensitive than LSC. The data showed that the HPLC-MSC technique is especially useful for detection and quantification of low levels of radioactivity in biological matrices.

Although both the Scintiplate in combination with the MSC or the Lumaplate with the TopCount offers improved radiochemical detection limits and counting efficiency over conventional online radiochemical techniques, the throughput is diminished compared to conventional online radio flow-through techniques because the off-line approaches involves drying of multiple 96-well plates (necessary for 45–60 min metabolite profiling experiments with 10–20 s per well) and then off-line counting. The 96-well plate approaches, however, are significantly streamlined process over the conventional fraction collection into scintillation vials and LSC.

5.5.4 Microplate Imager

Most recently, a MI technique was evaluated as an alternative to MSC systems. MSC systems, described above, use several photomultiplier tubes (PMTs) in conjunction with solid scintillant to detect β-particles–related light. The MI is an ultra-high-throughput technique capable of detecting all forms of fluorescence, absorbance, and luminescence including photons emitted from radioactive material. The MI detection technique involves a back-illuminated charged-coupled device (CCD) camera operating at −100°C coupled to an optimized telecentric optical lens. Unlike the PMT-based detection technique used in the MSC, which is capable of measuring radioactivity in one well at a time, the MI technique is capable of imaging an entire 96-, 384-, or 1536-well plate for the radioactivity without addition of liquid scintillation cocktail. An important difference between the PMT detection (MSC, all the flow-through radioactivity detectors, and liquid scintillation detectors) and the detection technique used in the MI is that the PMTs measure photons while the latter does not. Therefore, the data from PMT-based detectors are expressed as CPM and the MI data are expressed using relative light unit termed "analog-to-digital units" (ADU).

Dear et al. [253] evaluated MI for detecting and quantifying NCEs and their metabolites following UPLC separation and fraction collection into 96- and 384-well Lumaplates. Metabolite profiles obtained using MI technique was compared with those from MSC and the results demonstrated similar sensitivity and robustness between two techniques. The LOD and the limit of quantification for the MI technique were established at 3 and 7 dpm, respectively. A comparison of reconstructed radiochromatograms of a rat bile sample from a ^{14}C-study, following UPLC separation, fraction collection into 96-well lumaplates, and off-line counting for 40 and 5 min per plate using respectively MSC (TopCount) and MI (ViewLux) is shown in Figure 5.10. The reconstructed radiochromatograms show nine prominent metabolites (M1–M9) and unchanged parent drug (P). The metabolite profiles from both detectors are quantitatively and qualitatively similar. The advantages of using MI technique include higher-throughput (5 vs 40 min) and reduced plate cost (Figure 5.11). Although the MI technique is capable of imaging a 1536-well plate, compatible fraction collection techniques are not yet available to leverage these advantages to further improve the throughput.

5.5.5 Accelerator Mass Spectrometer

AMS is an extremely sensitive technique that offers 10^3–10^9-fold increases in sensitivity over LSC methods. AMS was originally used for radiocarbon dating of historical artifacts [254]. AMS measures individual atoms of rare isotopes (^{3}H, ^{14}C, ^{26}Al, ^{36}Cl, ^{41}Ca, and many isotopes of heavy elements), rather than relying on the detection of radioactive decay events. AMS separates elemental isotopes in a sample through differences in mass, charge, and energy. This technique has been mostly utilized for the detection of ^{14}C isotopes. AMS measurements can be up to 1 million times more sensitive than decay counting methods for ^{14}C [255,256]. AMS is used

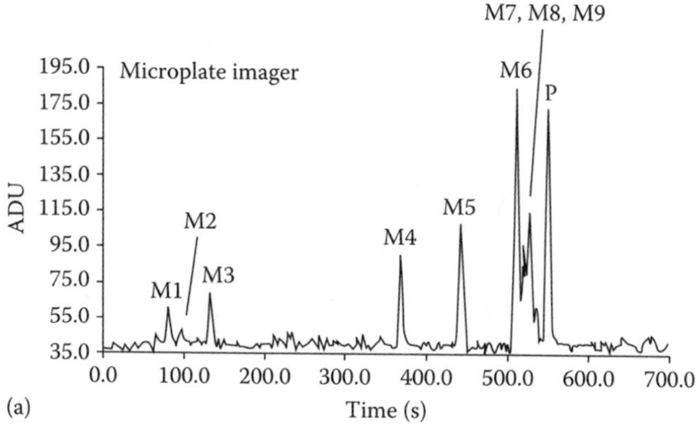

(a)

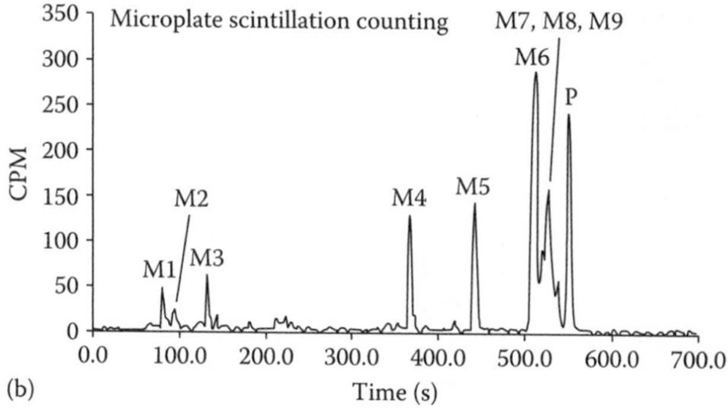

(b)

FIGURE 5.10 Comparison of reconstructed radiochromatograms obtained using a ViewLux MI (a) and TopCount MSC (b). Fraction collected at a speed of 2 s/well, into shallow-96-well Lumaplates. (Reprinted from Dear, G.J. et al., *J. Chromatogr. B: Anal. Technol. Biomed. Life Sci.*, 868, 49, 2008. With permission.)

to measure the isotope ratio (^{12}C:^{13}C:^{14}C) in graphitized samples. A cesium ion beam is directed at the graphite cathodes where negative carbon ions are produced. The ions are extracted from the ion source and directed via a magnet into an accelerator where they gain very high energies. In the accelerator, the negative ions collide with an inert gas, which has an electron stripping effect. The carbon ions, now stripped of electrons, gain a positive charge. The high-energy positive carbon ions are focused and filtered by a magnet, a quadrupole, and an electrostatic analyzer. Relatively abundant ^{12}C and ^{13}C ions are quantified in Faraday cups while rare ^{14}C ions are quantified in a gas ionization detector.

A number of recent studies have demonstrated the potential utility of AMS in pharmaceutical research [257] that includes mass balance and PK studies [258,259], microdosing [260], and covalent binding of ^{14}C-labeled active metabolites to DNA and to protein *in vivo* [261]. The ultrasensitivity of AMS technique allows the use of low chemical and/or radioisotope doses (microdose) in humans. Due to the small amount of radioactivity given to humans, administration would result in negligible radioactivity exposure,

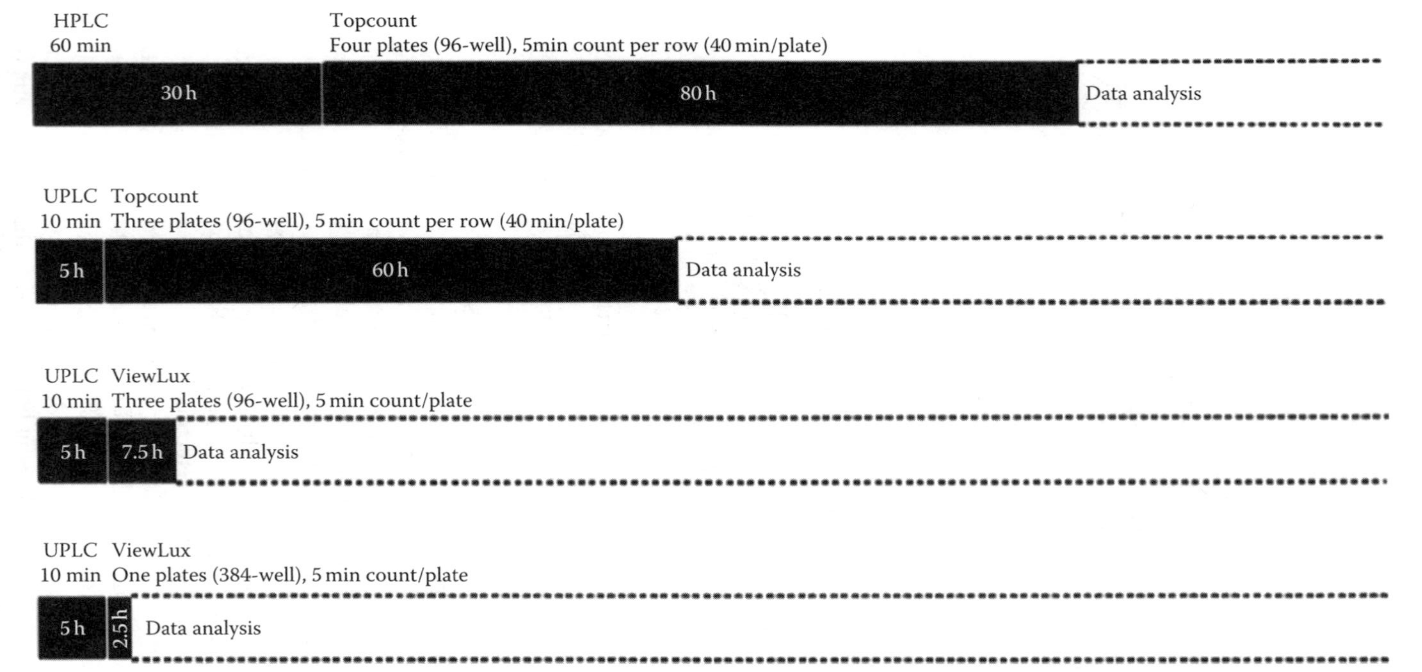

FIGURE 5.11 Metabolite profiling throughput improvements associated with transitioning from HPLC-TopCount to UPLC-ViewLux. (Reprinted from Dear, G.J. et al., *J. Chromatogr. B: Anal. Technol. Biomed. Life Sci.*, 868, 49, 2008. With permission.)

and the studies can be performed more safely in humans. Another advantage of a low radioactive dose is that radioactive waste generation is minimal. However, the method does require a contamination-free laboratory and is a destructive technique that does not allow generation of structural elucidation data. For more on AMS, see Chapter 14.

5.6 QUANTIFYING METABOLITES IN THE ABSENCE OF REFERENCE STANDARDS OR RADIOLABEL

The recent publication of the STDM guidance by the US FDA has clearly outlined the importance of detecting and characterizing human metabolites as early as possible during the course of discovery and development of an NCE. The STDM guidance also recommends detecting and characterizing human metabolites under the steady state conditions or after multiple dosing, which in turn makes the analysis of human MAD study samples for metabolites ever more important [24].

During the discovery and development of a drug, it is preferred that the quantitative bioanalytical method involves only the measurement of the parent NCE. If metabolites of an NCE need to be quantified as per the STDM guidance, the appropriate course of action involves complete identification of the metabolites in question, synthesis of the metabolite standards, obtaining certified standards, generating calibration curves for each metabolite and monitoring each metabolite using LC–MS/MS methods. To avoid developing a complex quantitative assay that is capable of monitoring all the metabolites and to be cost and resource efficient, it is prudent to obtain quantitative information during metabolite profiling studies. Although mass spectrometry–based techniques are eminently suited for detecting and characterizing metabolites, quantifying the metabolites in the absence of reference standards or radiolabel NCE administration is not straightforward. Difficulties in quantification stem from unpredicted LC–MS response differences between an NCE and its metabolites, nonuniform extraction efficiencies, highly variable interferences from the endogenous matrix and ionization efficiency differences.

To overcome some of the limitations mentioned above, several approaches have been developed (Figure 5.9). Among them, one option is to correlate the radioactive response peak area of the metabolite and the parent to the corresponding LC–MS or LC–MS/MS peak area responses [226,227]. Since the radioactivity response is nondiscriminating with respect to molecular weight and chemical structure, and is relatively immune to most matrix effects, response ratios between a parent NCE and its metabolite or between two metabolites would provide reliable quantitative estimates. LC–MS peak area response differences then can be corrected to match the radioactivity response ratios. Once a response correction ratio is established for a particular NCE and a metabolite present in a particular matrix, then the correction factor can be used for samples from nonradiolabeled studies such as the SAD and MAD studies or other discovery metabolism studies. Similar to radioactivity, UV and/or DAD also have been used to derive nondiscriminating responses between an NCE and its metabolites for the purpose of obtaining quantitative estimates [157,262]. Another approach is to use the ^{1}H-NMR to derive a response correction factor between an NCE and its biologically and nonbiologically generated metabolite, and this approach was recently demonstrated by Espina et al. [34].

Another novel approach that could be used to quantify metabolites in the absence of reference standards or radiolabel was described by Ramanathan et al. [78]. In this approach, recently introduced NSI technique and two HPLC systems were coupled with a Q-TOF mass spectrometer to obtain normalized LC–MS response of drugs and their metabolites. As shown in Figure 5.12, one HPLC system performs the separation of a drug and its metabolites, while the other system adds solvent postcolumn with an exact reverse of the mobile phase composition so that the final composition entering the NSI source is isocratic throughout the entire HPLC run.

To validate the response normalizing technique, data were obtained from four different structural classes of compounds (VCV, desloratadine [DL], tolbutamide, and cocaine) and their metabolites. The data demonstrated that by maintaining the solvent composition unchanged across the HPLC run, the influence of the solvent environment on the ionization efficiency is minimized. In comparison to responses obtained from radiochromatograms, conventional LC-ESI-MS overestimated the responses for the parent NCE between 6- and 20-fold. The response normalization modification resulted in nearly uniform LC-NSI-MS response for all compounds evaluated. During the course of drug development, if the AUC for a metabolite has to be determined to satisfy the STDM guidance, then a pooled plasma sample can be analyzed with and without response normalization to estimate the response correction factor as well as the amount of the metabolite. Once the response correction factor is known then the AUC of the parent NCE obtained using a validated quantitative bioanalytical technique can be used to roughly estimate the AUC of the metabolite.

5.7 SPECIAL STRATEGIES USED FOR METABOLITE DETECTION/CHARACTERIZATION

Once a metabolite is detected in a biological matrix and its molecular weight is deciphered using LC–MS techniques, the next steps involve structural characterization using LC–MS/MS, MS^n, and accurate mass measurement strategies. Most often MS/MS and MS^n experiments help to narrow down the metabolic modification region and distinguishing between the sites modification, such as whether an oxidation is occurring on a carbon or nitrogen, may require isolation of the metabolite followed by NMR. Furthermore, elemental compositions determined using accurate mass measurements can only allow one to decide whether the observed *m/z* is consistent with the proposed structure or not. However, some of the LC–MS and LC–MS/MS–based special strategies described below allows one to refine the proposed site of modifications and assist in identification of the metabolites. These strategies are potentially useful before investing time and resources for NMR characterization and synthesis of the metabolites.

5.7.1 Stable Isotope Labeling

Stable isotope labeling in conjunction with MS has been widely used for the identification of drug metabolites since late 1970s. In addition, stable isotope–labeled drugs

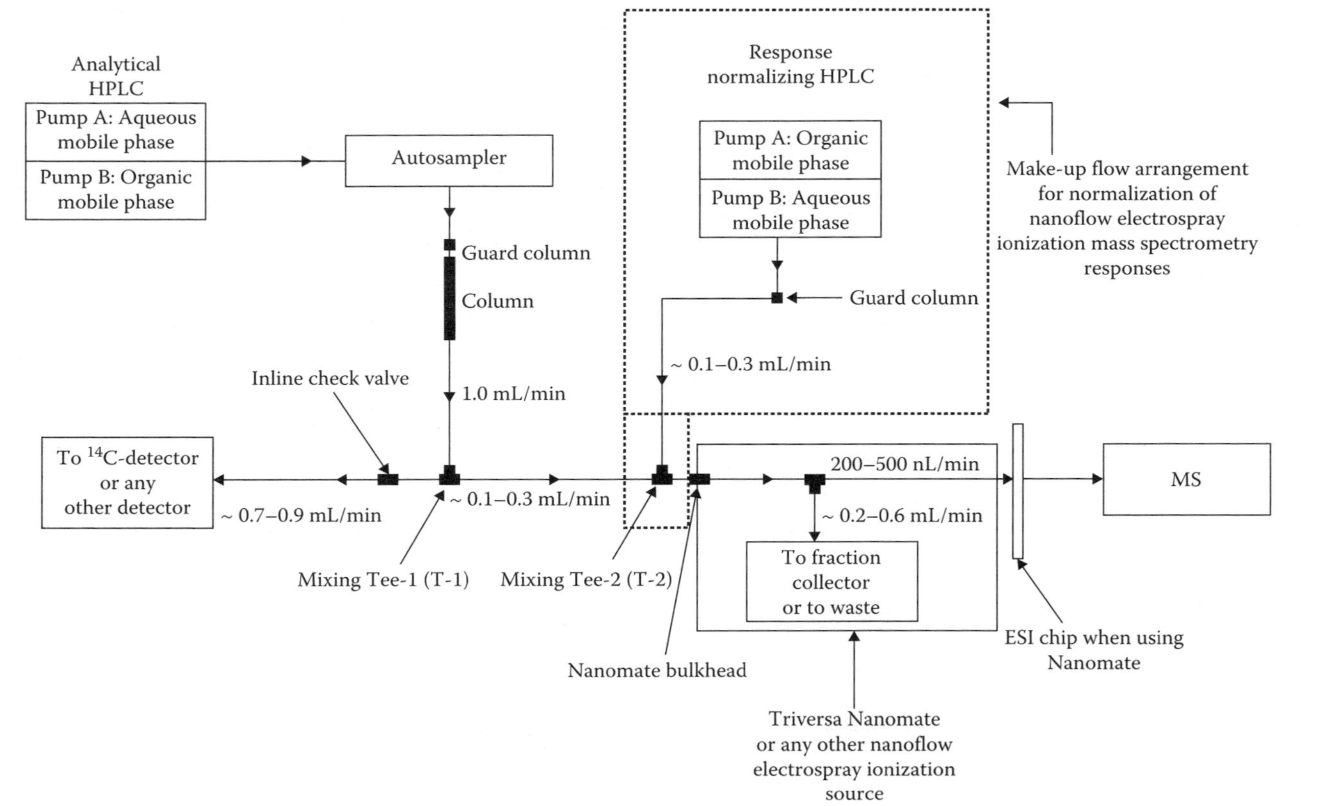

FIGURE 5.12 Response normalized liquid chromatography–nanospray ionization mass spectrometry set up used for quantifying metabolites in the absence of reference standards. (Reprinted from Ramanathan, R. et al., *J. Am. Soc. Mass Spectrom.*, 18, 1891, 2007. With permission.)

have been employed as internal standards for quantitative LC–MS work, in studies of absorption, bioavailability, distribution, biotransformation pathway determination, excretion, drug interactions, and pharmacological changes during pregnancy, etc. [263]. The stable isotopes can be separated by MS, such as ^{2}H from ^{1}H, ^{13}C from ^{12}C, ^{15}N from ^{14}N, ^{18}O from ^{16}O, ^{34}S from ^{32}S, and ^{81}Br from ^{79}Br. The isotopic clusters generated from the mixture of natural and synthetically introduced isotopes can effectively be used to track and identify the drug-related compounds in the complex biological matrices. In order to identify potential metabolites some computer programs have been developed, in which each scan of a chromatographic run is searched to find out the characteristic isotope cluster of the drug-related compounds. Use of stable isotope–labeled drugs can be a good alternative to the use of radiolabeled drugs in metabolism studies [264–266].

Recently, stable isotope–labeled compounds in combination with mass defect filter (MDF) technique and high-resolution accurate mass spectrometry were utilized to detect glutathione (GSH) conjugates generated in HLM with the use of MetWorks software [267]. In that study, high-resolution accurate mass analysis with isotope pattern triggered data-dependent product ion scans was employed for simultaneous detection and identification of GSH adducts and reactive intermediates within a single analysis using a LTQ-Orbitrap. In this method, a 2:1 ratio of unlabeled to stable isotope–labeled GSH was used as a nucleophilic chemical tag to trap reactive intermediate species. First, GSH conjugates were generated from incubating the test compounds (acetaminophen, tienilic acid, clozapine, ticlopidine, mifepristone) with HLM in the presence of NADPH-regenerating system and a 2:1 ratio of glutathione and ($^{13}C_2{}^{15}N$-Gly)-glutathione. The isotopic mass difference between the unlabeled and stable isotope–labeled GSH adducts was 3 Da. The isotopic doublets of chemically tagged GSH conjugates of five compounds were detected by isotope search of mass defect filtered and control subtracted full-scan MS data using MetWorks software, which utilized accurate mass difference of the chemical tags and a ratio corresponding to the conjugate composition of test compound + chemical tag −2H. Then, the reactive intermediate structures were elucidated using chemical formulae of both the protonated molecules and their product ions generated from high-resolution accurate mass measurements with data-dependent scans obtained from a single analysis. Some recent examples of the use of stable isotopes in metabolite identification have been presented by Martin et al. [268], Baillie et al. [269], Chowdhury et al. [265], Cuyckens et al. [270,271], and Mutlib [266].

Chlorine and bromine have a unique natural isotopic pattern that can easily be used to detect the drug-related compounds in complex biological mixtures. Therefore, drugs containing Cl or Br may not be needed to be stable labeled [109,272]. Furthermore the isotopic peaks can be individually subjected to tandem mass spectrometry experiments to decipher an origin of a certain fragment and/or site of modification. For example, separate tandem mass spectrometry experiments performed using the molecular ion peaks containing ^{35}Cl and ^{37}Cl allowed structural characterization of some of the metabolites following administration of loratadine to nonclinical species [109].

5.7.2 Hydrogen–Deuterium Exchange

HDX experiments in combination with LC–MS analysis can be used to generate valuable information for drug metabolites with the analysis of the MS data before and after this chemical reaction. Labile hydrogens present in some of the functional groups of an NCE and its metabolites, such as –OH, –SH, –NRH, $-NH_2$, $-CO_2H$, can exchange with deuterium when exposed to deuterated solvents or reagents [272]. In general, biotransformation of drugs produces metabolites with polar functional groups such as hydroxyls or blocks polar functionalities, which results in changes in the number of exchangeable hydrogens. With the determination of number of exchangeable hydrogens additional information is generated to facilitate the structural elucidation of the drug metabolite. LC–MS/MS analysis following online HDX using deuterated HPLC solvents such as D_2O has been widely applied to the structural elucidation of drug metabolites [272–275].

A recent example of application of LC–MS/MS following online HDX and stable isotope–labeling techniques was presented by Ni et al. [276] for characterization of metabolites of brimonidine, a highly selective α_2-anrenergic receptor agonist. A mixture of brimonidine and D_4-brimonidine was incubated in rat and HLM. In addition, the mixture was orally administered to rats. Then, *in vitro* (hepatic microsomal) and *in vivo* (rat urinary) metabolites of brimonidine were characterized by LC–MS/MS following HDX using deuterated mobile phase. Brimonidine has two exchangeable hydrogens attached to the nitrogens of the imidozoline ring. The hydroxyquinoxaline metabolite of brimonidine and dehydro-hydroxybrimonidine had three exchangeable hydrogens. The additional labile hydrogen detected in those metabolites indicated the addition of oxygen to carbon of the quinozaline ring, rather than nitrogen. Ni et al. [276] also discussed the effectiveness of stable isotope–labeling technique in structure elucidation of drug metabolites. In conclusion, LC–MS/MS following HDX and stable-isotope–labeling techniques generated additional information useful in characterization of metabolites of brimonidine.

Most recently, the utility of online HDX-based LC–MS for distinguishing isomeric N-oxides and hydroxylamines from oxidation at carbon was discussed [105]. Most often additional information from LC–MS-based HDX such as whether an oxidation is taking place on nitrogen or carbon can be very helpful for further characterization of a metabolite by NMR or narrow down the synthetic routes and save resource and time during the drug discovery and development process [104].

5.7.3 Chemical Derivatization

The chemical derivatization technique in combination with LC–MS/MS has also been used as an effective tool to facilitate the identification of metabolites, especially for unusual or unstable metabolites [277–279]. The MS detection sensitivity for poorly ionizable compounds can be increased by introduction of an easily ionizable functional group. Derivatization can reduce the polarity of very polar small molecular weight metabolites that elute very early together with many endogenous materials in the HPLC columns. As a result, the derivatized metabolite can be chromatographically separated from the endogenous material in the sample. Balani et al. [277] derivatized highly polar and extremely unstable human urinary metabolites of caspofungin (a potent antifungal agent) with ethyl chloroformate or acetic anhydride. With the increase in molecular weights of these metabolites and the improvement in their chromatographic retention, these metabolites were identified by LC–MS/MS.

Chemical derivatization has also been used in structural elucidation of conjugated metabolites [278,280,281]. In a recent study, Salomonsson et al. [278] successfully applied chemical derivatization of isomeric drug conjugates with 1,2-dimethylimidazole-4-sulfonyl chloride (DMISC) to determine the site of conjugation. Morphine has one phenol group and one aliphatic hydroxyl group where glucuronidation could take place. Two glucuronide conjugates of morphine were indistinguishable by MS analysis. DMISC reacted selectively with one of the metabolites that contain a free phenol. As a result it was possible to differentiate glucuronide conjugates by treatment with DMISC followed by MS analysis.

Derivatization is also useful to detect volatile metabolites. Liu et al. [282] described a specific, rapid, and sensitive *in situ* derivatization solid-phase microextraction (SPME) method for determination of volatile trichloroethylene (TCE) metabolites, trichloroacetic acid (TCA), dichloroacetic acid (DCA), and trichloroethanol (TCOH), in rat blood. The metabolites were derivatized to their ethyl esters with acidic ethanol, extracted by SPME and then analyzed by gas chromatography/negative chemical ionization mass spectrometry (GC-NCI-MS). After validation, the method was successfully applied to investigate the toxicokinetic behavior of TCE metabolites following an oral dose of TCE. Some of the common derivatization reagents include acetyl chloride and *N*-methyl-*N*-(*t*-butyldimethylsilyl)trofluoroacetamine (MTBSTFA) for phenols and aliphatic alcohols and amines, dansyl chloride and diazomethane for phenols, dansyl chloride for amines, acidic ethanol and diazomethane for carboxylic acids, and hydrazine for aldehydes.

5.7.4 Online Electrochemical Techniques

The potential of electrochemical detection (ECD) techniques for metabolite generation, detection, and quantification [283] should not be underestimated because the technique is especially useful for reactive metabolites [284,285]. ECD can be used as standalone or as a parallel or sequential method with LC–MS systems. Online ECD techniques and their utility in metabolite identification have been recently reviewed by Gamache et al. [286]. Most of the documented applications involve the use of the electrochemical cell for oxidizing an NCE to reactive metabolites and trapping and detecting them by online LC–MS and LC–MS/MS techniques.

5.8 APPLICATIONS

Applications of the mass analyzer types for discovery and development stage metabolite detection, characterization, and identification along with each analyzer type's limitations and advantages are discussed in the following section.

5.8.1 Triple Quadrupole Applications

Triple quadrupole mass spectrometers can perform tandem mass scan experiments in various modes including product ion (MS/MS), precursor ion, and neutral loss scan and SRM experiments, but they cannot be used for sequential MS^n experiments. The high sensitivity and specificity, in the SRM mode, have made triple quadrupole mass spectrometers a logical choice for metabolic stability experiments performed at relevant substrate concentrations. Despite the sensitivity of the triple quadrupole mass spectrometers, when an NCE or an NCE series exhibit unacceptable PK properties, metabolite identification studies are often initiated as follow-up studies in a separate set of experiments using incubation concentrations higher than the K_m of an NCE. Incubations at higher concentrations are required because conventional metabolite identification experiments required operation of the triple quadrupole mass spectrometer in the full-scan mode, which results in poor duty cycle and diminished sensitivity [287,288].

Poon et al. [288] and Tiller and Romanyshyn [287] demonstrated the application of integrated quantitative/qualitative approaches (within the same injection cycle), based on the combination of a fast gradient HPLC method, a short column, and a triple quadrupole mass spectrometer, for shortening the lead optimization process. A schematic of the integrated approach developed by these researchers is shown Figure 5.13. The integrated triple quadrupole approach was limited to targeting

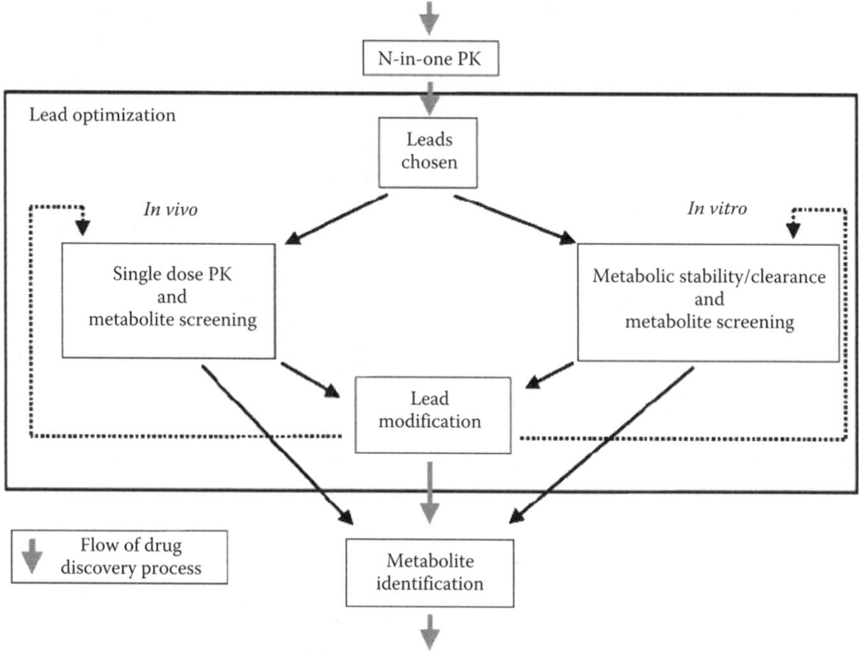

FIGURE 5.13 An integrated quantitative/qualitative metabolite screening approach. (Reprinted from Poon, G.K. et al., *Rapid Commun. Mass Spectrom.*, 13, 1943, 1999. With permission.)

some of the common expected metabolites such as (M+16), (M+32), and (M−14) and (M+16+176), (M+32+176), and (M−14+176) following incubations with microsomes and hepatocytes, respectively [287,288]. With conventional triple quadrupole mass spectrometers, each SRM transition can take anywhere from 50 to 300 ms, which in turn limits the number of SRM transitions that can be performed over a typical 10–30 s wide HPLC peak. The metabolite searching process was based on the NCE's (parent drug's) fragmentation offset by +16, +32, +176, or −14 for metabolites rather than being based on the MS/MS product ions of the metabolites. Even though the conventional triple quadrupole–based integrated quantitative and qualitative approaches have somewhat shortened the lead optimization timelines, the approach does not provide complete information, such as unique metabolites, often necessary for reliable decision-making. Recent MS developments such as the API 5500 and API 5000, which are capable of scanning fast without sacrificing resolution have reopened the door for utilizing conventional triple quadrupole mass spectrometers for obtaining metabolite information while performing quantification of drugs.

5.8.2 Time-of-Flight Applications

Historically, TOF mass spectrometers have had limited utility as a quantitation instruments due to their narrow dynamic range. However, over the last decade, TOF mass spectrometers have also been evaluated for quantification assays to exploit their efficient duty cycle and higher resolution capabilities over conventional triple quadrupole mass spectrometers. In contrast to QMFs, which have to filter or pass ions sequentially to the detector, TOF mass spectrometers are capable of detecting the entire mass range simultaneously. Simultaneous detection of a wide mass range leads to a full-scan duty cycle in the range of 10%–50% and results in an improved sensitivity. Whereas in conventional triple quadrupole mass spectrometers sequential scanning results in a duty cycle of less than 1% in the full-scan mode. When considering an optimum duty cycle, SIM and/or SRM are the most efficient scan types with QMF-based mass spectrometers, and full-scans are ideal for TOF mass spectrometers. Therefore, when using TOF mass spectrometers for quantitation of an NCE, information (*m/z*, elemental composition, etc.) about co-eluting metabolites, degradants, and impurities is available at no expense [289].

Nagele and Fandino [290] used an Agilent rapid resolution LC system in combination with an ESI-TOF mass spectrometer to detect and characterize metabolites while performing metabolic stability experiments. The integrated quantitative and qualitative method was tested using buspirone as a model compound. To maintain the throughput of the quantitative assay, a 2-position/10-port column switching valve was used. A short 50 mm sub-2 μm column, under low chromatographic resolution mode, was used to determine the metabolic stability of buspirone and for detecting major metabolites. A 150 mm sub-2 μm column was used in the high chromatographic resolution mode for the separation of several lower-level metabolites from major metabolites for detection by TOF mass spectrometer. Using the integrated quantitative and qualitative method, empirical formulas of the metabolites were confirmed with single digit parts-per-million errors.

Pelander et al. [291] used LC-TOF-MS in conjunction with metabolite prediction (Lhasa Meteor), metabolite detection (Bruker MetaboliteDetect), and fragmentation prediction (ACD/MS Fragmentar) programs to streamline detection of metabolites of the antipsychotic drug quetiapine in human urine. The nominal resolution of the TOF mass spectrometer was set at 10,000 and fragmentation for structural elucidation was achieved by ESI source CID. The integrated work flow included the following steps: (1) prediction of metabolites using Meteor software; (2) acquire accurate mass data using LC-TOF; (3) detect predicted metabolites using MetaboliteDetect software; (4) predict fragmentation using ACD/MS Fragmentar; (5) fragment the metabolites using ESI source CID and measure fragment ions *m/z* using accurate mass; (6) assign molecular formulas of the fragments; and (7) add metabolite exact mass and retention time to a database. Out of the 14 metabolites predicted by the Meteor software, the metabolite detection software detected 8 in the LC-TOF mass spectrum of the urine. In addition, five hydroxyl metabolites, not predicted by Meteor, were also detected in the mass spectra. Metabolites were confirmed with mass accuracies in the range of 2.4 ppm and by matching isotopic patterns generated by SigmaFit program. Isotopic pattern was used in addition to source CID fragmentation patterns, accurate mass data, and retention time information to confirm the metabolites.

To take advantage of the higher mass resolution and efficient duty cycle capabilities in the MS/MS mode, TOF mass analyzers have been coupled with QMFs as well as quadrupole ion traps. When coupling a TOF mass analyzer with a triple quadrupole spectrometer, the TOF mass analyzer is used in place of the Q3 mass analyzer. These hybrid systems are broadly referred to as Q-TOF mass spectrometers. In a typical Q-TOF quantification experiment, the precursor or the parent ion is isolated in the first quadrupole (Q1) and fragmented in the collision cell (q2). All the fragments are transferred to the pusher (accelerator) region and then detected after traversing through the reflectron. During SRM quantification experiments in the conventional triple quadrupole mass spectrometers, a single fragment ion is allowed to pass through Q3 to the detector. In Q-TOF quantification experiments, full-scan MS/MS spectra are available at each chromatographic time point. Therefore, when using a Q-TOF mass spectrometer for quantification of an NCE, information about metabolites become available at no expense. To date, Q-TOF mass spectrometer improvements associated with the dynamic range, signal-to-noise ratio, and resolution [287] coupled with the explosive growth of the ultra-high pressure chromatography (UPLC) have made the Q-TOF an ideal mass spectrometer for integrated quantitative and qualitative bioanalytical studies.

Tiller et al. [289] reported the use of accurate LC–MS/MS data from a Q-TOF system as a first-line approach for metabolite identification studies in the discovery phase of the drug development. In this approach, first, the MS/MS spectrum of the parent compound under conditions of high mass resolution is obtained in order to support the proposed fragmentation pattern by elemental composition data. In the next step of the accurate mass approach to metabolite identification, the *in vitro* samples of drug candidates from incubations in liver microsomal preparations or hepatocytes are analyzed by LC–MS^E on the Q-TOF mass spectrometer. Data are processed automatically and the report generated is used to assign structures to various metabolites detected with the use of metabolite identification software.

Kammerer et al. [292] used the fragmentation patterns of available standards of ML3403 (p38 MAP kinase inhibitor) elucidated by accurate mass measurements of precursor and fragment ions using ESI-Qq-TOF mass spectrometer to identify unknown metabolites. Tong et al. [293] showed that the high sensitivity of a QqTOF mass spectrometer in full-scan mode, in combination with the formation of alkali metal adduct ions in the mass spectra at lower ion source temperature, facilitated the identification of the true molecular ion of three major unstable metabolites of Lonafarnib (chemotherapeutic agent; selective inhibitor of farnesyl protein transferase) that were produced from incubation in human cDNA-expressed recombinant CYP3A4 enzymes. In their study, accurate mass measurement, stable isotope (^{2}H) incorporation, MS/MS, and NMR spectroscopic studies were used for the complete structural elucidation of the metabolites. In another example, ultra-high-performance liquid chromatography-TOF mass spectrometry was used to identify metabolites of an anticancer agent ARQ501 (3,4-dihydro-2,2-dimethyl-2*H*-naphthol(1,2-*b*)pyran-5,6-dione) generated in mouse, rat, dog, monkey, and human blood [294]. Some other applications of Q-TOF mass spectrometry to metabolism studies can be found in various recent publications [290,295–299].

Hybrid mass spectrometers resulting from the combination of either QIT or LIT mass analyzers and a TOF mass analyzer are broadly referred to as IT-TOF. An ion trap and a TOF mass analyzer are linked via a quadrupole collision cell (q2) to construct an IT-TOF mass spectrometer. IT-TOF introduced during the early 1990s achieved mass resolution of less than 10,000 because of the utilization of the ion trap as part of the TOF accelerator (axial trap-TOF) and the requirement of 1 mtorr range collision gas for focusing (to minimize kinetic energy spread) the ions exiting the ion trap. Popularity of the IT-TOF increased following the introduction of the orthogonal trap-TOF, in which the ions exiting the IT are introduced into the TOF accelerator and the ions are subsequently reaccelerated in the direction orthogonal to the IT ion exit path [300,301].

Although orthogonal-trap-TOF (oA-IT-TOF) mass spectrometers alleviated some of the limitations of the axial-trap-TOFs; oA-IT-TOF was limited in utility because of the mismatch between the exit (from the IT) and the arrival (into the accelerator region) times of the ions. In other words, the high-mass ions, which are reaching the accelerator region at a later time and the low-mass ions, which have already passed the accelerator region are missed during each specific acceleration event. This ion exit/arrival mismatch imposed a limit on the m/z range of the ions that can be accelerated and detected during each acceleration event [302]. Although several approaches were evaluated to alleviate the ion exit/arrival mismatch, modification of the oA-IT-TOF by addition of a collisional damping chamber [303] resulted in high mass resolution and a high mass accuracy over wide m/z range. The advantage of using an IT-TOF over a Q-TOF mass spectrometer is the ability to perform MS^n with high mass accuracy. Therefore, the utility of the IT-TOF mass spectrometers are in the area of drug and metabolite structural elucidation [304,305]. Fandino et al. [296] used an IT-TOF mass spectrometer for detection of several minor unexpected metabolites of buspirone that co-eluted with major expected metabolites present in

S9 fractions. Unexpected metabolites were detected using automated MS/MS fragmentation assignment software, data-dependent methods, and accurate mass measurements in the TOF mass spectrometer.

5.8.3 Quadrupole Ion Trap (3D) Applications

During the early 2000s, QIT were the mass spectrometers of choice for performing quantification of drugs and simultaneous identification of their metabolites. QITs were preferred for this task due to superior sensitivity, speed (duty cycle), ease of use, compact footprint, and cost [306–308]. Kantharaj et al. [63,309,310] and Cai et al. [160,161] demonstrated the use of QIT mass spectrometer for simultaneously acquiring drug stability and metabolite structural information. To establish the method, Kantharaj et al. [63,309,310] separately incubated four model compounds (verapamil, propranolol, cisapride, and flunarazine) with HLM for 15 min and determined the metabolic stability by comparing the peak area of the parent compound at time 0 with that obtained after 15 min of incubation. Since the mass spectrometer was operated in the full-scan mode (m/z 100–1000) rather than SRM or SIM mode, information about all drug-derived materials (metabolites, degradants, etc.) as well as additives and dosing components present in each sample was preserved. Full-scan data, in turn, allowed the authors to extract information about the most common Phase I metabolites associated with each test compound.

Superior full-scan duty cycle of the ion trap allowed Kantharaj et al. [63,309,310] to acquire "on-the-fly" MS/MS data for all the ions formed over a predetermined threshold (data-dependent MS/MS scan). For discovery compounds, which showed more than 20% metabolic turnover after 15 min of incubation, metabolite profiling evaluation was performed by interpreting the data-dependent MS/MS scans. Interpretation of the MS/MS scans involved comparison of the parent ion fragments with that of the metabolites. Once the structural elucidation was complete, data were discussed with the discovery teams and/or medicinal chemists to modify the structure for metabolic stability.

Typically, when a conventional triple quadrupole is operated in the full-scan mode (Q1 or Q3 scan), the sensitivity drops dramatically in comparison to SRM mode of operation. In contrast, QIT provides high sensitivity in the full-scan mode of operation (Table 5.4). Upon testing over 170 polar discovery compounds with molecular weights ranging from 300 to 650 Da, Cai et al. [160,161] found that the QIT had over 10-fold higher sensitivity than a conventional triple quadrupole (Perkin Elmer Sciex API 300) in full-scan analysis over a mass range of 100–800 Da. For α-1a antagonist compounds tested, Kantharaj et al. [310] achieved LODs in the range of 20–100 pg on the QIT, while QMF, for the same set of compounds, provided LODs in the range of 250–1200 pg.

The high sensitivity of the QIT in the full-scan mode prompted Cai et al. [160,161] to push the limits of QIT by simultaneously quantifying multiple drugs from cassette incubations and quantitatively monitoring metabolites from each drug. In an effort to increase the analytical throughput of the metabolic stability assays, compounds were individually incubated with dog microsomes for 60 min, quenched using acetonitrile, and the supernatants from four to five compounds were pooled to form cassette

TABLE 5.4
Comparison of Full Scan MS, Full-Scan MS/MS, and SRM Duty Cycles in QMF and QIT Mass Spectrometers

Scan Type	QIT	QMF	Sensitivity
Full scan (Q1 or Q3, or 100–1100 Da) Total scan time = 1 s	A group of ions (*m/z* 100–1100) will be trapped for 750 ms and then emptied within next 250 ms. Trap will repeat the same process 20 times over a 20 s wide peak	In any given time, ions of one *m/z* are present and allowed to pass through a QMF to the detector and all other ions are rejected. So, each *m/z* will be allowed to pass for 1 ms	750 ms/1 ms = 750 QIT is 750 times more sensitive
Full-scan product ion (MS/MS 100–500) Total scan time = 1 s	Same as above, a group of ions are trapped for 750 ms; isolated, fragmented, and detected within a second. So, each *m/z* is collected for 750 ms	Each *m/z* is sequentially scanned to the detector for 2.5 ms (1 s/400 Da = 2.5 ms/Da)	750 ms/2.5 ms = 300 QIT is 300 times more sensitive
SRM	A group of ions will be collected for 750 ms and then isolated, and fragmented. Duty cycle is less than 75%	Q1 and Q3 are set to pass a single *m/z*. Duty cycle is 100% because no scanning is required	A QMF provides an order of magnitude better LOQ for quantification experiments than a trap

groups before LC–MS analysis. The percent metabolized for each compound in a cassette was calculated by measuring the XIC peak intensities for the parent analyte from each 0 and 60 min incubation. The cassette analysis approach in combination with QIT mass spectrometry allowed Cai et al. [160,161] to evaluate 76 potential development candidates in 16 cassettes rather than individually. Not only did the cassette/QIT approach decreased the analysis time by fivefold in comparison to the traditional triple quadrupole–based approach, but also the results from the full-scan MS provided information about the potential metabolites. This in turn avoided the need for reanalysis of selected incubates for metabolite identification. Once the cassette/QIT approach was validated for the *in vitro* metabolic stability studies, Cai et al. [160,161] extended the technique to study plasma samples collected following administration of α-1a antagonist compounds to mice. Each test compound was administered separately at 10 mg/kg dose and plasma samples collected at selected time points for PK analysis. Analysis of the plasma revealed that the concentrations of the test compounds were similar at 0 min and at 2 h postdose and ranged from 36 to 1062 ng/mL.

Although 3D traps have been evaluated for simultaneous quantification of drugs and identification of their metabolites, the mainstay of the 3D traps within the pharmaceutical industry, during the early to mid-2000s, was in the area of qualitative bioanalysis, especially in the area of metabolite/unknown characterization. Therefore, there are several excellent examples showing the application of QIT in drug development metabolite profiling and identifications studies. Some of the recent applications include detection and characterization of metabolites following administration of loratadine or desloratadine to mice, rats, monkeys, and humans [109,232,233,311]. Another good example was presented by Christopher et al. [312–314] in identification of metabolites of dasatinib (Sprycel, a multiple kinase inhibitor) in *in vitro* and *in vivo* biotransformation studies. In support of NDA, the *in vitro* metabolism of ^{14}C-dasatinib in liver microsomes and hepatocytes (rat, monkey, and human) and the routes of excretion and extent of metabolism of ^{14}C-dasatinib after administration to rats and monkeys (species used in long-term toxicology evaluation) were investigated. *In vitro* and *in vivo* biotransformation profiles were determined by radiochromatographic profiling. The study also shows the use of the radioisotope detector and fraction collection/off-line LSC techniques for analysis of various samples with different levels of radioactivity. The high levels of radioactivity in metabolites of ^{14}C-dasatinib in microsomal and hepatocyte incubations allowed the use of a flow-through radioisotope detector (500 μL flow cell) (β-RAM model 3, IN/US) coupled with HPLC to determine the percent distribution of ^{14}C-labeled metabolites. The percentage of each radioactive metabolite was obtained by integration of the metabolite peak.

For *in vivo* samples with low radioactivity levels, HPLC fractions were collected in 96-well Packard Lumaplates (Perkin Elmer Life Sciences) over 0.25 min intervals. After drying, the plates were counted with a Top Count microplate scintillation analyzer to quantify radioactivity in each fraction. Radiochromatograms were obtained by plotting the net counts per minute values obtained from the Top Count vs time after injection using Microsoft Excel software. The metabolites were quantified based on the percentage of total radioactivity in each peak relative to the entire radiochromatogram. By splitting the HPLC flow between the radioisotope detector (or fraction collector) and mass spectrometer, it was possible to determine the HPLC metabolite profiles and to identify metabolites simultaneously. Biotransformation pathways of dasatinib involves conjugative and oxidative metabolism. Since conjugated metabolites, such as glucuronides and sulfates, and N-oxide metabolites typically undergo further metabolism by gut flora in the GI track [315,316] the studies also included administration of ^{14}C-dasatinib to BDC animals to understand its overall disposition, particularly for conjugated metabolites excreted in bile. Using LC–MS/MS more than 30 metabolites were identified in *in vitro* incubations with liver microsomes and hepatocytes from rats, monkeys, and humans and in plasma, urine, bile, and fecal samples collected after administration of a single oral or i.v. dose of ^{14}C-desatinib to rats and monkeys.

Although 3D traps have been extensively used, during the early to mid-2000s, for structural elucidation of metabolites, overall a slower scan rate compared to TOF mass analyzers, in combination with limited ion capacity and trapping efficiency are the limitations associated with the QITs for becoming the mass analyzer of choice for quantitative/qualitative bioanalysis. Most importantly, 3D traps can only simulate SRM by acquiring full-scan MS data, true SRM scan modes can only be performed using triple quadrupole type mass spectrometers. The major limitation of ion traps for metabolite identifications is that they are not capable of performing precursor ion and neutral loss scanning that is a convenient tool for detecting unknown metabolites in complex biological matrices. Neutral loss scan is a useful scanning technique especially for detection of Phase II metabolites, such as glucuronide (loss of 176 Da) and sulfate (loss of 80 Da) conjugates. Precursor ion scanning can be used to search for metabolites with common product ions that can be predicted from the fragmentation patterns of the NCE. In addition, there are several limitations associated with MS/MS fragmentation in a 3D ion trap. These limitations include unavailability of fragment ions in the low mass region and availability of fragment ions produced via lowest energy pathways due respectively to inherent low-mass cutoff and resonant excitation–based low-energy multiple collisions [160,161]. Recently, Hakala et al. [317] compared different mass spectrometric techniques including a 3D ion trap mass spectrometer to profile and characterize tramadol and its urinary metabolites and showed that 3D ion traps are less suitable for the profiling of tramadol and its metabolites due to the low-mass cutoff of the ion trap.

5.8.4 Linear Ion Trap (2D)/Hybrid Tandem Mass Spectrometry Applications

Commercial LITs were introduced in 2002 as either a stand-alone mass spectrometer (LTQ) [318] or as part of a triple quadrupole (Q-Trap) [319] or in 2005 as part of hybrid tandem mass spectrometers (LTQ-Orbitrap and LTQ-FTICR) [88,90]. Application of LTQ-FTICR for metabolism studies has been reviewed by Shipkova et al. [90]. In comparison to other mass analyzer types, FTICR-based mass spectrometers are not very popular for metabolite identification studies due to availability of less expensive and more user-friendly LTQ-Orbitrap and Q-TOF-based systems. Another limitation associated with the FTICR-based hybrid mass spectrometers is the "TOF effect," which results in efficient trapping of only the high-mass ions [90].

With the combined features of an ion trap and a triple quadrupole mass analyzer, the recently developed Q-Trap offers the capability of performing neutral loss and precursor scan experiments in addition to MS^n fragmentation [320–324]. Availability of triple quadrupole and ion trap functionalities prompted several groups to use the Q-Trap as a powerful mass analyzer for simultaneous metabolite identification and quantification of drugs in biological matrices. Simultaneous parent NCE quantification and metabolite identification experiments are necessary during early stages of drug discovery to streamline processes and expedite the candidate selection process. Some of the unique scan functions available with the Q-Trap mass spectrometer have been reviewed [325,326].

King et al. [164] used Q-Trap to simultaneously quantify an NCE and collect information about circulating metabolites, dosing vehicle, interfering matrix components, and co-eluting metabolites. The ability to operate the LIT in the enhanced MS (EMS) mode with a scan speed of 4000 Th/s allowed the combined SRM transitions (parent/IS) and the full scans to be completed in 0.31 s. The quantification data from the combined analysis (SRM/EMS) were compared with the data from SRM only analysis to show that the quantification data from both methods were generally acceptable in terms of sensitivity, accuracy, and precision. Although some loss of precision was observed at the lowest concentration, the combined SRM/EMS mode of operation generated information on the co-eluting metabolites and the PEG dosing vehicle. This in turn allowed modification of the quantification method to improve the quality of the assay. Additionally, information about circulating metabolites from early PK studies provided insight into the metabolic hot spots and allowed the modification of the structure to optimize the PK of the lead compound.

To improve upon the approach employed by King et al. [164], Li et al. [327] used SRM-triggered information-dependent acquisition (IDA) to acquire both parent drug quantification data and qualitative metabolite MS/MS data. To validate the IDA approach, Li et al. [327] tested both the conventional triple quadrupole mode as well as in the ion trap mode and showed that the cycle time improved from 2.78 to 1.14 s with the IDA approach. The longer cycle time in the triple quadrupole mode of operation would have resulted in possibly missing some the metabolites.

Xia et al. [328] demonstrated the utility of IDA approach to collect maximum amount of information with the minimum number of analytical runs during the course of identification of *in vitro* formed metabolites of gemfibrozil. The Q-Trap tandem mass spectrum contained fragment ions at *m/z* 113 and 85 and they were absent in the MS/MS spectrum generated using a 3D iontrap due to inherent low mass cutoff. With the 3D trap, fragment ions with *m/z* values lower than approximately one-third of the precursor ion were not detected.

Yao et al. [329] demonstrated a survey scan method involving multiple ion monitoring (MIM; a term coined by the authors) to detect and characterize predicted metabolites. Under MIM experimental setup, both Q1 and Q3 were used to monitor the predicted metabolites precursor ions while maintaining the q2 under low collision energy conditions (CE = 5 eV) to minimize fragmentation. MIM survey scans were used to trigger enhanced product ion (EPI) scans in the ion trap mode. Since metabolite fragment ions are not being filtered in Q3 during the execution of MIM (SRM involves filtering a precursor and product ions using Q1 and Q3, respectively), irrespective of fragmentation, screening for 100 possible metabolites was possible with the use of 10 ms dwell time and a 5 ms pause time. Overall, analysis of metabolites formed following incubation of ritonavir in rat hepatocytes showed that MIM-SRM-EPI was better than SRM-EPI in detecting metabolites that undergo both predictable and unpredictable fragmentation pathways.

Many similar applications of quadrupole linear ion trap instruments have been reported [320–322,329,330]. As discussed above, the Q-Trap is a triple quadrupole mass spectrometer capable of performing QMF type and 2D ion trapping experiments. This mass spectrometer can be operated exclusively in the QMF mode, as with a conventional QMF, or it can be operated exclusively in the ion trapping mode similar to a conventional 2D ion trap mass spectrometer. Advantages of using a Q-Trap mass spectrometer over a conventional QMF mass spectrometer come into play when one is attempting to perform both quantitative and qualitative metabolite detection/identification experiments from a single injection rather than separate dedicated injections for quantification of the parent drug and qualitative metabolite profiling.

Improvements over 3D traps with respect to trapping efficiency, ion capacity, sensitivity, scan speed, and MS/MS low-mass cutoff have made the stand-alone 2D ion traps (LTQs) the mass analyzer of choice in discovery and development metabolite identification groups. Unlike the Q-Trap, which can be used for qualitative and quantitative applications, mainstay of the LTQ is in the qualitative arena. The fast scanning capability of the LTQ mass spectrometer allowed Fitch et al. [331] to perform positive and negative polarity switching experiments to identify metabolites of R09237 following hepatocytes incubations with unlabeled and ^{14}C-radiolabeled form of the NCE. The comparison of the data obtained using each polarity showed that neither ESI mode is universal for detecting all the metabolites. Additionally, the fast scanning and MS^n capabilities were must for detecting and characterizing all the major Phase II metabolites. This example clearly demonstrates that all mass analyzer types are not capable of switching polarity on-the-fly to detect all the metabolites. Although polarity switching is feasible with triple quadrupoles, poor full-scan MS/MS duty cycle limit on-the-fly polarity switching experiments.

The recently introduced hybrid linear ion trap-Orbitrap mass spectrometer combines the high trapping capacity and MS^n scanning functions of a linear ion trap with the accurate mass measurement capability of the Orbitrap [332–336]. The LTQ-Orbitrap provides high-resolution, high mass accuracy, and good dynamic range. In the LTQ-Orbitrap, external calibration is used to obtain high mass accuracy (<3 ppm) of all measured spectra obtained from a single experimental analysis. With the capability of fast data-dependent acquisition of accurate MS^n data, the LTQ-Orbitrap has evolved into one of the most valuable MS instrument for metabolite identification [73,93,337,338]. All these features of the LTQ-Orbitrap allow high-resolution/high mass accuracy measurements on precursor and product ions and help in assigning fragment ion structure for all expected and unusual metabolites of NCEs.

Lim et al. [93] investigated the use of LTQ-Orbitrap to identify human liver microsomal metabolites of carvedilol, using parent mass list triggered data-dependent multiple stage accurate mass analysis. In this study, LTQ-Orbitrap was used for accurate mass measurements at a resolving power of 60,000 to obtain mass accuracy <2 ppm for all full-scan and MS^n spectra in the external calibration mode. The chemical formulas obtained from high-resolution accurate mass measurements were used to confirm structures of the product ions proposed by Mass Frontier.

In 2003, Zhang et al. [339] developed a MDF technique for the selective detection of drug metabolites. MDF is a postacquisition data filtering technique based on the mass defect of the parent drug and its metabolites. In order to use MDF, it is required to use high mass accuracy data from high-resolution analyses (e.g., LTQ-Orbitrap, LTQ-FTICR, Q-TOF). This technique is used to distinguish the metabolite ions from the interfering ions, including the isobaric ions, in the matrix using the similarity between the mass defect values of a drug and its metabolites. The mass defect of a compound is a characteristic related to its empirical formula. MDF technique is especially very useful to detect uncommon metabolites that are formed via unusual metabolic pathways. Metabolite ions and isobaric interference ions that are separated by approximately 50 mDa or more in a mass range of 200–1000 Da can be resolved by accurate mass LC–MS analysis. Since, mass defects of oxidative and conjugative metabolites are generally within 50 mDa relative to that of the parent drug, it is possible to set a mass defect window approximately ±50 mDa from that of the parent drug [339]. As a result, the ions with decimal portion sufficiently close to that of the parent drug can be retained and the ions with mass defect lying outside the defined window can be eliminated. Some examples of mass defect shifts for some common metabolites are −5 mDa for monooxygenation, −10 mDa for dioxygenation, −16 mDa for dehydrogenation, −16 mDa for demethylation, +11 mDa for addition of water, +16 mDa for methylation, +32 mDa for glucuronidation, and −43 mDa for sulfation [339,340].

Zhu et al. [340] studied the effectiveness, selectivity, and sensitivity of the LC–MS-based MDF approach for identification of drug metabolites that have mass defects either similar to or significantly different from those of the parent drug. MDF data were also compared with those obtained from the neutral loss and precursor ion scan experiments. Nefazodone was used as a model compound in that study. Nefazodone metabolism involves CYP-mediated hydroxylation, N-dealkylation, and oxidation products. In addition to all the known metabolites of nefazodone, two new metabolites that were not observed in the previous *in vivo* and *in vitro* studies [337,341,342] were detected using the MDF approach. For example, in the mass spectrum of metabolite M7 of nefazodone, many interference ions that have mass defects outside the filter windows were observed. However, after MDF processing, those interference ions were eliminated and abundance of the protonated molecule ion of M7 was increased. Although the interference ions with the mass defects falling within the filter windows are still observed in the MDF-processed spectrum, they can easily be excluded based on their molecular formulae. When conventional precursor ion and neutral loss scanning methods were used instead of MDF, the number of nefazodone metabolites detected in the same samples was less than that detected with the MDF method. Overall, the results showed that the MDF technique is applicable to detect and identify both common and uncommon metabolites, including metabolites whose molecular weights significantly differ from that of the parent NCE.

In another study, Zhu et al. [343] applied the MDF technique to screen and identify glutathione (GSH) trapped reactive metabolites and demonstrated the advantages of MDF over the neutral loss scanning technique. In general, the MDF technique was more sensitive in detecting *in vitro* GSH adducts when compared to the neutral loss scanning technique. The effectiveness of the MDF technique in the analysis of the GSH adducts that does not undergo fragmentation to generate a neutral loss of 129 Da was higher than the neutral loss scanning technique. The MDF approach was more selective in detecting drug–GSH adducts in rat bile containing endogenous GSH conjugates that have similar neutral loss fragmentation pathways to those of drug–GSH adducts. Since its development in 2003, many applications of MDF have been reported [338,343–346]. MDF is now being offered as a data-processing tool in software packages of various mass spectrometer manufacturers.

Most recently, Zhang et al. [347] demonstrated the combined use of a novel accurate mass-based background subtraction approach and MDF for detecting and characterizing troglitazone metabolites in rat plasma, bile, and urine. The standard

Xcalibur background subtraction algorithm and the more advanced algorithm available through MetWorks were less successful in subtracting many of the background and matrix ions present in the control samples. The novel algorithm was designed such that chromatographic fluctuations between the control and the analyte samples were dealt with before background subtraction. The combined use of the novel background subtraction algorithm with MDF allowed the detection and characterization of novel troglitazone metabolites as well as several previously reported metabolites. An improved variation of the software capable of extracting high-resolution metabolite LC–MS data with retention-time-shift-tolerant background-subtraction and noise-reduction algorithm was discussed by Zhu et al. [348].

5.9 FUTURE DIRECTIONS AND SUMMARY

Metabolite identification strategies and procedures at different stages of drug discovery and development have been described. As described, metabolite identification work processes can originate from various ADME studies or studies dedicated to metabolite structural elucidation. LC–MS-based techniques have now become a routine tool for the rapid identification of metabolites in biological samples. The process become even more sensitive, rugged, and capable of providing rapid turnaround when high-resolution TOF, Orbitrap mass spectrometers or fast scanning hybrid linear ion traps are involved.

As LC–MS technology advances, techniques once thought of as specialist techniques have now become more generalist techniques. The timing of this shift has allowed LC–MS to play a central role as part of the solution for the throughput needs of the biopharmaceutical industry. Overall, technological advances in mass spectrometer ionization sources and analyzers combined with the availability of more efficient sample processing, separation techniques, automated software tools, and powerful computer systems have paved the way for LC–MS to evolve as an indispensable tool to the biopharmaceutical industry, especially in the area of DM and PK. For the future, one could expect fully automated LC–MS systems to replace most manual and semiautomatic approaches and to produce metabolite identification information using integrated qualitative and quantitative approaches. MS techniques being developed for quantifying [349] and characterizing metabolites from microdosing and/or FIH studies are still in their infancy and have the potential to reduce cost and make safer drugs for humans.

REFERENCES

1. Kola, I. and Landis, J., Can the pharmaceutical industry reduce attrition rates?, *Nat. Rev. Drug Discov.*, 3(8), 711, 2004.
2. PhPMA, *The Pursuit of High Performance through Research and Development: Understanding Pharmaceutical Research and Development Cost Drivers*, Accenture, Editor, Washington, D.C., 1, 2007.
3. Tall, A.R., Yvan-Charvet, L., and Wang, N., The failure of torcetrapib: Was it the molecule or the mechanism?, *Arterioscler. Thromb. Vasc. Biol.*, 27(2), 257, 2007.
4. Agovino, T., Merck faces huge fallout over Vioxx suits, *Washington Post*, 2004.

5. Embi, P.J. et al., Responding rapidly to FDA drug withdrawals: Design and application of a new approach for a consumer health website, *J. Med. Internet Res.*, 8(3), e16, 2006.
6. Harris, G., FDA failing in drug safety, official asserts, in *New York Times*, 2004.
7. PhRMA, *U.S. Prescription Drug Safety: The Best in the World*, Pharmaceutical Research and Manufacturers of America, Washington, D.C., 2008, http://www.phrma.org/key_facts/
8. Guengerich, F.P. and MacDonald, J.S., Applying mechanisms of chemical toxicity to predict drug safety, *Chem. Res. Toxicol.*, 20(3), 344, 2007.
9. Murphy, P.J., The development of drug metabolism research as expressed in the publications of ASPET: Part 3, 1984–2008, *Drug Metab. Dispos.*, 36(10), 1977, 2008.
10. Murphy, P.J., The development of drug metabolism research as expressed in the publications of ASPET: Part 2, 1959–1983, *Drug Metab. Dispos.*, 36(6), 981, 2008.
11. Murphy, P.J., The development of drug metabolism research as expressed in the publications of ASPET: Part 1, 1909–1958, *Drug Metab. Dispos.*, 36(1), 1, 2008.
12. Gombar, V.K. et al., In silico metabolism studies in drug discovery: Prediction of metabolic stability, *Curr. Comput. Aid. Drug Des.*, 2(2), 177, 2006.
13. Gombar, V.K., Silver, I.S., and Zhao, Z., Role of ADME characteristics in drug discovery and their in silico evaluation: In silico screening of chemicals for their metabolic stability, *Curr. Topics Med. Chem.* (Sharjah, United Arab Emirates), 3(11), 1205, 2003.
14. Ramanathan, D.M. and LeLacheur, R.M., Evolving role of mass spectrometry in drug discovery and development, in *Mass Spectrometry in Drug Metabolism and Pharmacokinetics*, Ramanathan, R. (ed.), John Wiley & Sons, Hoboken, NJ, 1, 2008.
15. Prakash, C., Shaffer, C.L., and Nedderman, A., Analytical strategies for identifying drug metabolites, *Mass Spectrom. Rev.*, 26(3), 340, 2007.
16. Ma, S., Chowdhury, S.K., and Alton, K.B., Application of mass spectrometry for metabolite identification, *Curr. Drug Metab.*, 7(5), 503, 2006.
17. Lee, M.S. and Kerns, E.H., LC/MS applications in drug development, *Mass Spectrom. Rev.*, 18(3–4), 187, 1999.
18. Jemal, M. and Xia, Y.Q., LC-MS development strategies for quantitative bioanalysis, *Curr. Drug Metab.*, 7(5), 491, 2006.
19. Srinivas, N.R., Applicability of bioanalysis of multiple analytes in drug discovery and development: Review of select case studies including assay development considerations, *Biomed. Chromatogr.*, 20(5), 383, 2006.
20. Srinivas, N.R., Changing need for bioanalysis during drug development, *Biomed. Chromatogr.*, 22(3), 235, 2008.
21. Bakhtiar, R. and Majumdar, T.K., Tracking problems and possible solutions in the quantitative determination of small molecule drugs and metabolites in biological fluids using liquid chromatography-mass spectrometry, *J. Pharmacol. Toxicol. Meth.*, 55(3), 227, 2007.
22. Xu, R.N. et al., Recent advances in high-throughput quantitative bioanalysis by LC-MS/MS, *J. Pharm. Biomed. Anal.*, 44(2), 342, 2007.
23. Zhou, S. et al., Critical review of development, validation, and transfer for high throughput bioanalytical LC-MS/MS methods, *Curr. Pharm. Anal.*, 1(1), 3, 2005.
24. Food and Drug Administration, (US FDA), *Guidance for Industry: Safety Testing of Drug Metabolites*, 1, 2008. http://www.fda.gov/downloads/Drugs/GuidanceCompliance RegulatoryInformation/Guidances/UCM079266.pdf
25. Baillie, T.A. et al., Drug metabolites in safety testing, *Toxicol. Appl. Pharmacol.*, 182(3), 188, 2002.
26. Ramanathan, R. et al. Quantitative assessment of metabolites early in the clinical program: Potential for using LC-nanospray-MS, in *Eastern Analytical Symposium and Exposition*, Somerset, NJ, 2007.

27. Chowdhury, S.K., Early assessment of human metabolism: Why, how, challenges and opportunities, in *Proceedings of the 55th ASMS Conference on Mass Spectrometry and Allied Topics*, ASMS, Indianapolis, IN, 2007.
28. Tiller, P.R. et al., Fractional mass filtering as a means to assess circulating metabolites in early human clinical studies, *Rapid Commun. Mass Spectrom.*, 22(22), 3510, 2008.
29. Smith, D.A. and Obach, R.S., Metabolites in safety testing (MIST): Considerations of mechanisms of toxicity with dose, abundance, and duration of treatment, *Chem. Res. Toxicol.*, 22(2), 267, 2009.
30. Guengerich, F.P., Introduction: Human metabolites in safety testing (MIST) issue, *Chem. Res. Toxicol.*, 22(2), 237, 2009.
31. Anderson, S., Luffer-Atlas, D., and Knadler, M.P., Predicting circulating human metabolites: How good are we?, *Chem. Res. Toxicol.*, 22(2), 243, 2009.
32. Powley, M.W. et al., Safety assessment of drug metabolites: Implications of regulatory guidance and potential application of genetically engineered mouse models that express human P450s, *Chem. Res. Toxicol.*, 22(2), 257, 2009.
33. Leclercq, L. et al., Which human metabolites have we MIST? Retrospective analysis, practical aspects, and perspectives for metabolite identification and quantification in pharmaceutical development, *Chem. Res. Toxicol.*, 22(2), 280, 2009.
34. Espina, R. et al., Nuclear magnetic resonance spectroscopy as a quantitative tool to determine the concentrations of biologically produced metabolites: Implications in metabolites in safety testing, *Chem. Res. Toxicol.*, 22(2), 299, 2009.
35. Vishwanathan, K. et al., Obtaining exposures of metabolites in preclinical species through plasma pooling and quantitative NMR: Addressing metabolites in safety testing (MIST) guidance without using radiolabeled compounds and chemically synthesized metabolite standards, *Chem. Res. Toxicol.*, 22(2), 311, 2009.
36. Baillie, T.A., Approaches to the assessment of stable and chemically reactive drug metabolites in early clinical trials, *Chem. Res. Toxicol.*, 22(2), 263, 2009.
37. Ramanathan, R. et al., Early assessment of metabolites in clinical studies: Potential applications of NanoSpray ionization mass spectrometry, in *10th Annual Symposium on Chemical and Pharmaceutical Structure Analysis*, Langhorne, PA, 2007.
38. Marathe, P.H., Shyu, W.C., and Humphreys, W.G., The use of radiolabeled compounds for ADME studies in discovery and exploratory development, *Curr. Pharm. Des.*, 10(24), 2991, 2004.
39. Naidong, W. et al., Liquid chromatography/tandem mass spectrometric bioanalysis using normal-phase columns with aqueous/organic mobile phases—A novel approach of eliminating evaporation and reconstitution steps in 96-well SPE, *Rapid Commun. Mass Spectrom.*, 16(20), 1965, 2002.
40. Castro-Perez, J.M., Current and future trends in the application of HPLC-MS to metabolite-identification studies, *Drug Discov. Today*, 12(5–6), 249, 2007.
41. Vishwanathan, K. et al., Plasma pooling to obtain AUCs of metabolites present in preclinical species: Addressing MIST guidance using non-radiolabeled compounds, in *ISSX Meeting*, San Diego, CA, 2008.
42. Hamilton, R.A., Garnett, W.R., and Kline, B.J., Determination of mean valproic acid serum level by assay of a single pooled sample, *Clin. Pharmacol. Ther.*, 29(3), 408, 1981.
43. Hop, C.E. et al., Plasma-pooling methods to increase throughput for in vivo pharmacokinetic screening, *J. Pharm. Sci.*, 87(7), 901, 1998.
44. Guillarme, D. et al., Recent developments in liquid chromatography—Impact on qualitative and quantitative performance, *J. Chromatogr., A*, 1149(1), 20, 2007.
45. Hsieh, Y. et al., Direct plasma analysis of drug compounds using monolithic column liquid chromatography and tandem mass spectrometry, *Anal. Chem.*, 75(8), 1812, 2003.

46. Hsieh, Y. et al., Simultaneous determination of a drug candidate and its metabolite in rat plasma samples using ultrafast monolithic column high-performance liquid chromatography/tandem mass spectrometry, *Rapid Commun. Mass Spectrom.*, 16(10), 944, 2002.
47. van de Merbel, N.C. and Poelman, H., Experiences with monolithic LC phases in quantitative bioanalysis, *J. Pharm. Biomed. Anal.*, 33(3), 495, 2003.
48. Naxing, X.R. et al., A monolithic-phase based on-line extraction approach for determination of pharmaceutical components in human plasma by HPLC-MS/MS and a comparison with liquid-liquid extraction, *J. Pharm. Biomed. Anal.*, 40(3), 728, 2006.
49. Dear, G., Plumb, R., and Mallett, D., Use of monolithic silica columns to increase analytical throughput for metabolite identification by liquid chromatography/tandem mass spectrometry, *Rapid Commun. Mass Spectrom.*, 15(2), 152, 2001.
50. El Deeb, S., Preu, L., and Watzig, H., A strategy to develop fast RP-HPLC methods using monolithic silica columns, *J. Sep. Sci.*, 30(13), 1993, 2007.
51. El Deeb, S., Preu, L., and Watzig, H., Evaluation of monolithic HPLC columns for various pharmaceutical separations: Method transfer from conventional phases and batch to batch repeatability, *J. Pharm. Biomed. Anal.*, 44(1), 85, 2007.
52. Aboul-Enein, H.Y., Ali, I., and Hoenen, H., Rapid determination of haloperidol and its metabolites in human plasma by HPLC using monolithic silica column and solid-phase extraction, *Biomed. Chromatogr.*, 20(8), 760, 2006.
53. Plumb, R. et al., Direct analysis of pharmaceutical compounds in human plasma with chromatographic resolution using an alkyl-bonded silica rod column, *Rapid Commun. Mass Spectrom.*, 15(12), 986, 2001.
54. Johnson, K.A. and Plumb, R., Investigating the human metabolism of acetaminophen using UPLC and exact mass oa-TOF MS, *J. Pharm. Biomed. Anal.*, 39(3–4), 805, 2005.
55. Castro-Perez, J. et al., Increasing throughput and information content for in vitro drug metabolism experiments using ultra-performance liquid chromatography coupled to a quadrupole time-of-flight mass spectrometer, *Rapid Commun. Mass Spectrom.*, 19(6), 843, 2005.
56. Wang, G. et al., Ultra-performance liquid chromatography/tandem mass spectrometric determination of testosterone and its metabolites in in vitro samples, *Rapid Commun. Mass Spectrom.*, 20(14), 2215, 2006.
57. Xue, Y.J. et al., A 96-well single-pot protein precipitation, liquid chromatography/tandem mass spectrometry (LC/MS/MS) method for the determination of muraglitazar, a novel diabetes drug, in human plasma, *J. Chromatogr. B Analyt. Technol. Biomed. Life Sci.*, 831(1–2), 213, 2006.
58. Hsieh, Y. and Chen, J., Simultaneous determination of nicotinic acid and its metabolites using hydrophilic interaction chromatography with tandem mass spectrometry, *Rapid Commun. Mass Spectrom.*, 19(21), 3031, 2005.
59. Horning, E.C. et al., Atmospheric pressure ionization (API) mass spectrometry. Solvent-mediated ionization of samples introduced in solution and in a liquid chromatograph effluent stream, *J. Chromatogr. Sci.*, 12(11), 725, 1974.
60. Kebarle, P., A brief overview of the present status of the mechanisms involved in electrospray mass spectrometry, *J. Mass Spectrom.*, 35(7), 804, 2000.
61. Bruins, A.P., Liquid chromatography-mass spectrometry with ionspray and electrospray interfaces in pharmaceutical and biomedical research, *J. Chromatogr.*, 554(1–2), 39, 1991.
62. Sheen, J.F. and Her, G.R., Application of pentafluorophenyl hydrazine derivatives to the analysis of nabumetone and testosterone in human plasma by liquid chromatography-atmospheric pressure chemical ionization-tandem mass spectrometry, *Anal. Bioanal. Chem.*, 380(7–8), 891, 2004.

63. Kantharaj, E. et al., The use of liquid chromatography-atmospheric pressure chemical ionization mass spectrometry to explore the in vitro metabolism of cyanoalkyl piperidine derivatives, *Biomed. Chromatogr.*, 19(3), 245, 2005.
64. Robb, D.B., Covey, T.R., and Bruins, A.P., Atmospheric pressure photoionization (APPI): A new ionization technique for LC/MS, *Adv. Mass Spectrom.*, 15, 391, 2001.
65. Hsieh, Y., APPI: A new ionization source for LC-MS/MS assays, in *Using Mass Spectrometry for Drug Metabolism Studies*, Korfmacher, W.A. (ed.), CRC Press, Boca Raton, FL, 253, 2005.
66. Rosenberg, E., The potential of organic (electrospray- and atmospheric pressure chemical ionisation) mass spectrometric techniques coupled to liquid-phase separation for speciation analysis, *J. Chromatogr. A*, 1000(1–2), 841, 2003.
67. Dams, R. et al., Matrix effect in bio-analysis of illicit drugs with LC-MS/MS: Influence of ionization type, sample preparation, and biofluid, *J. Am. Soc. Mass Spectrom.*, 14(11), 1290, 2003.
68. Dams, R. et al., LC-atmospheric pressure chemical ionization-MS/ MS analysis of multiple illicit drugs, methadone, and their metabolites in oral fluid following protein precipitation, *Anal. Chem.*, 75(4), 798, 2003.
69. Hsieh, Y. et al., Quantitative screening and matrix effect studies of drug discovery compounds in monkey plasma using fast-gradient liquid chromatography/tandem mass spectrometry, *Rapid Commun. Mass Spectrom.*, 15(24), 2481, 2001.
70. Brewer, E. and Henion, J., Atmospheric pressure ionization LC/MS/MS techniques for drug disposition studies, *J. Pharm. Sci.*, 87(4), 395, 1998.
71. Robb, D.B., Covey, T.R., and Bruins, A.P., Atmospheric pressure photoionization: An ionization method for liquid chromatography-mass spectrometry, *Anal. Chem.*, 72(15), 3653, 2000.
72. Cai, Y. et al., Advantages of atmospheric pressure photoionization mass spectrometry in support of drug discovery, *Rapid Commun. Mass Spectrom.*, 19(12), 1717, 2005.
73. Erve, J.C., Demaio, W., and Talaat, R.E., Rapid metabolite identification with sub parts-per-million mass accuracy from biological matrices by direct infusion nanoelectrospray ionization after clean-up on a ZipTip and LTQ/Orbitrap mass spectrometry, *Rapid Commun. Mass Spectrom.*, 22(19), 3015, 2008.
74. Wilm, M. and Mann, M., Analytical properties of the nanoelectrospray ion source, *Anal. Chem.*, 68(1), 1, 1996.
75. Corkery, L.J. et al., Automated nanospray using chip-based emitters for the quantitative analysis of pharmaceutical compounds, *J. Am. Soc. Mass Spectrom.*, 16(3), 363, 2005.
76. Hop, C.E., Use of nano-electrospray for metabolite identification and quantitative absorption, distribution, metabolism and excretion studies, *Curr. Drug Metab.*, 7(5), 557, 2006.
77. Hop, C.E., Chen, Y., and Yu, L.J., Uniformity of ionization response of structurally diverse analytes using a chip-based nanoelectrospray ionization source, *Rapid Commun. Mass Spectrom.*, 19(21), 3139, 2005.
78. Ramanathan, R. et al., Response normalized liquid chromatography nanospray ionization mass spectrometry, *J. Am. Soc. Mass Spectrom.*, 18(10), 1891, 2007.
79. Seymour, J.L. et al., Comparison of DART ionization to traditional LC/MS methods for the analysis of metabolites in human urine, in *Proceedings of the 55th ASMS Conference on Mass Spectrometry and Allied Topics*, ASMS, Indianapolis, IN, 2007.
80. Yu, S. et al., DART for bioanalysis: Where are we now?, in *Proceedings of the 55th ASMS Conference on Mass Spectrometry and Allied Topics*, ASMS, Indianapolis, IN, 2007.
81. Williams, J.P. et al., The use of recently described ionisation techniques for the rapid analysis of some common drugs and samples of biological origin, *Rapid Commun. Mass Spectrom.*, 20(9), 1447, 2006.

82. Venter, A., Sojka, P.E., and Cooks, R.G., Droplet dynamics and ionization mechanisms in desorption electrospray ionization mass spectrometry, *Anal. Chem.*, 78(24), 8549, 2006.
83. Williams, J.P. et al., Polarity switching accurate mass measurement of pharmaceutical samples using desorption electrospray ionization and a dual ion source interfaced to an orthogonal acceleration time-of-flight mass spectrometer, *Anal. Chem.*, 78(21), 7440, 2006.
84. Kauppila, T.J. et al., Desorption electrospray ionization mass spectrometry for the analysis of pharmaceuticals and metabolites, *Rapid Commun. Mass Spectrom.*, 20(3), 387, 2006.
85. Cody, R.B., Laramee, J.A., and Durst, H.D., Versatile new ion source for the analysis of materials in open air under ambient conditions, *Anal. Chem.*, 77(8), 2297, 2005.
86. Takats, Z., Wiseman, J.M., and Cooks, R.G., Ambient mass spectrometry using desorption electrospray ionization (DESI): Instrumentation, mechanisms and applications in forensics, chemistry, and biology, *J. Mass Spectrom.*, 40(10), 1261, 2005.
87. Wiseman, J.M. et al., Direct characterization of enzyme-substrate complexes by using electrosonic spray ionization mass spectrometry, *Angew. Chem. Int. Ed. Engl.*, 44(6), 913, 2005.
88. Sanders, M. et al., Utility of the hybrid LTQ-FTMS for drug metabolism applications, *Curr. Drug Metab.*, 7(5), 547, 2006.
89. Peterman, S.M., Dufresne, C.P., and Horning, S., The use of a hybrid linear trap/FT-ICR mass spectrometer for on-line high resolution/high mass accuracy bottom-up sequencing, *J. Biomol. Tech.*, 16(2), 112, 2005.
90. Shipkova, P.A., Josephs, J.L., and Sanders, M., Changing role of FTMS in drug metabolism, in *Mass Spectrometry in Drug Metabolism and Pharmacokinetics*, Ramanathan, R. (ed.), John Wiley & Sons, Hoboken, NJ, 191, 2008.
91. Thompson, C.M. et al., A comparison of accurate mass techniques for the structural elucidation of fluconazole, *Rapid Commun. Mass Spectrom.*, 17(24), 2804, 2003.
92. Shipkova, P.A. et al., Use of a hybrid linear ion trap/Fourier transform mass spectrometer for determining metabolic stability and metabolite characterization, in *Proceedings of the 52nd ASMS Conference on Mass Spectrometry and Allied Topics*, ASMS, Nashville, TN, 2004.
93. Lim, H.K. et al., Metabolite identification by data-dependent accurate mass spectrometric analysis at resolving power of 60,000 in external calibration mode using an LTQ/Orbitrap, *Rapid Commun. Mass Spectrom.*, 21(12), 1821, 2007.
94. Samuel, K. et al., Addressing the metabolic activation potential of new leads in drug discovery: A case study using ion trap mass spectrometry and tritium labeling techniques, *J. Mass Spectrom.*, 38(2), 211, 2003.
95. Tozuka, Z. et al., Strategy for structural elucidation of drugs and drug metabolites using (MS)n fragmentation in an electrospray ion trap, *J. Mass Spectrom.*, 38(8), 793, 2003.
96. Nebert, D.W. et al., The P450 gene superfamily: Recommended nomenclature, *DNA*, 6(1), 1, 1987.
97. Nebert, D.W. et al., The P450 superfamily: Updated listing of all genes and recommended nomenclature for the chromosomal loci, *DNA*, 8(1), 1, 1989.
98. Nelson, D.R. et al., P450 superfamily: Update on new sequences, gene mapping, accession numbers and nomenclature, *Pharmacogenetics*, 6(1), 1, 1996.
99. Williams, J.A. et al., Drug-drug interactions for UDP-glucuronosyltransferase substrates: A pharmacokinetic explanation for typically observed low exposure (AUCi/AUC) ratios, *Drug Metab. Dispos.*, 32(11), 1201, 2004.
100. Shimada, T. et al., Interindividual variations in human liver cytochrome P-450 enzymes involved in the oxidation of drugs, carcinogens and toxic chemicals: Studies with liver microsomes of 30 Japanese and 30 Caucasians, *J. Pharmacol. Exp. Ther.*, 270(1), 414, 1994.

101. Paine, M.F. et al., The human intestinal cytochrome P450 "pie", *Drug Metab. Dispos.*, 34(5), 880, 2006.
102. Ramanathan, R. et al., Liquid chromatography/mass spectrometry methods for distinguishing N-oxides from hydroxylated compounds, *Anal. Chem.*, 72(6), 1352, 2000.
103. Peiris, D.M. et al., Distinguishing N-oxide and hydroxyl compounds: Impact of heated capillary/heated ion transfer tube in inducing atmospheric pressure ionization source decompositions, *J. Mass Spectrom.*, 39(6), 600, 2004.
104. Ramanathan, R., Chowdhury, S.K., and Alton, K.B., Oxidative metabolites of drugs and xenobiotics: LC-MS methods to identify and characterize in biological matrices, in *Identification and Quantification of Drugs, Metabolites and Metabolizing Enzymes by LC-MS*, Chowdhury, S.K. (ed.), Elsevier, Amsterdam, the Netherlands, 225, 2005.
105. Penner, N.A. et al., LC–MS methods with hydrogen/deuterium exchange for identification of hydroxylamine, N-oxide, and hydroxylated analogs of desloratadine, in *Mass Spectrometry in Drug Metabolism and Pharmacokinetics*, Ramanathan, R. (ed.), John Wiley & Sons, Hoboken, NJ, 297, 2008.
106. Burchell, B., Brierley, C.H., and Rance, D., Specificity of human UDP-glucuronosyltransferases and xenobiotic glucuronidation, *Life Sci*, 57(20), 1819, 1995.
107. Nobilis, M. et al., Identification and determination of phase II nabumetone metabolites by high-performance liquid chromatography with photodiode array and mass spectrometric detection, *J. Chromatogr. A*, 1031(1–2), 229, 2004.
108. Taylor, J.I., Grace, P.B., and Bingham, S.A., Optimization of conditions for the enzymatic hydrolysis of phytoestrogen conjugates in urine and plasma, *Anal. Biochem.*, 341(2), 220, 2005.
109. Ramanathan, R. et al., Metabolism and excretion of loratadine in male and female mice, rats and monkeys, *Xenobiotica*, 35(2), 155, 2005.
110. Pirnay, S.O. et al., Selection and optimization of hydrolysis conditions for the quantification of urinary metabolites of MDMA, *J. Anal. Toxicol.*, 30(8), 563, 2006.
111. Zenser, T.V., Lakshmi, V.M., and Davis, B.B., Human and *Escherichia coli* beta-glucuronidase hydrolysis of glucuronide conjugates of benzidine and 4-aminobiphenyl, and their hydroxy metabolites, *Drug Metab. Dispos.*, 27(9), 1064, 1999.
112. Faed, E.M., Properties of acyl glucuronides. Implications for studies of the pharmacokinetics and metabolism of acidic drugs, *Drug Metab. Rev.*, 15, 1213, 1984.
113. Patrick, J.E. et al., Disposition of the selective cholesterol absorption inhibitor ezetimibe in healthy male subjects, *Drug Metab. Dispos.*, 30(4), 430, 2002.
114. Ishii, Y. et al., Induction of two UDP-glucuronosyltransferase isoforms sensitive to phenobarbital that are involved in morphine glucuronidation. Production of isoform-selective antipeptide antibodies toward UGT1.1r and UGT2B1, *Drug Metab. Dispos.*, 25, 163, 1997.
115. Hawkins, D.R., Taylor, J.B., and Triggle, D.J., Comprehensive expert systems to predict drug metabolism, in *Comprehensive Medicinal Chemistry II*, Elsevier, Oxford, U.K., 795, 2007.
116. Wishart, D.S., Improving early drug discovery through ADME modelling: An overview, *Drugs R. D.*, 8(6), 349, 2007.
117. Yamashita, F. and Hashida, M., In silico approaches for predicting ADME properties of drugs, *Drug Metab. Pharmacokinet.*, 19(5), 327, 2004.
118. Caron, G., Ermondi, G., and Testa, B., Predicting the oxidative metabolism of statins: An application of the MetaSite algorithm, *Pharm. Res.*, 24(3), 480, 2007.
119. Yu, J. et al., In silico prediction of drug binding to CYP2D6: Identification of a new metabolite of metoclopramide, *Drug Metab. Dispos.*, 34(8), 1386, 2006.
120. van de Waterbeemd, H. and Gifford, E., ADMET in silico modelling: Towards prediction paradise?, *Nat. Rev. Drug Discov.*, 2(3), 192, 2003.

121. Zhou, D. et al., Comparison of methods for the prediction of the metabolic sites for CYP3A4-mediated metabolic reactions, *Drug Metab. Dispos.*, 34(6), 976, 2006.
122. Nassar, A.E. and Adams, P.E., Metabolite characterization in drug discovery utilizing robotic liquid-handling, quadruple time-of-flight mass spectrometry and in-silico prediction, *Curr. Drug Metab.*, 4(4), 259, 2003.
123. Klopman, G., Dimayuga, M., and Talafous, J., META. 1. A program for the evaluation of metabolic transformation of chemicals, *J. Chem. Inf. Comput. Sci.*, 34(6), 1320, 1994.
124. Talafous, J. et al., META. 2. A dictionary model of mammalian xenobiotic metabolism, *J. Chem. Inf. Comput. Sci.*, 34(6), 1326, 1994.
125. Testa, B. et al., Predicting drug metabolism—An evaluation of the expert system METEOR, *Chem. Biodivers.*, 2(7), 872, 2005.
126. Ekins, S. et al., A combined approach to drug metabolism and toxicity assessment, *Drug Metab. Dispos.*, 34(3), 495, 2006.
127. Ekins, S. and Swaan, P.W., Development of computational models for enzymes, transporters, channels, and receptors relevant to ADME/Tox, *Rev. Comp. Chem.*, 20, 333, 2004.
128. Desta, Z. et al., The gastroprokinetic and antiemetic drug metoclopramide is a substrate and inhibitor of cytochrome P450 2D6, *Drug Metab. Dispos.*, 30(3), 336, 2002.
129. Heinonen, M. et al., FiD: A software for ab initio structural identification of product ions from tandem mass spectrometric data, *Rapid Commun. Mass Spectrom.*, 22(19), 3043, 2008.
130. Ives, S., Gjervig-Jensen, K., and McSweeney, N., Speed up of the identification of the metabolites of Citalopram using LC/MS/MS and in silico prediction, in *Proceedings of the 55th ASMS Conference on Mass Spectrometry and Allied Topics*, ASMS, Indianapolis, IN, 2007.
131. Testa, B. et al., Predicting drug metabolism—An evaluation of the expert system METEOR, *Chem. Biodivers.*, 2(7), 872, 2005.
132. Sweeney, D.L., Small molecules as mathematical partitions, *Anal. Chem.*, 75(20), 5362, 2003.
133. Bayliss, M.A. et al., Rapid metabolite identification using advanced algorithms for mass spectral interpretation, in *Proceedings of the 55th ASMS Conference on Mass Spectrometry and Allied Topics*, ASMS, Indianapolis, IN, 2007.
134. Cruciani, G. et al., MetaSite: Understanding metabolism in human cytochromes from the perspective of the chemist, *J. Med. Chem.*, 48(22), 6970, 2005.
135. Williams, A., Applications of computer software for the interpretation and management of mass spectrometry data in pharmaceutical science, *Curr. Top. Med. Chem.*, 2(1), 99, 2002.
136. Ito, K. and Houston, J.B., Comparison of the use of liver models for predicting drug clearance using in vitro kinetic data from hepatic microsomes and isolated hepatocytes, *Pharm. Res.*, 21(5), 785, 2004.
137. Ito, K. and Houston, J.B., Prediction of human drug clearance from in vitro and preclinical data using physiologically based and empirical approaches, *Pharm. Res.*, 22(1), 103, 2005.
138. Leclercq, L. et al., IsoScore: Automated localization of biotransformations by mass spectrometry using product ion scoring of virtual regioisomers, *Rapid Commun. Mass Spectrom.*, 23(1), 39, 2008.
139. Korfmacher, W.A. et al., Development of an automated mass spectrometry system for the quantitative analysis of liver microsomal incubation samples: A tool for rapid screening of new compounds for metabolic stability, *Rapid Commun. Mass Spectrom.*, 13(10), 901, 1999.

140. Drexler, D.M. et al., An automated high throughput liquid chromatography-mass spectrometry process to assess the metabolic stability of drug candidates, *Assay Drug Dev. Technol.*, 5(2), 247, 2007.
141. Rajanikanth, M., Madhusudanan, K.P., and Gupta, R.C., Simultaneous quantitative analysis of three drugs by high-performance liquid chromatography/electrospray ionization mass spectrometry and its application to cassette in vitro metabolic stability studies, *Rapid Commun. Mass Spectrom.*, 17(18), 2063, 2003.
142. Baker, T. et al., High-throughput in vitro metabolic stability determinations of pharmaceuticals using automated sample preparation and multiple-compound HPLC/MS/MS, *Adv. Mass Spectrom.*, 15, 657, 2001.
143. Thompson, T.N., Early ADME in support of drug discovery: The role of metabolic stability studies, *Curr. Drug Metab.*, 1(3), 215, 2000.
144. Wan, H. et al., Single run measurements of drug-protein binding by high-performance frontal analysis capillary electrophoresis and mass spectrometry, *Rapid Commun. Mass Spectrom.*, 19(12), 1603, 2005.
145. Trainor, G.L., The importance of plasma protein binding in drug discovery, *Expert Opin. Drug Discov.*, 2(1), 51, 2007.
146. Fung, E.N., Chen, Y.H., and Lau, Y.Y., Semi-automatic high-throughput determination of plasma protein binding using a 96-well plate filtrate assembly and fast liquid chromatography-tandem mass spectrometry, *J. Chromatogr. B Analyt. Technol. Biomed. Life Sci.*, 795(2), 187, 2003.
147. Li, Y. et al., Increasing the throughput and productivity of Caco-2 cell permeability assays using liquid chromatography-mass spectrometry: Application to resveratrol absorption and metabolism, *Comb. Chem. High Throughput Screen.*, 6(8), 757, 2003.
148. Smalley, J. et al., Development of an on-line extraction turbulent flow chromatography tandem mass spectrometry method for cassette analysis of Caco-2 cell based bi-directional assay samples, *J. Chromatogr. B Analyt. Technol. Biomed. Life Sci.*, 830(2), 270, 2006.
149. van Breemen, R.B. and Li, Y., Caco-2 cell permeability assays to measure drug absorption, *Expert Opin. Drug Metab. Toxicol.*, 1(2), 175, 2005.
150. Whalen, K., Gobey, J., and Janiszewski, J., A centralized approach to tandem mass spectrometry method development for high-throughput ADME screening, *Rapid Commun. Mass Spectrom.*, 20(10), 1497, 2006.
151. Whalen, K.M. et al., AutoScan: An automated workstation for rapid determination of mass and tandem mass spectrometry conditions for quantitative bioanalytical mass spectrometry, *Rapid Commun. Mass Spectrom.*, 14(21), 2074, 2000.
152. Janiszewski, J.S. et al., A high-capacity LC/MS system for the bioanalysis of samples generated from plate-based metabolic screening, *Anal. Chem.*, 73(7), 1495, 2001.
153. Hop, C.E. et al., High throughput ADME screening: Practical considerations, impact on the portfolio and enabler of in silico ADME models, *Curr. Drug Metab.*, 9(9), 847, 2008.
154. Janiszewski, J.S., Liston, T.E., and Cole, M.J., Perspectives on bioanalytical mass spectrometry and automation in drug discovery, *Curr. Drug Metab.*, 9(9), 986, 2008.
155. Josephs, J.L. et al., A high quality high throughput LC/MS strategy for profiling of drug candidates with applications to structural integrity, permeability, stability, and related assays, in *Proceedings of the 50th ASMS Conference on Mass Spectrometry and Allied Topics*, ASMS, Orlando, FL, 2002.
156. Josephs, J.L. et al., A comprehensive strategy for the characterization and optimization of metabolic profiles of compounds using a hybrid linear ion trap/FTMS, in *Proceedings of the 53rd ASMS Conference on Mass Spectrometry and Allied Topics*, ASMS, San Antonio, TX, 2005.

157. Josephs, J.L. et al., A rapid approach to quantitative in vivo metabolite profiling without the need for authentic standards or labeled compounds, in *Proceedings of the 56th ASMS Conference on Mass Spectrometry and Allied Topics*, ASMS, Denver, CO, 2008.
158. White, R.E. and Manitpisitkul, P., Pharmacokinetic theory of cassette dosing in drug discovery screening, *Drug Metab. Dispos.*, 29(7), 957, 2001.
159. Beaudry, F. et al., In vivo pharmacokinetic screening in cassette dosing experiments; the use of on-line Pprospekt liquid chromatography/atmospheric pressure chemical ionization tandem mass spectrometry technology in drug discovery, *Rapid Commun. Mass Spectrom.*, 12(17), 1216, 1998.
160. Cai, Z. et al., High-throughput analysis in drug discovery: Application of liquid chromatography/ion-trap mass spectrometry for simultaneous cassette analysis of alpha-1a antagonists and their metabolites in mouse plasma, *Rapid Commun. Mass Spectrom.*, 15(8), 546, 2001.
161. Cai, Z., Sinhababu, A.K., and Harrelson, S., Simultaneous quantitative cassette analysis of drugs and detection of their metabolites by high performance liquid chromatography/ion trap mass spectrometry, *Rapid Commun. Mass Spectrom.*, 14(18), 1637, 2000.
162. Sadagopan, N., Pabst, B., and Cohen, L.H., Evaluation of online extraction/mass spectrometry for in vivo cassette analysis, *J. Chromatogr. B Analyt. Technol. Biomed. Life Sci.*, 820(1), 59, 2005.
163. Korfmacher, W.A. et al., Cassette-accelerated rapid rat screen: A systematic procedure for the dosing and liquid chromatography/atmospheric pressure ionization tandem mass spectrometric analysis of new chemical entities as part of new drug discovery, *Rapid Commun. Mass Spectrom.*, 15(5), 335, 2001.
164. King, R.C., Gundersdorf, R., and Fernandez-Metzler, C.L., Collection of selected reaction monitoring and full scan data on a time scale suitable for target compound quantitative analysis by liquid chromatography/tandem mass spectrometry, *Rapid Commun. Mass Spectrom.*, 17(21), 2413, 2003.
165. Ranasinghe, A. et al., Integrated approach to API quantitation and rapid in-vivo metabolic profiling in early discovery PK assays, in *Proceedings of the 57th ASMS Conference on Mass Spectrometry and Allied Topics*, ASMS, Philadelphia, PA, 2009.
166. Bateman, K. et al., Quantitative-qualitative data acquisition using a benchtop orbitrap mass spectrometer. *J. Am. Soc. Mass Spectrom.*, 20(8), 1441–1450, 2009.
167. Saljoughian, M. and Williams, P.G., Recent developments in tritium incorporation for radiotracer studies, *Curr. Pharm. Des.*, 6(10), 1029, 2000.
168. Lasser, K.E. et al., Timing of new black box warnings and withdrawals for prescription medications, *JAMA*, 287(17), 2215, 2002.
169. Wagner, A.K. et al., FDA drug prescribing warnings: Is the black box half empty or half full? *Pharmacoepidemiol. Drug Saf.*, 15(6), 369, 2006.
170. Lasser, K.E. et al., Adherence to black box warnings for prescription medications in outpatients, *Arch. Intern. Med.*, 166(3), 338, 2006.
171. Walgren, J.L., Mitchell, M.D., and Thompson, D.C., Role of metabolism in drug-induced idiosyncratic hepatotoxicity, *Crit. Rev. Toxicol.*, 35(4), 325, 2005.
172. Uetrecht, J., Idiosyncratic drug reactions: Past, present, and future, *Chem. Res. Toxicol.*, 21(1), 84, 2008.
173. Evans, D.C. et al., Drug-protein adducts: An industry perspective on minimizing the potential for drug bioactivation in drug discovery and development, *Chem. Res. Toxicol.*, 17(1), 3, 2004.
174. Argoti, D. et al., Cyanide trapping of iminium ion reactive intermediates followed by detection and structure identification using liquid chromatography-tandem mass spectrometry (LC-MS/MS), *Chem. Res. Toxicol.*, 18(10), 1537, 2005.

175. Grotz, D.E., Clarke, N.A., and Cox, K.A., The utility of in vitro screening for the assessment of electrophilic metabolite formation early in drug discovery using HPLC-MS/MS, in *Identification and Quantification of Drugs, Metabolites and Metabolizing Enzymes by LC-MS*, Chowdhury, S.K. (ed.), Elsevier, Amsterdam, the Netherlands, 159, 2005.
176. Castro-Perez, J.M., Applications of quadrupole time-of-flight mass spectrometry in reactive metabolite screening, in *Mass Spectrometry in Drug Metabolism and Pharmacokinetics*, Ramanathan, R. (ed.), John Wiley & Sons, Hoboken, NJ, 159, 2008.
177. Staack, R.F. and Hopfgartner, G., New analytical strategies in studying drug metabolism, *Anal. Bioanal. Chem.*, 388(7), 1365, 2007.
178. Dieckhaus, C.M. et al., Negative ion tandem mass spectrometry for the detection of glutathione conjugates, *Chem. Res. Toxicol.*, 18(4), 630, 2005.
179. Yan, Z., Caldwell, G.W., and Maher, N., Unbiased high-throughput screening of reactive metabolites on the linear ion trap mass spectrometer using polarity switch and mass tag triggered data-dependent acquisition, *Anal. Chem.*, 80(16), 6410, 2008.
180. Mutlib, A. et al., Application of stable isotope labeled glutathione and rapid scanning mass spectrometers in detecting and characterizing reactive metabolites, *Rapid Commun. Mass Spectrom.*, 19(23), 3482, 2005.
181. Meneses-Lorente, G. et al., A quantitative high-throughput trapping assay as a measurement of potential for bioactivation, *Anal. Biochem.*, 351(2), 266, 2006.
182. Gorrod, J.W. and Aislaitner, G., The metabolism of alicyclic amines to reactive iminium ion intermediates, *Eur. J. Drug Metab. Pharmacokinet.*, 19(3), 209, 1994.
183. Kalgutkar, A.S. et al., A comprehensive listing of bioactivation pathways of organic functional groups, *Curr. Drug Metab.*, 6(3), 161, 2005.
184. Lewicki, K. et al., Development of a fluorescence-based microtiter plate method for the measurement of glutathione in yeast, *Talanta*, 70(4), 876, 2006.
185. Guengerich, F.P., Principles of covalent binding of reactive metabolites and examples of activation of bis-electrophiles by conjugation, *Arch. Biochem. Biophys.*, 433(2), 369, 2005.
186. Gan, J. et al., In vitro Screening of 50 Highly prescribed drugs for thiol adduct formation comparison of potential for drug-induced toxicity and extent of adduct formation, *Chem. Res. Toxicol.*, 22(4), 690–698, 2009.
187. Ma, S. and Subramanian, R., Detecting and characterizing reactive metabolites by liquid chromatography/tandem mass spectrometry, *J. Mass Spectrom.*, 41(9), 1121, 2006.
188. Ma, S. and Zhu, M., Recent advances in applications of liquid chromatography-tandem mass spectrometry to the analysis of reactive drug metabolites, *Chem. Biol. Interact.*, 179(1), 25, 2009.
189. Williams, J.A. et al., Reaction phenotyping in drug discovery: Moving forward with confidence?, *Curr. Drug Metab.*, 4(6), 527, 2003.
190. Bell, L. et al., Evaluation of fluorescence- and mass spectrometry-based CYP inhibition assays for use in drug discovery, *J. Biomol. Screen*, 13(5), 343, 2008.
191. Yin, H. et al., Automated high throughput human CYP isoform activity assay using SPE-LC/MS method: Application in CYP inhibition evaluation, *Xenobiotica*, 30(2), 141, 2000.
192. Ramanathan, R. et al., Application of semi-automated metabolite identification software in the drug discovery process for rapid identification of metabolites and the cytochrome P450 enzymes responsible for their formation, *J. Pharm. Biomed. Anal.*, 28(5), 945, 2002.
193. Rodrigues, A.D., Integrated cytochrome P450 reaction phenotyping: Attempting to bridge the gap between cDNA-expressed cytochromes P450 and native human liver microsomes, *Biochem. Pharmacol.*, 57(5), 465, 1999.

194. Williams, J.A. et al., In vitro ADME phenotyping in drug discovery: Current challenges and future solutions, *Curr. Opin. Drug Discov. Develop.*, 8(1), 78, 2005.
195. Crespi, C.L. and Miller, V.P., The use of heterologously expressed drug metabolizing enzymes—State of the art and prospects for the future, *Pharmacol. Ther.*, 84(2), 121, 1999.
196. Hurst, S., Williams, J.A., and Hansel, S., Reaction phenotyping, in *Drug Metabolism in Drug Design and Development*, Zhang, D., Zhu, M., and Humphreys, W.G. (eds.), John Wiley & Sons, Hoboken, NJ, 477, 2008.
197. Ghosal, A. et al., Identification of human liver cytochrome P450 enzymes involved in biotransformation of vicriviroc, a CCR5 receptor antagonist, *Drug Metab. Dispos.*, 35(12), 2186, 2007.
198. Harper, T.W. and Brassil, P.J., Reaction phenotyping: Current industry efforts to identify enzymes responsible for metabolizing drug candidates, *AAPS J.*, 10(1), 200, 2008.
199. Lu, A.Y., Wang, R.W., and Lin, J.H., Cytochrome P450 in vitro reaction phenotyping: A re-evaluation of approaches used for P450 isoform identification, *Drug Metab. Dispos.*, 31(4), 345, 2003.
200. Bjornsson, T.D. et al., The conduct of in vitro and in vivo drug-drug interaction studies: A PhRMA perspective, *J. Clin. Pharmacol.*, 43(5), 443, 2003.
201. Bjornsson, T.D. et al., The conduct of in vitro and in vivo drug-drug interaction studies: A Pharmaceutical Research and Manufacturers of America (PhRMA) perspective, *Drug Metab. Dispos.*, 31(7), 815, 2003.
202. Marier, J.F. et al., Metabolism and disposition of resveratrol in rats: Extent of absorption, glucuronidation, and enterohepatic recirculation evidenced by a linked-rat model, *J. Pharmacol. Exp. Ther.*, 302(1), 369, 2002.
203. Dickinson, R.G. et al., Disposition of valproic acid in the rat: Dose-dependent metabolism, distribution, enterohepatic recirculation and choleretic effect, *J. Pharmacol. Exp. Ther.*, 211(3), 583, 1979.
204. Teichert, J. et al., Synthesis and characterization of some new phase II metabolites of the alkylator bendamustine and their identification in human bile, urine, and plasma from patients with cholangiocarcinoma, *Drug Metab. Dispos.*, 33(7), 984, 2005.
205. Ghibellini, G. et al., A novel method for the determination of biliary clearance in humans, *AAPS J.*, 6(4), e33, 2004.
206. Ghibellini, G., Leslie, E.M., and Brouwer, K.L., Methods to evaluate biliary excretion of drugs in humans: An updated review, *Mol. Pharm.*, 3(3), 198, 2006.
207. Davit, B. et al., FDA evaluations using in vitro metabolism to predict and interpret in vivo metabolic drug-drug interactions: Impact on labeling, *J. Clin. Pharmacol.*, 39(9), 899, 1999.
208. Zhang, H., Sinz, M.W., and Rodrigues, A.D., Metabolism-mediated drug–drug interactions, in *Drug Metabolism in Drug Design and Development*, Zhang, D., Zhu, M., and Humphreys, W.G. (eds.), John Wiley & Sons, Hoboken, NJ, 113, 2008.
209. Iyer, R. and Zhang, D., Role of drug metabolism in drug development, in *Drug Metabolism in Drug Design and Development*, Zhang, D., Zhu, M., and Humphreys, W.G. (eds.), John Wiley & Sons, Hoboken, NJ, 261, 2008.
210. Roffey, S.J. et al., What is the objective of the mass balance study? A retrospective analysis of data in animal and human excretion studies employing radiolabeled drugs, *Drug Metab. Rev.*, 39(1), 17, 2007.
211. Food and Drug Administration (US FDA), *Guidance for Industry, Investigators, and Reviewers—Exploratory IND Studies*, 1, 2006. http://www.fda.gov/downloads/Drugs/GuidanceComplianceRegulatoryInformation/Guidances/UCM078933.pdf
212. European Medicines Agency (EMEA), *Position Paper on Non-clinical Safety Studies to Support Clinical Trials with a Single Microdose*, 1, 2004. http://www.emea.europa.eu/pdfs/human/swp/259902en.pdf

213. Lappin, G. et al., Use of microdosing to predict pharmacokinetics at the therapeutic dose: Experience with 5 drugs, *Clin. Pharmacol. Ther.*, 80(3), 203, 2006.
214. Sarapa, N. et al., The application of accelerator mass spectrometry to absolute bioavailability studies in humans: Simultaneous administration of an intravenous microdose of 14C-nelfinavir mesylate solution and oral nelfinavir to healthy volunteers, *J. Clin. Pharmacol.*, 45(10), 1198, 2005.
215. Smith, D.A., Johnson, D.E., and Park, B.K., Editorial overview: Use of microdosing to probe pharmacokinetics in humans—Is it too much for too little? *Curr. Opin. Drug Dis. Dev.*, 639, 2003.
216. Buchholz, L.M. et al., LC-MS-MS method development in support of sub-therapeutic preclinical pharmacokinetics, in *Proceedings of the 53rd ASMS Conference on Mass Spectrometry and Allied Topics*, ASMS, San Antonio, TX, 2005.
217. Ni, J. et al., Microdosing assessment to evaluate pharmacokinetics and drug metabolism in rats using liquid chromatography-tandem mass spectrometry, *Pharm. Res.*, 25(7), 1572, 2008.
218. El-Kattan, A., Holliman, C., and Cohen, L.H., Quantitative bioanalysis in drug discovery and development: Principles and applications, in *Mass Spectrometry in Drug Metabolism and Pharmacokinetics*, Ramanathan, R. (ed.), John Wiley & Sons, Hoboken, NJ, 87, 2008.
219. Lam, W.W. et al., Applications of high-sensitivity mass spectrometry and radioactivity detection techniques in drug metabolism studies, in *Mass Spectrometry in Drug Metabolism and Pharmacokinetics*, Ramanathan, R. (ed.), John Wiley & Sons, Hoboken, NJ, 253, 2008.
220. Seto, C. et al., Quantification and identification of metabolites at microdosing levels, in *Proceedings of the 57th ASMS Conference on Mass Spectrometry and Allied Topics*, ASMS, Philadelphia, PA, 2009.
221. Mahmood, I., Application of allometric principles for the prediction of pharmacokinetics in human and veterinary drug development, *Adv. Drug Deliv. Rev.*, 59(11), 1177, 2007.
222. Gomase, V.S. and Tagore, S., Species scaling and extrapolation, *Curr. Drug Metab.*, 9(3), 193, 2008.
223. Waibler, Z. et al., Toward experimental assessment of receptor occupancy: TGN1412 revisited, *J. Allergy Clin. Immunol.*, 122(5), 890, 2008.
224. Liedert, B. et al., Safety of phase I clinical trials with monoclonal antibodies in Germany—The regulatory requirements viewed in the aftermath of the TGN1412 disaster, *Int. J. Clin. Pharmacol. Ther.*, 45(1), 1, 2007.
225. Bhogal, N. and Combes, R., TGN1412: Time to change the paradigm for the testing of new pharmaceuticals, *Altern. Lab Anim.*, 34(2), 225, 2006.
226. Yu, C. et al., A rapid method for quantitatively estimating metabolites in human plasma in the absence of synthetic standards using a combination of liquid chromatography/mass spectrometry and radiometric detection, *Rapid Commun. Mass Spectrom.*, 21(4), 497, 2007.
227. Zhang, D. et al., LC-MS/MS-based approach for obtaining exposure estimates of metabolites in early clinical trials using radioactive metabolites as reference standards, *Drug Metab. Lett.*, 1(4), 1293, 2007.
228. Deroubaix, X. and Coquette, A., The ins and outs of human ADME studies, *Business Briefing: Pharmatech 2004*, 1, 2004.
229. Li, Q. et al., The distribution pattern of intravenous [(14)C] artesunate in rat tissues by quantitative whole-body autoradiography and tissue dissection techniques, *J. Pharm. Biomed. Anal.*, 48(3), 876, 2008.
230. Richter, W.F., Starke, V., and Whitby, B., The distribution pattern of radioactivity across different tissues in quantitative whole-body autoradiography (QWBA) studies, *Eur. J. Pharm. Sci.*, 28(1–2), 155, 2006.

231. Masse, F.X. and Miller, T., Exposure to radiation and informed consent, *Irb*, 7(4), 1, 1985.
232. Ramanathan, R. et al., Disposition of loratadine in healthy volunteers, *Xenobiotica*, 37(7), 753, 2007.
233. Ramanathan, R. et al., Disposition of desloratadine in healthy volunteers, *Xenobiotica*, 37(7), 770, 2007.
234. Mallikaarjun, S. et al., Effects of hepatic or renal impairment on the pharmacokinetics of aripiprazole, *Clin. Pharmacokinet.*, 47(8), 533, 2008.
235. Jemal, M., Ouyang, Z., and Powell, M.L., A strategy for a post-method-validation use of incurred biological samples for establishing the acceptability of a liquid chromatography/tandem mass-spectrometric method for quantitation of drugs in biological samples, *Rapid Commun. Mass Spectrom.*, 16(16), 1538, 2002.
236. Jemal, M., High-throughput quantitative bioanalysis by LC/MS/MS, *Biomed. Chromatogr.*, 14(6), 422, 2000.
237. Onorato, J.M. et al., Selected reaction monitoring LC-MS determination of idoxifene and its pyrrolidinone metabolite in human plasma using robotic high-throughput, sequential sample injection, *Anal. Chem.*, 73(1), 119, 2001.
238. Poquette, M.A., Lensmeyer, G.L., and Doran, T.C., Effective use of liquid chromatography-mass spectrometry (LC/MS) in the routine clinical laboratory for monitoring sirolimus, tacrolimus, and cyclosporine, *Ther. Drug Monit.*, 27(2), 144, 2005.
239. Vengurlekar, S.S. et al., A sensitive LC-MS/MS assay for the determination of dextromethorphan and metabolites in human urine—Application for drug interaction studies assessing potential CYP3A and CYP2D6 inhibition, *J. Pharm. Biomed. Anal.*, 30(1), 113, 2002.
240. Egnash, L.A. and Ramanathan, R., Comparison of heterogeneous and homogeneous radioactivity flow detectors for simultaneous profiling and LC-MS/MS characterization of metabolites, *J. Pharm. Biomed. Anal.*, 27(1–2), 271, 2002.
241. Cuyckens, F. et al., Improved liquid chromatography—Online radioactivity detection for metabolite profiling, *J. Chromatogr. A*, 1209(1–2), 128, 2008.
242. Nassar, A.E., Bjorge, S.M., and Lee, D.Y., On-line liquid chromatography-accurate radioisotope counting coupled with a radioactivity detector and mass spectrometer for metabolite identification in drug discovery and development, *Anal. Chem.*, 75(4), 785, 2003.
243. Nassar, A.E. and Lee, D.Y., Novel approach to performing metabolite identification in drug metabolism, *J. Chromatogr. Sci.*, 45(3), 113, 2007.
244. Nassar, A.E. et al., Liquid chromatography-accurate radioisotope counting and microplate scintillation counter technologies in drug metabolism studies, *J. Chromatogr. Sci.*, 42(7), 348, 2004.
245. Lee, D.Y., Anderson, J.J., and Ryan, D.L., LC-ARC: A novel sensitive in-line detection system/method for radio-HPLC, in *7th International Symposium of IIS*, Dresden, Germany, 2000.
246. Boernsen, K.O., Floeckher, J.M., and Bruin, G.J.M., Use of a microplate scintillation counter as a radioactivity detector for miniaturized separation techniques in drug metabolism, *Anal. Chem.*, 72, 3956, 2000.
247. Kiffe, M., Jehle, A., and Ruembeli, R., Combination of high-performance liquid chromatography and microplate scintillation counting for crop and animal metabolism studies: A comparison with classical on-line and thin-layer chromatography radioactivity detection, *Anal. Chem.*, 75(4), 723, 2003.
248. Zhu, M., Zhang, D., and Skiles, G.L., Quantification and structural elucidation of low quantities of radiolabeled metabolites using microplate scintillation counting techniques in conjunction with LC-MS, in *Identification and Quantification of Drugs, Metabolites and Metabolizing Enzymes by LC-MS*, Chowdhury, S.K. (ed.), Elsevier, Amsterdam, the Netherlands, 195, 2005.

249. Nedderman, A.N. et al., The use of 96-well Scintiplates to facilitate definitive metabolism studies for drug candidates, *J. Pharm. Biomed. Anal.*, 34(3), 607, 2004.
250. Zhang, D. et al., Comparative metabolism of radiolabeled muraglitazar in animals and humans by quantitative and qualitative metabolite profiling, *Drug Metab. Dispos.*, 35(1), 150, 2007.
251. Bruin, G.J. et al., A microplate solid scintillation counter as a radioactivity detector for high performance liquid chromatography in drug metabolism: Validation and applications, *J. Chromatogr. A*, 1133(1–2), 184, 2006.
252. Dear, G.J. et al., TopCount coupled to ultra-performance liquid chromatography for the profiling of radiolabeled drug metabolites in complex biological samples, *J. Chromatogr., B: Anal. Technol. Biomed. Life Sci.*, 844(1), 96, 2006.
253. Dear, G.J. et al., Utilizing a −100°C microplate CCD imager, yttrium silicate coated 384-microplates and ultra-performance liquid chromatography for improved profiling of radiolabeled drug metabolites in complex biological samples, *J. Chromatogr. B: Anal. Technol. Biomed. Life Sci.*, 868(1–2), 49, 2008.
254. Turteltaub, K.W. and Vogel, J.S., Bioanalytical applications of accelerator mass spectrometry for pharmaceutical research, *Curr. Pharm. Des.*, 6(10), 991, 2000.
255. Bennett, C.L. et al., Radiocarbon dating using electrostatic accelerators: Negative ions provide the key, *Science*, 198, 508, 1977.
256. Nelson, D.E., Korteling, R.G., and Scott, W.R., Carbon-14: Direct detection at natural concentrations, *Science*, 198, 507, 1977.
257. Vogel, G. and Browse, J., Preparation of radioactively labeled synthetic sn-1,2-diacylglycerols for studies of lipid metabolism, *Anal. Biochem.*, 224(1), 61, 1995.
258. Garner, R.C. et al., Evaluation of accelerator mass spectrometry in a human mass balance and pharmacokinetic study—Experience with 14C-labeled (R)-6-[amino(4-chlorophenyl)(1-methyl-1H-imidazol-5-yl)methyl]-4-(3-chlorophenyl)-1-methyl-2(1H)-quinolinone (R115777), a farnesyl transferase inhibitor, *Drug Metab. Dispos.*, 30(7), 823, 2002.
259. Garner, R.C. et al., Comparison of the absorption of micronized (Daflon 500 mg) and nonmicronized 14C-diosmin tablets after oral administration to healthy volunteers by accelerator mass spectrometry and liquid scintillation counting, *J. Pharm. Sci.*, 91(1), 32, 2002.
260. Lappin, G. and Garner, R.C., Ultra-sensitive detection of radiolabelled drugs and their metabolites using accelerator mass spectrometry, in *Bioanalytical Separations: Handbook of Analytical Separation*, Wilson, I.D. (ed.), Vol. 4, Elsevier, Amsterdam, the Netherlands, 331, 2003.
261. Turteltaub, K.W. and Dingley, K.H., Application of accelerated mass spectrometry (AMS) in DNA adduct quantification and identification, *Toxicol. Lett.*, 102–103, 435, 1998.
262. Ranasinghe, A. et al., Is standard-free quantitation of metabolites possible? in *Proceedings of the 56th ASMS Conference on Mass Spectrometry and Allied Topics*, ASMS, Denver, CO, 2008.
263. Browne, T.R., Stable isotopes in pharmacology studies: Present and future, *J. Clin. Pharmacol.*, 26(6), 485, 1986.
264. Abramson, F.P. et al., Replacing 14C with stable isotopes in drug metabolism studies, *Drug Metab. Dispos.*, 24(7), 697, 1996.
265. Chowdhury, S.K. et al., Detection and characterization of highly polar metabolites by LC-MS: Proper selection of LC column and use of stable isotope labeled drug to study metabolism of rabavirin in rats, in *Identification and Quantification of Drugs, Metabolites and Metabolizing Enzymes by LC-MS*, Chowdhury, S.K. (ed.), Elsevier, Amsterdam, the Netherlands, 277, 2005.
266. Mutlib, A.E., Application of stable isotope-labeled compounds in metabolism and in metabolism-mediated toxicity studies, *Chem. Res. Toxicol.*, 21(9), 1672, 2008.

267. Lim, H.K. et al., A generic method to detect electrophilic intermediates using isotopic pattern triggered data-dependent high-resolution accurate mass spectrometry, *Rapid Commun. Mass Spectrom.*, 22(8), 1295, 2008.
268. Martin, S.W. et al., Pharmacokinetics and metabolism of the novel anticonvulsant agent N-(2,6-dimethylphenyl)-5-methyl-3-isoxazolecarboxamide (D2624) in rats and humans, *Drug Metab. Dispos.*, 25(1), 40, 1997.
269. Baillie, T.A. et al., Mechanistic studies on the reversible metabolism of rofecoxib to 5-hydroxyrofecoxib in the rat: Evidence for transient ring opening of a substituted 2-furanone derivative using stable isotope-labeling techniques, *Drug Metab. Dispos.*, 29(12), 1614, 2001.
270. Cuyckens, F. et al., Extracting relevant data out of the MS background, in *Proceedings of the 56th ASMS Conference on Mass Spectrometry and Allied Topics*, ASMS, Denver, CO, 2008.
271. Cuyckens, F. et al., Use of the bromine isotope ratio in HPLC-ICP-MS and HPLC-ESI-MS analysis of a new drug in development, *Anal. Bioanal. Chem.*, 390(7), 1717, 2008.
272. Lam, W. and Ramanathan, R., In electrospray ionization source hydrogen/deuterium exchange LC-MS and LC-MS/MS for characterization of metabolites, *J. Am. Soc. Mass Spectrom.*, 13(4), 345, 2002.
273. Liu, D.Q. et al., Use of on-line hydrogen/deuterium exchange to facilitate metabolite identification, *Rapid Commun. Mass Spectrom.*, 15(19), 1832, 2001.
274. Chen, G. et al., Structural characterization of in vitro rat liver microsomal metabolites of antihistamine desloratadine using LTQ-Orbitrap hybrid mass spectrometer in combination with online hydrogen/deuterium exchange HR-LC/MS, *J. Mass Spectrom.*, 44(2), 203, 2009.
275. Liu, D.Q. and Hop, C.E., Strategies for characterization of drug metabolites using liquid chromatography-tandem mass spectrometry in conjunction with chemical derivatization and on-line H/D exchange approaches, *J. Pharm. Biomed. Anal.*, 37(1), 1, 2005.
276. Ni, J. et al., Characterization of benzimidazole and other oxidative and conjugative metabolites of brimonidine in vitro and in rats in vivo using on-line HID exchange LC-MS/MS and stable-isotope tracer techniques, *Xenobiotica*, 37(2), 205, 2007.
277. Balani, S.K. et al., Metabolites of caspofungin acetate, a potent antifungal agent, in human plasma and urine, *Drug Metab. Dispos.*, 28(11), 1274, 2000.
278. Salomonsson, M.L., Bondesson, U., and Hedeland, M., Structural evaluation of the glucuronides of morphine and formoterol using chemical derivatization with 1,2-dimethylimidazole-4-sulfonyl chloride and liquid chromatography/ion trap mass spectrometry, *Rapid Commun. Mass Spectrom.*, 22(17), 2685, 2008.
279. Miao, Z., Kamel, A., and Prakash, C., Characterization of a novel metabolite intermediate of ziprasidone in hepatic cytosolic fractions of rat, dog, and human by ESI-MS/MS, hydrogen/deuterium exchange, and chemical derivatization, *Drug Metab. Dispos.*, 33(7), 879, 2005.
280. Kondo, T. et al., Characterization of conjugated metabolites of a new angiotensin II receptor antagonist, candesartan cilexetil, in rats by liquid chromatography/electrospray tandem mass spectrometry following chemical derivatization, *J. Mass Spectrom.*, 31(8), 873, 1996.
281. Cui, D. and Harvison, P.J., Determination of the site of glucuronidation in an N-(3,5-dichlorophenyl)succinimide metabolite by electrospray ionization tandem mass spectrometry following derivatization to picolinyl esters, *Rapid Commun. Mass Spectrom.*, 14(21), 1985, 2000.
282. Liu, Y. et al., In situ derivatization/solid-phase microextraction coupled with gas chromatography/negative chemical ionization mass spectrometry for the determination of trichloroethylene metabolites in rat blood, *Rapid Commun. Mass Spectrom.*, 22(7), 1023, 2008.

283. Gu, Y. et al., Comparison of HPLC with electrochemical detection and LC-MS/MS for the separation and validation of artesunate and dihydroartemisinin in animal and human plasma, *J. Chromatogr. B Analyt. Technol. Biomed. Life Sci.*, 867(2), 213, 2008.
284. Lohmann, W. and Karst, U., Generation and identification of reactive metabolites by electrochemistry and immobilized enzymes coupled on-line to liquid chromatography/mass spectrometry, *Anal. Chem.*, 79(17), 6831, 2007.
285. King, R. et al., On-line electrochemical oxidation used with HPLC-MS for the study of reactive drug intermediates, in *Proceedings of the 52nd ASMS Conference on Mass Spectrometry and Allied Topics*, Nashville, TN, 2004.
286. Gamache, P.H. et al., Online electrochemical–LC–MS techniques for profiling and characterizing metabolites and degradants, in *Mass Spectrometry in Drug Metabolism and Pharmacokinetics*, Ramanathan, R. (ed.), John Wiley & Sons, Hoboken, NJ, 275, 2008.
287. Tiller, P.R. and Romanyshyn, L.A., Liquid chromatography/tandem mass spectrometric quantification with metabolite screening as a strategy to enhance the early drug discovery process, *Rapid Commun. Mass Spectrom.*, 16(12), 1225, 2002.
288. Poon, G.K. et al., Integrating qualitative and quantitative liquid chromatography/tandem mass spectrometric analysis to support drug discovery, *Rapid Commun. Mass Spectrom.*, 13(19), 1943, 1999.
289. Tiller, P.R. et al., High-throughput, accurate mass liquid chromatography/tandem mass spectrometry on a quadrupole time-of-flight system as a "first-line" approach for metabolite identification studies, *Rapid Commun. Mass Spectrom.*, 22(7), 1053, 2008.
290. Nagele, E. and Fandino, A.S., Simultaneous determination of metabolic stability and identification of buspirone metabolites using multiple column fast liquid chromatography time-of-flight mass spectrometry, *J. Chromatogr. A*, 1156(1–2), 196, 2007.
291. Pelander, A., Tyrkko, E., and Ojanpera, I., In silico methods for predicting metabolism and mass fragmentation applied to quetiapine in liquid chromatography/time-of-flight mass spectrometry urine drug screening, *Rapid Commun. Mass Spectrom.*, 23(4), 506, 2009.
292. Kammerer, B. et al., In vitro metabolite identification of ML3403, a 4-pyridinylimidazole-type p38 MAP kinase inhibitor by LC-Qq-TOF-MS and LC-SPE-cryo-NMR/MS, *Xenobiotica*, 37(3), 280, 2007.
293. Tong, W. et al., Identification of unstable metabolites of Lonafarnib using liquid chromatography-quadrupole time-of-flight mass spectrometry, stable isotope incorporation and ion source temperature alteration, *J. Mass Spectrom.*, 41(11), 1430, 2006.
294. Miao, X.S. et al., Identification of the in vitro metabolites of 3,4-dihydro-2,2-dimethyl-2H-naphthol[1,2-b]pyran-5,6-dione (ARQ 501; beta-lapachone) in whole blood, *Drug Metab. Dispos.*, 36(4), 641, 2008.
295. Timischl, B., Dettmer, K., and Oefner, P.J., Hyphenated mass spectrometry in the analysis of the central carbon metabolism, *Anal. Bioanal. Chem.*, 391(3), 895, 2008.
296. Fandino, A.S., Nagele, E., and Perkins, P.D., Automated software-guided identification of new buspirone metabolites using capillary LC coupled to ion trap and TOF mass spectrometry, *J. Mass Spectrom.*, 41(2), 248, 2006.
297. Sleno, L. et al., Investigating the in vitro metabolism of fipexide: Characterization of reactive metabolites using liquid chromatography/mass spectrometry, *Rapid Commun. Mass Spectrom.*, 21(14), 2301, 2007.
298. Wang, Y. et al., Metabolite identification of arbidol in human urine by the study of CID fragmentation pathways using HPLC coupled with ion trap mass spectrometry, *J. Mass Spectrom.*, 43(8), 1099, 2008.

299. Ren, L. et al., Characterization of the in vivo and in vitro metabolic profile of PAC-1 using liquid chromatography-mass spectrometry, *J. Chromatogr. B Analyt. Technol. Biomed. Life Sci.*, 876(1), 47, 2008.
300. Zhang, N. et al., Quantification and rapid metabolite identification in drug discovery using API time-of-flight LC/MS, *Anal. Chem.*, 72(4), 800, 2000.
301. Decaestecker, T.N. et al., Evaluation of automated single mass spectrometry to tandem mass spectrometry function switching for comprehensive drug profiling analysis using a quadrupole time-of-flight mass spectrometer, *Rapid Commun. Mass Spectrom.*, 14(19), 1787, 2000.
302. Weaver, P.J., Laures, A.M.F., and Wolff, J.C., Investigation of the advanced functionalities of a hybrid quadrupole orthogonal acceleration time-of-flight mass spectrometer, *Rapid Commun. Mass Spectrom.*, 21(15), 2415, 2007.
303. Michael, S.M., Chien, M., and Lubman, D.M., An ion trap storage/time-of-flight mass spectrometer, *Rev. Sci. Instrum.*, 63, 4277, 1992.
304. Hashimoto, Y. et al., Orthogonal trap time-of-flight mass spectrometer using a collisional damping chamber, *Rapid Commun. Mass Spectrom.*, 19(2), 221, 2005.
305. Hopfgartner, G. et al., A statistical based approach in metabolite identification by high mass accuracy MS^n analysis, in *Proceedings of the 55th ASMS Conference on Mass Spectrometry and Allied Topics*, ASMS, Indianapolis, IN, 2007.
306. Zurek, G. et al., Rapid metabolite identification using accurate and MS^n mass spectra in combination with smart processing tools, *LC-GC Europe*, 24, 2004.
307. Zurek, G. et al., Novel strategies for metabolite identification using HPLC-ion trap mass spectrometry, *Chem. Mag.*, (3), 34, 2003.
308. Hopfgartner, G. and Varesio, E., Application of mass spectrometry for quantitative and qualitative analysis in life sciences, *Chimia*, 59(6), 321, 2005.
309. Kantharaj, E. et al., Simultaneous measurement of metabolic stability and metabolite identification of 7-methoxymethylthiazolo[3,2-a]pyrimidin-5-one derivatives in human liver microsomes using liquid chromatography/ion-trap mass spectrometry, *Rapid Commun. Mass Spectrom.*, 19(8), 1069, 2005.
310. Kantharaj, E. et al., Simultaneous measurement of drug metabolic stability and identification of metabolites using ion-trap mass spectrometry, *Rapid Commun. Mass Spectrom.*, 17(23), 2661, 2003.
311. Ramanathan, R. et al., Metabolism and excretion of desloratadine (Clarinex) in mice, rats, monkeys, and humans, *Drug Metab. Rev.*, 38(Suppl. 2), 160, 2006.
312. Wang, L. et al., Identification of the human enzymes involved in the oxidative metabolism of dasatinib: An effective approach for determining metabolite formation kinetics, *Drug Metab. Dispos.*, 36(9), 1828, 2008.
313. Christopher, L.J. et al., Biotransformation of [14C]dasatinib: In vitro studies in rat, monkey, and human and disposition after administration to rats and monkeys, *Drug Metab. Dispos.*, 36(7), 1341, 2008.
314. Christopher, L.J. et al., Metabolism and disposition of dasatinib after oral administration to humans, *Drug Metab. Dispos.*, 36(7), 1357, 2008.
315. Slatter, J.G. et al., Pharmacokinetics, metabolism, and excretion of irinotecan (CPT-11) following I.V. infusion of [(14)C]CPT-11 in cancer patients, *Drug Metab. Dispos.*, 28(4), 423, 2000.
316. Mitchell, A.E. et al., Quantitative profiling of tissue- and gender-related expression of glutathione S-transferase isoenzymes in the mouse, *Biochem. J.*, 325 (Pt 1), 207, 1997.
317. Hakala, K.S., Kostiainen, R., and Ketola, R.A., Feasibility of different mass spectrometric techniques and programs for automated metabolite profiling of tramadol in human urine, *Rapid Commun. Mass Spectrom.*, 20(14), 2081, 2006.
318. Schwartz, J.C., Senko, M.W., and Syka, J.E.P., A two-dimensional quadrupole ion trap mass spectrometer, *J. Am. Soc. Mass Spectrom.*, 13, 659, 2002.

319. Hager, J.W., A new linear ion trap mass spectrometer, *Rapid Commun. Mass Spectrom.*, 16, 512, 2002.
320. Le Blanc, J.C. et al., Unique scanning capabilities of a new hybrid linear ion trap mass spectrometer (Q TRAP) used for high sensitivity proteomics applications, *Proteomics*, 3(6), 859, 2003.
321. Hager, J.W. and Le Blanc, J.C., High-performance liquid chromatography-tandem mass spectrometry with a new quadrupole/linear ion trap instrument, *J. Chromatogr. A*, 1020(1), 3, 2003.
322. Hager, J.W. and Yves Le Blanc, J.C., Product ion scanning using a Q-q-Q linear ion trap (Q TRAP) mass spectrometer, *Rapid Commun. Mass Spectrom.*, 17(10), 1056, 2003.
323. Hopfgartner, G. et al., Triple quadrupole linear ion trap mass spectrometer for the analysis of small molecules and macromolecules, *J. Mass Spectrom.*, 39(8), 845, 2004.
324. King, R. and Fernandez-Metzler, C., The use of Qtrap technology in drug metabolism, *Curr. Drug Metab.*, 7(5), 541, 2006.
325. Hopfgartner, G. and Zell, M., Q Trap MS: A new tool for metabolite identification, in *Using Mass Spectrometry for Drug Metabolism Studies*, Korfmacher, W.A. (ed.), CRC Press, Boca Raton, FL, 277, 2005.
326. Jones, E.B., Quadrupole, triple-quadrupole, and hybrid linear ion trap mass spectrometers for metabolite analysis, in *Mass Spectrometry in Drug Metabolism and Pharmacokinetics*, Ramanathan, R. (ed.), John Wiley & Sons, Hoboken, NJ, 123, 2008.
327. Li, A.C. et al., Simultaneously quantifying parent drugs and screening for metabolites in plasma pharmacokinetic samples using selected reaction monitoring information-dependent acquisition on a QTrap instrument, *Rapid Commun. Mass Spectrom.*, 19(14), 1943, 2005.
328. Xia, Y.Q. et al., Use of a quadrupole linear ion trap mass spectrometer in metabolite identification and bioanalysis, *Rapid Commun. Mass Spectrom.*, 17(11), 1137, 2003.
329. Yao, M. et al., Rapid screening and characterization of drug metabolites using a multiple ion monitoring-dependent MS/MS acquisition method on a hybrid triple quadrupole-linear ion trap mass spectrometer, *J. Mass Spectrom.*, 43(10), 1364, 2008.
330. Zheng, J. et al., Screening and identification of GSH-trapped reactive metabolites using hybrid triple quadruple linear ion trap mass spectrometry, *Chem. Res. Toxicol.*, 20, 757, 2007.
331. Fitch, W.L. et al., Application of polarity switching in the identification of the metabolites of RO9237, *Rapid Commun. Mass Spectrom.*, 21(10), 1661, 2007.
332. Hu, Q. et al., The Orbitrap: A new mass spectrometer, *J. Mass Spectrom.*, 40(4), 430, 2005.
333. Makarov, A. et al., Performance evaluation of a hybrid linear ion trap/orbitrap mass spectrometer, *Anal. Chem.*, 78(7), 2113, 2006.
334. Makarov, A. et al., Dynamic range of mass accuracy in LTQ Orbitrap hybrid mass spectrometer, *J. Am. Soc. Mass Spectrom.*, 17(7), 977, 2006.
335. Erickson, B., Linear ion trap/Orbitrap mass spectrometer, *Anal. Chem.*, 78(7), 2089, 2006.
336. Perry, R.H., Cooks, R.G., and Noll, R.J., Orbitrap mass spectrometry: Instrumentation, ion motion and applications, *Mass Spectrom. Rev.*, 27, 661, 2008.
337. Peterman, S.M. et al., Application of a linear ion trap/orbitrap mass spectrometer in metabolite characterization studies: Examination of the human liver microsomal metabolism of the non-tricyclic anti-depressant nefazodone using data-dependent accurate mass measurements, *J. Am. Soc. Mass Spectrom.*, 17(3), 363, 2006.

338. Ruan, Q. et al., An integrated method for metabolite detection and identification using a linear ion trap/Orbitrap mass spectrometer and multiple data processing techniques: Application to indinavir metabolite detection, *J. Mass Spectrom.*, 43(2), 251, 2008.
339. Zhang, H., Zhang, D., and Ray, K., A software filter to remove interference ions from drug metabolites in accurate mass liquid chromatography/mass spectrometric analyses, *J. Mass Spectrom.*, 38(10), 1110, 2003.
340. Zhu, M. et al., Detection and characterization of metabolites in biological matrices using mass defect filtering of liquid chromatography/high resolution mass spectrometry data, *Drug Metab. Dispos.*, 34(10), 1722, 2006.
341. Mayol, R.F. et al., Characterization of the metabolites of the antidepressant drug nefazodone in human urine and plasma, *Drug Metab. Dispos.*, 22(2), 304, 1994.
342. Kalgutkar, A.S. et al., Metabolic activation of the nontricyclic antidepressant trazodone to electrophilic quinone-imine and epoxide intermediates in human liver microsomes and recombinant P4503A4, *Chem. Biol. Interact.*, 155(1–2), 10, 2005.
343. Zhu, M. et al., Detection and structural characterization of glutathione-trapped reactive metabolites using liquid chromatography-high-resolution mass spectrometry and mass defect filtering, *Anal. Chem.*, 79(21), 8333, 2007.
344. Bateman, K.P. et al., MSE with mass defect filtering for in vitro and in vivo metabolite identification, *Rapid Commun. Mass Spectrom.*, 21(9), 1485, 2007.
345. Zhang, H. and Ray, K., Profiling impurities in a taxol formulation: Application of a mass-dependent mass defect filter to remove polymeric excipient interferences, in *Proceedings of the 55th ASMS Conference on Mass Spectrometry and Allied Topics*, ASMS, Indianapolis, IN, 2007.
346. Zhang, H. et al., Mass defect profiles of biological matrices and the general applicability of mass defect filtering for metabolite detection, *Rapid Commun. Mass Spectrom.*, 22(13), 2082, 2008.
347. Zhang, H. et al., An algorithm for thorough background subtraction from high-resolution LC/MS data: Application to the detection of troglitazone metabolites in rat plasma, bile, and urine, *J. Mass Spectrom.*, 43(9), 1191, 2008.
348. Zhu, P. et al., A retention-time-shift-tolerant background-subtraction and noise-reduction algorithm (BgS-NoRA) for extraction of drug metabolites in LC-MS Data, in *Proceedings of the 57th ASMS Conference on Mass Spectrometry and Allied Topics*, ASMS, Philadelphia, PA, 2009.
349. Wright, P., Miao, Z., and Shilliday, B., Metabolite quantitation: Detector technology and MIST implications, *Bioanalysis*, 1(4), 831, 2009.

6 Reactive Metabolite Screening and Covalent-Binding Assays

Gérard Hopfgartner

CONTENTS

6.1 INTRODUCTION

Drug metabolism follows the biotransformation of endogenous compounds and xenobiotics into more polar compounds that can then be eliminated from the body. While most metabolites are less active than the parent drug, bioactivation can generate reactive metabolites that can covalently bind to macromolecules and lead to unexpected toxicities [1]. The understanding of the metabolic fate and the pharmacokinetics of a drug and its metabolites plays a major role in pharmaceutical drug discovery and development [2–4]. Adverse drug reactions (ADRs) have become a major reason for drug failure in the clinical testing phase. Severe adverse effects are believed to be one of the leading causes of death in the United States, and it has been estimated that in 1994 over 2 million hospitalized patients had serious ADR, and 106,000 had fatal outcomes [5]. The effects are critical because they occur mainly once a drug is in the market and are then difficult to predict [5–7]. Adverse effects can be classified in two categories: the first as pharmacology-related toxicity and the second as not related to a pharmacological effect. For the second category, often the compounds form reactive metabolites that are thought to cause toxicity by covalently altering cellular macromolecules such as proteins and DNA [1,5,8]. These reactive metabolites (Figure 6.1) can be generated by metabolic phase I reactions (oxidation, reduction), for example, catalyzed by cytochrome P450s (CYP), flavin monooxygenases (FMO), and peroxidases such as myeloperoxidase (MPO), cyclooxygenases (COX), and also by phase II reactions (conjugation), for example, catalyzed by UDP glucuronyltransferases (UGT) or sulfotransferases (SULT) [8]. One of the demonstrated hypotheses (hapten hypothesis) is that these metabolites covalently bind to proteins and the immunogenic adduct leads to idiosyncratic drug reactions (IDRs) [5,9,10]. Therefore, knowledge of the formation of reactive metabolites is considered critical and has led to their screening in the early phase of drug development.

Reactive metabolites can be divided into two categories: the first one includes radicals that can by a cascade mechanism generate further radicals and the second one includes electrophiles that are classified as hard or soft. The low concentration and the unstable nature of these metabolites make direct detection challenging. The most commonly used strategy to detect these intermediates is to screen either *in vitro* or *in vivo* their trapping products using nucleophilic compounds such as glutathione, cyanide, methoxylamine, β-mercaptoethanol, and their protein adducts using mass spectrometry approaches. Several reviews have already have been recently published on the analysis of reactive metabolites by mass spectrometry [11,12]. Liquid chromatography coupled to mass spectrometry (LC–MS) plays a major role in drug metabolism studies and faces two major challenges [12,13]. The first challenge is to fish out minor low molecular weight metabolites in the presence of a significant background of endogenous compounds. The second challenge is that the structural identification is often solely based on mass spectrometric techniques. In the past, when LC–MS was used for drug metabolism studies, typically these were performed on triple quadrupole instruments.

Recently, significant progress has been made in the field of mass spectrometry and a large variety of instruments are now available including two-dimensional and three-dimensional ion traps (ITs), quadrupole time-of-flight (QqTOF), triple quadrupole linear ion trap (QqQ_{LIT}), Fourier transform ion cyclotron resonance (FT-ICR), and Orbitrap systems. Each instrument has strengths and limitations, and a similar metabolic investigation can be solved successfully using selective scan modes, MS^n, or high-resolution/accurate mass strategies. The aim of the present chapter is to

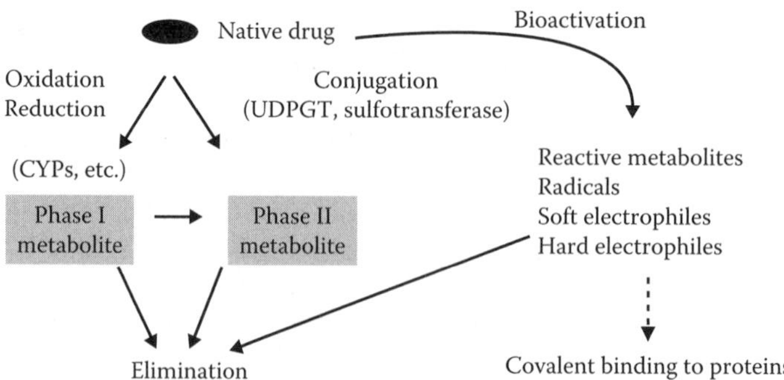

FIGURE 6.1 Common metabolic pathway of pharmaceuticals.

present a comprehensive status of the current strategies to detect reactive metabolites and their covalent adducts based on mass spectrometry.

6.2 IDENTIFICATION OF REACTIVE METABOLITES BY TRAPPING TECHNIQUES AND MASS SPECTROMETRIC DETECTION

6.2.1 *In Vitro* and *In Vivo* Reactive Metabolite Profiling

Due to the short lifetimes of electrophilic reactive intermediates, the strategy of trapping these intermediates *in situ* is widely used. The most commonly used *in vitro* assay to screen for reactive metabolites is liver microsome. Although CYP 450 enzymes play an important role in generating reactive metabolites, other bioactivation pathways should be kept in mind when choosing the *in vitro* test systems, that is, the addition of different cofactors and the use of different *in vitro* systems, for example, hepatocytes or neutrophils. Different nucleophiles are used as trapping agents such as thiols (e.g., glutathione [GSH], *N*-acetylcysteine), amines (semicarbazide and methoxylamine), and cyanide anions. In general "soft nucleophiles" tend to react with "soft electrophiles" and "hard nucleophiles" tend to react with "hard electrophiles." The most widely used trapping agent, GSH, a soft nucleophile, has been shown to react with quinoneimines, nitrenium ions, arene oxides, quinones, imine methides, and Michael acceptors [14]. Hard electrophiles such as iminium species and aldehydes and ketones will react with cyanide or amines, respectively. The formed adducts can be subsequently characterized for structure elucidation and quantitated.

6.2.2 Glutathione Adducts

An indirect approach to monitor the formation of soft electrophilic reactive metabolites is to detect, by LC–MS, their adducts with the endogenous tripeptide, γ-glutamylcysteinyl-glycine (GSH). With liver being the target organ, the abundance of GSH conjugates formed in microsomal incubations is determined by several factors (content of CYP enzyme, substrate concentration, extent of bioactivation, trapping efficiency of GSH) and the amounts of formed adducts are not necessarily proportional to the levels of the corresponding reactive metabolite. Traditionally, the formation of glutathione adducts has been monitored by triple quadrupole mass spectrometry using the constant neutral loss (NL) scanning of 129 Da, which is based on loss from the protonated molecule of pyroglutamic acid, derived from the glutamate residue of the conjugate (Figure 6.2) [15]. However, this approach suffers from several drawbacks: (1) detection of false positives due to possible loss of 129 Da from various endogenous analytes; (2) the difficulty of finding a generic collision energy in a collision-induced dissociation (CID) that works for most compounds; (3) the capture by glutathione favors the formation of doubly charged precursors [8]; and (4) not all classes of glutathione adduct afford the 129 Da NL. A broader screening of all classes of glutathione adducts can be performed by using a precursor ion (PI) mode scan of *m/z* 272 in the negative ion mode as has been demonstrated by Dieckhaus et al. [16]. The negatively charged fragment observed at *m/z* 272 corresponds to the elimination of H_2S to the deprotonated glutamyl dehydroalanyl glycin (Figure 6.2). Figure 6.3 illustrates the

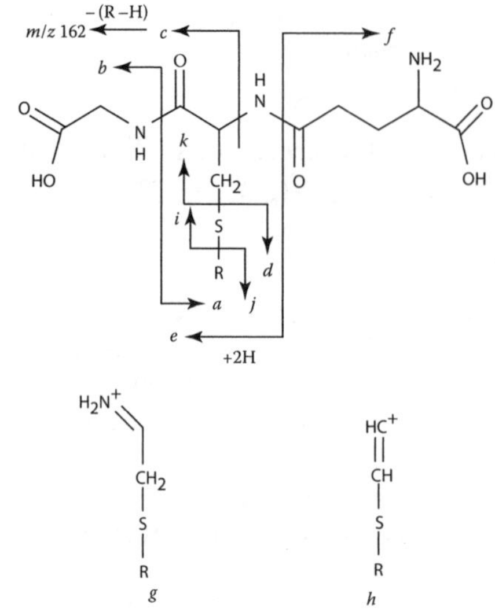

(a)

Fragment ion	m/z
a	$MH^+ - 75$
b	76
c	$MH^+ - 146$
c − (R − H)	162
c − (H_2O)	$MH^+ - 164$
d	$MH^+ - 273$
e	$MH^+ - 129$
e − H_2O	$MH^+ - 147$
f	130
g	$MH^+ - 232$
g–H_2O	$MH^+ - 250$
h	$MH^+ - 249$
i	308
j	$MH^+ - 305/307$
k	274

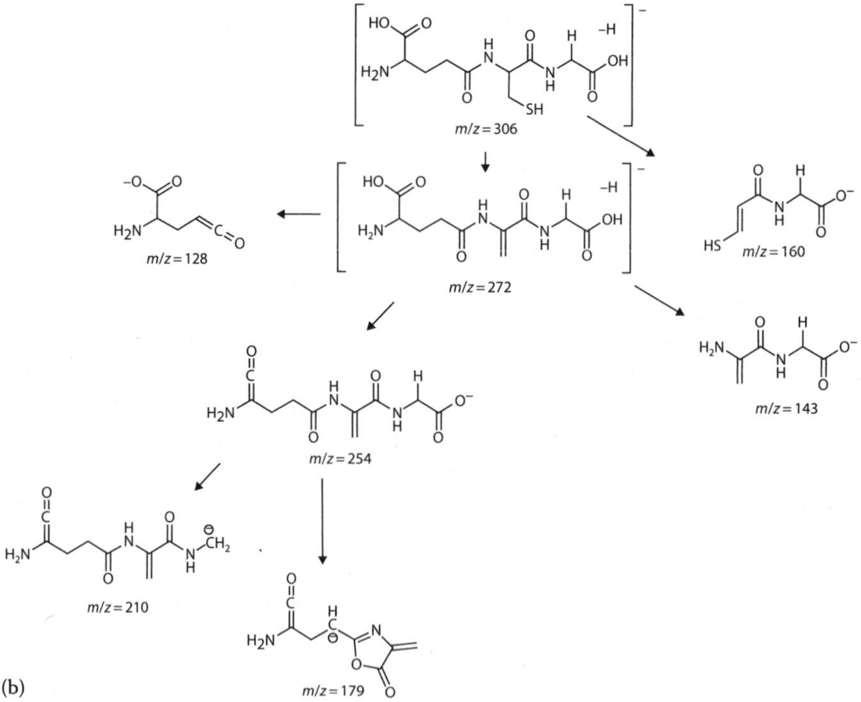

(b)

FIGURE 6.2 (a) Characteristic fragments from glutathione conjugate in positive mode. (From Ma, S. and Subramanian, R., *J. Mass Spectrom.*, 41, 1121, 2006. With permission.) (b) Characteristic fragments from glutathione conjugate in negative mode. (From Dieckhaus, C.M. et al., *Chem. Res. Toxicol.*, 18, 630, 2005. With permission.)

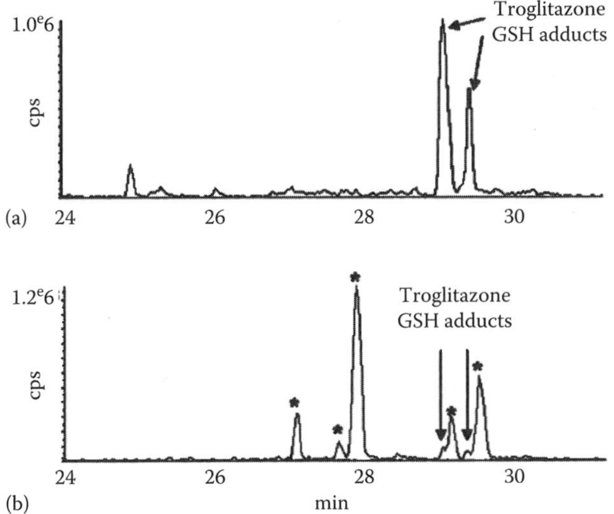

FIGURE 6.3 LC–MS/MS analysis of bile from a rat dosed with troglitazone (a) using the negative ion precursor of *m/z* 272 scan and (b) the positive ion NL of 129 Da scan. (From Dieckhaus, C.M. et al., *Chem. Res. Toxicol.*, 18, 630, 2005. With permission.)

liquid chromatography–tandem mass spectrometry (LC–MS/MS) analysis of dog bile after oral administration of troglitazone. The NL of 129 Da in the positive ion mode does not allow one to clearly identify the adducts, while in the negative ion mode the glutathione adduct can be seen by using the *m/z* 272 PI mode. However, negative mode product ion spectra are rather noninformative because they show mainly fragments of the glutathione. Therefore, in most cases multiple experiments need to be run in both positive and negative modes in order to determine the structure of the metabolite. While in the early days mainly triple quadrupole instruments were available for drug metabolism studies, with the development of data-dependent software tools and new high-resolution instruments, novel strategies have been developed to screen for reactive metabolites. For example, to overcome the limited selectivity of the NL of the pyroglutamique moiety, Castro-Perez et al. [17] developed a screening procedure using exact mass measurements and MS/MS on a hybrid QqTOF mass spectrometer. Exact NL detection is achieved via sequential low- and high-energy MS acquisitions (typically 5 and 20 eV). In this mode the quadrupole mass analyzer is operated in the wide band mass mode. After detection of the loss of the pyroglutamic acid moiety (129.0426 Da), using a window of ±20 mDa on the high-energy scan, MS/MS is carried out on the parent mass of interest to confirm the common NL.

Another strategy for the detection of glutathione adducts is to rely on cluster analysis approaches in which a labeled form and a nonlabeled form of glutathione are used at a defined ratio. Various labels (Table 6.1) have been proposed and Mutlib et al. [18] used a deuterium labeled in the glutamate moiety (mass difference 3 Da) and switched later to a GSH-labeled $^{13}C_2$-^{15}N in the glycine moiety (mass difference 3 Da) [19]. If the data-dependent acquisition mode is positive in the survey scan, then the

TABLE 6.1
Structures of Various GSH Derivatives

Name	Structure
Glutathione (GSH) MW = 307.3 u	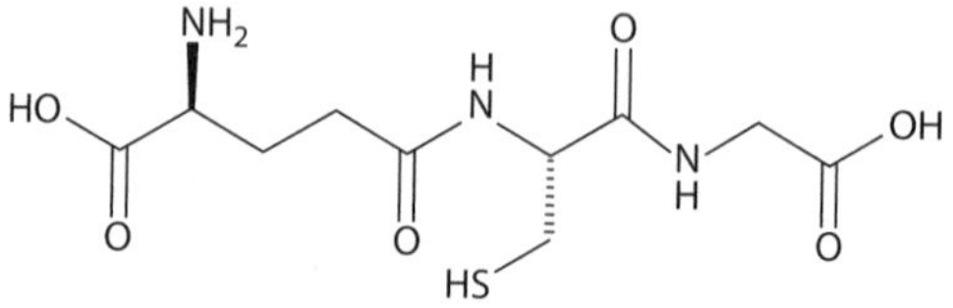
Ethyl ester glutathione (EE-GSH) MW = 334.4 u	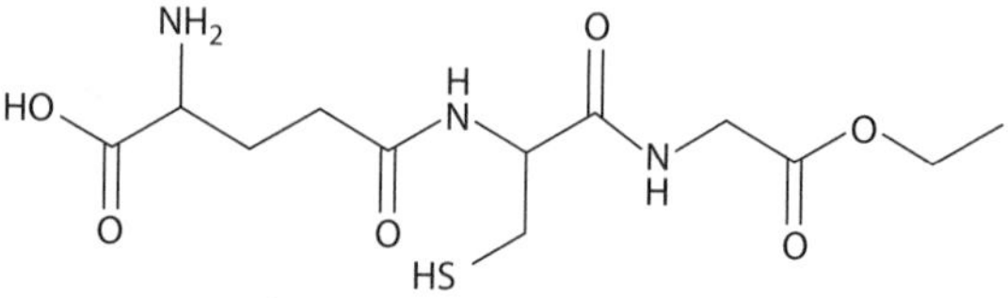
Deuterated-glutathione (^{2}H-GSH) MW = 310.4 u	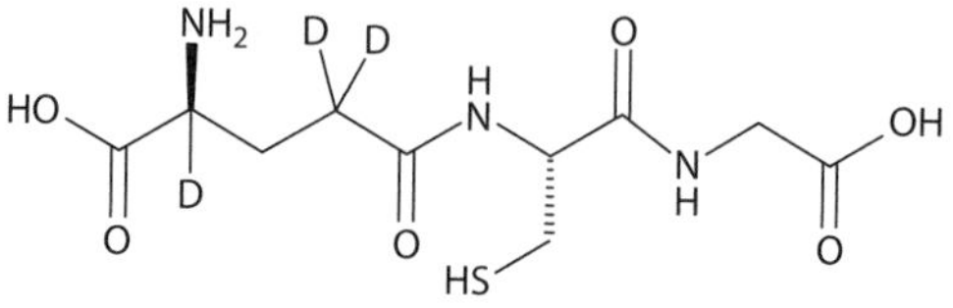
$^{13}C_2$^{15}N–Glutathione ($^{13}C_2$^{15}N-GSH) MW = 310.3 u	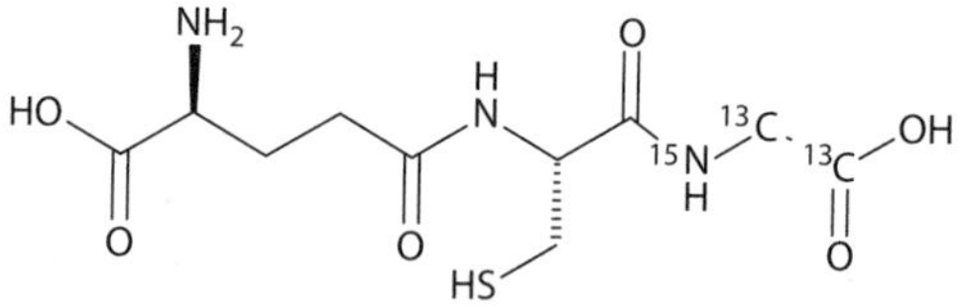
Dansyl-glutathione (*d*-GSH) MW = 540.6 u	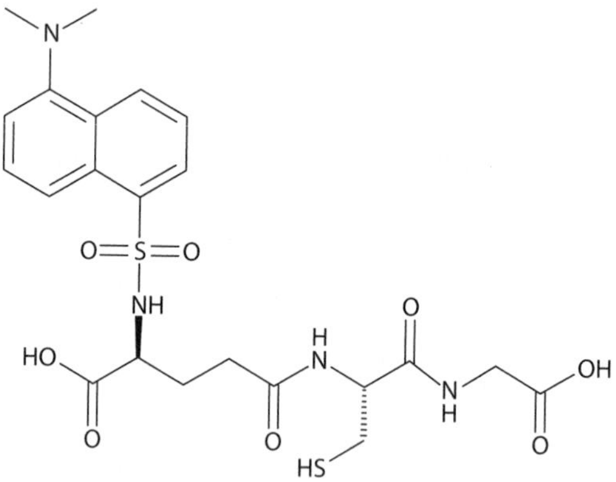
Quaternary amine glutathione (QA-GSH) MW = 433.6 u	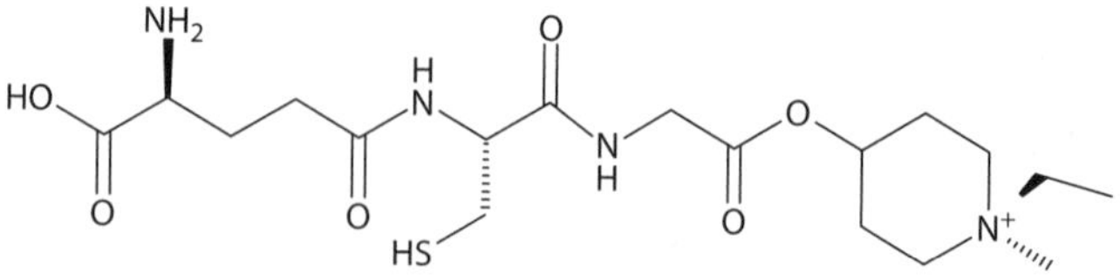

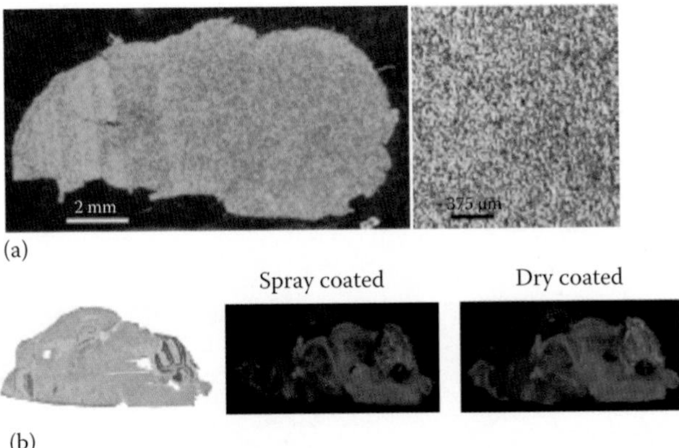

FIGURE 12.4 Developments in sample preparation have led to matrix coating procedures that induce minimal delocalization. (a) Scanned optical image of a dry-coated sagittal mouse brain section on a gold-coated MALDI plate, with a 4× magnification of the DHB layer shown on the right. (b) Comparison of phospholipid images obtained from a sagittal mouse brain either spray-coated (middle) or dry-coated (right) with DHB. (Adapted from Puolitaival, S.M. et al., *J. Am. Soc. Mass Spectrom.*, 19, 882, 2008. With permission.)

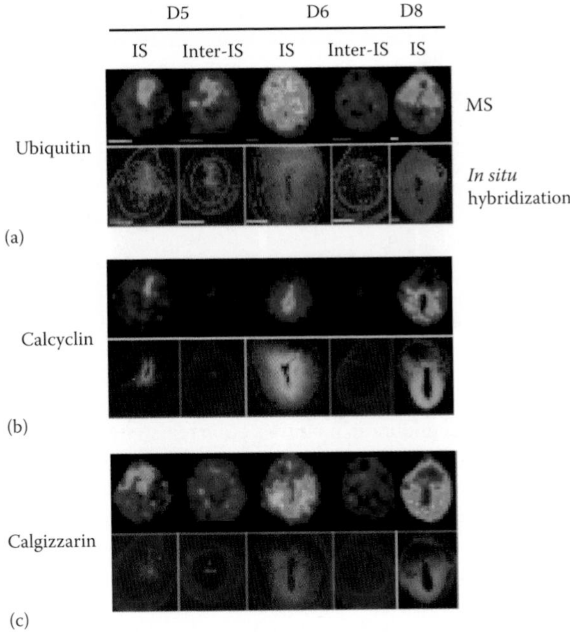

FIGURE 12.5 Imaging MS shows spatial localization for selected proteins of interest on days 5, 6, and 8 of pregnancy. Spatial localization of (a) ubiquitin, (b) calcyclin, and (c) calgizzarin by imaging MS (upper panels) and respective mRNAs by *in situ* hybridization (lower panels) in the implantation site (IS) or interimplantation site (inter-IS) of the mouse uterus. Bar, 700 µm. (Adapted from Burnum, K.E. et al., *Endocrinology*, 149, 3274, 2008. With permission.)

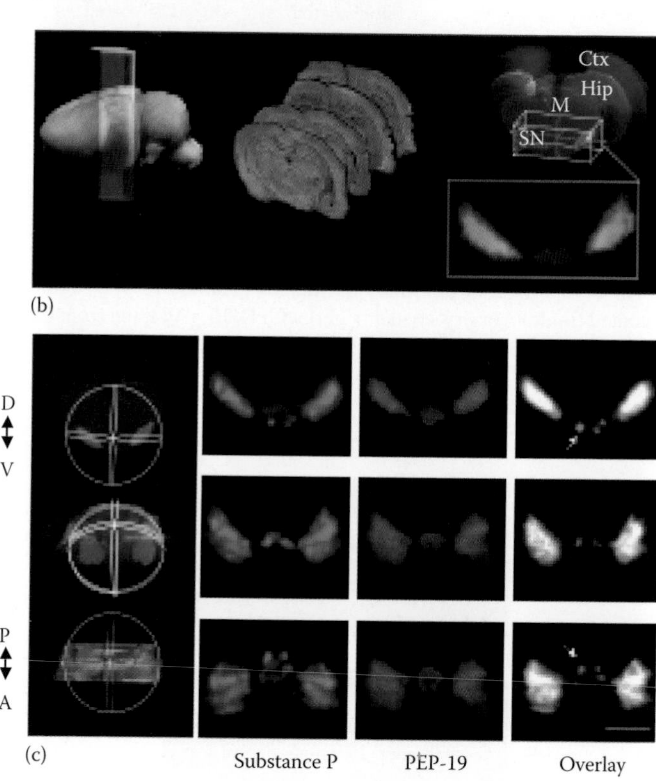

FIGURE 12.6 3D volume reconstructions of one peptide and one protein as detected by MALDI IMS of the rat ventral midbrain. (a) Intrasection registration of pixelated IMS images is performed so that each individual pixel matches its matrix deposit. (b) The intersection registration uses cresyl violet stained images to create a *z*-stack for alignment using Amira software. The 3D surface reconstructions of cortex (Ctx), hippocampus (Hip), midbrain (M), and substantia nigra (SN) visualize the alignment. The box outlines the volume analyzed by MALDI IMS. (c) MALDI IMS images of substance P (*m/z* 1347.8 $[M + H]^+$) and PEP-19 (*m/z* 6717 $[M + H]^+$) were registered using the same factors obtained for intra- and intersection registration. Virtual *z*-stacks were created in Image J software and 3D volume reconstructions of substance P and PEP-19 were rotated for frontal and dorsal views (top and bottom rows, respectively). Overlap of substance P and PEP-19 localization was assessed by combining *z*-stacks. Differential localization was observed in the interpeduncular nucleus. Anterior (A), posterior (P), ventral (V), dorsal (D). Scale bars, 2 mm. (Reproduced from Andersson, M. et al., *Nat. Methods*, 5, 101, 2008. With permission.)

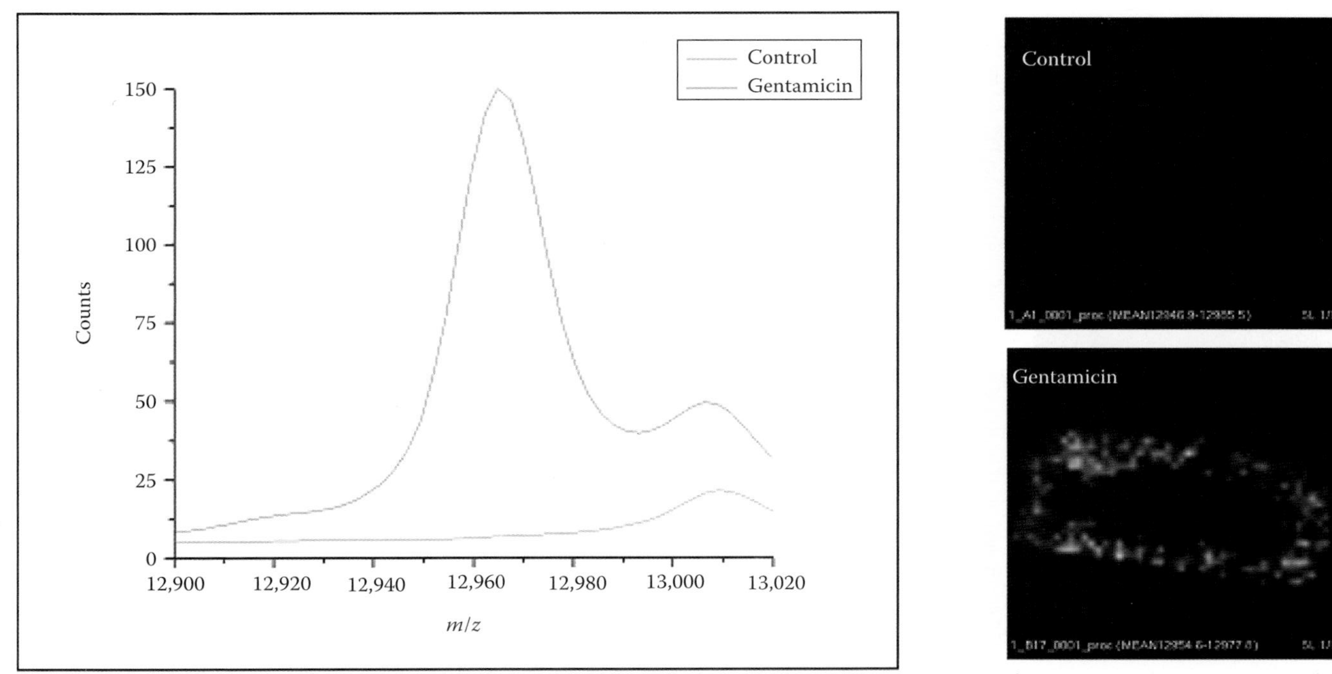

FIGURE 12.7 Differential imaging analysis of a control vs. a gentamicin-treated rat kidney section. Average spectrum (cortex) and intensity distribution of m/z 12,959 on a control and a treated kidney section. (Adapted from Meistermann, H. et al., *Mol. Cell. Proteomics*, 5, 1876, 2006. With permission.)

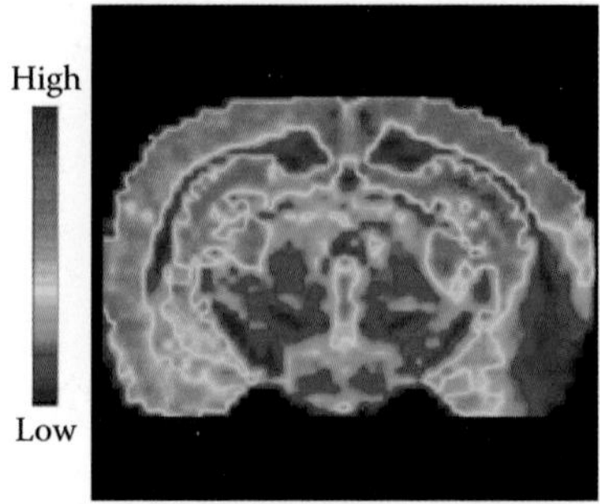

FIGURE 12.8 On-tissue digestion aids the detection of high-molecular-weight species. Distribution of ion detected at *m/z* 1496.75 corresponding to a tryptic peptide of the 71 kDa protein, synapsin I. (Reproduced from Groseclose, M.R. et al., *J. Mass Spectrom.*, 42, 254, 2007. With permission.)

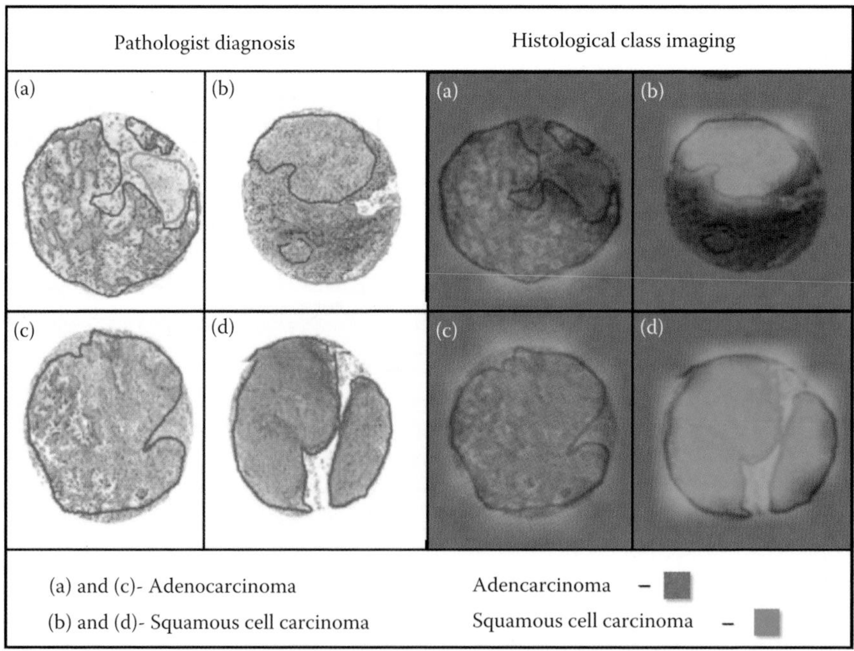

FIGURE 12.9 On-tissue digestion of lung tumor TMA allows for high-throughput diagnostic analyses. Shown is a visual representation of the statistical classification of four biopsies compared to the marking and diagnosis based on histology. (Reproduced from Groseclose, M.R. et al., *Proteomics*, 8, 3715, 2008. With permission.)

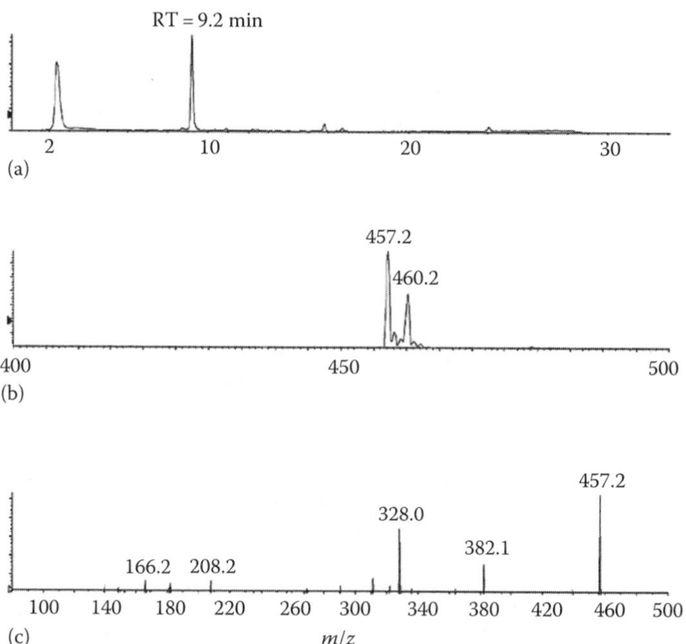

FIGURE 6.4 Detection and characterization of the GSH adduct of acetaminophen using the API 4000 Q Trap instrument. A mixture of labeled and nonlabeled GSH was used to trap the reactive metabolite of acetaminophen (100 mM) incubated with rat liver microsomes. (a) The ion chromatogram obtained from the NL experiment. (b) The precursor protonated molecular ions detected for the peak at 9.2 min. (c) An MS/MS spectrum of the precursor ion at *m/z* 457. (From Mutlib, A. et al., *Rapid Commun. Mass Spectrom.*, 19, 3482, 2005. With permission.)

spectral signature can be used to look for glutathione adducts and one can confirm the findings by MS/MS. The LC–MS analysis of acetaminophen incubated in rat liver microsomes is depicted in Figure 6.4, where neutral, loss, enhanced product ion (EPI) spectrum, and MS^3 spectra could be obtained in a single run. The positive identification of the glutathione is shown in Figure 6.4b, where a characteristic doublet at *m/z* 457 and *m/z* 460 specific for the label is observed. Another advantage of the method is the possibility to perform the experiments on any type of mass analyzer [20]. Using this approach, they were able to detect unresolved minor reactive metabolites using a triple quadrupole mass analyzer. Furthermore, the same group [21] developed an alternative approach to screen for glutathione adducts using a linear ion trap where PI and NL scan functions are not available. Their strategy is again based on the use of a mixture of nonlabeled and labeled forms of glutathione at a given isotopic ratio and a mass difference of 3 Da. This mass pattern is then applied to trigger in a data-dependent acquisition the dependent scan MS^2. In addition to improving the selectivity, a polarity switch is implemented and in a single LC–MS run both positive and negative ionization data are recorded (Figure 6.5). In this way, all the required information can be obtained in a single LC–MS/MS run. Linear ion traps work well

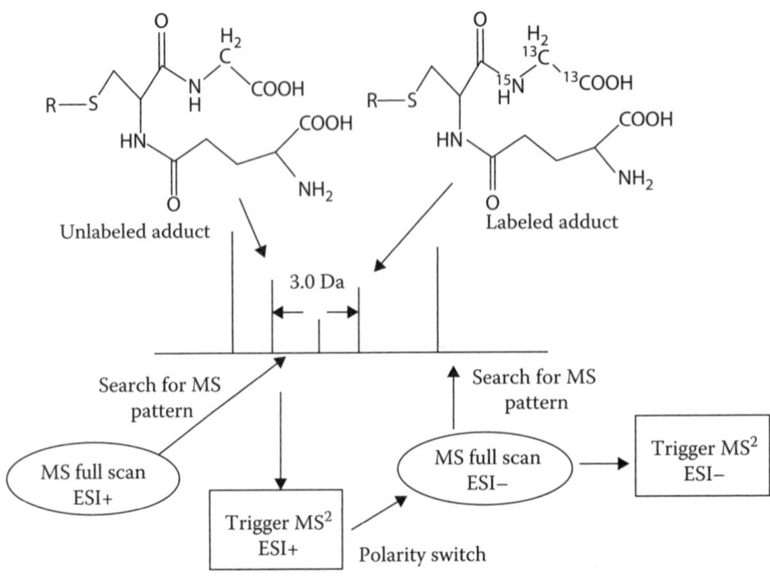

FIGURE 6.5 MS methodology of the mass tag-triggered data-dependent acquisition. R denotes the drug moiety. (From Yan, Z. et al., *Anal. Chem.*, 80, 6410, 2008. With permission.)

for this purpose because they have a short duty cycle and are quite efficient for fast data-dependent acquisition. Based on the $^{13}C_2{}^{15}N$-GSH adduct it was demonstrated that the combination of the isotope pattern-dependent scanning with the postacquisition data mining was very effective in detecting low levels of GSH adducts. To evaluate the performance of the assay, GSH adducts of carbamazepine, diclofenac, 4-ethylphenol, acetaminophen, *p*-cresol, and omeprazole were analyzed [22].

The application of an Orbitrap platform was also described for this purpose. In this case, one can use the excellent full scan sensitivity of the high-resolution accurate mass measurement in both the survey and data-dependent scan modes in order to identify the reactive intermediates and then provide data for their structure assignment [23]. A labeled form of GSH 2:1 ratio of 5 mM glutathione and [$^{13}C_2{}^{15}N$-Gly] glutathione was used. The high mass accuracy in the survey scan can be used to detect the GSH adduct. The method was successfully applied for the detection and characterization of GSH conjugates of acetaminophen, tienilic acid, clozapine, ticlopidine, and mifepristone.

High-resolution instruments such as QqTOF, FT-ICR, and Orbitrap instruments do not have true NL or selected reaction monitoring (SRM) capabilities, which limits their use for screening unexpected metabolites. Therefore, an approach was developed based on the use of high-resolution accurate mass data. With high-resolution data, it was demonstrated that common biotransformations [24] result in a limited shift of the mass defect of the protonated metabolite and a mass defect filter (MDF) can be applied to screen for metabolites. Table 6.2 shows that for common bioactivation steps leading to GSH adducts, the mass defects are not greater than 40 mDa and a ±40 mDa filter around the mass defect of the filter template should

TABLE 6.2
Mass Defect Shifts of GSH-Trapped Reactive Metabolites That Are Formed via Common P450-Mediated Bioactivation Reactions

Mass Shift of the Drug Moiety	GSH Adduct Composition	Mass Defect Shift of GSH Adduct[a]	Functional Group	Common Metabolic Activation Reaction
−34	P + GSH − HCl	+0.0389	β-Cl to a nitrogen or sulfur	Loss of Cl to form aziridinium or episulfonium
−32	P + GSH − S − 2H	+0.0279	Thiourea	R–NH–C(=S)–NH–R′→R–NH–C(SG)=N–R′
−30	P + GSH − 2N − 4H	−0.0218	Hydrazine	R-NH-NH_2→R–SG
−18	P + GSH + O − HCl	+0.0339	Aromatic chlorine	Formation of quinone imine or epoxide followed by GSH attack and loss of HCl
−14	P + GSH − CH_2–2H	0.0157	Aromatic ether	Demethylation followed by oxidation to quinone
−12	P + GSH − C − 2H	0	Methylenedioxy	Formation of quinone
2	P + GSH + O − HF	+0.0043	Aromatic fluoride	Epoxidation followed by GSH attack and loss of HF
0	P + GSH − 2H	0	Substituted phenol derivative	Formation of quinone methide
+1	P + GSH + O − NH − 2H	−0.016	Aromatic amine	Formation of quinone via oxidative deamination
+2	P + GSH	+0.0157	αβ-Unsaturated carbonyl	CH_2=CH–C(=0)–R→GS–CH_2–CH–C(=0)–R
+14	P + GSH + O − 4H	−0.0208	Benzylamine	Formation of hydroxylamine followed by oxidation to nitrile oxide
+16	P + GSH + O − 2H	−0.0051	Phenol	Formation of quinone
+18	P + GSH + O	+0.0106	Furan	Formation of epoxide followed by GSH attack
+28	P + GSH + CO − 2H	−0.0051	Benzylamine	Formation of isocyanate
+32	P + GSH + 2O − 2H	−0.0102	Benzene	Formation of quinone

Source: Zhu, M. et al., *Anal. Chem.*, 79, 8333, 2009. With permission.

[a] The mass defect shifts of GSH adducts are defined as the difference in the mass defects between a detected GSH adduct and the GSH adduct MDF template (MH^+ of the Drug + GSH − 2H).

be sufficient to detect the GSH adducts [25]. The performance of this approach was demonstrated with an analysis of seven model compounds (acetaminophen, carbamazepine, diclofenac, clozapine, *p*-cresol, 4-ethylphenol, and 3-methylindole). The process to identify GSH adducts included the following three steps: (1) full scan acquisition with high-resolution filtering; (2) MDF; and (3) structural elucidation of GSH adducts of product ion data acquired in the data-dependent mode or after a second LC–MS analysis. The benefit of the approach compared to the triple quadrupole was mainly sensitivity and the possibility to detect GSH adducts that do not show the 129 Da NL.

QqQ_{LIT}s are instruments that can be used in both the triple quadrupole mode and the ion trap acquisition mode. In particular, in a data-dependent experiment, the SRM mode can be used as a survey scan and the EPI as a dependent scan [26]. This combination is attractive for metabolite screening because the good sensitivity and selectivity of the SRM mode. Zheng et al. have applied this approach to screen for glutathione adducts [27] based on 114 SRM transitions from protonated molecules constructed on the basis of common bioactivation reactions predicted to occur in human liver microsomes. To be able to cover the maximum number of possibilities, two strategies were followed: (1) targeted SRM screening of predicted GSH adducts and (2) comprehensive SRM screening after PI and NL prescreening.

One of the major issues for detecting reactive metabolites by trapping with glutathione is the relatively low sensitivity of the assay. Another point is that it would of great importance to obtain some quantitative information about the amount of reactive metabolites formed. Soglia et al. [28] focused on an approach in which they optimized the sensitivity of the *in vitro* test method and the throughput of the LC–MS/MS analysis by switching to a more lipophilic glutathione analog, the glutathione ethyl ester (GSH-EE). The microsomal incubations were analyzed after semiautomated 96-well plate solid-phase extraction (SPE) using microbore liquid chromatography-micro-electrospray ionization-tandem mass spectrometry (μLC-μ-ESI-MS/MS). Twelve test compounds were analyzed using GSH and GSH-EE as trapping agents. The use GSH-EE improved the MS sensitivity of the assay by a factor of 10 and in the case of acetaminophen-GSH-EE the peak area increase was approximately 80-fold. The analyses were performed on a triple quadrupole instrument in the positive ion mode. By using a more lipophilic derivative, increased retention of the compounds was observed on reverse-phase chromatography, which also allowed more flexibility in the SPE sample preparation. In addition, the GSH-EE adducts can be detected in the negative ion mode. The CID in the negative mode generates fragments at *m/z* 316, 300, 282, 210, and 128 [29]. The fragment at *m/z* 300 certainly results from the loss of H_2S while the fragment at *m/z* 282 includes a subsequent loss of water. These ions are very specific for GSH-EE compounds and can be selected to perform PI mode scans to screen for GSH-EE adducts. The product ion spectra of GSH-EE are far more informative in the positive mode than in the negative ion mode. The use of PI (*m/z* 300) in negative mode as a survey scan and the EPI mode as a dependent scan is a useful combination [29]. As a general rule, triple quadrupole mass spectrometers suffer from relatively moderate sensitivity when used for qualitative analysis. One of the unique features of the quadrupole linear ion trap is their high sensitivity in the product ion scan mode. The major difference from ion traps is that the CID is performed in a collision cell and is often more informative than typical ion trap CID [26]. In this configuration the limited sensitivity of the PI mode is less critical because its selectivity in the negative mode is sufficient to trigger the mass of the PI to perform a sensitive product ion spectrum in the positive mode.

Fluorescence detection has been applied to quantify glutathione adducts. Gan et al. developed a quantitative method using a dansylated GSH (dGSH) derivative, with the fluorescent dansyl group being added to the free amino group of GSH [30]. A concentration of 1 mM dGSH was selected as a compromise between trapping efficiencies and the fluorescence background. In the present LC–MS setup the fluorescence detector was placed in line before the MS system. One limitation of the approach is that it cannot be applied for compound that causes fluorescence interferences. It was also shown that despite the introduction of the bulky dansyl group, dansyl GSH did not serve as a cofactor for glutathione *S*-transferase (GST) and no CYP 450 inhibition was observed, The analysis time of one sample lasted about 50 min which currently limits the applicability of the assay for high throughput but the application of fast LC or ultrahigh-performance liquid chromatography (UHPLC) could be beneficial for the method.

Ideally, one would like to perform quantitation directly with mass spectrometric detection using a GSH adduct that provides standardized response independent of the nature of the parent drug. One approach proposed by Soglia et al. [31] is the use of a quaternary ammonium GSH (QA-GSH) analog with a fixed positive charge where the MS response factor is directly related to the quaternary ammonium. Comparison of equimolar amounts of the parent standards and the corresponding QA-GSH standards showed that conjugation with QA-GSH resulted in improved detection capability and a similar MS response. MS signal responses of QA-GSH conjugate reference compounds were within 3.3-fold of one another whereas the responses of the corresponding parent standard differed as much as 19-fold. Furthermore, only a 1.5-fold difference in slope of the standard curves of QA-GSH standards was observed. The quantification methodology was as follows: addition of QA-GSH standards (one is single charged and one is double charged), as internal standards, to the microsomal incubation sample prior to analysis. LC–MS/MS analysis was performed using capillary liquid chromatography coupled to microelectrospray ionization-tandem mass spectrometry. In a first step the QA-GSH conjugate of interest has to be identified. This was done either by calculation of the *m/z* of M^+ and MH^{2+} ions of the QA-GSH conjugate of interest for multiple single reacting monitoring analysis or by PI scanning of *m/z* 144 (corresponding to the 4-hydroxy-*N*-methylethyl piperidinium ion). Once the *m/z* of the M^+ or MH^{2+} ion was verified a MS method was used to monitor four independent scan events (MH^{2+} and M^+ of QA-GSH adduct of interest and the MH^{2+} and M^+ of the internal standards) during LC–MS/MS analysis.

6.2.3 Cyanide Adducts

During metabolic activation hard electrophilic metabolites are also generated, for example, iminium ions derived from compounds with an alicyclic amine structure. Cyanide, a hard nucleophile, has been used as trapping agent for detection of this category of reactive metabolites. The complementarities of both trapping screens (using GSH and cyanide) could be demonstrated for several compounds [32]. Gorrod et al. described the use of radiolabeled [^{14}C]cyanide in Ref. [33]. Meneses-Lorente et al. extended this approach into a semi-automated high-throughput assay [34]. Following incubations of cold test compounds with liver microsomes in the presence of [^{14}C]cyanide, the unreacted trapping agent was removed by SPE and the amount of radiolabeled conjugate was determined by liquid scintillation counting. Evans et al. mentioned the use of a 1:1 mixture of cyanide and stable isotopically labeled cyanide ($^{13}C^{15}N^-$). They reported that detection of the adducts was greatly facilitated by the presence of prominent isotopic doublets, which differed in mass by 2 Da (monoadduct) or 4 Da (bis-adduct) and that the MS/MS spectra were characterized by a NL of 27/29 (HCN/$H^{13}C^{15}N$) [14]. Argoti et al. developed a LC–MS/MS method for screening iminium ion formation in liver microsomes using cyanide [32]. The microsomal incubation mixtures were fortified with KCN or $K^{13}C^{15}N$. The cyanide conjugates were screened for by constant NL scanning of 27 (HCN) or 29 ($H^{13}C^{15}N$). In addition, GSH trapping experiments were conducted. Fourteen alicylclic amine compounds were analyzed using GSH and cyanide trapping screenings. Comparison of the results demonstrated the importance of cyanide trapping for detection and identification of iminium ion intermediates and the complementarities of GSH and cyanide trapping experiments.

6.2.4 Other Peptides

Reichardt et al. proposed a model using a dipeptide trapping agent for monitoring adduct formation of xenobiotics and for structural characterization of the molecular structures [35]. Lysine–Tyrosine (Lys–Tyr) was chosen as model peptide. This dipeptide was selected to minimize the steric hindrances and dipole effects, factors that are known to have an influence of the binding characteristics to proteins. Lysine was selected due to its high basicity and degree of ionization, the ε-amino group of the lysine residue belongs to the most commonly involved and characteristic groups for protein–xenobiotic interaction *in vivo*. Tyrosine was chosen as second amino acid as its aromatic ring is easily detectable in UV-light and because of its much lower reactivity in order to favor reactions at the lysine residue. Test compounds, including aldehydes and other electrophilic compounds, were incubated with the dipeptide at 37°C and pH 7.4 (phosphate buffer). After centrifugation, the adducts were analyzed by flow injection analysis MS (FIA–MS). Besides adduct formation of aldehyde groups via Schiff bases, covalent adducts of other electrophilic compounds, such as toluene-2,4-diisocyanate, 2,4-dinitro-1-fluorobenzene, 2,4,6-trinitrobenzene sulfonic acid, dansyl chloride, and phthalic acid anhydride with the lysine moiety could also be detected. Quantitation of peptide reactivity was performed by determination of the amount of peptide with apparently unchanged molecular structure remaining after a specific duration of incubation by HPLC–UV and fluorescence monitoring. A factor of reactivity defined as the amount of xenobiotic necessary to allow a reaction with 50% of the peptide in a given solution was introduced.

Wang et al. used a slightly different dipeptide lysine–phenylalanine (Lys–Phe) for studying the reactivity of acyl glucuronides to proteins [36]. Acyl glucuronides are a commonly formed metabolite for compounds with a carboxylic acid moiety. Although, acyl glucuronides are reactive electrophiles they are relatively stable compared to other reactive metabolites and can circulate in the body and are excreted intact in urine or bile [37,38]. The chemical reactivity is suspected to be responsible to the toxicity observed for carboxylic acids. Covalent binding to proteins may occur via two different mechanisms [39]: first, via transacylation, that is, nucleophilic attack of the acyl carbon of the glucuronide, for example, by a nucleophilic group of a protein, resulting in liberation of the glucuronic acid and acylated protein or second, via glycation. Following acyl migration, the resulting positional isomers can transiently exist in the open chain form of the sugar ring, thereby exposing the aldehyde group, which can form Schiff base, for example, with ε-amine group of a lysine residue of a protein. In this case, the glucuronic acid moiety is retained in the adduct, forming a bridge between the aglycone and the target protein [38]. The approach of Wang et al. [36] aimed at determining the reactivity of acyl glucuronides toward the test peptide, whereby the formation of adducts via the Schiff base mechanism was monitored. It included two incubation steps. First, formation of acyl glucuronides by incubation of the carboxylic acid containing compound with human liver microsomes fortified with uridine diphosphate glucuronic acid (UDPGA) and second, formation of adducts by addition of the model peptide into the first step incubation supernatant. After dilution of an aliquot with methanol:water 50:50, the samples were analyzed by LC–MS.

Another approach was proposed by Mitchell et al. [40] using neither an LC–MS nor radiolabeled drug for detecting reactive metabolites. A custom trapping 11 amino acid peptide (ECGHDRKAHYK) was designed for use with surface enhanced laser desorption/ionization time-of-flight (SELDI–TOF). The peptide contains cystein, some nucleophilic amino acid residues, and a charged residue for binding to a weak cation exchange. The uniqueness of the assay is that the peptide (50 pmol) is bound to a weak cation exchange chip and the covalent adduct can be detected by desorptive ionization. The peptide is able to trap efficiently the reactive metabolites even in the presence of competing proteins (2 mg/mL). The assay was found to be suitable for the following activation systems: rat and human liver microsomes, purified recombinant P450 enzymes, horseradish peroxidase (HRP), MPO, and alcohol dehydrogenase (ADH). Typically the test compound is added at a concentration of 200 μM in the activation system and 50 μM of trapping peptide is used. Incubations are carried out at 37°C for 15 min. One microliter of the sample is then spotted onto the weak cation exchange chip and incubated for 20 min. After several buffer wash steps the matrix (a-cyano) is added. Figure 6.6 illustrates the peptide adduct formation for butylated hydroxytoluene using the rat liver microsomes activation system.

6.2.5 Cobalamine

Highly reactive hydrophilic low molecular weight epoxy metabolites of 1,3-butadiene are particularly difficult to analyze. Gas chromatography techniques were used for analyses following solvent extraction or by head space GC. However, both procedures were not optimal for hydrophilic low molecular weight compounds. The supernucleophile Cob(I)alamin, a compound with high nucleophilic strength, was proposed as an analytical tool for characterization of reactive metabolites. When Cobalamin (Co(III)) is reduced to Cob(I)alamin, the upper ligand is replaced by a free pair of electrons, which react rapidly with electrophilic compounds such as oxiranes to form alkyl cobalamin complexes [41]. These polar alkyl cobalamin complexes of high molecular weight are amenable for analysis by LC–MS. Haglund et al. developed a

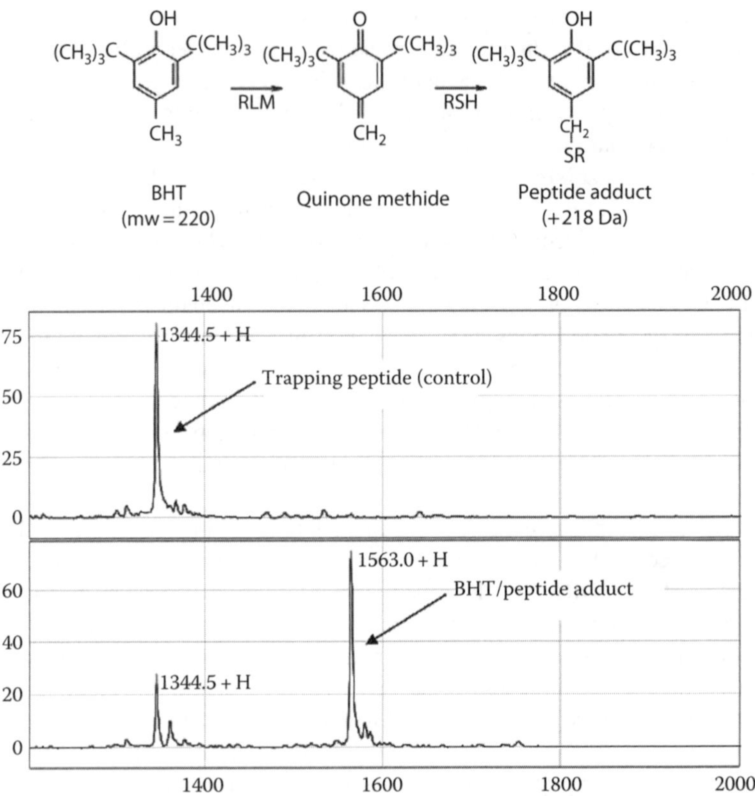

FIGURE 6.6 Peptide adduct formed from BHT using the rat liver microsomal activation system. The upper panel is the control reaction (–NADPH) and shows only the trapping peptide, while the lower panel is the complete reaction (BHT + NADPH + RLM) and shows both the peptide and the BHT–peptide adduct. (From Mitchell, M.D. et al., *Chem. Res. Toxicol.*, 21, 859, 2008. With permission.)

validated capillary LC–ESI–MS/MS method using the SRM mode with a column switching setup for sensitive and accurate determination of reactive metabolites trapped with Cob(I)alamin [42]. The performance of this approach was demonstrated by using diepoxybutane, a metabolite of 1,3-butadiene (MW, 86.1 Da), as test compound. The intermediate metabolites epoxybutene was incubated with S9 fraction fortified with NADPH for CYP450 catalyzed bioactivation to diepoxybutane. Aliquots of the incubation sample were mixed with Cob(I)alamin solution before analysis of the formed alkyl cobalamin complexes.

6.2.6 Electrochemical System for Metabolite Generation

A purely instrumental approach based on using electrochemistry to catalyze reactions and to mimic CYP activity has been proposed as another way to study the formation of metabolites [43,44]. The electrochemical cell (EC) is placed in front of the mass spectrometer and the analysis can be performed in the flow injection (EC–FIA–MS) mode or combined with liquid chromatography (EC–LC–MS) (Figure 6.7). The online reactor is either an EC flow-through cell connected to a potentiostat (EC/LC/MS system) or functionalized magnetic particles that are fixed with two magnets in a capillary (HRP/LC/MS system). In loading position, the reaction mixture from the EC cell or the HRP particles is collected in the 10 μL loop [45]. After switching the valve, the loop is emptied by the HPLC pumps and the content is injected on the column. Figure 6.7b shows the chromatograms of amodiaquine (AQ) before and after electrochemical oxidation. By using the electrochemical system, the reactive amodiaquine quinoneimine (AQQI) could be observed, whereas it was not observed in the microsomal incubation study. While direct detection of reactive metabolites is of interest, it can be challenging either *in vitro* or *in vivo* because the electrophilic metabolites formed can spontaneously react with endogenous nucleophiles. EC–MS and EC–LC–MS studies on the metabolism of clozapine and acetaminophen were

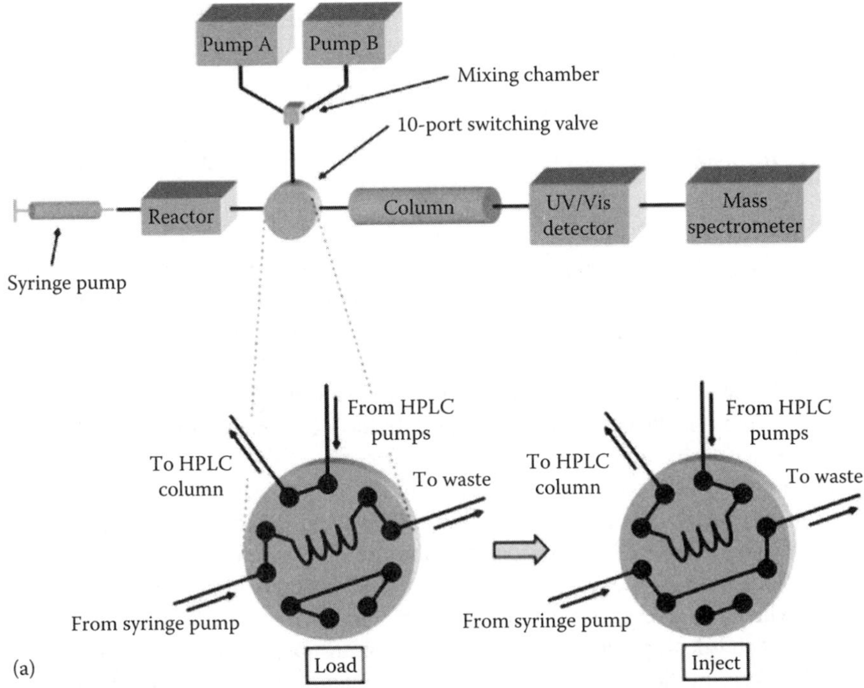

FIGURE 6.7 (a) Schematic setup of the systems for the online generation, separation, and identification of metabolites. (b) ESI(+)-TIC chromatograms (combined from SIM measurements) of (i) AQ and metabolized AQ using (ii) online electrochemistry at 700 mV vs. Pd/H_2, (iii) online HRP, (iv) off-line microsomes, and (v) off-line microsomes in the presence of GSH. Peak assignment: 1, AQ (*m/z* 356); 2, DESAQ (*m/z* 328); 3, AQQI (*m/z* 354); 4, DESAQQI (*m/z* 326); 5, AQ aldehyde* (*m/z* 299); 6, AQQI N-oxide* (*m/z* 370); 7, AQ N-oxide* (*m/z* 372); 8, bis-DESAQ (*m/z* 300); 9, GS-AQ (*m/z* 661); 10, GS-DESAQ (*m/z* 633). Structures with * are tentatively identified. (From Lohmann, W. and Karst, U., *Anal. Chem.*, 79, 6831, 2007. With permission.)

(*continued*)

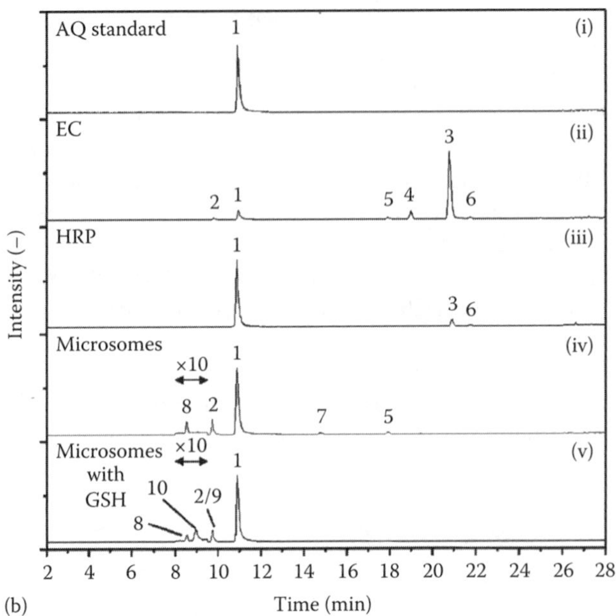

FIGURE 6.7 (continued)

reported in the absence and presence of trapping agents, for example, GSH [46,47]. As for acetaminophen, the known oxidative metabolic detoxification pathway could be mimicked, that is, GSH and *N*-acetylcysteine adducts of the electrochemically generated reactive acetaminophen metabolite, *N*-acetyl-*p*-benzoquinoneimine (NAPQI), could be identified. However, in contrast to *in vivo* experiments, different isomers of these adducts were observed [47]. In the study on clozapine, several of the known oxidative metabolic pathways were observed.

Jurva et al. performed comprehensive studies on the comparability of EC–MS and cytochrome P 450 catalyzed reactions [48,49] and found that EC–MS successfully mimicked those CYP reactions that are initiated by a one-electron oxidation, such as *N*-dealkylation, *S*-oxidation, *P*-oxidation, alcohol oxidation, and dehydrogenation, whereas the reactions that are initiated via direct hydrogen atom abstraction, for example, *O*-dealkylation and hydroxylation of unsubstituted aromatic rings, were not mimicked. In the latter cases the oxidation potentials were too high for electrochemical oxidation in aqueous solutions. A further disadvantage is that EC lacks the stereospecifity of the reactions, in contrast to CYP catalyzed reactions [50].

6.3 COVALENT BINDING TO PROTEINS

Despite the fact that no direct correlation of drug–protein adducts with organ toxicity has been shown, the investigation of these adducts are of prime importance in the drug selection process. One approach to monitor the potential of a drug to generate reactive metabolites is to determine the amount of drug binding to liver proteins, which requires radiolabeled drug. Once the measurement has been performed, the challenge becomes to define how much covalent binding is acceptable for a good clinical candidate. Evans et al. [14] suggested a conservative threshold of 50 pmol drug equivalent/mg total liver proteins for a 60 min reaction. In order to determine the amount of protein adducts the drug is generally incubated in microsomes or hepatocytes or the liver of the animal is collected up to 24 h after administration. Figure 6.8a illustrates the centrifugation-based scheme where the proteins were precipitated and several washing steps were performed on the pellets to remove all noncovalently bound material. The pellet is then resolubilized with a tissue solubilizer and measured by liquid scintillation counting. The procedure is labor intensive and difficult to automate. Day et al. [51] have described a semi-automated method in which a Brandel cell harvester is used to collect and wash the proteins as shown in Figure 6.8b. The authors concluded that centrifugation and the Brandel cell harvester system provided similar results. The described approach may be useful to perform a ranking

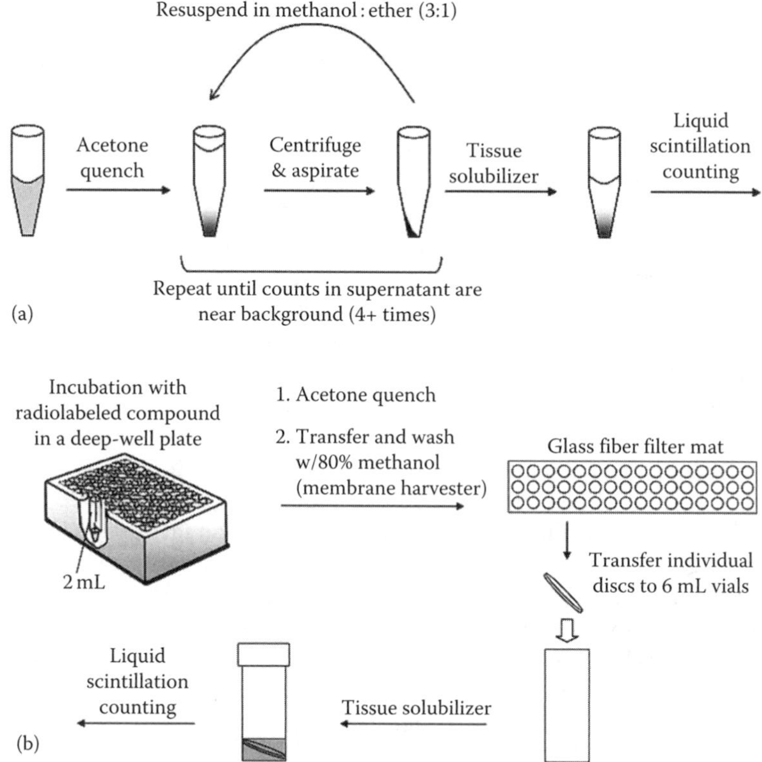

FIGURE 6.8 (a) Schematic of the traditional centrifugation-based method for measuring covalent protein binding. (b) Schematic of the filter method. Precipitated protein, containing covalently bound drug, was transferred by vacuum to a filter mat where 80% methanol was used to remove unbound radioactivity. Individual filters were punched from the mat into vials where solubilization of protein and liquid scintillation counting were done. (From Day, S.H. et al., *J. Pharmacol. Toxicol. Methods*, 52, 278, 2005. With permission.)

of the investigated compounds compared to others, however no knowledge on which proteins the metabolites were bound to is obtained. Molecular weight information of the covalent metabolite protein complex can be obtained using a 1D-gel and radioactivity detection. Quantitative whole-body autoradiography (QWBA) allows one to obtain a picture of the distribution of the compound and metabolites in tissues using ^{14}C-labeled parent compound. Takakusa et al. [52] performed a study where they compared tissue distribution retention properties of ^{14}C-labeled NCEs with covalent binding of several drugs associated with IDR or not. They found that the radioactivity measured in liver 78 or 168 h postdose correlated well with the rat *in vivo* covalent binding yield of the drugs.

Masubuchi et al. [53] have investigated the relationship between the *in vitro* formation rate of GSH conjugates and the covalent binding to protein for 10 test compounds known to form reactive intermediates. They found that (1) there is a correlation between binding rate of reactive metabolites and GSH formation; (2) the covalent-binding rate of human liver microsomal proteins correlates with that of rat liver microsomal proteins; (3) GSH formation can be used as surrogate marker; and (4) the *in vitro* covalent-binding rate correlates with the *in vivo* binding rate in the case where a predictive mode using the unbound fraction of the AUC is applied.

Identification of the proteins involved in covalent binding is also of interest. Most approaches are based on a proteomic workflow [54] where after administration of the radiolabeled drug the protein extract (liver, plasma) is separated on a 2D-Gel electrophoresis followed by trypsin digestion and mass spectrometric identification either with MALDI peptide mass fingerprint or LC–MS/MS assay. A major challenge of trypsin digestion is that, in most cases, the peptide coverage is rarely 100% and only a partial picture of modified peptides can be obtained. Also the modification of the peptides may generate a signal reduction of enhancement of the specific peptide. There is still little limited information about which proteins are the target for the reactive metabolites and in particular, which mechanism occurs showing the need for more comprehensive studies. A reactive metabolites target protein database compilation has been described by Hanzlik et al. [55]. Little is known concerning the microsomal protein targets of reactive metabolites. Identification of the proteins that target various modifications resulting in a toxic response could help to predict toxicity with better specificity [56]. For example the protein involved in the binding of the acyl glucuronide of (AcMPAG) mycophenolic was investigated using a proteomic approach [57]. The kidney proteins were separated by 2D-gel electrophoresis and the covalent product adduct were detected by Western blotting with an antibody specific for MPA/AcMPAG. Twenty-one proteins were identified as potential targets of AcMPAG.

The adduct formation of the reactive metabolites of paracetamol, AQ, and clozapine with the proteins β-lactoglobulin A and human serum albumin was studied using an electrochemical flow-through cell coupled to an online LC–MS system [58]. The formed adducts were directly characterized with time-of-flight mass spectrometry. The adduct formation with the electrophilic metabolites of paracetamol (NAPQI), amodiaquine (AQQI), and clozapine (CLZox) are with β-lactoglobulin A as illustrated in Figure 6.9. The mass shift of the respective adduct can be clearly read out from the spectrum, which becomes more challenging with the heterogeneity of

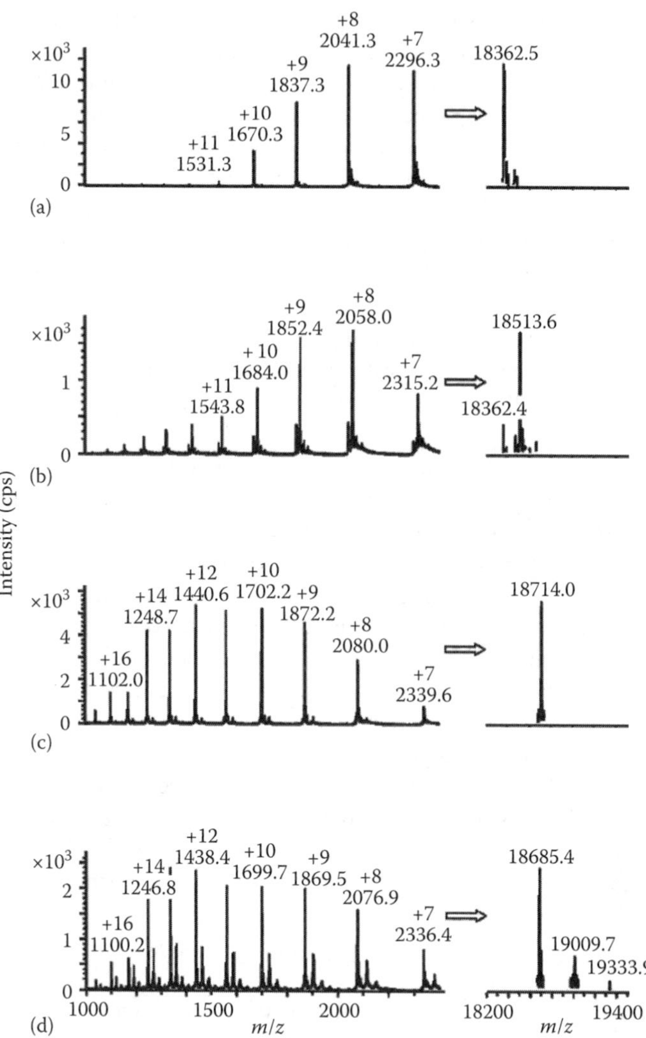

FIGURE 6.9 Time-of-flight mass spectra and deconvolution results of (a) unmodified β-LGA, (b) β-LGA after reaction with NAPQI, (c) β-LGA after reaction with AQQI, and (d) β-LGA after reaction with CLZox. The mass spectra of the modified protein were obtained after online reaction of the unmodified protein with electrochemically generated reactive metabolites. (From Lohmann, W. et al., *Anal. Chem.*, 80, 9714, 2008. With permission.)

larger proteins. To identify the site of modification, the protein was digested using trypsin and the formed peptides were analyzed by FT-ICR–MS. The potential of the approach was also investigated with human serum albumin.

Most screening procedures for the investigation of covalent binding to proteins require the synthesis of radiolabeled parent drug which is relatively expensive and is time-consuming. Therefore, a strategy based on unlabeled drug is of great interest. The use of ECs is certainly a major step in this direction to investigate the behavior of various drug candidates but may not completely reflect the biological system. Hepatocytes and microsomes remain the most relevant systems. Sleno et al. [59] described a procedure to visualize the comprehensive binding of reactive metabolites to proteins. Using this approach they were able to determine a reactive metabolite cystein adduct of fipexide in microsomes without radiolabeled parent drug using pronase digestion. Fipexide is a nootropic drug that has been withdrawn from the market due to serious toxic effects. Fipexide hydrolyses to two potential toxic species: 3,4-methylenedioxybenzylpiperazine (MDBP) and 4-chlorophenoxyacetic acid (4-CPA). *In vitro* experiments showed that the metabolism of fipexide leads to reactive metabolites [60]. Adducts with GSH, GSH ethyl ester, and *N*-acetylcysteine were observed with the demethylenated fipexide. Catechol intermediates are known to be precursor to the electrophilic orthoquinone which can subsequently undergo a nucleophilic attack by the cystein of peptides or proteins. 4-CPA was shown to form an acyl glucuronide and acyl-CoA thioester.

Direct visualization of protein adducts may be very challenging in complex mixtures such as microsomes or hepatocytes. Proteomic strategies such as two-dimensional LC–MS/MS after tryptic digestion of the protein pellets to fish out the modified peptides may be considered. Due the low concentration of protein adducts, the partial tryptic coverage of the proteins and the low throughput this setup may not be suitable to screen for covalent protein adducts *in vitro* or *in vivo*. Pronase digests proteins to single amino acids. This approach was used to digest the microsomal peptides after incubation of fipexide and the resulting amino acids adducts were separated by LC–MS/MS. Based on the *in vitro* knowledge of the formed *in vitro* reactive metabolites, a data-dependent experiment using the SRM mode as a survey scan and an EPI or targeted EPI could be performed to visualize the cystein adduct (Figure 6.10). It could clearly establish that the fipexide *o*-quinone metabolite covalently binds to microsomal proteins. The strategy is compound independent and brings additional information to the classical trapping experiments.

6.4 CONCLUSIONS

Most of the current developed mass spectrometric strategies allow one to efficiently screen for the potential of new drug candidates to form reactive metabolites and understand the mechanism of bioactivation. Further development in the field of hybrid mass spectrometry will provide to the scientist even more powerful instruments to perform drug metabolism investigations. Surprisingly, most of the current efforts in the investigation of reactive metabolites are concentrated solely on the characterization of GSH adducts. Currently, no direct correlation between the presence of these metabolites and potential adverse effects can be made despite the fact that they will affect certain biological processes. Visualization and structure elucidation of reactive metabolites is of prime importance but new tools are needed to understand their real impact. Binding to proteins has been established but still limited knowledge is available on which protein and under which circumstances the adducts are formed. Differential proteomics studies may be of interest because they would help to identify which protein is affected by the presence of reactive metabolites and would allow one to obtain a more global picture. In a similar way

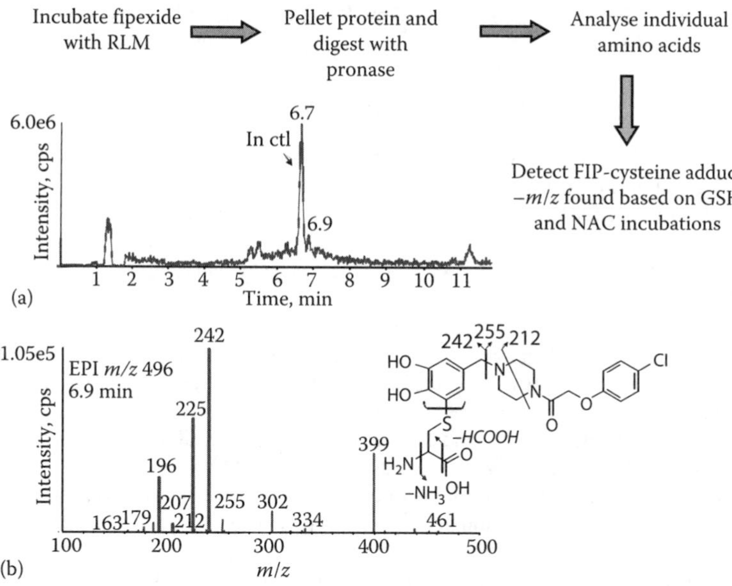

FIGURE 6.10 EPI chromatogram (a) of *m/z* 496 from pronase digestion of RLM after incubation with fipexide for 18 h. Note that peak at 6.7 min is also seen in the control incubation and therefore not due to fipexide metabolism. The MS/MS spectrum of the protonated adduct (b) yields several product ions indicative of the Cys-adduct of the demethylenated fipexide metabolite. Some important fragment ions are rationalized in the structure shown (note that the position of the cysteine on the ring is unknown).

metabolomic approaches may allow to find specific toxicity markers and in more general help to understand which compounds are up- or downregulated and which biological pathways are affected.

ACKNOWLEDGMENT

The author is grateful to Roland Staack for fruitful input on the topic.

REFERENCES

1. Liebler, D. C. and Guengerich, F. P., Elucidating mechanisms of drug-induced toxicity, *Nat. Rev. Drug Discov.*, 4 (5), 410–20, 2005.
2. Korfmacher, W. A., editor, *Using Mass Spectrometry for Drug Metabolism Studies*, CRC Press, Boca Raton, FL, 2005.
3. Nassar, A. E., Talaat, R. E., and Kamel, A. M., The impact of recent innovations in the use of liquid chromatography-mass spectrometry in support of drug metabolism studies: Are we all the way there yet? *Curr. Opin. Drug Discov. Devel.*, 9 (1), 61, 2006.
4. Baillie, T. A., Metabolism and toxicity of drugs. Two decades of progress in industrial drug metabolism, *Chem. Res. Toxicol.*, 21 (1), 129, 2008.
5. Walgren, J. L., Mitchell, M. D., and Thompson, D. C., Role of metabolism in drug-induced idiosyncratic hepatotoxicity, *Crit. Rev. Toxicol.*, 35 (4), 325, 2005.

6. Lazarou, J., Pomeranz, B. H., and Corey, P. N., Incidence of adverse drug reactions in hospitalized patients: A meta-analysis of prospective studies, *JAMA*, 279 (15), 1200, 1998.
7. Navarro, V. J. and Senior, J. R., Drug-related hepatotoxicity, *N. Engl. J. Med.*, 354 (7), 731, 2006.
8. Kalgutkar, A. S. et al., A comprehensive listing of bioactivation pathways of organic functional groups, *Curr. Drug Metab.*, 6 (3), 161, 2005.
9. Kaplowitz, N., Idiosyncratic drug hepatotoxicity, *Nat. Rev. Drug Discov.*, 4 (6), 489, 2005.
10. Uetrecht, J., Idiosyncratic drug reactions: Past, present, and future, *Chem. Res. Toxicol.*, 21 (1), 84, 2008.
11. Ma, S. and Subramanian, R., Detecting and characterizing reactive metabolites by liquid chromatography/tandem mass spectrometry, *J. Mass Spectrom.*, 41 (9), 1121, 2006.
12. Staack, R. F. and Hopfgartner, G., New analytical strategies in studying drug metabolism, *Anal. Bioanal. Chem.*, 388 (7), 1365, 2007.
13. Prakash, C., Shaffer, C. L., and Nedderman, A., Analytical strategies for identifying drug metabolites, *Mass Spectrom. Rev.*, 26 (3), 340, 2007.
14. Evans, D. C. et al., Drug-protein adducts: An industry perspective on minimizing the potential for drug bioactivation in drug discovery and development, *Chem. Res. Toxicol.*, 17 (1), 3, 2004.
15. Baillie, T. A. and Davis, M. R., Mass spectrometry in the analysis of glutathione conjugates, *Biol. Mass Spectrom.*, 22 (6), 319, 1993.
16. Dieckhaus, C. M. et al., Negative ion tandem mass spectrometry for the detection of glutathione conjugates, *Chem. Res. Toxicol.*, 18 (4), 630, 2005.
17. Castro-Perez, J. et al., A high-throughput liquid chromatography/tandem mass spectrometry method for screening glutathione conjugates using exact mass neutral loss acquisition, *Rapid Commun. Mass Spectrom.*, 19 (6), 798, 2005.
18. Mutlib, A. et al., Formation of unusual glutamate conjugates of 1-[3-(aminomethyl) phenyl]-*N*-[3-fluoro-2′-(methylsulfonyl)-[1,1′-biphenyl]-4-yl]-3-(trifluoromethyl)-1H-pyrazole-5-carboxamide (DPC 423) and its analogs: the role of gamma-glutamyltranspeptidase in the biotransformation of benzylamines, *Drug Metab. Dispos.*, 29 (10), 1296, 2001.
19. Mutlib, A. et al., Application of stable isotope labeled glutathione and rapid scanning mass spectrometers in detecting and characterizing reactive metabolites, *Rapid Commun. Mass Spectrom.*, 19 (23), 3482, 2005.
20. Yan, Z. and Caldwell, G. W., Stable-isotope trapping and high-throughput screenings of reactive metabolites using the isotope MS signature, *Anal. Chem.*, 76 (23), 6835, 2004.
21. Yan, Z., Caldwell, G. W., and Maher, N., Unbiased high-throughput screening of reactive metabolites on the linear ion trap mass spectrometer using polarity switch and mass tag triggered data-dependent acquisition, *Anal. Chem.*, 80 (16), 6410, 2008.
22. Ma, L. et al., Rapid screening of glutathione-trapped reactive metabolites by linear ion trap mass spectrometry with isotope pattern-dependent scanning and postacquisition data mining, *Chem. Res. Toxicol.*, 21 (7), 1477, 2008.
23. Lim, H. K. et al., A generic method to detect electrophilic intermediates using isotopic pattern triggered data-dependent high-resolution accurate mass spectrometry, *Rapid Commun. Mass Spectrom.*, 22 (8), 1295, 2008.
24. Zhang, H., Zhang, D., and Ray, K., A software filter to remove interference ions from drug metabolites in accurate mass liquid chromatography/mass spectrometric analyses, *J. Mass Spectrom.*, 38 (10), 1110, 2003.
25. Zhu, M. et al., Detection and structural characterization of glutathione-trapped reactive metabolites using liquid chromatography-high-resolution mass spectrometry and mass defect filtering, *Anal. Chem.*, 79 (21), 8333, 2007.

26. Hopfgartner, G. et al., Triple quadrupole linear ion trap mass spectrometer for the analysis of small molecules and macromolecules, *J. Mass Spectrom.*, 39 (8), 845, 2004.
27. Zheng, J. et al., Screening and identification of GSH-trapped reactive metabolites using hybrid triple quadruple linear ion trap mass spectrometry, *Chem. Res. Toxicol.*, 20 (5), 757, 2007.
28. Soglia, J. R. et al., The development of a higher throughput reactive intermediate screening assay incorporating micro-bore liquid chromatography-micro-electrospray ionization-tandem mass spectrometry and glutathione ethyl ester as an in vitro conjugating agent, *J. Pharm. Biomed. Anal.*, 36 (1), 105, 2004.
29. Wen, B. and Fitch, W. L., Screening and characterization of reactive metabolites using glutathione ethyl ester in combination with Q-trap mass spectrometry, *J. Mass Spectrom.*, 44(1), 90, 2009.
30. Gan, J. et al., Dansyl glutathione as a trapping agent for the quantitative estimation and identification of reactive metabolites, *Chem. Res. Toxicol.*, 18 (5), 896, 2005.
31. Soglia, J. R. et al., A semiquantitative method for the determination of reactive metabolite conjugate levels in vitro utilizing liquid chromatography-tandem mass spectrometry and novel quaternary ammonium glutathione analogues, *Chem. Res. Toxicol.*, 19 (3), 480, 2006.
32. Argoti, D. et al., Cyanide trapping of iminium ion reactive intermediates followed by detection and structure identification using liquid chromatography-tandem mass spectrometry (LC-MS/MS), *Chem. Res. Toxicol.*, 18 (10), 1537, 2005.
33. Gorrod, J. W., Whittlesea, C. M., and Lam, S. P., Trapping of reactive intermediates by incorporation of 14C-sodium cyanide during microsomal oxidation, *Adv. Exp. Med. Biol.*, 283, 657, 1991.
34. Meneses-Lorente, G. et al., A quantitative high-throughput trapping assay as a measurement of potential for bioactivation, *Anal. Biochem.*, 351 (2), 266, 2006.
35. Reichardt, P. et al., Identification and quantification of in vitro adduct formation between protein reactive xenobiotics and a lysine-containing model peptide, *Environ. Toxicol.*, 18 (1), 29, 2003.
36. Wang, J. et al., A novel approach for predicting acyl glucuronide reactivity via Schiff base formation: development of rapidly formed peptide adducts for LC/MS/MS measurements, *Chem. Res. Toxicol.*, 17 (9), 1206, 2004.
37. Bailey, M. J. and Dickinson, R. G., Acyl glucuronide reactivity in perspective: Biological consequences, *Chem. Biol. Interact.*, 145 (2), 117, 2003.
38. Wainhaus, S., Acyl glucuronides: assays and issues, in *Using Mass Spectrometry for Drug Metabolism Studies*, Korfmacher, Ed., CRC Press, Boca Raton, FL, p. 175, 2005.
39. Boelsterli, U. A., Xenobiotic acyl glucuronides and acyl CoA thioesters as protein-reactive metabolites with the potential to cause idiosyncratic drug reactions, *Curr. Drug Metab.*, 3 (4), 439, 2002.
40. Mitchell, M. D. et al., Peptide-based in vitro assay for the detection of reactive metabolites, *Chem. Res. Toxicol.*, 21 (4), 859, 2008.
41. Fred, C. et al., Characterization of alkyl-cobalamins formed on trapping of epoxide metabolites of 1,3-butadiene, *J. Sep. Sci.*, 27 (7-8), 607, 2004.
42. Haglund, J. et al., Cobalamin as an analytical tool for analysis of oxirane metabolites of 1,3-butadiene: development and validation of the method, *J. Chromatogr. A*, 1119 (1-2), 246, 2006.
43. Permentier, H. P., Bruins, A. P., and Bischoff, R., Electrochemistry-mass spectrometry in drug metabolism and protein research, *Mini Rev. Med. Chem.*, 8 (1), 46, 2008.
44. Karst, U., Electrochemistry/mass spectrometry (EC/MS)-a new tool to study drug metabolism and reaction mechanisms, *Angew. Chem. Int. Ed. Engl.*, 43 (19), 2476, 2004.

45. Lohmann, W. and Karst, U., Generation and identification of reactive metabolites by electrochemistry and immobilized enzymes coupled on-line to liquid chromatography/mass spectrometry, *Anal. Chem.*, 79 (17), 6831, 2007.
46. van Leeuwen, S. M. et al., Prediction of clozapine metabolism by on-line electrochemistry/liquid chromatography/mass spectrometry, *Anal. Bioanal. Chem.*, 382 (3), 742, 2005.
47. Lohmann, W. and Karst, U., Simulation of the detoxification of paracetamol using on-line electrochemistry/liquid chromatography/mass spectrometry, *Anal. Bioanal. Chem.*, 386, 1701, 2006.
48. Jurva, U., Wikstrom, H. V., and Bruins, A. P., Electrochemically assisted Fenton reaction: reaction of hydroxyl radicals with xenobiotics followed by on-line analysis with high-performance liquid chromatography/tandem mass spectrometry, *Rapid Commun. Mass Spectrom.*, 16 (20), 1934, 2002.
49. Jurva, U. et al., Comparison between electrochemistry/mass spectrometry and cytochrome P450 catalyzed oxidation reactions, *Rapid Commun. Mass Spectrom.*, 17 (8), 800, 2003.
50. Jurva, U., Wikstrom, H. V., and Bruins, A. P., In vitro mimicry of metabolic oxidation reactions by electrochemistry/mass spectrometry, *Rapid Commun. Mass Spectrom.*, 14 (6), 529, 2000.
51. Day, S. H. et al., A semi-automated method for measuring the potential for protein covalent binding in drug discovery, *J. Pharmacol. Toxicol. Methods*, 52 (2), 278, 2005.
52. Takakusa, H. et al., Covalent binding and tissue distribution/retention assessment of drugs associated with idiosyncratic drug toxicity, *Drug Metab. Dispos.*, 36 (9), 1770, 2008.
54. Masubuchi, N., Makino, C., and Murayama, N., Prediction of in vivo potential for metabolic activation of drugs into chemically reactive intermediate: Correlation of in vitro and in vivo generation of reactive intermediates and in vitro glutathione conjugate formation in rats and humans, *Chem. Res. Toxicol.*, 20 (3), 455, 2007.
54. Qiu, Y., Benet, L. Z., and Burlingame, A. L., Identification of the hepatic protein targets of reactive metabolites of acetaminophen in vivo in mice using two-dimensional gel electrophoresis and mass spectrometry, *J. Biol. Chem.*, 273 (28), 17940, 1998.
55. Hanzlik, R. P. et al., The reactive metabolite target protein database (TPDB)–a web-accessible resource, *BMC Bioinformatics*, 8, 95, 2007.
56. Shin, N. Y. et al., Protein targets of reactive electrophiles in human liver microsomes, *Chem. Res. Toxicol.*, 20 (6), 859, 2007.
57. Asif, A. R. et al., Proteins identified as targets of the acyl glucuronide metabolite of mycophenolic acid in kidney tissue from mycophenolate mofetil treated rats, *Biochimie*, 89 (3), 393, 2007.
58. Lohmann, W., Hayen, H., and Karst, U., Covalent protein modification by reactive drug metabolites using online electrochemistry/liquid chromatography/mass spectrometry, *Anal. Chem.*, 80 (24), 9714, 2008.
59. Sleno, L., Varesio, E., and Hopfgartner, G., Determining protein adducts of fipexide: Mass spectrometry based assay for confirming the involvement of its reactive metabolite in covalent binding, *Rapid Commun. Mass Spectrom.*, 21 (24), 4149, 2007.
60. Sleno, L. et al., Investigating the in vitro metabolism of fipexide: Characterization of reactive metabolites using liquid chromatography/mass spectrometry, *Rapid Commun. Mass Spectrom.*, 21 (14), 2301, 2007.

7 Fast Metabolite Screening in a Discovery Setting

Xiaoying Xu

CONTENTS

7.1 INTRODUCTION

Recent data indicate that the discovery and development of a new drug costs around \$1 billion, and it may take approximately 10 years for the drug to reach the marketplace [1]. Considering these staggering numbers, it is critical that efforts are made to reduce attrition of drug candidates during the various stages of drug discovery and development. One of the sources of attrition can be inappropriate drug disposition characteristics. Drugs are xenobiotics to living organisms, which therefore biotransform them to less toxic, less active, and more hydrophilic forms and so enhance their excretion in urine. However, biotransformation can also lead to some unwanted consequences, such as rapid clearance of the drug from the body, formation of active metabolites, drug–drug interactions due to enzyme induction or competition, and formation of reactive or other toxic metabolites [2].

Metabolite identification is crucial to the drug-discovery process because it can be used to investigate the Phase I metabolites that are likely to be formed *in vivo*, the differences between species in drug metabolism, the major circulating metabolites of an administered drug, Phase I and Phase II metabolic pathways, and pharmacologically active or toxic metabolites, and it can also help to determine the effects of metabolizing enzyme inhibition and/or induction. The ability to produce this information early in the discovery phase is becoming increasingly important as a basis for judging whether or not a drug candidate merits further development. Metabolite identification enables early identification of potential metabolic liabilities or issues, provides a metabolism perspective that guides the synthetic route with the aim of either blocking or enhancing metabolism to optimize the pharmacokinetic and safety profiles of newly synthesized drug candidates and assists in the prediction of the metabolic pathways of potential drug candidates chosen for development [3].

The information required to determine the metabolic fate of a new chemical entity (NCE) includes detection of metabolites, structure characterization, and quantitative analysis. Recent efforts among researchers have focused on developing faster methods for metabolite identification. Liquid chromatography–mass spectrometry (LC–MS) has become an ideal and widely used method in the analysis of metabolites owing to its superior specificity, sensitivity, and efficiency [4].

The goal of this chapter is to discuss the challenges involved in incorporating traditionally time-consuming and complex assays into the fast-paced, high-throughput environment of drug discovery. The success of metabolite identification in this setting depends not only on the implementation of cutting-edge technology but also on devising strategies based on critical thinking—to answer the questions that are crucial to a particular program as it progresses compounds from general screening to the final selection of a discovery recommendation.

7.2 REVIEW OF RECENT LITERATURE

7.2.1 Drug Discovery and Development Stages

With the advent of combinatorial chemistry and high-throughput screening, the number of new candidate structures emerging from the discovery cycle has increased significantly. This has created a demand for earlier and faster determination of the absorption, distribution, metabolism, excretion (ADME) properties of these molecules. Based on pure chemical structures, some software packages [5–7] can predict its potential metabolites and toxicity profiles in a high-throughput manner even before the chemical synthesis of a molecule. In this presynthesis stage of discovery, although *in silico* studies cannot replace conventional *in vitro* and *in vivo* testing, they are great tools that are used in conjunction with chemical synthesis efforts to select, eliminate, or modify the chemical structure of a drug candidate.

In the early drug-discovery stage, compounds are made available for screening. At this stage, relatively small quantities of drugs are available (milligram amounts) and radiolabeled standard is typically not available. The popular studies including metabolic stability, drug–drug interaction, and enzyme kinetic studies are based on the quantitative analysis of a parent drug or a few of its metabolites. The key is high throughput. The determination of metabolite profiles is usually performed for a limited number of lead molecules *in vitro* and *in vivo*. The goal is to identify the hot spots in a structure series to guide the modification of the structure to minimize the metabolic liability. Early feedback about metabolically labile sites or potentially toxic metabolites is crucial to the discovery team in order to direct future synthesis pathways. Many discovery programs aim to obtain a general picture of the metabolic fate of NCEs at this stage.

As the compound reaches the late discovery and candidate selection stage, the focus is to determine its major metabolic pathways, metabolic difference between species, and to identify potential pharmacologically active or toxic metabolites. Because of the complexity, comprehensive metabolite characterization studies have been typically conducted at this stage with radiolabeled standard. Identification of circulating metabolites is also important at this stage to explain the pharmacokinetic or the pharmacodynamic profile. An NCE may show efficacy that is inconsistent with what is predicted based upon the known concentration of the parent drug. These inconsistencies could be due to the presence of active metabolites. The knowledge of these metabolites will also dictate how the analysis of samples will be conducted in the development and clinical studies.

When the drug candidate goes to preclinical and clinical development, the goals of drug metabolism studies are to identify unambiguously all significant metabolites observed in humans, and confirm that they are present in the animal species employed for preclinical safety evaluation.

As the drug candidate moves down in the pipeline, the requirements for metabolite identification differ at the different stages of drug discovery and development, leading to the introduction of "fit for purpose" analytical strategies that provide the appropriate level of information for the purpose at hand, while simultaneously minimizing resource expenditures [8]. In order to accomplish the needs, more specialized instrumentation is used to fulfill the task at a later stage. The relative merits of current and potential strategies for dealing with metabolite characterization at the various stages of drug discovery and development with its recommended instrumentation and studies are listed in Table 7.1 [9].

7.2.2 Instrumentation

In the 1970s and 1980s, gas chromatography–mass spectrometry (GC–MS) was the most commonly used method for small-molecule analysis and is still used today for the detection of many metabolic disorders [10]. Due to its convoluted sample preparation that involves metabolite extraction and derivatization to improve volatility, lengthy analysis time, and the limits on the size and type of molecule that can be analyzed, LC–MS has become a more popular choice for these analyses and studies. LC–MS is used at various stages for identifying compounds and their metabolites either to confirm the structure of a known compound or to identify unknown metabolites of drug candidates. The two most commonly employed atmospheric pressure ionization techniques in LC–MS interface are electrospray ionization (ESI) and atmospheric pressure chemical ionization (APCI). They provide efficient ionization for very different type of molecules including polar, labile, and high molecular mass drugs and metabolites. The utility of using ESI lies in its ability to generate gas-phase ions directly from the liquid phase, which establishes the technique as a convenient mass analysis platform for both liquid chromatography and automated sample analysis. APCI generates ions by using a corona discharge with a heated

TABLE 7.1
Tools That Help in the Search and Identification of Metabolites in Drug Metabolism

Stage	Tools	Uses
Presynthesis compound for drug discovery	*In silico*	Assistance with chemical synthesis efforts to select or eliminate compounds
Available compounds for screening and early drug discovery	*In silico* LC–MS/MS	Assistant with synthetic efforts to block or enhance metabolism Identification of simple and major metabolites, for example, dealkylations and conjugations such as glucuronide
Late drug discovery and candidate selection	LC–MS/MS QTOF (high resolution and exact mass measurement) MS^3 H–D exchange	Determination of metabolite differences between species Identification of potential pharmacologically active or toxic metabolites
Preclinical and clinical development	LC–MS/MS QTOF (high resolution and exact mass measurement) MS^3 H–D exchange Radioactivity detector LC–MS–NMR	Determination of the percentage of metabolite formed *in vitro* or *in vivo* Synthesis of metabolites for toxicology testing Comparison of human pathways Drug–drug interactions

Source: Adapted from Nassar, A.F. and Talaat, R.E., *Drug Discov. Today*, 9, 317, 2004. With permission.
Abbreviations: H–D, hydrogen–deuterium; LC, liquid chromatography; MS, mass spectrometry; MS–MS, tandem mass spectrometry; MS^3, ion trap; NMR, nuclear magnetic resonance; QTOF, quadrupole time-of-flight.

nebulizer where the solvent vapor can act as the reagent gas providing chemical ionization reactions. The recent atmospheric pressure photoionization (APPI) has expended the applicability of API techniques toward less polar compounds [11]. In the dopant-assisted APPI method, the major processes leading to ionization in APPI are proton transfer and charge exchange in the positive ion mode. Its sensitivity is dependent on ion-molecule reactions in the gas phase that are largely governed by the proton affinity of the analytes [12].

A wide range of mass spectrometers have been used for metabolite identification [13,14]. The most frequently used instruments are single- and triple-quadrupole (QQQ), three-dimensional (3D) ion trap, linear ion trap (LIT), and quadrupole-time-of-flight (Q-TOF) (Figure 7.1). Fourier-transform mass spectrometers (FTMS) have been used as well typically at a later stage. The new hybrid technology of QTrap and LTQ-Orbitrap provides unique features that bring in a lot of excitement in the drug metabolism area [15].

In the current pharmaceutical industry, most work in metabolite analysis is still carried out by using QQQ mass spectrometers due to its durability and low cost. Their tandem mass spectrometric (MS/MS) scan types are highly helpful in the

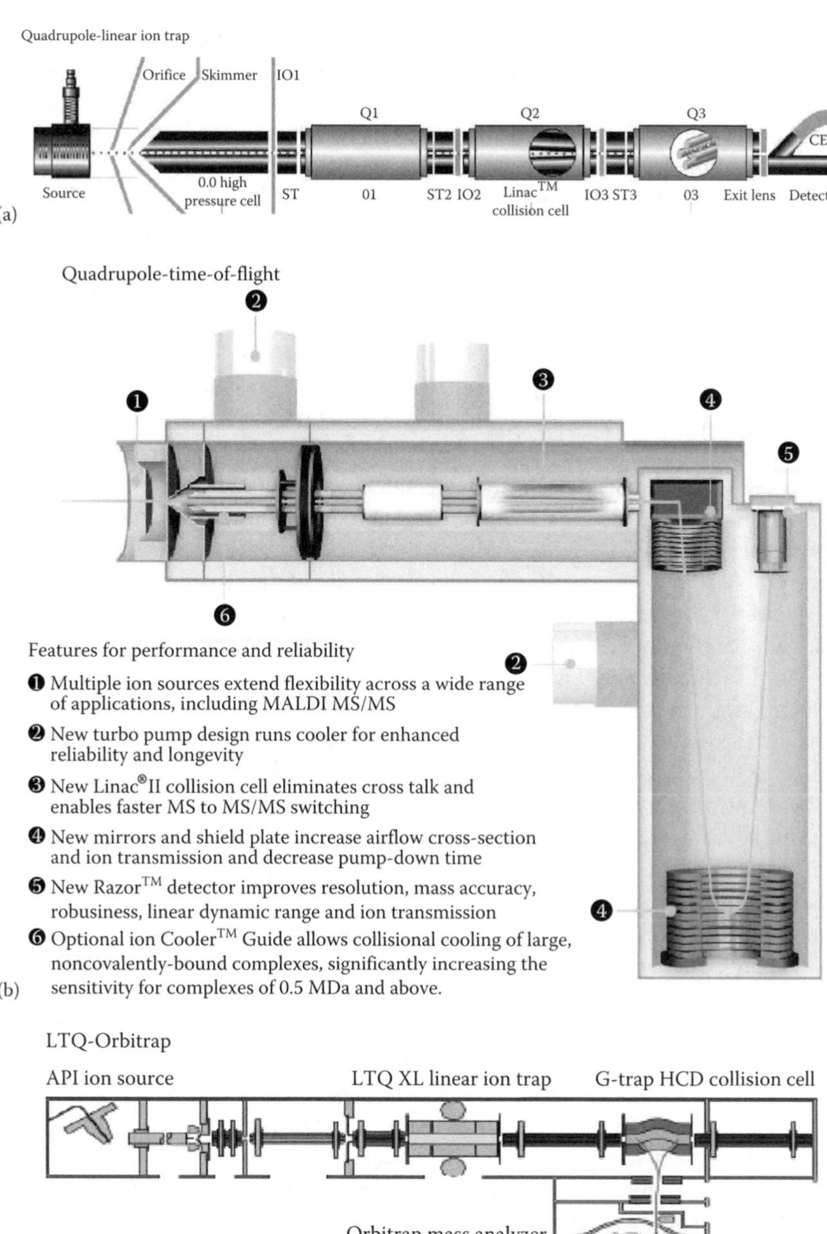

FIGURE 7.1 Different types of mass spectrometers that are used for various drug metabolism assays. (Courtesy of Applied Biosystems MDS Analytical Technologies and ThermoFisher, Foster City, CA. With permission.)

identification of metabolites and provide the required specificity and sensitivity. With single and triple quadrupole mass spectrometers, MS data can be obtained by scanning the first quadrupole and detecting the mass-separated ions. Product ion MS/MS data can be obtained with a QQQ by mass-selecting the ions of interest with the first quadrupole, fragmenting these ions in the collision cell and mass separating the ions with the third quadrupole for subsequent detection [1]. The tandem MS capabilities of a QQQ are useful in providing a quick look at the metabolic profile of an NCE and offer the best chance of identifying novel or unexpected metabolites. QQQ has the unique constant neutral loss and precursor ion MS/MS features, which has been widely used and they require no prior knowledge of the metabolites and may only require a minor structural similarity to the NCE [16,17]. However, QQQ has its major disadvantage of poor full-scan sensitivity and slow scan rates, which makes it difficult to capture low abundance metabolites.

Often, a single MS/MS experiment is insufficient to narrow down the site of modification. An ion trap mass spectrometer is unique in that it can perform multiple MS/MS experiments. In a 3D ion trap system, all ions are at the center of the ion trap and a spectrum is obtained by sequentially ejecting ions out of the trap. MS^n is possible, which provides more fragmentation information about the structure of a metabolite [18]. However, it suffers poor trapping efficiency and space charge effects on mass peak width and mass measurement accuracy. A geometrical variation of the 3D ion trap MS, LIT has garnered a lot of interest. It can provide very high ion acceptance and more efficient trapping. The detection sensitivity in a LIT is at least two orders of magnitude higher than that in a 3D ion trap [19,20].

As QQQ and LIT posses their own unique features for drug metabolism studies, a combined instrument hybrid linear ion trap-triple quadrupole mass spectrometer (QTrap) offers the advantages of both the ion trap and the QQQ and provides great flexibility and excellent performance. The ability of the third quadrupole to act as both a mass filter and a scanning ion trap gives QTrap a wide range of capabilities. The instrument retains all of the scans of the typical QQQ, and it also includes many of the scans available in the 3D ion trap. The system can perform the common QQQ scans for precursor ions and constant neutral loses, selected reaction monitoring (SRM) and selected ion monitoring (SIM). In addition, the system can perform trap isolation and fragmentation to yield trap-like product ions or as a means of forming second-generation product ions for MS^3 studies. The system also has the ability to look at ions stored for different periods of time in the trap. It allows selective detection and high speed–high sensitivity collection of product ions without space charge and mass range limitations of the 3D trap. However, one limitation is its mass resolving power and mass accuracy capability [21].

Q-TOF technology has provided rugged high-performance accurate mass (AM) and high resolution capabilities that are useful in the evaluation of metabolites present in the complex matrices. Its advantages are sensitivity, scan speed, and mass accuracy. The high mass resolution enables the separation of metabolites from nominally isobaric background ions. The mass accuracy is usually within 5 ppm, which makes it possible to assign a specific molecular formula to the metabolite of interest. However, it cannot distinguish isomers that have the same exact mass. Its fast scan speed can provide highly sensitive detection of low-level metabolites, for example, circulating metabolites in plasma [22,23].

Other instruments, such as FTMS, are capable of achieving the highest resolution and mass accuracies (<1 ppm), thus providing accurate determination of unknown metabolites more reliably than any other MS available today. However, there are very few reports on pharmaceutical FTMS applications specifically for drug metabolism analyses due to its high cost, difficulty to use, and low throughput [24]. The big step forward for drug metabolism and small molecule analyses came with the introduction of the hybrid FTMS (e.g., LTQ–FTMS). The hybrid systems are compatible with standard HPLC flow rates, have high-throughput and automation compatibility, and are data-dependent MS^n with high mass accuracy and mass resolution [25–27].

The recently introduced LTQ-Orbitrap is another hybrid detection instrument, where the superconducting magnet and the ICR cell in an FTICR instrument are replaced by an electrostatic trap providing performance close to that of an FTICR instrument [25,28]. The LTQ-Orbitrap shows excellent performance for small molecule MS/MS. It can achieve high mass resolution up to 150,000 and mass accuracy (1 ppm) [19,29]. In most cases, the sensitivity of Orbitrap detection is superior to the FTICR detection of small molecules. Low mass transmission does not appear to be an issue, as spectra are very similar to those from the LTQ. Mass accuracies did not appear to be as precise as those for the LTQ–FTMS; however, with the vast majority of mass measurements falling within 3 ppm, its performance is well suited for structure elucidation.

In the end, each of the instruments described here provides unique capabilities that are highly useful in metabolite structure elucidation [30]. The most effective and comprehensive metabolite characterization experiments utilize a combination of these instruments at different stages of the drug discovery and development in order to answer a specific question.

7.3 CURRENT USES AND TECHNOLOGY

Regardless of which drug stage we are in, the first step for metabolite profiling is detection. The second step is structure elucidation and finally quantitation. The definitive metabolic analysis typically uses radioactive labeling (^{14}C, ^{3}H) of a parent drug and detection of its metabolites by LC–MS along with a radioactivity flow detector [31]. However, synthesis and purification of radioactive compounds are expensive and time-consuming, radiation is a potential health risk for humans and the requirements for handling radioactive material and wastes make the use of radiolabeled compounds very costly. For these reasons, radioactively labeled compounds are rarely used in the early drug discovery phase, and the early metabolite profiling studies almost totally rely on various MS techniques [32]. This chapter focuses on the recent practices and the new technologies utilized in metabolite profiling in early drug discovery phase.

7.3.1 Sample Preparation

In the early stage of drug discovery, the metabolic pathways of a given compound are typically not known, and the samples do not contain radiolabeled compounds, any cleanup or manipulation of the samples could result in loss of metabolites. Since the matrices for metabolite characterization are typically very dirty, some forms of sample cleanup are always needed. The simplest and most effective way is protein precipitation followed by centrifugation. Liquid–liquid extraction (LLE) or solid-phase extraction (SPE) can be utilized when there is some *a priori* knowledge of the possible metabolites [16].

7.3.2 Metabolite Identification Methods

Traditional metabolite identification has primarily relied on using the QQQ. A typical procedure starts with screening the test and control samples over the full mass range in both positive and negative ionization modes. By comparing the total ion chromatogram (TIC) from the test and control samples, all predicted or unexpected drug-related materials that are only presented in the test samples can be determined. However, this approach usually suffers by its low sensitivity. The reconstructed ion chromatogram (RIC) of potential metabolites based on common biotransformation reactions and their associated mass shifts are more useful, and it could reveal the presence of many drug-related ions hidden under background matrix ions [33,34]. Other survey scans, such as precursor ion scan (PIS) or constant neutral loss (CNLS) experiments are commonly used in the next step. In PIS, the second mass analyzer is set to only pass the ions with a selected *m/z* value, while the first mass analyzer is scanned over a defined *m/z* range. In CNLS, both mass analyzers are scanned while the *m/z* difference between the mass spectrometers is kept constant [19]. Utilizing these two techniques can identify families of metabolites quickly. The last step includes product ion scan or multiple stage MS^n; an ion of a given *m/z* value is selected, activated, and fragmented, which provides structural information. This traditional approach suffers from iterative analysis and low throughput and typically requires large amounts of samples and significant user intervention. Under the currently fast-paced drug discovery timelines, it becomes a less-desired approach. However, the recent commercial availability of a QQQ with enhanced mass resolution and AM capability (TSQ Quantum AM QQQ, ThermoFinnigan, San Jose, CA, United States) has provided attractive features. Jemal et al. demonstrated the feasibility of using the Quantum AM QQQ to identify nefazodone and its two metabolites in a liver microsome incubation. In the study, the CNLS and PIS were able to identify most of the metabolites of nefazodone. The subsequent AM SIM and enhanced resolution AM SRM experiments gave mass accuracy of better than ±0.003 u for the masses of the precursor and product ions of nefazodone and all the metabolites [35].

When LC–MS/MS is used for the quantitation of drugs, the quantitation experiments have reached a level that one can term as high throughput in that sample processing is automated and analysis is done using QQQ with fast run times of less than 5 min. Metabolite identification of the drug candidates typically is done in a separate experiment that require longer HPLC run times. Recently, quantitation of drug candidates with simultaneous structural characterization of their metabolites has become more fully realized and may be done within the time frame of a single high-throughput quantitative LC–MS/MS experiment without having to reanalyze the plasma samples. Chen et al. [36] utilized SRM-triggered product ion scan with reversed energy ramp (RER) scanning to simultaneously identify and quantify propranolol and its metabolites in rat plasma on a Finnigan TSQ Quantum Ultra. With the quality-enhanced data-dependent (QED) scanning and RER techniques, not only were five metabolites of propranolol characterized with high-quality product ion spectra, but time course profiles of propranolol and its metabolites were also obtained simultaneously [36]. Quantitative data of the parent obtained with SRM-dependent scan approach is very similar to those with SRM only approach. Spectra acquired with RER scans have richer fragment information which facilitates structure elucidation of metabolites. Cai et al. [37] tested a similar approach on a Bruker Esquire ion trap MS system (Bruker-Franzen, Bremen, Germany). The α-1a antagonists were dosed in the mouse. Plasma samples were collected at various times. The major metabolites in the mouse plasma samples were detected simultaneously through the interpretation of full-scan mass spectra. Varoglu and colleagues [38] ran a microsomal incubation assay on an Orbitrap LC–MS system. The AM measurements coupled with the ability to create extracted ion chromatogram of very narrow 0.01–0.02 Da mass ranges provided chromatograms with very high signal-to-noise ratios to confidently detect the parent compound and its metabolites. In addition, MS–MS transitions were not necessary to observe analytes of interest in the samples and the need for individual compound method development was eliminated.

More applications in simultaneously quantifying parent drugs and screening metabolites in a single sample injection for either *in vitro* or *in vivo* samples were reported by various groups using a hybrid API 4000 QTrap LC–MS/MS system (Applied Biosystems, Foster City, CA, United States). Gao et al. [39] used a survey experiment consisting of monitoring SRM transitions for the internal standard, the parent, and 48 SRM transitions designed to cover the most common Phase I and II biotransformations in a hepatocyte incubation sample. An information-dependent acquisition (IDA) method was employed to trigger product ion scans above the SRM signal threshold. The high sensitivity and specificity of SRM enabled the detection of minor metabolites at trace levels in biological matrices. Coeluting metabolites could be detected and identified due to the specificity of their respective SRM transitions, which also enabled the usage of a rapid chromatographic method compared to conventional metabolite identification methods where a longer chromatographic method was required for better separation. Three biotransformations of a lead compound had been identified through enhanced product ion (EPI) scans and the respective SRM transitions of those metabolites were selected for semiquantitation. The study integrated the metabolite identification, the semiquantitative analysis of parent disappearance, and the formation of the metabolites along the time course by a rapid LC–MS/MS analysis.

Li et al. [40–42] used a QTrap system to incorporate both the conventional SRM-only acquisition of parent compounds and the SRM-triggered IDA of potential metabolites within the same scan cycle during the same LC–MS/MS run in plasma sample analysis. The fast scanning capability of the LIT allowed for the IDA of metabolite MS/MS spectra <1 s/scan, in addition to the collection of adequate data points for SRM-only channels. A SRM survey scan containing 30–150 SRM channels was established by running a script in Analyst software. If the intensities in these survey channels exceeded a predefined threshold, EPI scan spectra were automatically collected from the most intense peak (Figures 7.2 and 7.3). The information of the product ion scan spectrum can be readily enhanced by spreading the collision energy for the collision-induced dissociation (CID). The final spectrum without a low mass cutoff is obtained by combining the spectra collected at three different collision energies with a predefined spread energy. The author further confirmed that quantifying parent compounds using SRM + SRM–IDA approach was not negatively impacted by the addition of survey SRM channels and the IDA–EPI scans and generated equivalent quantitative results for parent compounds to those obtained by conventional SRM-only method. The data files generated using this approach were only about twice as large as those from the SRM-only method, which was of less concern in processing speed of the quantitative data. The advantage of this approach is that the SRM transitions can be generated through the script in Analyst software to cover the common metabolites. This list can be further improved by addition of possible metabolites or removal of unlikely transitions based on the knowledge of the metabolite predictions and the fragment pattern of the parent drug [40,41]. The disadvantage of this target scanning method is that it is unable to detect unexpected metabolites compared to the trap mode full MS scan (EMS)-triggered IDA or data-dependent MS/MS although full data-dependent qualitative analysis was hindered in practice because of the high background signal from the matrix. The importance of adequate chromatographic separation should also be emphasized when this approach was applied. If multiple metabolites elute at the same retention time, potential misses can happen because EPI will be performed only on the most intense peak.

In addition to this simultaneous quantitative and qualitative function, the QTrap can be used under full-scan MS in the ion trap mode and/or CNLS, PSI as survey scans to trigger product ion scan (MS^2) and MS^3 experiments to obtain structural information of drug metabolites "on-the-fly." Xia et al. [43] reported five metabolites of gemfibrozil were detected in a single injection. Bramwell and colleagues [44] utilized different scan functions in the QTrap to identify GSH conjugates in human liver microsome incubations.

With the capabilities of EMS (Q3 scan in the ion trap mode), EPI scan (MS^2 in the ion trap mode), MS^3, SRM, CNLS, PSI, and IDA, the QTrap system can be employed as a complementary tool to the conventional metabolite profiling with the capability to meet the fast turnaround demands in drug discovery. It can also stand alone to aid the selection and design of the structural scaffold at the lead selection stage or be incorporated to the overall metabolic profile strategy to evaluate preclinical candidates. However, this approach is only intended to be an "added bonus" during parent compound quantitation, and it is not a replacement for the full profiling of metabolites in a given sample because of the highly specific nature of SRM, there definitely exists the possibility that the survey channels could miss important metabolite pathways.

In a recent report, Tiller et al. [45] described increased throughput for drug metabolite identification studies in the early discovery phase with a record of 21 diverse structural classes of NCEs assayed in one day using AM LC–MS/MS on a Waters QTOF Premier MS system (Waters, Milford, MA, United States) and targeted data analysis procedures. An MS^E experiment [46], utilizing dynamic range enhancement

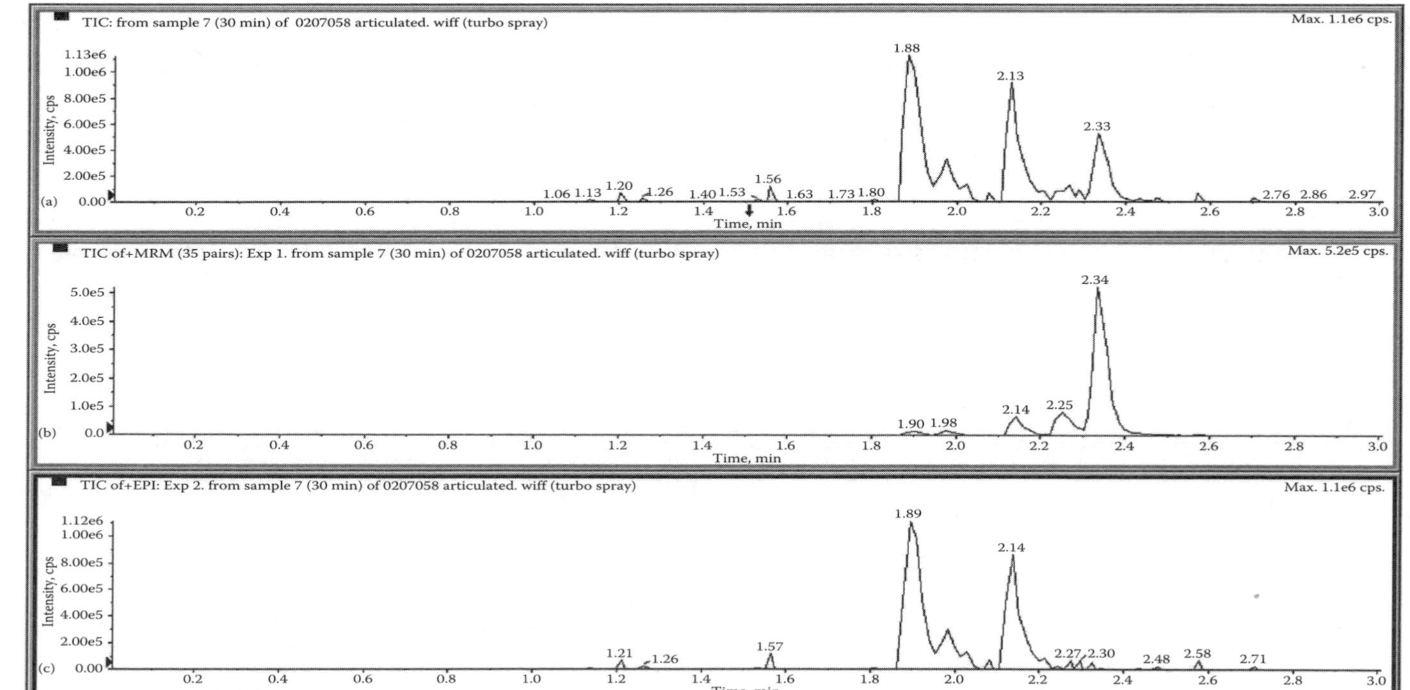

FIGURE 7.2 RICs obtained from a 30 min incubation sample of bufuralol using the SRM + SRM–IDA–EPI approach. (a) TIC, (b) reconstructed TIC of all the SRM channels, and (c) reconstructed TIC of the EPI scans. (Adapted from Shou, W.Z. et al., *J. Mass Spectrom.*, 40, 1347, 2005. With permission.)

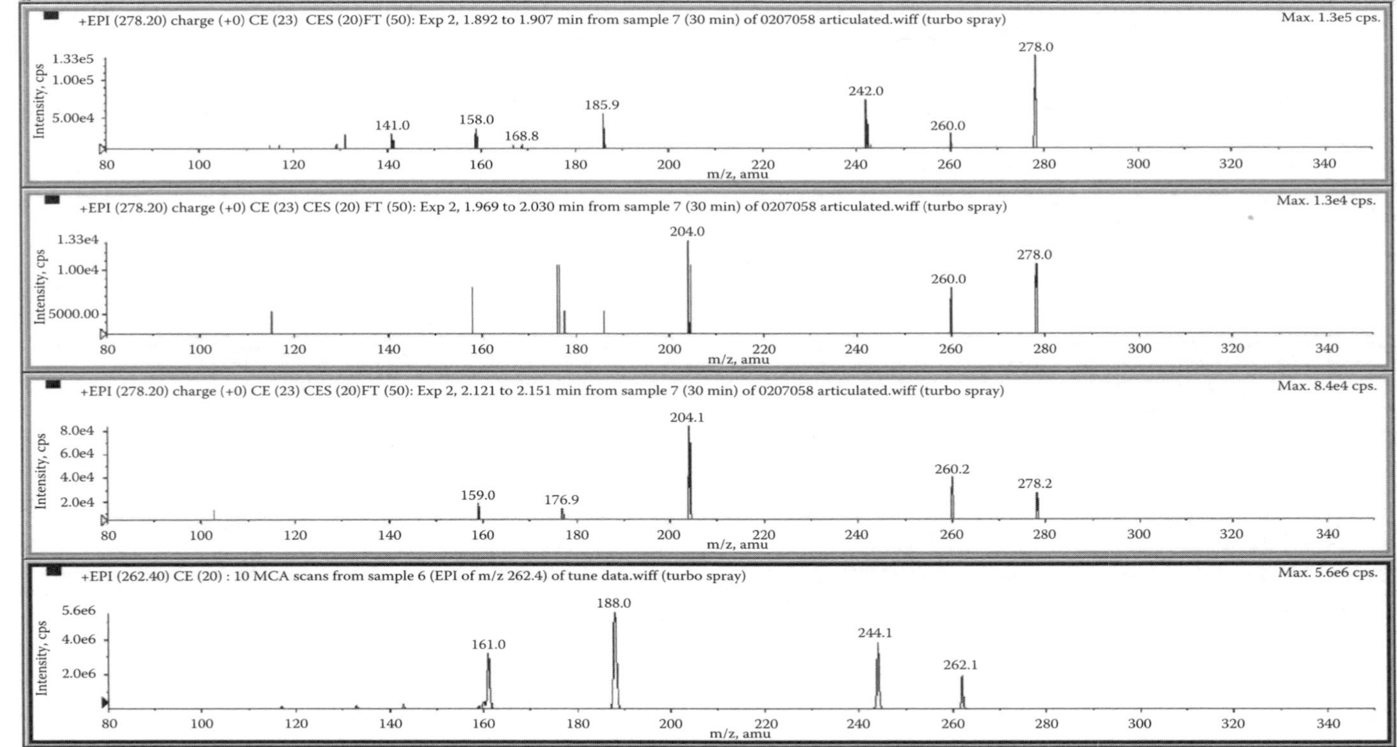

FIGURE 7.3 EPI spectra of bufuralol and its three metabolites. (Adapted from Shou, W.Z. et al., *J. Mass Spectrom.*, 40, 1347, 2005. With permission.)

(DRE) with two scan functions, enables the almost simultaneous acquisition of both LC–MS and fragmentation data from a single experiment, although since no precursor ion selection occurs in the mass spectrometer good chromatographic separation is a critical factor in affording fragment ions that are derived predominantly from the analyte of interest. MS^E works well when applied to relatively simple matrices, such as microsomal and hepatocytes incubations, but it does not work well with complex matrices such as bile samples. The targeted data analysis was carried out with the aid of Metabolynx software. The availability of elemental composition data on precursor and all fragment ions in each spectrum greatly enhanced confidence in the ion structure assignments, while computer-based algorithms for defining sites of biotransformation based upon mass shifts of diagnostic fragment ions facilitated identification of positions of metabolic transformation in drug candidates. This approach typically requires as little as 1 h of acquisition time (during an overnight run) and about 1 h of data interpretation time by a scientist with a savings of 13 h per molecule (7 h of acquisition time and 6 h of data interpretation time) to the traditional method. With this capacity, AM MS/MS technology can cease to play its traditional function as a confirmatory technique in metabolite structure elucidation, but will increasingly assume a "first-line" role in the detection and characterization of drug metabolites in early drug discovery.

Another "all-in-one" strategy proposed by Wrona et al. [47] used the QTOF system to assay *in vitro* samples in a drug discovery setting. Full-scan MS and MS/MS data was acquired using collision energy switching without the preselection of precursor ions. Data was collected using two scan functions. In the first function, Q1 was scanned using a normal low collision energy that provided for transmission of intact ions to the analyzer and detected with high resolution (8000 FWHM) and mass accuracy (<10 ppm). The second scan functions scanned Q1 in the same mass range with a high collision energy that fragmented all ions. In this way, two mass chromatograms were generated, one with information on intact molecules and one with fragment ion information. A variety of data-processing algorithms then were used to extract metabolite information from these data. These included using narrow window extracted ion chromatograms for expected biotransformations, extracted ion chromatograms for the product ions of the parent compounds, and/or expected modification of these product ions, and neutral loss chromatograms. The advantage of this approach is that a single injection using a generic acquisition method provided a data set from which single ion, precursor, product, and neutral loss chromatograms could be extracted. No implicit assumptions about the biotransformations were made prior to data acquisition.

A directed search for metabolites that are predicted on the basis of the structure of the parent compound offers the greatest probability of success in detecting drug-related materials in the presence of large extraneous background. Another novel approach, knowledge-based predictions of metabolic pathways, is first derived from a commercial database, the output from which is used to formulate a list-dependent LC/MS^n data acquisition protocol. Reza Anari et al. used indinavir as a model drug, a substructure similarity search on a MDL metabolism database (MDL Information Systems, Inc.) with a similarity index of 60% yielded 188 hits, pointing to the possible operation of two hydrolytic, two *N*-dealkylation, three *N*-glucuronidation, one

N-methylation, and several aromatic and aliphatic oxidation pathways. The overall biotransformation spreadsheet included all classical primary metabolic pathways for xenobiotic biotransformation as well as various multistage oxidative metabolic reactions (Figure 7.4). The calculated mass-to-charge ratios for all plausible metabolites of indinavir were exported to the Xcalibur LCQ software in order to set up a list-dependent instrument method. Integration of this information with data-dependent LC/MS^n analysis using a LCQ DECA XP ion trap mass spectrometer (ThermoFinnigan, San Jose, CA, United States) led to the identification of 18 metabolites of indinavir. This result was accomplished with only a single LC/MS^n run, representing significant savings in instrument use and operator time, and afforded an accurate view of the complex metabolic profiling of the drug, as shown in the flowchart in Figure 7.5 [48].

Regardless of which approach is utilized, a prerequisite for successful LC–MS/MS metabolite identification is good chromatographic separation. Although significant advances have occurred in LC modes and stationary phases, the separation of isobaric metabolites can require long run times. Ultraperformance liquid chromatography (UPLC) has enabled these analyses to be carried out at higher speed with better sensitivity, increase sample throughput by reducing analytical run time without sacrificing chromatographic integrity (see Chapter 8) [49]. Using UPLC, Wainhaus and colleagues demonstrated the powerful high speed separation of two hydroxyl analogs from its parent within a 7 min, while a normal HPLC would require at least 20 min to perform the sample separation [50]. Other researchers applied UPLC systems in their fast metabolite identification strategy because of its excellent chromatographic resolving power [45,47]. By improving the chromatographic resolution, increased peak capacity can be achieved with a reduction in the number of co-eluting peaks leading to superior separation, which can significantly reduce ion suppression and improve the MS sensitivity [49].

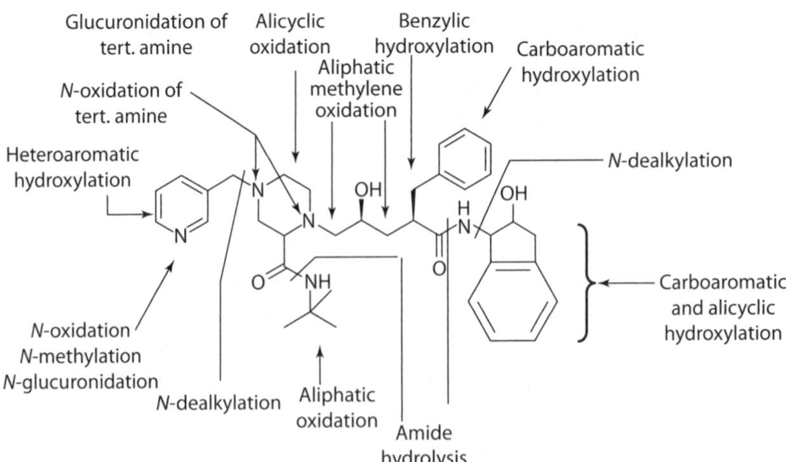

FIGURE 7.4 Knowledge-based metabolic prediction of indinavir. (Adapted from Reza Anari, M. et al., *Anal. Chem.*, 76, 823, 2004. With permission.)

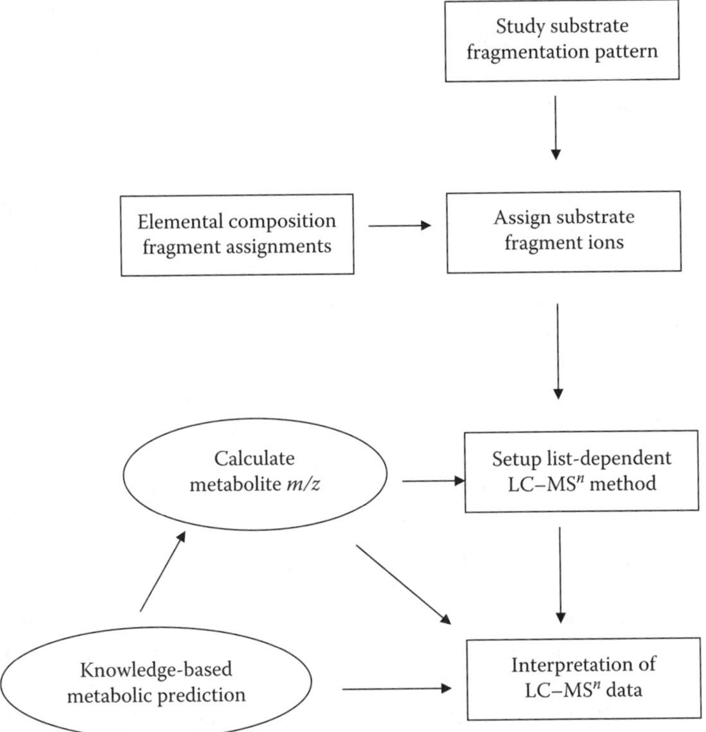

FIGURE 7.5 Metabolite identification strategy based on integration of knowledge-based metabolic predictions with liquid chromatography list-dependent tandem mass spectrometry. (Adapted from Reza Anari, M. et al., *Anal. Chem.*, 76, 823, 2004. With permission.)

In the end, the bottleneck occurs in the data processing and interpretation. Automated software algorithms, such as Waters MetabolLynx, Applied Biosystems Lightsight, and MetWorks from Thermo, which came with the different types of mass spectrometers, became useful tools for detecting biotransformations for expected or unexpected metabolites. Typically acquired results were reported via a "Data Browser" that enabled the chromatographic and mass spectroscopic evidence that supported each automated metabolic assignment [51,52].

The advent of all the new ion sources and mass analyzers, along with the recent development in data-acquisition and handling software, have significantly improved quality and detailed information acquired for identification of unknown structures. These methods can be readily used for rapid frontline identification of drug metabolite structures in early drug discovery.

7.3.3 Quantitation of Metabolites

Quantitative information on metabolites that have been identified to have pharmacological and/or toxicological activities is very important during the drug discovery and development process. One limitation of LC–MS for metabolite quantification is that different ESI response factors are observed for different metabolites, which makes ESI a poor quantitative tool. This problem is further influenced by gradient elution because the desolvation efficiency increases during the run due to increased organic solvent content. In addition, another challenge with metabolite quantitation is associated with the difference in extraction recovery between the metabolites and the parent compounds, particularly with commonly used selective sample extraction techniques such as SPE or LLE. Therefore, typically, metabolite standards are required or alternative radiolabeled compounds will be needed for quantification. However, in the early drug discovery stage, metabolite standards or radiolabeled compounds are usually not available, which makes alternative approaches an urgent need.

Since most drugs contain at least one nitrogen atom, one approach is utilizing a chemiluminescent nitrogen-specific detector (CLND). CLND gives a response corresponding to the number of nitrogens in the molecule independent of the compound structure, with a few exceptions, for example, compounds with nitrogens adjacent to each other like hydrazines may eliminate nitrogen gas during combustion in the detector which results in a lower response [53]. In HPLC–CLND, samples in the HPLC effluent are first combusted at high temperatures in an oxygen-rich furnace to convert all organic species to oxides of carbon, nitrogen, water, and other combustion products. Nitric oxide, generated from nitrogen-containing compounds, is then reacted with ozone to produce nitrogen dioxide in an excited state. The rapid relaxation of the excited nitrogen dioxide results in the release of a photon of light, which is then captured and amplified in a photomultiplier tube. The chemiluminescent response is proportional to the number of moles of nitrogen originally present in the analyte and the signal is independent of structure. Due to its low sensitivity and high background interference from endogenous nitrogen-containing compounds in biological matrices, CLND by itself is a poor choice for quantifying metabolites and typically is coupled with MS. In this setup, CLND serves as a calibrator to obtain the response factor ratio for a metabolite and its parent drug. For example, Deng and colleagues evaluated a HPLC–CLND/MS system with oxazepam and its metabolite temazepam. Taking advantage of the equimolar response feature of the CLND, a response factor ratio between oxazepam and temazepam on the mass spectrometer was obtained by comparing the peak areas generated on the CLND and a QQQ MS system. From the ratio, temazepam was quantified using the oxazepam standard curve. The difference between the concentration of temazepam obtained from the reconstructed standard curve and the concentration obtained directly from a real temazepam standard curve was within 13%, except for the least concentrated standard (31%) (Figure 7.6) [54]. One limitation of this technique is the low sensitivity of the CLND system, which requires a significant amount of metabolites (low micromole level) in the samples. Another limitation is that it is not applicable to compounds that do not contain nitrogen, or in those cases where the metabolites have no nitrogen atoms.

There has recently been some interest in nanoflow LC as a mean of normalizing the mass spectrometric response [55–58]. Nanospray is more than simply a reduction of ESI flow rate. ESI is an electrohydrodynamic process, with daughter droplets generated when Columb repulsion in a charged liquid (the parent droplet) overcomes

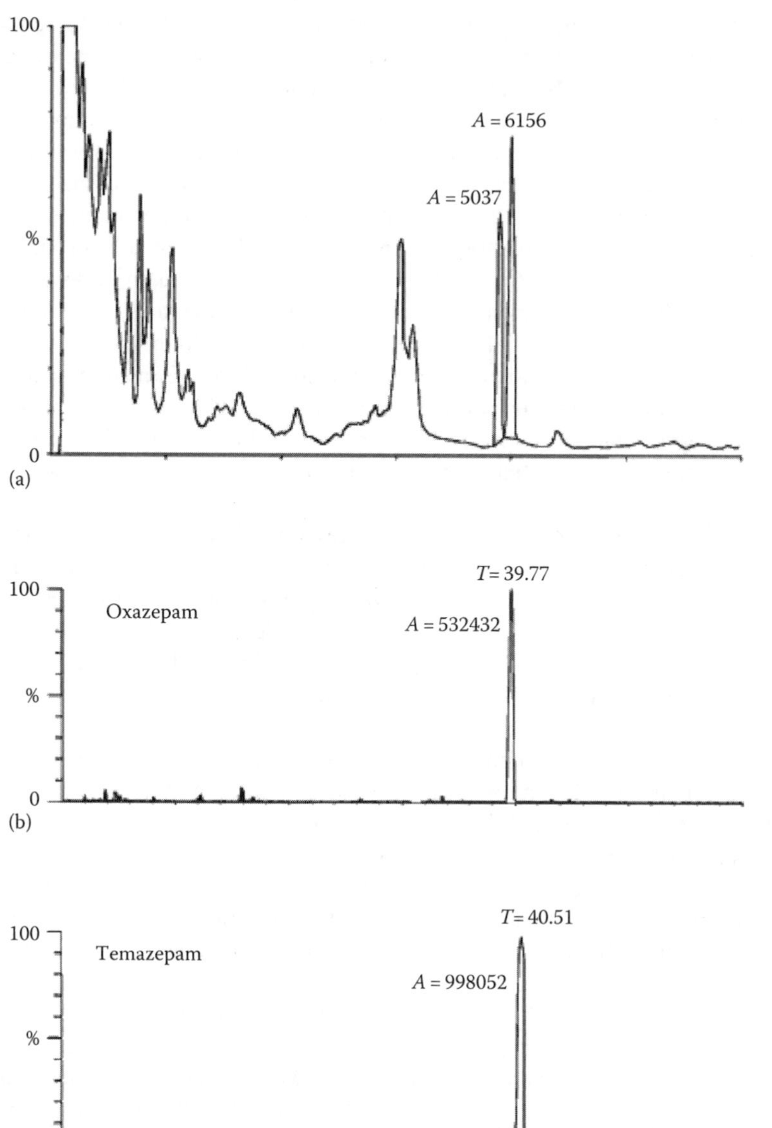

FIGURE 7.6 A sample containing 1 μM of oxazepam and an unknown concentration of temazepam in dog urine was injected into the HPLC/CLND/MS system. (a) CLND chromatogram, (b) and (c) MRM chromatograms for oxazepam and temazepam. (Adapted from Deng, Y. et al., *Rapid Commun. Mass Spectrom.*, 18, 1681, 2004. With permission.)

surface tension. Nanospray typically generates a smaller initial droplet size than its conventional ESI counterpart. The initial droplet size generated at an emitter orifice is a complex function of both intrinsic mobile phase properties and external experimental parameters, perhaps most notably including flow rate. The reduction of initial droplet size can impact the quality of ESI spectra. Smaller droplets will increase total available surface area, decrease diffusion time for a solvated species to the droplet surface, and decrease the number of columbic explosions required to yield a sufficiently small droplet suitable for ionization, which minimize ion suppression. Valaskovic and colleagues [58] used a custom-built, off-line, nanospray system to provide for direct, accurate measurement and control of flow rates in the low-nanoliter/min regime on extracted plasma samples. Nanospray exhibited a distinct trend toward equimolar response when flow rate was reduced from 25 nL/min to less than 10 nL/min. A more uniform response between the parent drug and the corresponding metabolites was obtained at flow rates of 10 nL/min or lower (Figure 7.7). The results indicated that the control of nanospray parameters, especially flow rate, would be a key consideration for success.

Hop et al. [56,59] demonstrated that ESI response factors could be normalized using nanoelectrospray. In this report, 25 compounds from 6 structurally distinct classes were tested using silicon chip-based nanoelectrospray devices (NanoMate) with <300 nL/min flow rate. The MS responses between these 25 compounds were relatively consistent (within 2.2-fold). This can be compared to standard LC–MS data which would have a 21-fold difference in response; therefore, it was a big improvement.

Ramanathan et al. [57] reported using nanoelectrospray in combination with a postcolumn makeup solution having the inverse composition to the analytical gradient. The set-up involves two HPLC systems, a chip-based nanospray ionization (NSI) source and a Q-TOF mass spectrometer. One HPLC unit performed the analytical separation, while the other unit added solvent post-column with an exact reverse of the mobile phase composition such that the final composition entering the NSI source was isocratic throughout the entire HPLC run. The data obtained from four different structural classes of compounds and their metabolites indicated that by maintaining the solvent composition unchanged across the HPLC run, the influence of the solvent environment on the ionization efficiency was minimized. Their examples showed excellent agreement between radioactivity and LC/MS data, and the response normalization modification resulted in nearly uniform response for all tested compounds.

In the light of the difficulty of integrating an LC system capable of delivering nanoflow rates reproducibly in a regular laboratory setting, Wainhaus explored a different approach to quantify metabolites without an authentic standard [50]. The tested model compound (BAS12439) and its series of hydroxyl group metabolites were evaluated under regular LC–MS/MS or UPLC–MS/MS setting under both ESI and APCI modes. In ESI, mass spectrometric response can vary dramatically with observed 2–8-fold response difference between parent and its metabolites across all fragments. However, when the same conditions were applied in APCI, only 1–4-fold of difference was observed. A twofold difference for minor molecular modification was common under APCI conditions. This phenomenon appeared to be consistent with a variety compounds. Within early discovery, these errors

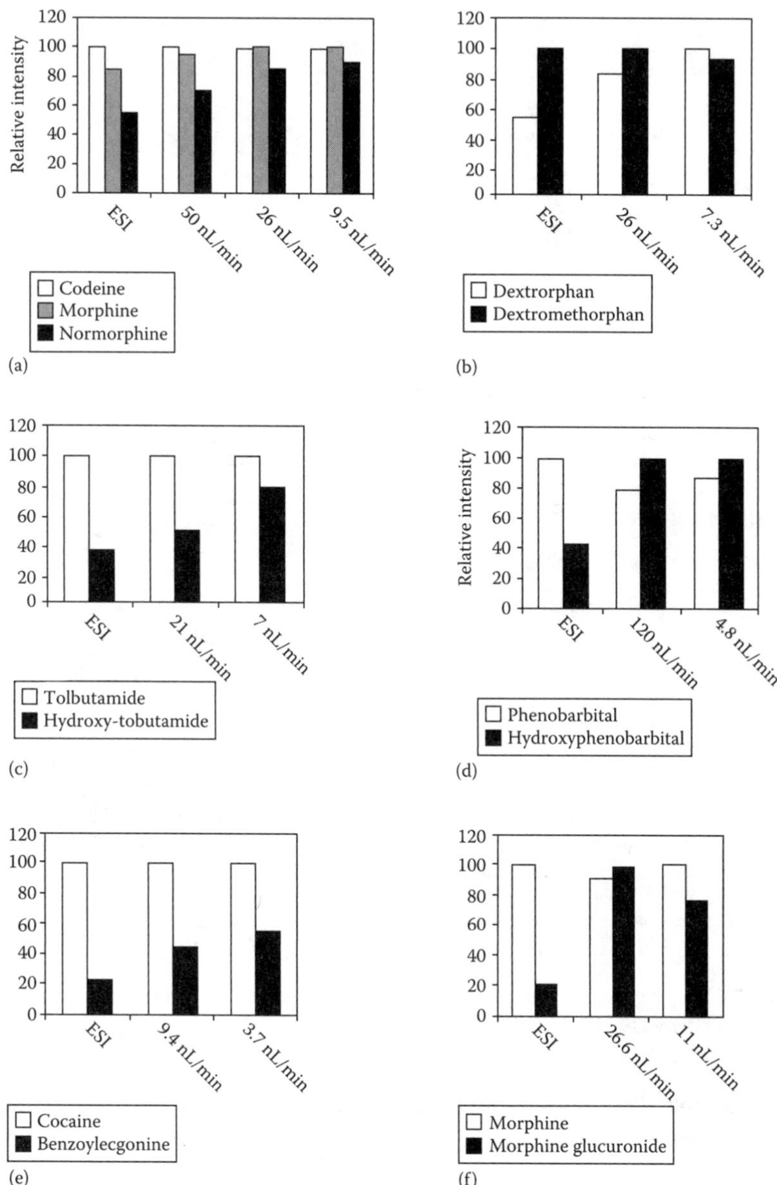

FIGURE 7.7 Results (a through f) of the equimolar response study for a variety of diverse compounds and their metabolites, respectively, for both conventional ESI (5 µL/min) and nanospray (flow given in nanoliter/min). Each bar graph shows the relative average signal intensities obtained from full-scan MS (average 12 scans). Each compound set was spiked into LLE plasma (2.5 µM final conc. each). Analysis was carried out by conventional ESI (5 µL/min) and ultralow flow nanospray (a 5 µm. i.d., short taper, metal-coated fused-silica emitter). *Note* that the commonly observed trend toward equimolar response as flow rate is reduced. (Adapted from Valaskovic, G.A. et al., *Rapid Commun. Mass Spectrom.*, 20, 1087, 2006. With permission.)

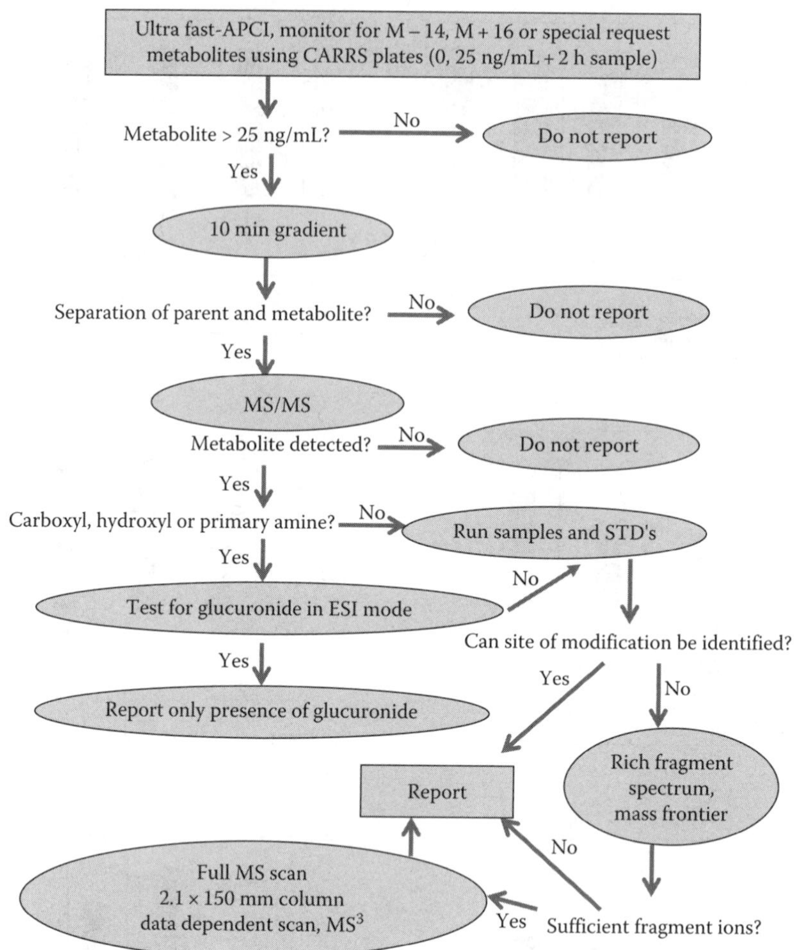

FIGURE 7.8 Systematic targeted metabolite screening procedure. (Adapted from Wainhaus, S. et al., *Am. Drug Discov.*, 2, 6, 2007. With permission.)

were acceptable for an early estimation of metabolite quantity and a new paradigm was established (Figure 7.8).

7.4 NEW EMERGING TECHNOLOGY

The field of metabolite identification and its impact for drug discovery is dynamic and advances in technology, and strategies are constantly evolving. There are significant challenges still be faced and exciting technologies on the horizon [60]. Traditionally, the incubations for *in vitro* drug metabolism studies were performed "off-line." Recently, several approaches using methods for "online metabolite generation" have been described to increase throughput in early drug discovery. The production of immobilized enzyme reactors (IMERs) enables online coupling of enzyme-catalyzed reactions with LC [61]. Microsomal incubations in a chip-based format coupled to ESI–MS has been demonstrated by Benetton et al. [61]. Two syringe pumps delivered microsomal protein, substrate, and cofactors in a buffer to the chip reaction region and were incubated at 37°C. Two configurations for desalting, protein removal and preconcentration before MS analysis enable fully integrated on-chip sample preparation. Staack et al. [62] described an approach for studying the metabolism of unlabeled drugs by online LC–MS/MS combined with chip-based infusion after fraction collection into 96-well plates. Owing to reduced sample consumption (typically 1–5 μL), the ability to infuse the sample for an extended period of time and acquisition time was no longer an issue, which resulted in the improved quality of MS/MS spectra and sensitivity. Chip-based nanoelectrospray had also proved suitable for the analysis of glutathione adducts with a 100 times more sensitivity than conventional LC–MS/MS method [63]. Another high-resolution screening methodology presented by Kool et al. [64] utilized postcolumn online profiling of estrogenic compound and its metabolites with affinity for the estrogen receptor α (ERα). Most of the 14 metabolites detected exhibited affinity for the ERα. This methodology appeared to be very sensitive and selective. It allows to quickly screen active metabolite formation in time, which offers great perspectives for applications in early drug discovery.

Another purely instrumental approach is the use of online electrochemistry coupling with liquid chromatography/mass spectrometry (EC–LC–MS) to mimic cytochrome P-450-catalyzed reactions. This configuration enabled generation of more data on the oxidation products. Studies on the metabolism of clozapine and acetaminophen in the absence and presence of trapping agent had been reported [65,66]. In the study, the reactive acetaminophen metabolite, *N*-acetyl-*p*-benzoquinoneimine, was identified. In another application, Toremifene was oxidized and detected in the online system as hydroxylated, *N*- and *O*-dealkylated, and dehydrogenated metabolites [67].

Metal and nonmetal atoms such as Cl, Br, I, S, or P can be detected by inductively coupled plasma mass spectrometry (ICP–MS) [68]. Because nonmetal atoms are often present in a drug and the sensitivity of this technique is element-independent, ICP–MS (GVI platform ICPMS instrument, GV instruments Ltd., Manchester, United Kingdom) in combination with HPLC–ESI–MS was used to investigate the metabolism of bromobenzoic acid in bile-cannulated rats [69]. The qualitative metabolic studies using the combination of a single separation combined with postchromatographic splitting of the eluent to both ICP–MS and ESI–MS had been performed. As isocratic HPLC method was used to maintain a constant response of ICP–MS to Br during the run. ^{79}Br and ^{81}Br were both measured by ICP–MS and the combined signal was used for profiling. Since ICP–MS is an element-detection technique, these profiles do not reveal any information on the structure of the metabolites. Therefore, the analysis of representative samples was undertaken using ESI–MS in parallel. The application of ICP–MS to bromine-specific detection of the metabolites of bromobenzoic acid highlighted the potential of using this technique in metabolite studies without a need of radiolabeling.

Last but not the least, matrix-assisted laser-desorption ionization (MALDI) imaging mass spectrometry (IMS) caught a lot of attentions in the drug metabolism area. Drug distribution and individual metabolite distributions within whole-body tissue sections were detected simultaneously at various time points following drug administration by Khatib-Shahidi and colleague [70,71]. The model compound olanzapine was administered in rats via oral administration. Its signal was detected primarily in the stomach at early time points and gradually localized into surrounding organs by 6–12 h. The emergence of its metabolites, *N*-desmethyl and 2-hydroxymethyl olanzapine, was also detected in tissue and correlated to the loss of parent drug signal. MALDI–MS/MS data of parent drug and its metabolites correlated well with published quantitative whole body autoradiography data. The advantage of IMS over traditional drug image techniques is the capability to detect a nonradiolabeled molecule with molecular specificity. Therefore, IMS has the ability to differentiate between intact drug and its metabolites which may be present within a single tissue section collected at various time points following drug administration (see Chapter 11).

7.5 CONCLUSIONS

Metabolite identification has proven to be of great value in drug discovery and development, however, identification and quantification of metabolites in complex matrices remain challenging tasks. Early knowledge of the metabolic fate of an NCE can redefine the focus of the chemistry, and efforts can result in the advancement of a superior drug product. Metabolite studies in the early drug discovery can provide insight into potential metabolic issues that would not otherwise have been brought to light until the NCE was well advanced into the clinical development. Structural information of metabolites can be obtained using the state-of-the-art LC–MS strategies available today. LC–MS has been widely accepted as the main tool in the identification, structure characterization and quantitative analysis of drug metabolites owing to its superior sensitivity, specificity, and efficiency. Advances in analytical technologies continue to improve the quality of the metabolite identification experiments with faster turnaround. Tools to automate data interpretation are valuable as well and combining these with the LC–MS data acquisition software should provide a powerful combination. The use of standards or radiolabeled compounds in quantitative analysis can be avoided, when online coupling of LC–MS to detection techniques that provide equimolar responses are used. All of these current and future advances in instrumentation and predictive software, when applied appropriately, will result in screening larger numbers of compounds, making discovery decisions more rapidly, developing better compounds, and delivering a safer and more efficacious drug product to the patient. The progress in microfluidics suggests that metabolite analysis will be carried out by miniaturized lab-on-a-chip techniques integrated with miniaturized mass spectrometers in the near future.

REFERENCES

1. Hop, C.E.C.A. and Prakash, C. Metabolite identification by LC-MS. In *Application in Drug Discovery and Development*, Chapter 6, Chowdhury, S.K. (ed.). Elsevier: the Netherlands, p. 123, 2005.
2. Parkinson, A. Biotransformation of xenobiotics. In *Casarett & Doull's Toxicology*, Klaassen, C.D. (ed.). McGraw-Hill: New York, pp. 133–224, 2001.

3. Li, C. et al., Integrated application of capillary HPLC/continuous-flow liquid secondary ion mass spectrometry to discovery stage metabolism studies, *Anal. Chem.*, 57, 2931, 1995.
4. Sinz, M.W. and Podoll, T., The mass spectrometer in drug metabolism, In *Mass Spectrometry in Drug Discovery*, Rossi, D.T. and Sinz, M.W. (eds.). Marcel Dekker: New York, pp. 271–335, 2002.
5. Wilson, A.G., White, A.C., and Mueller, R.A., Role of predictive metabolism and toxicity modeling in drug discovery—A summary of some recent advancements, *Curr. Opin. Drug Discov. Devel.*, 6, 123, 2003.
6. Greene, N., Computer systems for the prediction of toxicity: An update, *Adv. Drug Deliv. Rev.*, 54, 417, 2002.
7. Richard, A.M., Structure-based methods for predicting mutagenicity and carcinogenicity: Are we there yet? *Mutat. Res.*, 400, 493, 1998.
8. Anari, M.R., Tiller, P.R., and Baillie, T.A., *Progress in Pharmaceutical and Biomedical Analysis*, vol. 6, Chowdhury, S.K. (ed.). p. 183, 2005.
9. Nassar, A.F. and Talaat, R.E., Strategies for dealing with metabolite elucidation in drug discovery and development, *Drug Discov. Today*, 9, 317, 2004.
10. Want, E.J., Cravatt, B.F., and Siuzdak, G., The expending roles of mass spectrometry in metabolite profiling and characterization, *Chem. Biochem.*, 6, 1941, 2005.
11. Syage, J.A. and Evans, M.D., Photoionization mass spectrometry as a powerful new tool for drug discovery, *Spectroscopy*, 16, 14, 2001.
12. Hsieh, Y., APPI: A new ionization source for LC-MS/MS assays. In *Using Mass Spectrometry for Drug Metabolism Studies*, Chapter 9, Korfmacher, W., (ed.). CRC Press: Boca Raton, FL, p. 272, 2005.
13. Prakash, C., Shaffer, C.L., and Nedderman, A., Analytical strategies for identifying drug metabolites, *Mass Spectrom. Rev.*, 26, 340, 2007.
14. Kamel, A. and Prakash, C., High performance liquid chromatography/atmospheric pressure ionization/tandem mass spectrometry (HPLC/API/MS/MS) in drug metabolism and toxicology, *Curr. Drug. Metab.*, 7, 837, 2006.
15. Ackermann, B.L. et al., Current applications of liquid chromatography/mass spectrometry in pharmaceutical discovery after a decade of innovation, *Annu. Rev. Anal. Chem.*, 1, 357, 2008.
16. Cox, K., Special requirements for metabolite characterization. In *Using Mass Spectrometry for Drug Metabolism Studies*, Chapter 8, Korfmacher,W., (ed.). CRC Press: Boca Raton, FL, pp. 233–234, 2005.
17. Perchalski, A.D., Yost, R.A., and Wilder, B.J., Structural elucidation of drug metabolites by triple-quadrupole mass spectrometry, *Anal. Chem.*, 54, 1466, 1982.
18. Stafford, G.C. et al., Recent improvements in analytical applications of ion-trap technology, *Int. J. Mass Spectrom. Ion Proc.*, 60, 85, 1984.
19. Ma, S., Chowdhury, S.K., and Alton, K.B., Application of mass spectrometry for metabolite identification, *Curr. Drug Metab.*, 7, 503, 2006.
20. Douglas, D.J., Frank, A.J., and Mao, D., Linear ion traps in mass spectrometry, *Mass Spectrom. Rev.*, 24, 1, 2005.
21. King, R. and Fernandez-Metzler, C., The use of Qtrap technology in drug metabolism, *Curr. Drug Metab.*, 7, 541, 2006.
22. Hop, C.E.C.A. et al., Elucidation of fragmentation mechanisms involving transfer of three hydrogen atoms using a quadrupole time-of-flight mass spectrometer, *J. Mass Spectrom.*, 36, 575, 2001.
23. Baranczewski, P. et al., Introduction to early in vitro identification of metabolites of new chemical entities in drug discovery and development, *Pharmacolog. Rep.*, 58, 341, 2006.
24. Fountain, S.T., A mass spectrometry primer. In *Mass Spectrometry in Drug Discovery*, Chapter 3, Rossi, D.T. and Sinz, M.W. (eds.). Marcel Dekker: New York, p. 66, 2002.

25. Sanders, M. et al., Utility of the hybrid LTQ-FTMS for drug metabolism applications, *Curr. Drug Metab.*, 7, 547, 2006.
26. Shipkova, P.A. et al., Use of a hybrid linear ion trap/Fourier transform mass spectrometer for determining metabolic stability and metabolite characterization, *Proceeding of 52nd ASMS Conference on Mass Spectrometry and Allied Topics*. Nashville, TN, May 23–27, 2004.
27. Josephs, J.L. et al., The hybrid linear ion trap/Fourier transform mass spectrometer as a tool for investigating biotransformation pathways, *Proceedings of Conference on Small Molecules Science*. Bristol, RI, August 8–14, 2004.
28. Makarov, A., Electrostatic axially harmonic orbital trapping: A high-performance technique of mass analysis, *Anal. Chem.*, 72, 1156, 2000.
29. Hu, Q. et al., The Orbitrap: a new mass spectrometer, *J. Mass Spectrom.*, 40, 430, 2005.
30. Hughes, J.M. et al., A comparison of LC/MS mass analyzers for screening, confirmation and quantification of drugs in blood, *Proceeding of 56th ASMS Conference on Mass Spectrometry and Allied Topics*. Denver, CO, June 1–5, 2008.
31. Egnash, L.A. and Ramanathan, R., Comparison of heterogeneous and homogeneous radioactivity flow detectors for simultaneous profiling and LC-MS/MS characterization of metabolites, *J. Pharm. Biomed. Anal.*, 27, 271, 2002.
32. Kostiainen, R. et al., Liquid chromatography/atmospheric pressure ionization—mass spectrometry in drug metabolism studies, *J. Mass Spectrom.*, 38, 357, 2003.
33. Reza Anari, M. and Baillie, T.A., Bridging cheminformatic metabolite prediction and tandem mass spectrometry, *Drug Discov. Today*, 10, 711, 2005.
34. Clark, N.J. et al., Systematic LC/MS metabolite identification in drug discovery, *Anal. Chem.*, 73, 430A, 2001.
35. Jemal, M. et al., A strategy for metabolite identification using triple-quadrupole mass spectrometry with enhanced resolution and accurate mass capability, *Rapid. Commun. Mass Spectrom.*, 17, 2732, 2003.
36. Chen, Y. et al., High throughput strategies for metabolite identification in drug discovery pharmacokinetic studies, *Proceeding of 56st ASMS Conference on Mass Spectrometry and Allied Topics*. Denver, CO, June 1–5, 2008.
37. Cai, Z. et al., High-throughput analysis in drug discovery: Application of liquid chromatography/ion-trap mass spectrometry for simultaneous cassette analysis of α-1a antagonists and their metabolites in mouse plasma, *Rapid Commun. Mass Spectrom.*, 15, 546, 2001.
38. Varoglu, M. et al., Simultaneous quantitative and qualitative measurements of in vitro microsomal metabolism assays by Orbitrap LC-MS methods, *Proceeding of 56th ASMS Conference on Mass Spectrometry and Allied Topics*. Denver, CO, June 1–5, 2008.
39. Gao, H. et al., Method for rapid metabolite profiling of drug candidates in fresh hepatocytes using liquid chromatography coupled with a hybrid quadrupole linear ion trap, *Rapid Commun. Mass Spectrom.*, 21, 3683, 2007.
40. Li, A.C. et al., Simultaneously quantifying parent drugs and screening for metabolites in plasma pharmacokinetic samples using selected reaction monitoring information-dependent acquisition on a QTrap instrument, *Rapid Commun. Mass Spectrom.*, 19, 1943, 2005.
41. Shou, W.Z. et al., A novel approach to perform metabolite screening during the quantitative LC-MS/MS analyses of in vitro metabolic stability samples using a hybrid triple-quadrupole linear ion trap mass spectrometer, *J. Mass Spectrom.*, 40, 1347, 2005.
42. Li, A.C., Gohdes, M.A., and Shou, W.Z., "N-in-one" strategy for metabolite identification using a liquid chromatography/hybrid triple quadrupole linear ion trap instrument using multiple dependent product ion scans triggered with full mass scan, *Rapid Commun. Mass Spectrom.*, 21,1421, 2007.

43. Xia, Y.Q. et al., Use of a quadrupole linear ion trap mass spectrometer in metabolite identification and bioanalysis, *Rapid Commun. Mass Spectrom.*, 17, 1137, 2003.
44. Bramwell, C., Wingate, J., and Elicone, C., GSH conjugate metabolite identification with a hybrid triple-quadrupole-linear ion-trap instrument and automated software-driven method development, *Curr. Trends Mass Spectrom.*, May, 42, 2008.
45. Tiller, P.R. et al., High-throughput, accurate mass liquid chromatography/tandem mass spectrometry on a quadrupole time-of-flight system as a "first-line" approach for metabolite identification studies, *Rapid Commun. Mass Spectrom.*, 22, 1053, 2008.
46. Bateman, K.P. et al., MSE with mass defect filtering for in vitro and in vivo metabolite identification, *Rapid Commun. Mass Spectrom.* 21(9), 1485, 2007.
47. Wrona, M. et al., "All-in-One" analysis for metabolite identification using liquid chromatography/hybrid quadrupole time-of-flight mass spectrometry with collision energy switching, *Rapid Commun. Mass Spectrom.*, 19, 2597, 2005.
48. Reza Anari, M. et al., Integration of knowledge-based metabolic predictions with liquid chromatography data-dependent tandem mass spectrometry for drug metabolism studies: Application to studies on the biotransformation of indinavir, *Anal. Chem.*, 76, 823, 2004.
49. Castro-Perez, J. et al., Increasing throughput and information content for in vitro drug metabolism experiments using ultraperformance liquid chromatography coupled to a quadrupole time-of-flight mass spectrometer, *Rapid Commun. Mass Spectrom.*, 19, 843, 2005.
50. Wainhaus, S. et al., Ultra fast liquid chromatography-MS/MS for pharmacokinetic and metabolic profiling within drug discovery, *Am. Drug Discov.*, 2, 6, 2007.
51. Castro-Perez, J.M., Current and future trends in the application of HPLC-MS to metabolite-identification studies, *Drug Discov. Today*, 12, 249, 2007.
52. Mortishire-Smith, R.J. et al., Accelerated throughput metabolic route screening in early drug discovery using high-resolution liquid chromatography/quadrupole time-of-flight mass spectrometry and automated data analysis, *Rapid. Commun. Mass Spectrom.*, 19, 2659, 2005.
53. Taylor, E.W., Qian, M.G., and Dollinger, G.D., Simultaneous online characterization of small organic molecules derived from combinatorial libraries for identity, quantity and purity by reversed-phase HPLC with chemiluminescent nitrogen, UV, and mass spectrometric detection, *Anal. Chem.*, 70, 3339, 1998.
54. Deng, Y. et al., Quantitation of drug metabolites in the absence of pure metabolite standards by high-performance liquid chromatography coupled with a chemiluminescent nitrogen detector and mass spectrometer, *Rapid Commun. Mass Spectrom.*, 18, 1681, 2004.
55. Utley, L. et al., The evaluation of nanospray as a potential response calibrator for the estimation of metabolites levels in drug discovery, *Proceeding of 53rd Conference on Mass Spectrometry and Allied Topics.* San Antonio, TX, June 5–9, 2005.
56. Hop, C.E., Chen, Y., and Yu, L.J., Uniformity of ionization response of structurally diverse analytes using a chip-based nanoelectrospray ionization source, *Rapid Commun. Mass Spectrom.*, 19, 3139, 2005.
57. Ramanathan, R. et al., Response normalized liquid chromatography nanospray ionization mass spectrometry, *J. Am. Soc. Mass Spectrom.*, 18,1891, 2007.
58. Valaskovic, G.A. et al., Ultra-low flow nanospray for the normalization of conventional liquid chromatography/mass spectrometry through equimolar response: Standard-free quantitative estimation of metabolite levels in drug discovery, *Rapid Commun. Mass Spectrom.*, 20, 1087, 2006.
59. Hop, C.E.C.A., Use of nano-electrospray for metabolite identification and quantitative absorption, distribution, metabolism and excretion studies, *Curr. Drug Metab.*, 7, 557, 2006.

60. Staack, R.F. and Hopfgartner, G., New analytical strategies in studying drug metabolism, *Anal. Bioanal. Chem.*, 388, 1365–13, 2007.
61. Benetton, S. et al., Chip-based P450 drug metabolism coupled to electrospray ionization-mass spectrometry detection, *Anal. Chem.*, 75, 6430, 2003.
62. Staack, R.F., Varesio, E., Hopfgartner, G., The combination of liquid chromatography/tandem mass spectrometry and chip-based infusion for improved screening and characterization of drug metabolites, *Rapid Commun. Mass Spectrom.*, 19, 618, 2005.
63. Chen, Y., Yu, L.J., and Hop, C.E.C.A., Development of a high throughput in-vitro assay for glutathione conjugated metabolites using an automated chip based nanoelectrospray MS/MS method. *Proceeding of the 52nd ASMS Conference on Mass Spectrometry and Allied Topics*. Nashville, TN, May 23–27, 2004.
64. Kool, J. et al., Rapid on-line profiling of estrogen receptor binding metabolites of Tamoxifen, *J. Med. Chem.*, 49, 3287, 2006.
65. van Leeuwen, S.M. et al., Prediction of clozapine metabolism by on-line electrochemistry/liquid chromatography/mass spectrometry, *Anal. Bioanal. Chem.*, 382, 742, 2005.
66. Lohmann, W. and Karst, U., Simulation of the detoxification of paracetamol using online electrochemistry/liquid chromatography/mass spectrometry, *Anal. Bioanal. Chem.*, 386, 1701, 2006.
67. Lohmann, W. and Karst, U., Simulation of Phase I and II metabolism of toremifene using on-line electrochemistry/immobilized enzymes/liquid chromatography/mass spectrometry, *Proceedings of 56th ASMS Conference on Mass Spectrometry and Allied Topics*, Denver. CO, June 1–5, 2008.
68. Lobinski, R., Schaumloffel, D., and Szpunar, J., Mass spectrometry in bioinorganic analytical chemistry, *Mass Spectrom. Rev.*, 25, 255, 2006.
69. Jensen, B.P. et al., Investigation of the metabolic fate of 2-, 3- and 4-bromobenzoic acids in bile-duct-cannulated rats by inductively coupled plasma mass spectrometry and high-performance liquid chromatography/inductively coupled plasma mass spectrometry/electrospray mass spectrometry, *Rapid Commun. Mass Spectrom.*, 19, 519, 2005.
70. Khabit-Shahidi, S. et al., Detecting drug distribution in whole rat sagittal sections by imaging mass spectrometry, *Proceedings of 53nd ASMS Conference on Mass Spectrometry and Allied Topics*. San Antonio, TX, June 5–9, 2005.
71. Khabit-Shahidi, S. et al., Direct molecular analysis of whole-body animal tissue sections by imaging MALDI mass spectrometry, *Anal. Chem.*, 78, 6448, 2006.

8 Fast Chromatography with UPLC and Other Techniques

Sam Wainhaus

CONTENTS

8.1 INTRODUCTION

The coupling of liquid chromatography (LC) with mass spectrometry (MS) has undergone much evolution since its initial inception [1,2]. Atmospheric pressure ionization techniques such as electrospray ionization (ESI) and atmospheric pressure chemical ionization (APCI) opened the door for the ionization and analysis of non-volatile or thermally labile analytes. This technique revolutionized drug discovery and development allowing for dramatic improvements in sensitivity, selectivity, and speed. This area continues to grow, and significant advances have been and continue to be achieved in all three areas [3–5].

One of the key issues in LC–MS involves separation of the analytes of interest from each other and from the matrix in which they are present. The introduction of the high-performance liquid chromatography (HPLC) column prior to the ionization source accomplished this task. The use of different stationary phases coupled with an appropriate selection of mobile phases and gradient conditions allowed for excellent separation of analytes as well as separation from the solvent front and endogenous material that could negatively impact sensitivity and reliability [6]. The impact of HPLC–MS on drug discovery and development was dramatic in both qualitative and quantitative arenas. The identification of metabolites in biological matrices such as bile, urine, and plasma became more rigorous especially when coupled with online radioflow detection [7]. It is not uncommon to identify 30 or more metabolites for a given drug with HPLC–MS [8]. As quantitation of drugs in plasma became faster and more sensitive, questions such as to what extent a drug is absorbed or how rapidly it is cleared are now answered early in drug discovery. The ability to run a study in a single overnight run became commonplace. Running parallel to the use of this technique was the ever-improving HPLC column by a plethora of manufacturers. More reproducible stationary phases, smaller particle size, narrower column width, and shorter column length led to trade-offs in resolution, speed, and selectivity. HPLC pump manufacturers typically rated their pumps up to 4500 psi and were slow to keep up with the pressure demands of smaller particle size with its promise of increased number of theoretical plates and greater resolution. Thus, it was fairly common to hear the familiar ringing alarm of an overpressured pump when entering an HPLC–MS laboratory on the morning after an overnight run. Column heaters were employed to alleviate the backpressure and shorter columns were utilized (thereby diminishing some of the gains of the smaller particle size).

A variety of improvements with regard to HPLC–MS have occurred. The 2× 30 mm (4 μ) supplanted the longer wider column and 2–3 min run times were achieved with ballistic gradients [9]. Questions regarding matrix ion suppression and issues with coeluting endogenous and exogenous impurities resulted in issues with both discovery (non-GLP) and development (GLP) assays [10–12].

Ultraperformance liquid chromatography (UPLC) entered the scene as a viable tool in 2003 [13,14]. UPLC has two components: (1) a sub-2 μm particle size column and (2) a high-pressure pump rated at around 15,000 psi for handling the extreme backpressure generated by these columns. A third component required to fully utilize this technique is a fast scanning detector to capture sufficient points across these narrow (1–2 s) peakwidths. The concept of UPLC has been around for a long time. Jorgenson et al. has been a pioneer in the field and helped develop it to a mature technique [15–17]. The van Deemter curve clearly demonstrates that for conventional columns there is a significant decrease in the number of theoretical plates with an increase in linear velocity as shown in Figure 8.1 [18]. However, when the particle size decreases below 2 μm the number of theoretical plates is not diminished. A flow rate of 500 μL/min in a 2×30 mm sub-2 μm particle column typically generates a backpressure of ~10,000 psi. The Waters Corporation seized this opportunity and created a high-pressure pump (Aquity) and coupled it to their Micromass Premier fast scanning (5 ms) triple quadrupole mass spectrometer. Figure 8.2 shows the trade-offs between speed, resolution, and sensitivity for normal HPLC along with the situation for UPLC where no trade-off is required.

A critical question for each analyst to determine is "how fast do I really need to go and at what cost?" The needs of each analyst and environment are different and this dictates what sort of gradient is required and the resolution of the column. We will

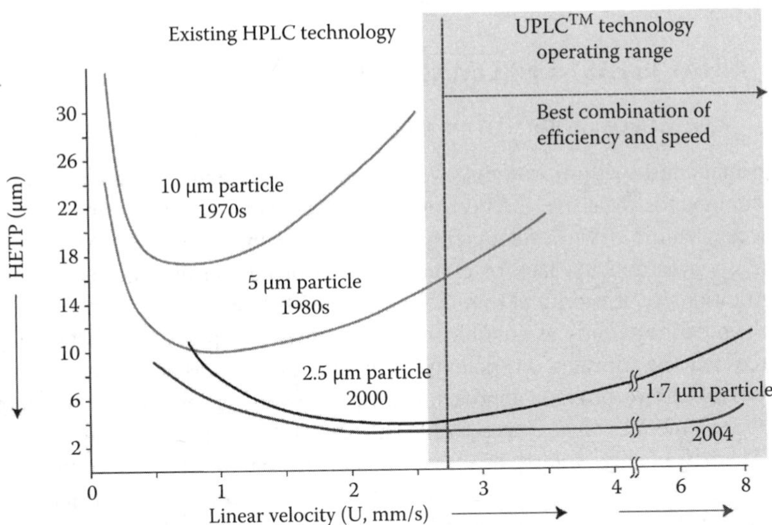

FIGURE 8.1 van Deemter plot showing the relationship between linear velocity (flow rate) and plate height. In general, the plate height increases with increasing velocity so that the number of plates (resolution) decreases. The exception to this is for sub-2 μm particles.

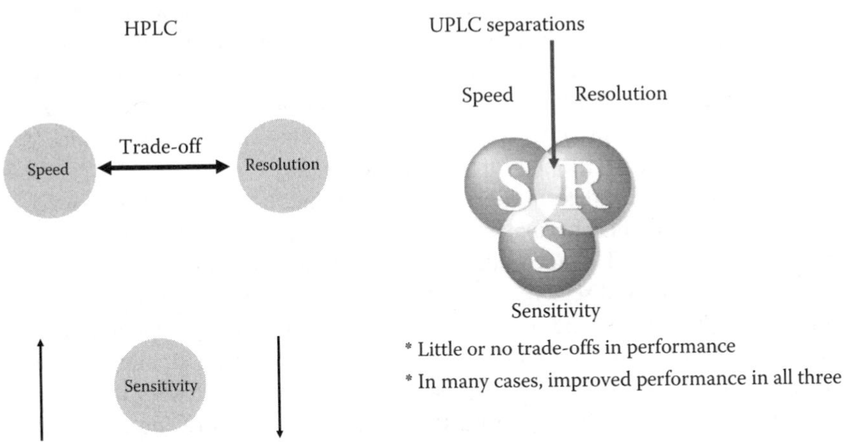

FIGURE 8.2 Relationship between resolution, sensitivity, and speed for HPLC and UPLC. UPLC does not result in trade-offs or compromises between these parameters.

begin to address these issues in a high-throughput drug discovery setting where the quantitation of a single analyte in a biological matrix is the main focus.

8.2 WHAT IS FAST CHROMATOGRAPHY?

8.2.1 Fast Chromatography in a Drug Discovery Environment

The primary task within drug discovery in the pharmaceutical industry is to bring forth compounds that have the best overall profile. This insures the highest chance for success within drug development, where new chemical entities (NCEs) fail for reasons such as toxicity, lack of efficacy, and poor human pharmacokinetics (PK) [19]. Additionally, it is critical to remove compounds that have no chance of success from the pipeline as early as possible in the drug discovery process so that money and resources may be allocated on "potential winners" and not wasted on the "losers." We therefore take a two-pronged approach in drug discovery: (1) find the best compounds and progress them as quickly as possible and (2) "kill" the "losers" as quickly as possible. There are many "filters" or screens that determine the fate of a compound and constitute a go/no go decision in early drug discovery. One of the most critical "filters" is *in vivo* rat PK. Clearly, if the requisite plasma concentrations for efficacy and drug safety are not achieved in rat then, in most cases, the drug has no chance of success. It stands to reason then, that obtaining this data in as short a timeframe as possible is critical within drug discovery. There are various limiting factors within the development of such a paradigm including resources, communication, and instrumentation.

We have developed a novel and streamlined approach that avoids classical "cassette dosing" and its potential pitfalls. The cassette-accelerated rapid rat screen (CARRS) is a high-throughput *in vivo* screen that provides the earliest PK information on 72 NCEs per week [20–24]. Six NCEs are selected per cassette for a total of 12 cassettes. Two rats are dosed with each NCE and plasma samples are collected at 0.5, 1, 2, 3, 4, and 6 h post-dose. Tissue samples may also be harvested at the 6 h time-point for specific programs. The total turnaround time for dosing, method development, analysis, and report submission is two weeks. In a recent study, Mei et al. found good correlation between CARRS and full rat PK studies [25].

The primary tool for determining drug concentrations in biological matrices is high-performance liquid chromatography combined with tandem mass spectrometry (HPLC–MS/MS). This technique provides a two-dimensional approach that is both highly sensitive and selective for the analyte of interest. Limits of quantitation in the 1–10 ng/mL range and a dynamic range over 3–5 orders of magnitude are now routine. Instrumentation has become more compact, rugged, and easy to use so that multiple systems are routinely found in many drug discovery and development metabolism and PK laboratories. Recent advances in HPLC column technology combined with fast scanning mass spectrometers has resulted in shorter cycle times and narrower peak widths [26].

In addition to providing quantitative information on the parent drug, HPLC–MS/MS is used to analyze biological samples within drug discovery and monitor for select metabolites. This serves two functions, to help chemists identify metabolic "hot spots" and help identify a new lead compound or prodrug.

The driving force of "fast chromatography" is the ability to diminish run times such that the end user realizes a significant time savings while sacrificing nothing. In the world of CARRS this means that an analyst can analyze an entire cassette (6 compounds ~120 injections) and report the data the same day or run numerous cassettes in a single overnight run. This represents a dramatic improvement in turnaround time. Ideally, one would be able to see an additional gain so that analyte and endogenous material, metabolites, chiral analytes, etc., may be separated faster.

8.2.2 Fast Chromatography: Ballistic Gradients

The use of short HPLC columns and ultrafast gradients has been explored in order to decrease HPLC–MS/MS analytical run times [27–30]. Typical cycle times ranged from 85 s to 2 min and used 2×30 mm, 4 μm HPLC columns (Table 8.1). The 1 min HPLC method is shown in Table 8.1. Mobile phases A and B are 0.01 M ammonium acetate in water/methanol (80:20) and methanol (100%), respectively. A ballistic gradient from 1% to 70% B was run over 0.01 min and held for 0.09 min. A linear gradient was then employed from 70% to 95% B for 0.6 min and held for 0.1 min. The column was then reequilibrated to 1% B over 0.2 min at a constant flow rate of 0.8 mL/min.

For the 2 min HPLC method, a linear gradient from 1% to 95% B was run over 1.2 min, held for 0.3 min and reequilibrated to 1% over 0.5 min at a constant flow rate of 0.8 mL/min. The injection volume was 10 μL for both methods. The mass resolution on the TSQ Quantum for both methods was 0.7 for both Q1 and Q3. The scan time was 0.1 s for the 2 min linear gradient and 0.05 s for the 1 min ballistic gradient. Figure 8.3 compares the chromatograms from these two methods for a typical compound (CPD X). The ballistic gradient results in a twofold decrease in retention time and makes sub-1 min run times a reality. This in turn opens the door for same day turnaround of data that has been accomplished used this method.

TABLE 8.1
Gradient Profiles for CARRS Assay Methods

1 min Ballistic Gradient		2 min Linear Gradient	
Time (min)	**%B**	**Time (min)**	**%B**
0	1	0	1
0.01	70	0.2	1
0.1	70	1.2	95
0.7	95	1.5	95
0.8	95	1.7	1
0.9	1	2	1
1.0	1		

Source: Dunn-Meynell, K.W. et al., *Rapid Commun. Mass Spectrom.*, 19, 2905, 2005. With permission.

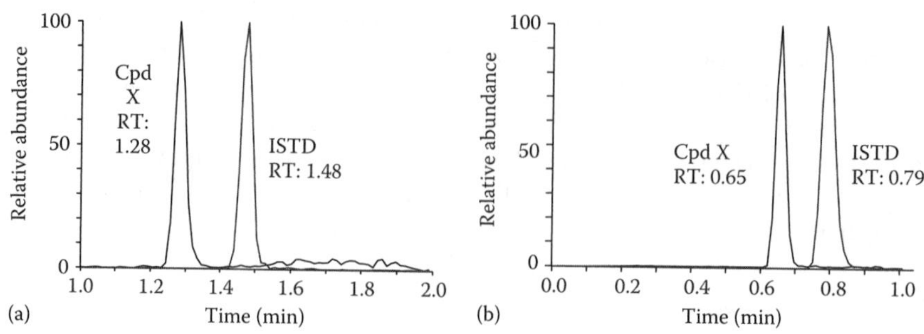

FIGURE 8.3 Comparison of (a) the 2 min linear gradient and (b) the 1 min ballistic gradient. (From Dunn-Meynell, K.W. et al., *Rapid Commun. Mass Spectrom.*, 19, 2905, 2005. With permission.)

The matrix ion suppression is a common concern when shortening the run time [31,32]. Fast isocratic methods have matrix effects that can last for most of the analysis time. The solvent front elutes later and lasts longer while fast gradients have solvent fronts that elute earlier. Additionally, sample preparation techniques can impact the extent of matrix ion suppression [33]. Protein precipitation can also increase the likelihood of matrix effects due to the limited sample cleanup. Liquid–liquid and solid-phase extraction result in cleaner samples that suffer from matrix effects to a smaller extent. The best way to eliminate matrix effects would be to develop analyte-specific sample extraction methods; however, this would also be impractical for higher-throughput assays. Figure 8.4 shows the matrix ion suppression traces of the 1 min ballistic gradient using mobile phase A vs. precipitated rat plasma; CPD X and the internal standard (ISTD) elute well past the suppression area.

There are several drawbacks with this approach. The loss of separation resolution at higher linear velocities results in issues ranging from matrix ion suppression to

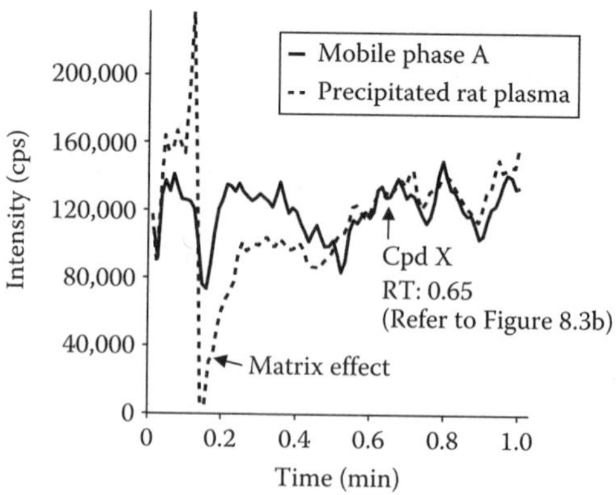

FIGURE 8.4 Matrix ion suppression effect in precipitated rat plasma vs. mobile phase A using the 1 min ballistic gradient. (From Dunn-Meynell, K.W. et al., *Rapid Commun. Mass Spectrom.*, 19, 2905, 2005. With permission.)

coeluting isobaric peaks. Additionally, this technique cannot be used for metabolite analysis. Finally, there is a question as to whether or not there is sufficient resolution to achieve meaningful separation for wide assortments of compounds given the small column length and ballistic nature of the gradient. Additionally, although this technique has served us well, the possibility of matrix ion suppression continues to be a factor and a method that improved the separation of the analyte from both the solvent front and sources of matrix ion suppression without the need for constant monitoring would be a most welcome technique in the area of high-throughput *in vivo* drug screening.

8.3 SMALL PARTICLES AND THE VAN DEEMTER EQUATION

Recently, advances in HPLC column technology have made small particle size columns (average particle size 1.7–2.1 μm) readily available [34,35]. These sub-2 μm particles dramatically increase column efficiency, which has a large impact on the speed of HPLC–MS/MS applications. The major drawback in using small particles is maintaining stable flow rates and the increase in column backpressure requiring the use of pumps with higher pressure ratings (15,000 psi). We have used these small particle columns in conjunction with high-pressure pumps to create a true ultrafast chromatographic approach to PK and metabolism. In order to fully appreciate the benefit of using small particle column packing material one must examine the van Deemter equation which is fundamental to chromatographic theory. According to the van Deemter equation:

$$H = A + \frac{B}{\mu} + C\mu$$

where

H is the column's plate height
μ is the mobile phase flow rate
A, B, and C are constants [36,37]

A smaller plate height equals a greater separation resolution, which is vital for separating the analyte of interest from its metabolites and from endogenous material in the matrix. The A and C constants are dependent on the particle size and therefore a decrease in particle size will result in a decrease in H. However, as the flow rate increases, the C term becomes more important and results in a loss of resolution above a certain flow rate as shown in Figure 8.1. The benefit of using 2 μm or smaller particles is that it allows a significant increase in flow rate before a loss of resolution. In the world of HPLC–MS/MS this translates into the equivalent of "having your cake and eating it too." In a typical HPLC–MS/MS analysis, there is a trade-off between speed, resolution, and sensitivity as shown in Figure 8.2. With the advent of small particle HPLC columns, it is possible to attain all three without sacrificing anything. UPLC performance is characterized by a dramatic reduction in retention times together with an improvement of the sensitivity without affecting peak resolution

8.4 INSTRUMENTATION

The key to large-scale use of sub-2 μm columns has been the prolific availability of very high-pressure (15,000 psi) pumps from a variety of vendors. The Waters Acquity was the first high-pressure pump on the market. Early issues with the sample organizer, needle, and mass spectrometer communication have been resolved. It has been a highly reliable pump in our laboratory for the past three years. There are now several very high-pressure pumps available on the market including the Thermo Accela, as well as models from CTC (Leap) Technologies and Agilent. The state of the art is such that all of these pumps are highly reliable with an associated cost of $30,000 to $70,000 depending upon the various bells and whistles.

8.5 ASSESSING HPLC VERSUS UPLC

In order to have a true comparison between conventional HPLC and ultrafast HPLC using small particles, it is necessary to compare them head to head using the same compounds. Our standard HPLC conditions are shown in Table 8.1. Mobile phase A is methanol/water (20:80), 0.010 M ammonium acetate and mobile phase B is methanol/water (90:10), 0.010 M ammonium acetate, 0.2% of 10% acetic acid. Under our "old" HPLC conditions, a typical cycle time of 2.5 min is easily achievable. The analyte of interest elutes at a retention time of 1.2–1.6 min with a run time of 2 min then, approximately 30 s is needed for the mass spectrometer to communicate with the autosampler in preparation for the next sample. In the CARRS paradigm, a cassette of plasma samples requires 108 injections including standards, samples, and blanks. Therefore, the total run time is 4.5 h and increases to 7 h for a plasma and tissue cassette. This precludes same day turnaround and does not allow for the most efficient use of automated method development software. UPLC can achieve 1–1.5 min cycle times which allows for same day turnaround of a plasma and tissue cassette and the completion of four to seven cassettes in an overnight run. This allows for rapid turnaround of data in a highly efficient manner.

Figure 8.5 shows the comparison between a typical HPLC (2 × 30 mm, 4 μm) run and ultrafast HPLC (2 × 30 mm, 1.7 μm) for BAS 12439 and its hydroxyl analog (CL-3472–1460). The ultrafast HPLC run used a gradient that ramps from 10% B to 100% B between 0.1 and 0.15 min. The flow rate was 0.9 mL/min resulting in a backpressure of approximately 14,000 psi. The HPLC run ramps from 10% B to 100% B between 0.2 and 0.5 min. A threefold improvement in speed and threefold decrease in peak width is observed. This is a relatively straightforward example of what is possible with UPLC and similar results have been shown in many laboratories for many different compounds. The key feature is the increase in speed while maintaining or improving sensitivity and quality. It is important to note that under these conditions the separation of the parent and hydroxyl analogs was not possible. Figure 8.6 shows another example of UPLC, using a Thermo Electron Accela pump and a sub-2 μm particle column, for imipramine, doxepin, clozapine, and timolol with retention times of 0.3–0.4 min and peak widths of 3–4 s FWHM. The ballistic gradient represents what is possible for UPLC. In practice, we use a less-aggressive approach that yields retention times of 0.5–0.8 min. This is sufficient for same day analysis as well as the analysis of seven 96 well plates in 24 h. Also shown in

Figure 8.6 is the corresponding CARRS standard curve (25, 250, 2500 ng/mL), which demonstrates good linearity across the region of interest. The low end of the standard curve is easily achieved with minimal method development. We employ

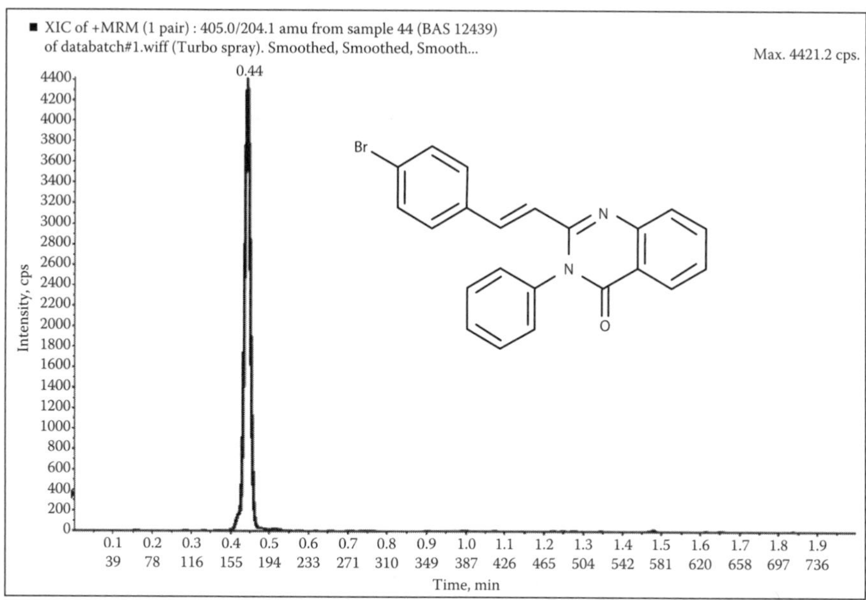

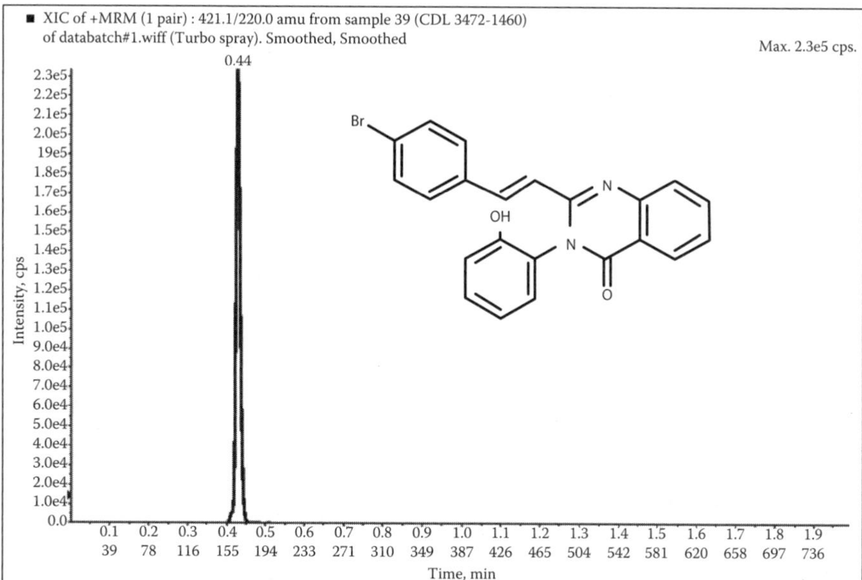

FIGURE 8.5 UPLC versus HPLC for a compound and its hydroxyl metabolite. A threefold improvement in speed is observed while improving resolution and sensitivity. UPLC (1.7 μm particle size), RT 0.44 min, peak width 2 s FWHM. HPLC (4 μm particle size), RT 1.3–1.5 min, peak width 6 s FWHM. (From Wainhaus, S. et al., *Amer. Drug Discov.*, 2, 6, 2007. With permission.)

(continued)

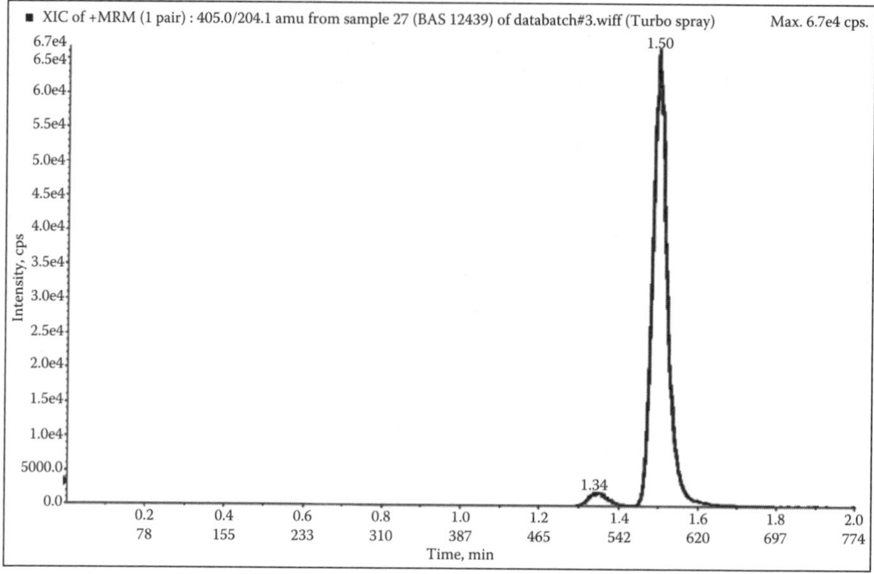

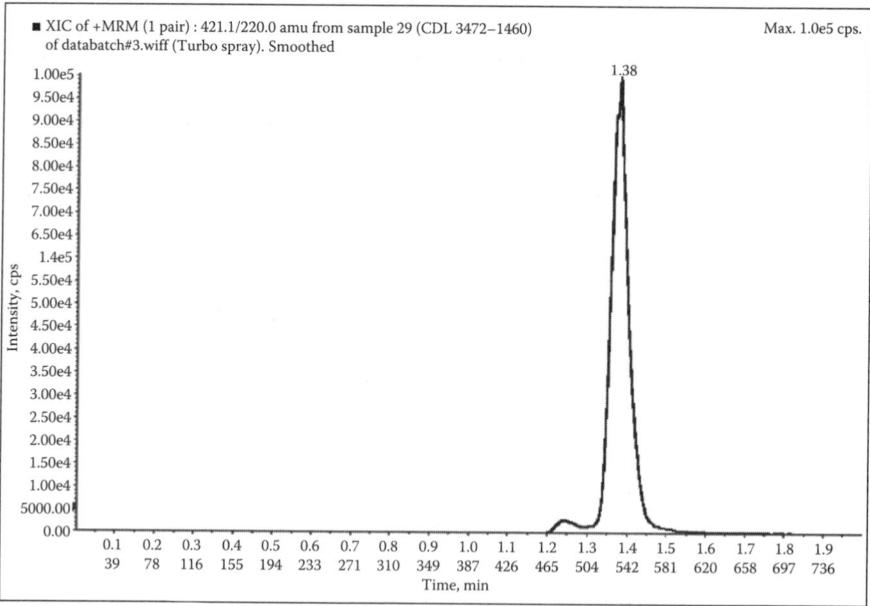

FIGURE 8.5 (continued)

generic gradients and automated method development for over 90% of our assays with little difficulty.

Table 8.2 shows a comparison of ultrafast, standard and ballistic HPLC for BAS 12439. The UPLC utilized a gradient that ramped from 40% B to 100% B in 0.2 min, normal HPLC ramped from 10% B to 100% B in 1 min and ballistic HPLC ramped

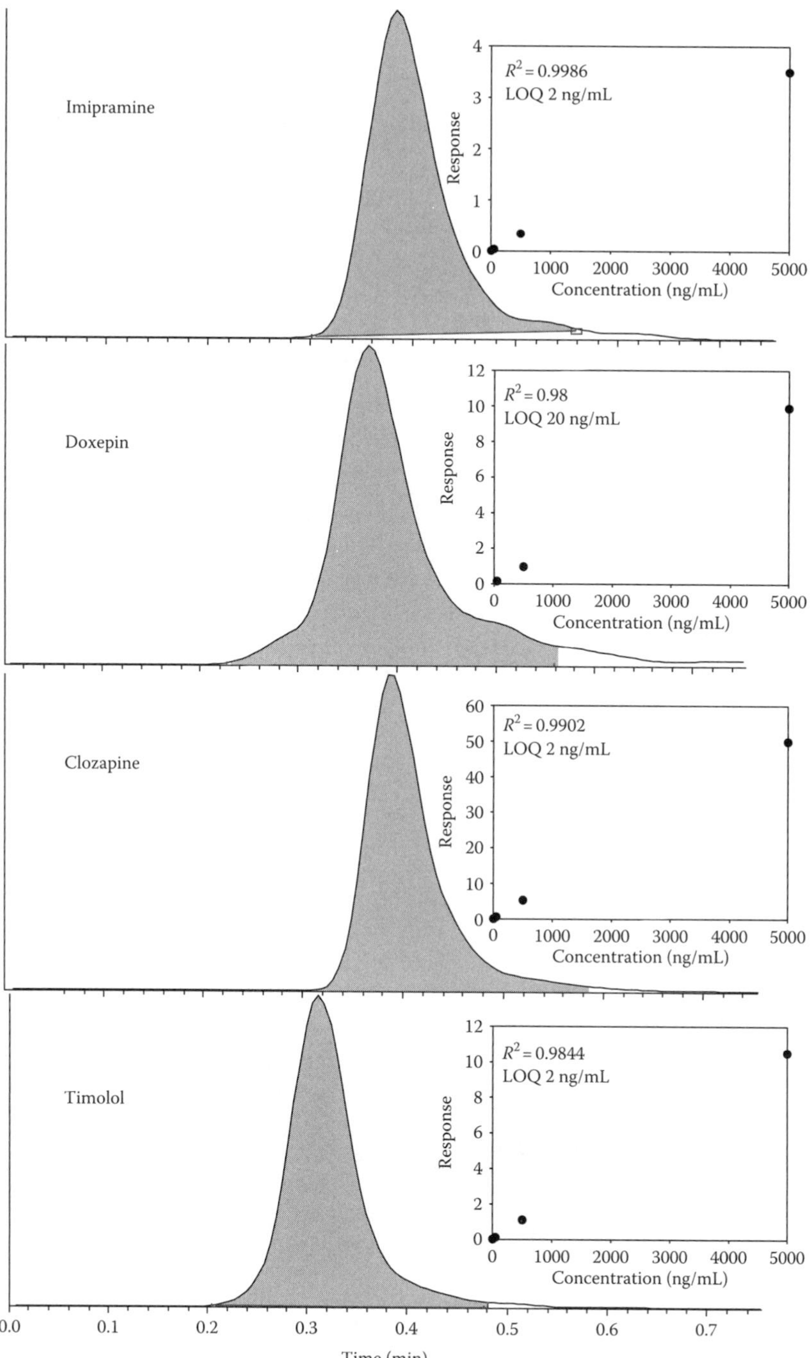

FIGURE 8.6 UPHPLC–MS/MS chromatograms obtained with the Accela/TSQ Quantum system for four generic drugs. The observed retention times are less than 25 s and generate linear standard curves across three orders of magnitude in the CARRS assay. (From Wainhaus, S. et al., *Amer. Drug Discov.*, 2, 6, 2007. With permission.)

TABLE 8.2
Comparison of Sensitivity and Linearity as a Function of Method

1.7 μm Particle Size (LOQ 0.04 ng/mL)		HPLC (LOQ 0.9 ng/mL)		Ballistic HPLC (LOQ 1.5 ng/mL)	
Conc. (ng/mL)	Response Ratio	Conc. (ng/mL)	Response Ratio	Conc. (ng/mL)	Response Ratio
0.1	0.00061	0.1	0.0022	0.1	0.00070
1	0.0062	1	0.011	1	0.010
10	0.069	10	0.093	10	0.12
100	0.63	100	1.0	100	1.3

Source: Dunn-Meynell, K.W. et al., *Rapid Commun. Mass Spectrom.*, 19, 2905, 2005. With permission.

from 10% B to 100% B in 0.1 min. The LOQ is defined as 40% of the lowest usable standard or three times the back-calculated blank (standard "0") value, whichever is greater. The results observed in Table 8.2 are typical for what is routinely observed. Linearity, dynamic range, and sensitivity are all under ultrafast HPLC conditions.

Transferring methods from HPLC to UPLC has become a popular topic and is generally quite straightforward [38–40].

8.6 UPLC REPRODUCIBILITY

We monitored the reproducibility for a 10 μL injection of loperamide and ketoconazole over a 12.5 h period in both rat plasma and homogenized rat brain (diluted 1:4 with water). The results are displayed in Figure 8.7 and demonstrate the overall improved reproducibility that has been observed for UPLC. This is important for determining potential matrix effects. Improved carryover has also been observed in most cases (<0.1%) precluding the need for multiple solvent blank injections [41].

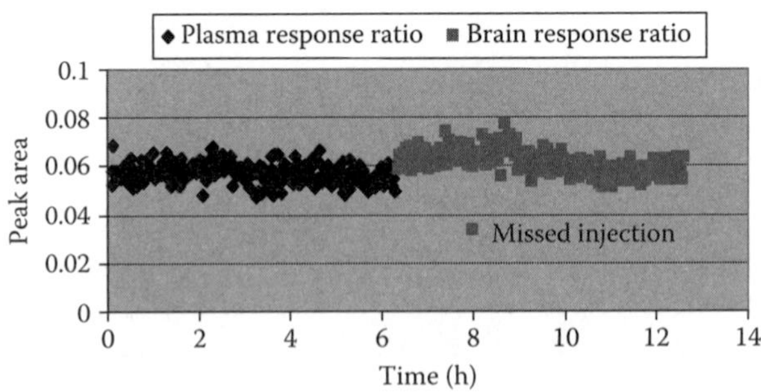

FIGURE 8.7 Internal standard reproducibility in plasma and tissue matrices demonstrated over 12 h for 2 min cycle times. This reproducibility time scale easily covers the time required for a typical CARRS assay. Ketoconazole/loperamide stability in rat plasma and brain. (From Wainhaus, S. et al., *Amer. Drug Discov.*, 2, 6, 2007. With permission.)

8.7 UPLC SEPARATION

One of the difficulties encountered when screening many compounds is interfering peaks due to endogenous impurities. If the analyte of interest coelutes with the impurity, quantitation becomes quite challenging. There are various remedies to this problem, such as increasing the LC run to obtain separation, selecting a different fragment ion, or improving the sample cleanup (solid-phase or liquid–liquid extraction). Unfortunately, these are all "after the fact" solutions and create logistical difficulties in a high-throughput arena. Ultrafast HPLC has proven to fix the problem a priori through its improved column efficiency. One example is shown in Figure 8.8 (analyte not shown). An endogenous peak was present at a retention time of 1.44 min during a normal HPLC run (2×30 mm, 5 μm C18). The analyte of interest coeluted with and was lost in this endogenous peak as shown in Figure 8.8. When we switched over to UPLC (2 × 30 mm, 1.7 μm C18), the endogenous peak eluted at 0.77 min, however, the analyte of interest eluted at 0.68 min and was sufficiently separated for quantitation. This is clearly a large benefit from UPLC, and has saved us on numerous occasions. We routinely operate with ballistic gradients and flow rates of 0.4–0.5 mL/min, taking full advantage of the speed, sensitivity, and selectivity of UPLC.

In 2008 the CARRS group at Schering-Plough developed UPLC–MS/MS methods for 3454 compounds and reported data that were critical for making decisions about progressing those compounds, as shown in Figure 8.9. We reported data on 75% of these compounds within five business days of sample receipt and 98% within 10 days. UPLC has been vital to our group for developing generic methods, separating the analyte of interest from coeluting interfering peaks with minimal or no effort, minimizing matrix ion suppression, decreasing carryover, improving sensitivity, and improving turnaround time.

UPLC–MS has been used for a plethora of analyses since its initial inception. Wang et al. showed that UPLC could be used to separate diastereomers in 5 min compared to more than 30 min in a conventional HPLC assay [42] and Leandro et al. used UPLC–MS/MS for a rapid determination and confirmation of 17 pesticide residues in baby food extracts [43].

8.8 UTILIZING UPLC FOR METABOLITE SCREENING

In addition to quantitation of parent drug, there is great utility from obtaining quantitative information about metabolites [44]. This is not straightforward and one must take great care about making quantitative determinations of metabolites when an authentic standard is not readily available [45]. Clearly, within drug discovery it is highly unusual to have anything but the authentic standard of the parent drug. Therefore, in order to obtain quantitative information about metabolites, one must first be sure that a metabolite is being observed. The first criterion is the MS/MS transition. For example, a hydroxyl metabolite will add *m/z* 16 to the parent molecular weight and may or may not add *m/z* 16 to the fragment molecular weight (depending on where the modification occurs on the molecule and which piece of the molecule is being observed in the MS/MS transition). If an MS/MS peak is observed that corresponds to a potential metabolite, further confirmation is necessary. Both

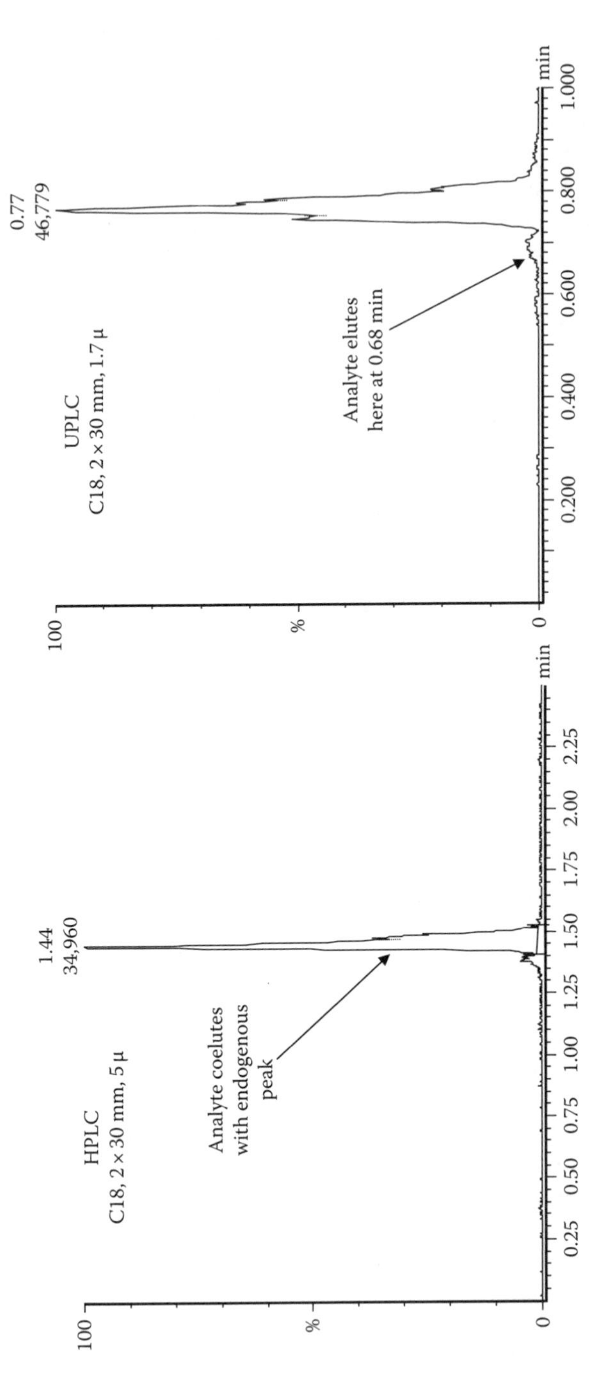

FIGURE 8.8 UPLC versus HPLC for separation of endogenous impurities from the analyte of interest. The analyte (not shown) coelutes with the endogenous impurity peak when standard HPLC is used. The two peaks are easily separated by UPLC. (From Wainhaus, S. et al., *Amer. Drug Discov.*, 2, 6, 2007. With permission.)

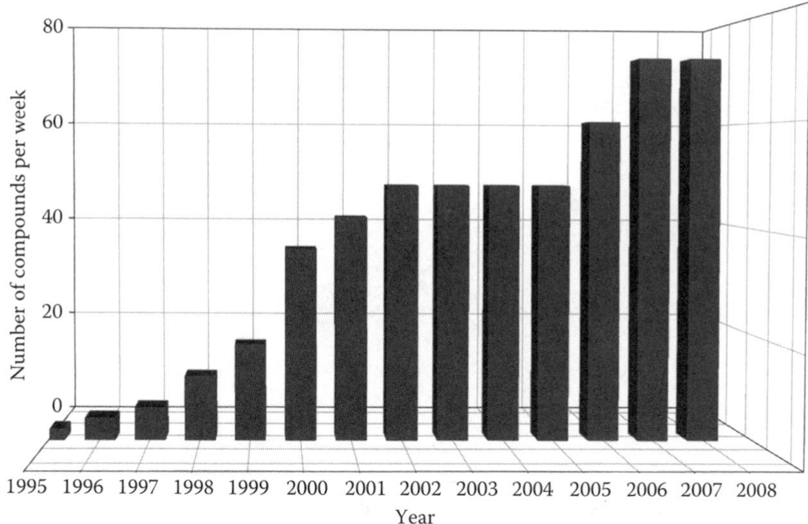

FIGURE 8.9 The number of CARRS compounds that have been assayed yearly since 2003 has mirrored advances in chromatography and mass spectrometry. The advent of UPLC has allowed us to routinely increase our capacity to 72 compounds per week with improved turnaround time and data quality. (From Wainhaus, S. et al., *Amer. Drug Discov.*, 2, 6, 2007. With permission.)

retention time and MS/MS of the metabolite can serve to authenticate the existence of the suspected metabolite. This is where ultrafast HPLC comes into play as shown in Figure 8.10 for BAS 12439 and two of its hydroxyl analogs, CL-3472-1460 and Chembridge 6141660. Figure 8.10a shows that UPLC can separate the parent as well as both M + 16 metabolites in 7 min while normal HPLC (Figure 8.10b) would require longer than 9.5 min (in practice this requires at least 20 min) to perform the same separation. This is very typical of UPLC and this fast separation represents a breakthrough in metabolite screening. The separation of metabolites from each other as well as the parent drug is vital for determining their authenticity.

UPLC has been widely utilized for metabolite analysis. Wang et al. developed a fast, sensitive, and UPLC–MS/MS method for the determination of testosterone (T) and its four metabolites, 6β-OH-T, 16α-OH-T, 16β-OH-T, and 2α-OH-T [46], *in vitro* samples. UPLC has been used to separate the glucuronides of midazolam which coelute under HPLC conditions as shown in Figure 8.11 [47]. UPLC has been used with time-of-flight mass spectrometry (UPLC/TOF MS) for a high-throughput and comprehensive analysis with minimal sample preparation [48]. Using this technique, a wide range of metabolites was investigated.

8.9 ALTERNATIVES TO UPLC

The benefits of UPLC are clear, however, these instruments may not be readily available to the pharmaceutical chemist. Various authors have explored a practical alternative that offers increased efficiencies, but at conventional HPLC pressure

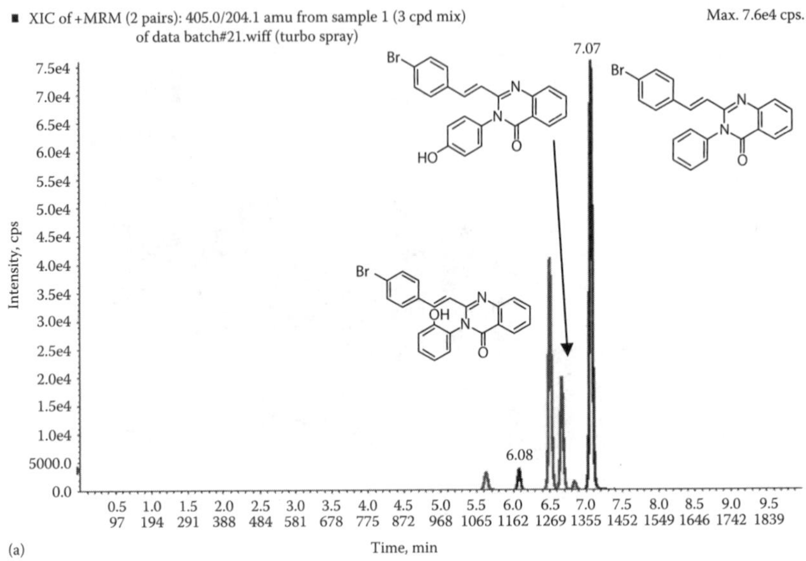

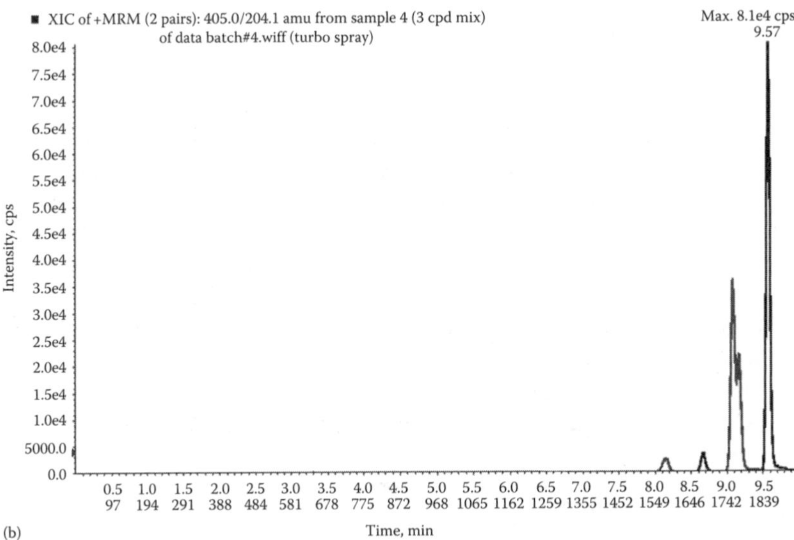

FIGURE 8.10 (a) UPLC versus (b) HPLC for a compound and two of its hydroxyl analogs. UPLC allows for the separation of hydroxyl analogs with little effort. This is critical for fast routine metabolite screening in drug discovery. (From Wainhaus, S. et al., *Amer. Drug Discov.*, 2, 6, 2007. With permission.)

limitations. These particles are called fused-core particles and are comprised of a 1.7 μm solid core encompassed by a 0.5 μm porous silica layer. Salisbury found that compared to the Waters Acquity particles, the fused-core particles achieved approximately 80% of the efficiency but with half the observed backpressure [49]. Hsieh et al. compare fast HPLC–MS/MS (2.7 μ) and UPLC–MS/MS (1.7 μ) using

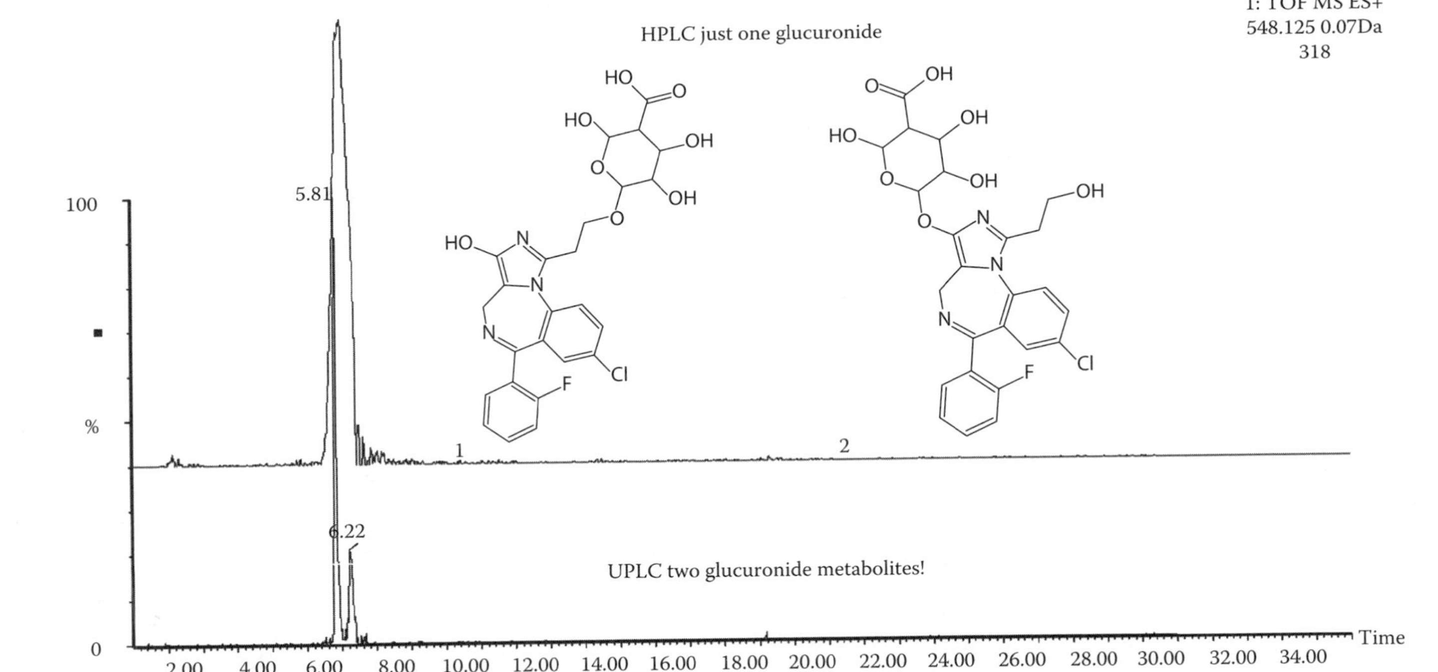

FIGURE 8.11 UPLC versus HPLC for analysis of glucuronide metabolites. UPLC clearly shows the existence of two glucuronide conjugates while HPLC only shows one. (From Plumb, R. et al., *Rapid Commun. Mass Spectrom.*, 18, 2331, 2004. With permission.)

fused-core silica particles for the determination of rimonabant in mouse plasma samples [50]. The fused-core silica particles allows for a shorter diffusional mass transfer path for solutes and are less affected in resolving power by increases in mobile-phase velocity compared with the sub-2 μm porous silica packings. The overall result is faster separations and higher sample throughput without the need for special high-pressure pumps. Cunliffe et al. described the use of fused-core silica particles as an alternative to sub-2 μm particles for routine bioanalysis [51]. Bones et al. discussed the use of monolithic HPLC columns for ultrafast LC without the backpressure issues encountered for the use of sub-2 μm particles [52].

High-speed gradient parallel HPLC–MS/MS has been successfully used with fully automated sample preparation for bioanalysis. These techniques typically use a MUX interface and can provide analysis times as low as 30 s per sample. This technique suffers from instrument complexity that have dramatic consequences is system ruggedness [53].

Although these techniques provide a cheap alternative to UPLC–MS, the replacement of standard HPLC pumps with high-pressure pumps continues to occur. The overall benefit and flexibility of UPLC–MS and the relatively inexpensive cost of the high-pressure pump make this an attractive technique.

Seemingly, the next frontier for fast bioanalysis is the removal of the column altogether. Techniques such as direct analysis in real time (DART) and desorption electrospray ionization (DESI) have shown great potential. These techniques are discussed in more detail in Chapter 13.

8.10 CONCLUSIONS

The advent of 1.7–2 μm reproducible HPLC stationary phases and high-pressure pumps has opened up a new world of UPLC, with a variety of applications ranging from high-speed chromatography resulting in greater sensitivity and selectivity to separation of previously unresolved HPLC peaks. We have incorporated this new technology into our daily work and improved turnaround time while still increasing the number samples and improving the quality of our data.

ACKNOWLEDGMENT

The author thanks Dr. Kate Yu and Dr. Marion Twohig for their help when UPLC–MS/MS was in its infancy.

REFERENCES

1. Baldwin, M.A. and McLafferty, F.W., Liquid chromatography-mass spectrometry interface. I. Direct introduction of liquid solutions into a chemical ionization mass spectrometer, *Org. Mass Spectrom.*, 7, 1353, 1973.
2. Bruins A.P, Covey, T.R., and Henion, J.D., Ion spray interface for combined liquid chromatography/atmospheric pressure ionization mass spectrometry, *Anal. Chem.*, 59, 2642, 1987.
3. Xu, R.N. et al., Recent advances in high-throughput quantitative bioanalysis by LC-MS/MS, *J. Pharm. Biomed. Anal.*, 44(2), 342, 2007.

4. Ma, S. and Zhu, M., Recent advances in applications of liquid chromatography-tandem mass spectrometry to the analysis of reactive drug metabolites, *Chem. Biol. Interact.*, 179(1), 25, 2009.
5. Kamel, A. and Prakash, C., High performance liquid chromatography/atmospheric pressure ionization/tandem mass spectrometry (HPLC/API/MS/MS) in drug metabolism and toxicology, *Curr. Drug Metab.*, 7(8), 837, 2006.
6. Korfmacher, W., Bioanalytical assays in a drug discovery environment, in *Using Mass Spectrometry for Drug Metabolism Studies*, W. Korfmacher, Editor. 2005, CRC Press: Boca Raton, FL. p. 1.
7. Ma, S., Chowdhury, S.K., and Alton, K.B., Application of mass spectrometry for metabolite identification, *Curr. Drug Metab.*, 7(5), 503, 2006.
8. Triolo, A. et al., In vivo metabolite detection and identification in drug discovery via LC-MS/MS with data-dependent scanning and postacquisition data mining, *J. Mass Spectrom.*, 40(12), 1572, 2005.
9. Dunn-Meynell, K.W., Wainhaus, S., and Korfmacher, W.A., Optimizing an ultrafast generic high-performance liquid chromatography/tandem mass spectrometry method for faster discovery pharmacokinetic sample throughput, *Rapid Commun. Mass Spectrom.*, 19(20), 2905, 2005.
10. Chambers, E. et al., Systematic and comprehensive strategy for reducing matrix effects in LC-MS/MS analyses, *J. Chromatogr. B Analyt. Technol. Biomed. Life Sci.*, 852(1–2), 22, 2007.
11. Xu, X. et al., A study of common discovery dosing formulation components and their potential for causing time-dependent matrix effects in high-performance liquid chromatography tandem mass spectrometry assays, *Rapid Commun. Mass Spectrom.*, 19(18), 2643, 2005.
12. Mei, H. et al., Investigation of matrix effects in bioanalytical high-performance liquid chromatography/tandem mass spectrometric assays: Application to drug discovery, *Rapid Commun. Mass Spectrom.*, 17(1), 97, 2003.
13. Yu, K. et al., Ultra-performance liquid chromatography/tandem mass spectrometric quantification of structurally diverse drug mixtures using an ESI-APCI multimode ionization source, *Rapid Commun. Mass Spectrom.*, 21(6), 893, 2007.
14. Tolley, L., Jorgenson, J.W., and Moseley, M.A., Very high pressure gradient LC/MS/MS, *Anal. Chem.*, 73(13), 2985, 2001.
15. MacNair, J.E., Lewis, K.C., and Jorgenson, J.W., Ultrahigh-pressure reversed-phase liquid chromatography in packed capillary columns, *Anal. Chem.*, 69(6), 983, 1997.
16. MacNair, J.E. et al., Rapid separation and characterization of protein and peptide mixtures using 1.5 microns diameter non-porous silica in packed capillary liquid chromatography/mass spectrometry, *Rapid Commun. Mass Spectrom.*, 11(12), 1279, 1997.
17. MacNair, J.E., Patel, K.D., and Jorgenson, J.W., Ultrahigh-pressure reversed-phase capillary liquid chromatography: Isocratic and gradient elution using columns packed with 1.0-micron particles, *Anal. Chem.*, 71(3), 700, 1999.
18. Jerkovich, A.D. et al., Linear velocity surge caused by mobile-phase compression as a source of band broadening in isocratic ultrahigh-pressure liquid chromatography, *Anal. Chem.*, 77(19), 6292, 2005.
19. Schuster, D., Laggner, C., and Langer, T., Why drugs fail–a study on side effects in new chemical entities, *Curr. Pharm. Des.*, 11(27), 3545, 2005.
20. Wainhaus, S. et al., Ultra fast liquid chromatography-MS/MS for pharmacokinetic and metabolic profiling within drug discovery, *Amer. Drug Discov.*, 2(1), 6, 2007.
21. Korfmacher, W.A. et al., Cassette-accelerated rapid rat screen: A systematic procedure for the dosing and liquid chromatography/atmospheric pressure ionization tandem mass spectrometric analysis of new chemical entities as part of new drug discovery, *Rapid Commun. Mass Spectrom.*, 15(5), 335, 2001.

22. Tiller, P.R., Romanyshyn, L.A., and Neue, U.D., Fast LC/MS in the analysis of small molecules, *Anal. Bioanal. Chem.*, 377(5), 788, 2003.
23. Korfmacher, W.A. et al., Development of an automated mass spectrometry system for the quantitative analysis of liver microsomal incubation samples: A tool for rapid screening of new compounds for metabolic stability, *Rapid Commun. Mass Spectrom.*, 13(10), 901, 1999.
24. Cox, K. et al., Novel procedure for rapid pharmacokinetic screening of discovery compounds in rats., *Drug Discov. Today*, 4(5), 232, 1999.
25. Mei, H., Korfmacher, W., and Morrison, R., Rapid in vivo oral screening in rats: Reliability, acceptance criteria, and filtering efficiency, *AAPS J.*, 8(3), E493, 2006.
26. Thorngren, J.O., Ostervall, F., and Garle, M., A high-throughput multicomponent screening method for diuretics, masking agents, central nervous system (CNS) stimulants and opiates in human urine by UPLC-MS/MS, *J. Mass Spectrom.*, 43(7), 980, 2008.
27. Romanyshyn, L., Tiller, P.R., and Hop, C.E., Bioanalytical applications of "fast chromatography" to high-throughput liquid chromatography/tandem mass spectrometric quantitation, *Rapid Commun. Mass Spectrom.*, 14(18), 1662, 2000.
28. Romanyshyn, L. et al., Ultra-fast gradient vs. fast isocratic chromatography in bioanalytical quantification by liquid chromatography/tandem mass spectrometry, *Rapid Commun. Mass Spectrom.*, 15(5), 313, 2001.
29. Romanyshyn, L.A. and Tiller, P.R., Ultra-short columns and ballistic gradients: Considerations for ultra-fast chromatographic liquid chromatographic-tandem mass spectrometric analysis, *J. Chromatogr. A*, 928(1), 41, 2001.
30. Hsieh, Y. and Korfmacher, W.A., Increasing speed and throughput when using HPLC-MS/MS systems for drug metabolism and pharmacokinetic screening, *Curr. Drug Metab.*, 7(5), 479, 2006.
31. Tiller, P.R. and Romanyshyn, L.A., Implications of matrix effects in ultra-fast gradient or fast isocratic liquid chromatography with mass spectrometry in drug discovery, *Rapid Commun. Mass Spectrom.*, 16(2), 92, 2002.
32. Hsieh, Y. et al., Quantitative screening and matrix effect studies of drug discovery compounds in monkey plasma using fast-gradient liquid chromatography/tandem mass spectrometry, *Rapid Commun. Mass Spectrom.*, 15(24), 2481, 2001.
33. Lagana, A. et al., Sample preparation for determination of macrocyclic lactone mycotoxins in fish tissue, based on on-line matrix solid-phase dispersion and solid-phase extraction cleanup followed by liquid chromatography/tandem mass spectrometry, *J. AOAC Int.*, 86(4), 729, 2003.
34. Wu, N. and Clausen, A.M., Fundamental and practical aspects of ultrahigh pressure liquid chromatography for fast separations, *J. Sep. Sci.*, 30(8), 1167, 2007.
35. Anspach, J.A., Maloney, T.D., and Colon, L.A., Ultrahigh-pressure liquid chromatography using a 1-mm id column packed with 1.5-microm porous particles, *J. Sep. Sci.*, 30(8), 1207, 2007.
36. van Deemter, J.J., Zuiderweg, F.J., and Klinkenberg, A., Longitudinal diffusion and resistance to mass transfer as causes of nonideality in chromatography, *Chem. Eng. Sci.*, 5, 271, 1956.
37. J.C. Giddings, *Dynamics of Chromatography*, Part 1, Marcel Dekker: New York 1965.
38. Guillarme, D. et al., Method transfer for fast liquid chromatography in pharmaceutical analysis: Application to short columns packed with small particle. Part II: Gradient experiments, *Eur. J. Pharm. Biopharm.*, 68(2), 430, 2008.
39. Guillarme, D. et al., Method transfer for fast liquid chromatography in pharmaceutical analysis: Application to short columns packed with small particle. Part I: Isocratic separation, *Eur. J. Pharm. Biopharm.*, 66(3), 475, 2007.
40. Russo, R. et al., Pharmaceutical applications on columns packed with sub-2 microm particles, *J. Chromatogr. Sci.*, 46(3), 199, 2008.

41. Yu, K. et al., High-throughput quantification for a drug mixture in rat plasma-a comparison of ultra performance liquid chromatography/tandem mass spectrometry with high-performance liquid chromatography/tandem mass spectrometry, *Rapid Commun. Mass Spectrom.*, 20(4), 544, 2006.
42. Wang, G. et al., Ultra-performance liquid chromatography/tandem mass spectrometric determination of diastereomers of SCH 503034 in monkey plasma, *J. Chromatogr. B Analyt. Technol. Biomed. Life Sci.*, 852(1–2), 92, 2007.
43. Leandro, C.C. et al., Comparison of ultra-performance liquid chromatography and high-performance liquid chromatography for the determination of priority pesticides in baby foods by tandem quadrupole mass spectrometry, *J. Chromatogr. A*, 1103(1), 94, 2006.
44. Thompson, T.N., Drug metabolism in vitro and in vivo results: How do these data support drug discovery? in *Using Mass Spectrometry for Drug Metabolism Studies*, W. Korfmacher, Editor. 2005, CRC Press: Boca Raton, FL. p. 35.
45. Roddy, T.P. et al., Mass spectrometric techniques for label-free high-throughput screening in drug discovery, *Anal. Chem.*, 79(21), 8207, 2007.
46. Wang, G. et al., Ultra-performance liquid chromatography/tandem mass spectrometric determination of testosterone and its metabolites in vitro samples, *Rapid Commun. Mass Spectrom.*, 20(14), 2215, 2006.
47. Plumb, R. et al., Ultra-performance liquid chromatography coupled to quadrupole-orthogonal time-of-flight mass spectrometry, *Rapid Commun. Mass Spectrom.*, 18(19), 2331, 2004.
48. Kaufmann, A. et al., Ultra-performance liquid chromatography coupled to time of flight mass spectrometry (UPLC-TOF): A novel tool for multiresidue screening of veterinary drugs in urine, *Anal. Chim. Acta.*, 586(1–2), 13, 2007.
49. Salisbury, J.J., Fused-core particles: A practical alternative to sub-2 micron particles, *J. Chromatogr. Sci.*, 46(10), 883, 2008.
50. Hsieh, Y., Duncan, C.J., and Brisson, J.M., Fused-core silica column high-performance liquid chromatography/tandem mass spectrometric determination of rimonabant in mouse plasma, *Anal. Chem.*, 79(15), 5668, 2007.
51. Cunliffe, J.M. and Maloney, T.D., Fused-core particle technology as an alternative to sub-2-microm particles to achieve high separation efficiency with low backpressure, *J. Sep. Sci.*, 30(18), 3104, 2007.
52. Bones, J. et al., Using coupled monolithic rods for ultra-high peak capacity LC and LC-MS under normal LC operating pressures, *Analyst*, 133(2), 180, 2008.
53. Morrison, D., Davies, A.E., and Watt, A.P., An evaluation of a four-channel multiplexed electrospray tandem mass spectrometry for higher throughput quantitative analysis, *Anal. Chem.*, 74(8), 1896, 2002.

9 Supercritical Fluid Chromatography–Mass Spectrometry

Yunsheng Hsieh

CONTENTS

9.1 INTRODUCTION

The search for higher-throughput methods for small molecular analysis has fueled many innovations on the development of high-speed chromatographic techniques [1–5]. According to the van Deemter equation, reducing particle size of the packing materials is a direct and effective way to enhance the column efficiency (e.g., see Figure 9.1). However, reducing the particle diameter will result in an increase in the column backpressure. One solution to overcome the high backpressure is to adopt a shorter column with small particles (~3 μm) for rapid separation [6]. Monolithic columns made from a single piece of porous silica gel can be operated at higher flow rates than conventional HPLC columns due to the higher permeability of monolithic silica rods making higher speed separation possible without a noticeable effect on chromatographic resolution [7]. Ultrahigh-pressure performance liquid (UHPLC) is a technical advance that primarily provides greater mobile-phase pressures to counter the high backpressure resulting from stationary phases with sub-2 μm particles. UHPLC offers theoretical advantages in chromatographic resolution, speed, and sensitivity over conventional HPLC systems (see Chapter 8) [4,8–10].

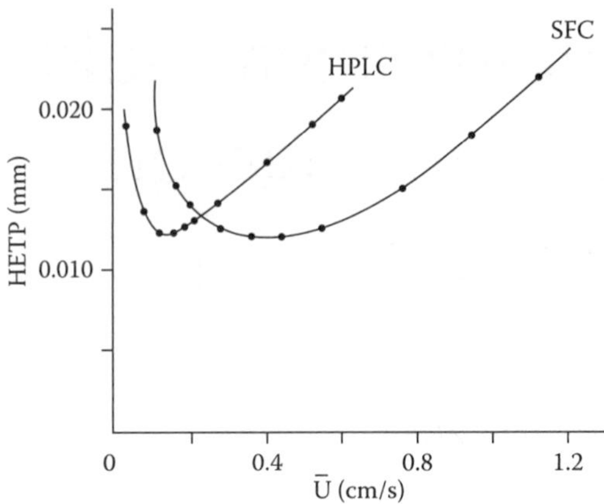

FIGURE 9.1 van Deemter plots for chromatographic data from HPLC and SFC elution of pyrene; HETP is height equivalent to a theoretical plate.

Supercritical fluid chromatography (SFC) is another practical fast chromatographic approach that uses supercritical fluid in place of organic solvents as the mobile phase to run the separation process. SFC is basically a hybrid of gas and liquid chromatography that eases the resolution of a mixture of compounds not conveniently resolved by either gas chromatography (GC) or liquid chromatography (LC) [11]. The mobile phases with a supercritical fluid have low viscosities and high diffusion coefficients compared to those for high-performance liquid chromatography (HPLC) and allow for high-efficiency separations. SFC uses supercritical fluid as the mobile phase, polar organic solvents as the modifiers in conjunction with acidic/basic compounds as additives to run the chromatographic process like in HPLC. In many applications, SFC-based methods are advantageous over HPLC-based methods as a separation tool in terms of efficiency and economical impact perspectives. Today, the availability of commercial hardware and atmospheric pressure ionization (API) interfaces with a mass spectrometer makes SFC even more applicable for pharmaceutical analysis in drug discovery and development fields. In this chapter, a variety of novel SFC–MS methods and technological advances are summarized.

9.2 SUPERCRITICAL FLUID CHROMATOGRAPHY

SFC is the application of a supercritical fluid, any substance at a temperature and pressure above its thermodynamic critical point (Figure 9.2) with both gas- and liquid-like abilities to diffuse through solids, and dissolve materials, respectively, as the mobile phase in the chromatographic process. The most widely used mobile phase for SFC is carbon dioxide because of its low critical pressure (73 atm), low critical temperature (31°C), inertness, low toxicity, and high purity at low cost [12,13]. Historically there were two approaches in developing modern SFC: the use of either the packed and microbore columns designed for HPLC application or the open-tubular capillary GC type columns [13,14]. The conventional packed HPLC

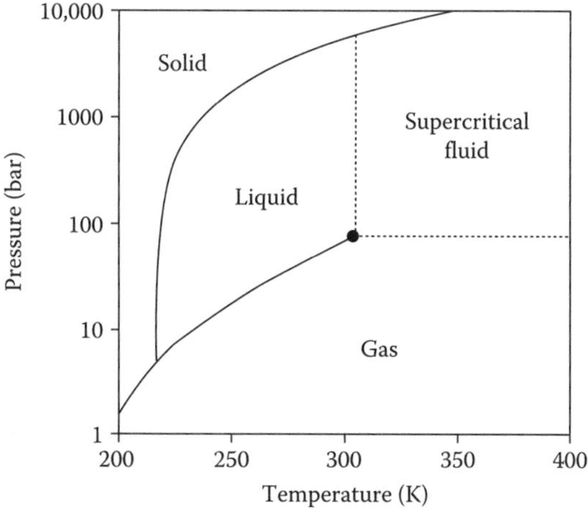

FIGURE 9.2 Carbon dioxide pressure–temperature phase diagram.

columns containing small particles (<10 μm) such as silica-based C-18, phenyl, and cyano packings could also be employed for SFC columns. However, the capillary SFC columns are smaller in diameter than the typical capillary GC columns [15]. In comparison with a packed SFC column, a capillary SFC column could provide larger number of theoretical plates for resolution of complex mixtures but lower sample throughput for the analysis of simple mixtures [13]. The use of packed capillary columns offers a compromise on the chromatographic performances between the two extremes of packed and open-tubular capillary columns.

Retention in SFC is basically governed by interaction between solute and the stationary phase and the solute volatility as well as the strength of the mobile phase. Both HPLC- and GC-like separation behaviors contribute to retention mechanism in SFC. As in both reversed-phase (RP) and normal-phase (NP) HPLC, retention rules in SFC are dependent on the nature of the stationary phase. For HPLC, the opposite polarity of the mobile phases is normally required for these two domains. However, for SFC, the same mobile phase can be employed with both polar and nonpolar stationary phases. The lack of polarity of neat CO_2 as a mobile phase makes it difficult to elute polar compounds on packed columns with high efficiency. Consequently, SFC with CO_2 requires the combination of more polar solvents such as methanol as a modifier to enhance the solvent strength of the mobile phase. The higher modifier content shortens the retention times of the analytes. The decrease of retention factor, k', of the analytes by increasing modifier concentration is due to the increase of solvent strength of the mobile phase and the deactivation of active sites on the surface of the packing material. As an example shown in Figure 9.3, the retention times of cytarabine (ara-C) and clofazimine on a packed silica column (100 × 4.6 mm, 10 μm) decreased as the methanol concentration increased [16]. The addition of a small concentration of additives such as TFA, ammonium acetate, or water was reported to improve the solvation power of methanol-modified CO_2 to further enhance chromatographic performance [17,18]. The effectiveness of an additive to disrupt specific

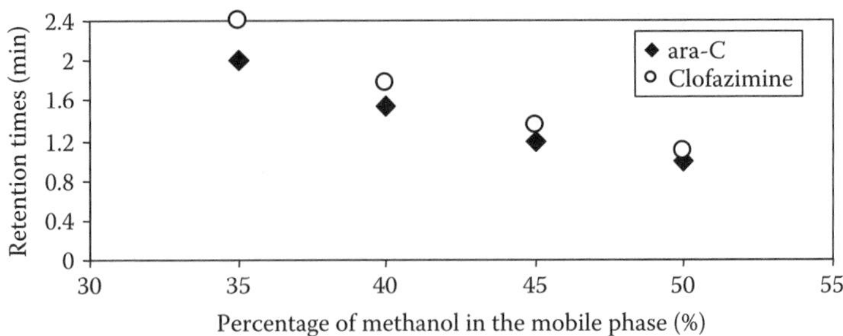

FIGURE 9.3 Effects of percentage of methanol in the mobile phase. (Adapted from Hsieh, Y. et al., *Anal. Chem.*, 79, 3856, 2007. With permission.)

binding through molecular interaction on retention is highly associated with the type of functional groups of the analytes. The retention in SFC is also dependent upon the temperature and the flow rate of the mobile phase. Temperature has several effects in SFC. In general, as temperature increases, density decreases at constant pressure resulting in an increase in chromatographic efficiency. Figure 9.4 shows that the retention times of both ara-C and clofazimine reduce as the flow rate of mobile phase increases from 2 to 6 mL/min [16]. Here, the gas-like properties of supercritical CO_2 allow for faster optimum flow rates and the use of longer columns than with liquids.

The SFC systems are grouped into two categories: analytical SFC for chemical analysis and preparative SFC for scale-up chemical synthesis and purification. On a fundamental level, the instrumentation for SFC consists of the following: (1) a fluid delivery system with high-pressure pumps to transport the sample in a mobile phase and to control the pressure; (2) the column in a thermostat-controlled oven where the separation process occurs; (3) a restrictor to maintain the high pressure in the column; (4) a detection system; and (5) a computer to control the system as well as to record the results (see Figure 9.5 as an example). In SFC the mobile phase

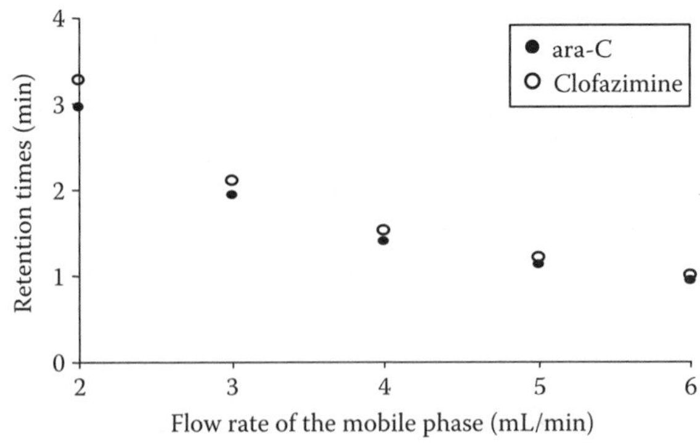

FIGURE 9.4 Influence of the mobile-phase flow rates on the retention times. (Adapted from Hsieh, Y. et al., *Anal. Chem.*, 79, 3856, 2007. With permission.)

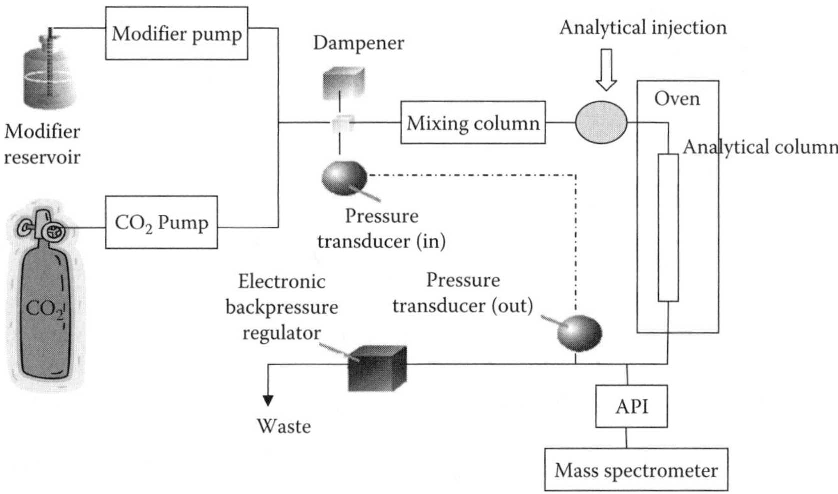

FIGURE 9.5 Schematic diagram of a packed column SFC–MS system.

is initially pumped through the supercritical region prior to the analytical column. For packed columns reciprocating pumps to allow easier mixing of the mobile phase are generally employed while for capillary SFC syringe pumps are used to provide consistent pressure for a neat mobile phase. For the analytical SFC system, the mixed mobile phase passes through a high-pressure injection component where the sample is loaded and carried through the column. Unlike normal liquids, supercritical fluids are highly compressible when they are pushed, the volume changes significantly, while the pressure changes only when the volume changes. In SFC, the fluid is chilled in order to be liquefied and these fluids remain compressible. Direct injection of large sample volumes might overwhelm the local mobile phase solvent strength at the head of the column causing distortion of chromatographic peak shape. Supercritical conditions are maintained by a pressure transducer, which is not required in HPLC. The detectors can be interfaced to the analytical columns through a "tee" between the column and the backpressure regulator.

9.3 INTEGRATION OF SFC TO MS

In the past, the majority of SFC equipment was developed using either UV [19] or flame ionization detection (FID) [20]. Similar to other chromatographic techniques, there is a trend toward hyphenation of SFC with a mass spectrometer for more sensitive and selective measurement of the analytes in a complex mixture. Thermospray (TS) and particle-beam (PB) systems, as examples, were early interfacing approaches for online coupling of chromatographic systems to a mass spectrometer. The exponential growth in HPLC–MS applications since the late 1990s is primarily due to the introduction of the API sources for HPLC–MS because the earlier interfacing techniques were not robust and had poor sensitivity. The integration of SFC to the mass analyzer for chemical analysis was not challenging due to the highly volatile property of CO_2, which can enhance volatilization during the ionization process in the various interface devices [21,22].

Earlier implementation of SFC–MS followed the evolution of both HPLC–MS and GC–MS interfaces [11,21,23–26]. As the API interfaces of HPLC–MS became mainstream analytical techniques in recent decades, they were also quickly employed for SFC–MS [21,23,26–37]. The atmospheric pressure chemical ionization (APCI) [27,33] and electrospray ionization (ESI) [36,37] sources are the most popular API interfaces for SFC–MS systems and allow for direct introduction of the effluent to the inlet of the mass spectrometer (Table 9.1). In some cases, the commercial API sources used for HPLC–MS system were proven to be applicable to the SFC–MS system with no modification [11,21,38–41]. However, some modification in the SFC–MS interface may be desired for SFC to achieve stable operation and enhanced ionization [22]. The ideal interfaces for SFC–MS would provide uniform pulse free flow, maintain chromatographic integrity, and ionize a wide range of analytes.

The mobile phase conditions of SFC might not only alter the chromatographic performances (Figure 9.3) but also the ionization efficiencies of the analytes when hyphenating to an API source prior to mass spectrometric detection (Figure 9.6). Addition of a proton-donating organic modifier such as methanol to the CO_2 mobile phase is generally necessary for ion generation. For the APCI, the effluent is vaporized through the heated nebulizer that consists of a pneumatic nebulizer and a heated quartz tube. The heated mixture of mobile phase and vapor is then introduced into the APCI reaction chamber. The plasma of air, solvent, and sample components then encounters a cloud of electrons emitted from the tip of a corona electrode pin held at 3–5 kV in atmosphere. The electric field is sufficiently strong to ionize nitrogen and solvent vapor by removal of an electron through electron impact ionization. The analytical ions could be produced by the proton transfer process with the reagent gas plasma if the acidity of the analytes is less than that of protonated charged clusters as shown in the following example [38]:

$$e^- + N_2 \rightarrow N_2^{+} + 2e^-$$

$$CH_3OH + e^- \rightarrow CH_3O^{+}H + 2e^-$$

$$CH_3O^{+}\ H + n(CH_3OH) \rightarrow [(CH_3OH)_n + H]^{+} + CH_3O$$

$$[(CH_3OH)_n + H]^{+} + M\ (\text{analyte}) \rightarrow n(CH_3OH) + [M + H]^{+}$$

In the gas phase, the proton transfer reaction occurs when the proton affinity (PA) of the neutral solvent molecules is greater than that of the donors. Ion source chemistry is of fundamental importance in the SFC–APCI system. The evaporation process to remove the solvent and to leave the ionized analytes in the gas phase plays an important role in all API interfaces. For HPLC–MS, higher mobile-phase flow rates normally result in appreciable negative impact on the ionization efficiency of the analytes. However, the tolerable input of mobile-phase flow rate into the APCI

TABLE 9.1
Representative SFC–API/MS Methods for Chemical Analysis

Compound	Interface	Ion Mode	Column	Mobile Phase	Mass Spectrometer	Reference
(R,S)-propranolol, (+)-pindolol	APCI	+	Chiralcel OD-H	Carbon dioxide/MeOH	Triple-quadrupole	[39]
Dextromethorphan	ESI	+	Cyano column	Carbon dioxide/MeOH	Triple-quadrupole	[11]
Hydrocortisone, testosterone, troleandomycin, picromycin, oleandomycin, erythromycin	APCI	+	Cyano column	Carbon dioxide/MeOH/IPA	Q-TOF	[42]
(R,S)-ketoprofen	ESI	+	Chirex 3005	Carbon dioxide/MeOH	Triple-quadrupole	[35]
R/S-warfarin	APCI	+	Chriralpak AD	Carbon dioxide/MeOH	Triple-quadrupole	[48]
Estrogen metabolites	APCI	+	Diol/Cyanopropyl/columns	Carbon dioxide/MeOH	Triple-quadrupole	[51]
Cytarabine, Clofazimine	APCI	+	Silica column	Carbon dioxide/MeOH/ ammonium acetate	Triple-quadrupole	[16]
Polypeptides	ESI	+	Ethylpyridine/aminopropyl columns	Carbon dioxide/MeOH/TFA	Single-quadrupole	[53]
Pyrimidine, caffeine, sulfanilamide	APCI	+	Zymor Zyrosil-Hybrid	Carbon dioxide/MeOH	Single-quadrupole/TOF	[30]
Primaquine	ESI	+	Chriralpak AD-H	Carbon dioxide/EtOH/ DEA	Triple-quadrupole	[28]

Note: IPA, isopropylamine; DEA, diethylamine.

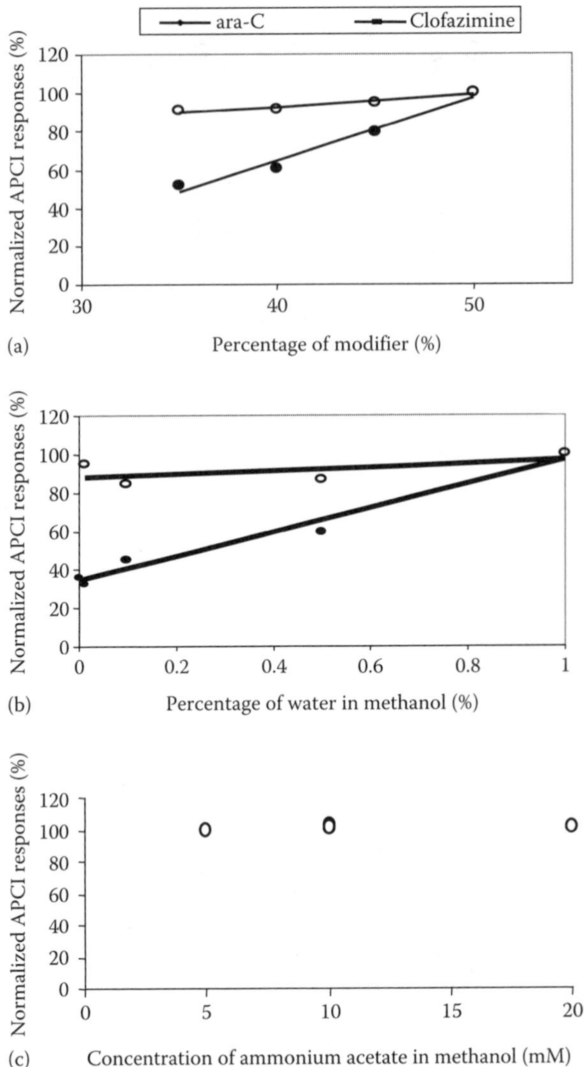

FIGURE 9.6 Effects of (a) percentage of methanol in the mobile phase, (b) percentage of water in methanol, and (c) percentage of ammonium acetate in methanol on the APCI responses of ara-C and clofazimine. (Adapted from Hsieh, Y. et al., *Anal. Chem.*, 79, 3856, 2007. With permission.)

source was enhanced in the presence of supercritical fluid due to its high volatility which efficiently assists the nebulization process [40]. APCI normally generates both $[M + H]^+$ and $[M - H]^-$ ions to make mass spectral interpretation trivial. The SFC–APCI–MS method was proved to be as durable, reliable, and user-friendly as the LC–MS method for screening of a large and diverse library of pharmaceutically relevant compounds [21]. In addition, the APCI source demanded less cleaning for the SFC–MS systems than the LC–MS systems [21]. The ionization efficiencies of the APCI and the ESI sources could be significantly different under the same SFC conditions. For the ESI source, a fused-silica inner capillary and a stainless-steel outer capillary are used for introducing the sample and the nebulizing nitrogen gas, respectively. The solvent and analytes are ionized through the combined action of applying a high electric field (3–5 kV) and the pneumatic nebulization. These charged droplets shrink due to evaporation leading to the formation of highly charged microdroplets as they are directed toward the mass spectrometer. In contrast to APCI, ESI occurs in the liquid phase and produces few fragmentation ions for most polar compounds [7].

There are several types of MS analyzers available for SFC–MS which have different strengths to meet the demands required by different analytical applications. The triple-quadrupole (QqQ) is a tandem mass spectrometer commonly used for conducting product ion scan (MS^2), precursor ion scan, neutral loss scan modes for structure elucidation, and selected reaction monitoring (SRM) for quantitative analyses of trace components. The integration of SFC and a QqQ tandem mass spectrometer allows less sample preparation for simultaneous determination of the compounds of interest in complex mixtures with good sensitivity and linearity [11]. The commercial time-of-flight (TOF) mass spectrometers can produce up to 500 average spectra per second, which are far in excess of the digital resolution required by the narrow chromatographic peaks. The use of TOF–MS detection allows the resolving power for high-speed SFC applications to be exploited in high-throughput analysis [30]. A chromatographic comparison of QqQ–MS and TOF–MS showed high-speed TOF total ion current data point sampling to be more indicative for fast SFC separations [30]. Similarly, a high-resolution hybrid mass spectrometer (Q-TOF) as a detector integrated into the SFC–APCI system for online accurate mass measurements was demonstrated to be very useful for the characterization of new chemical entities [42].

9.4 CHIRAL SFC–MS

According to the U.S. Food and Drug Administration's (FDA) policy statement for the development of new stereoisomeric drugs, in order to evaluate the pharmacokinetics of a single enantiomer or mixture of enantiomers, manufacturers should develop quantitative assays for individual enantiomers in *in vivo* samples early in drug development [43]. SFC, offering higher resolution per unit of time and faster column re-equilibration, is becoming the top choice in enantioselective chromatographic techniques such as gas chromatography (GC), capillary electrophoresis (CE), thin-layer chromatography (TLC), and HPLC [44]. In the drug discovery environment, there is a need to explore the biological responses of new chemical entities with respect to stereochemistry as part of lead characterization [45]. For the chromatographic resolution of chiral compounds, SFC was proven to provide at least two times faster separation than NP HPLC [44,45]. As an example, a chiral packed-column SFC system linked with APCI–MS was reported for the simultaneous determination of (R,S)-propranolol and (+)-pindolol in metabolic stability samples and mouse blood samples (Figure 9.7) [38,39]. A polysaccharide derivative column is commonly chosen for enantioseparation under NP conditions due to the

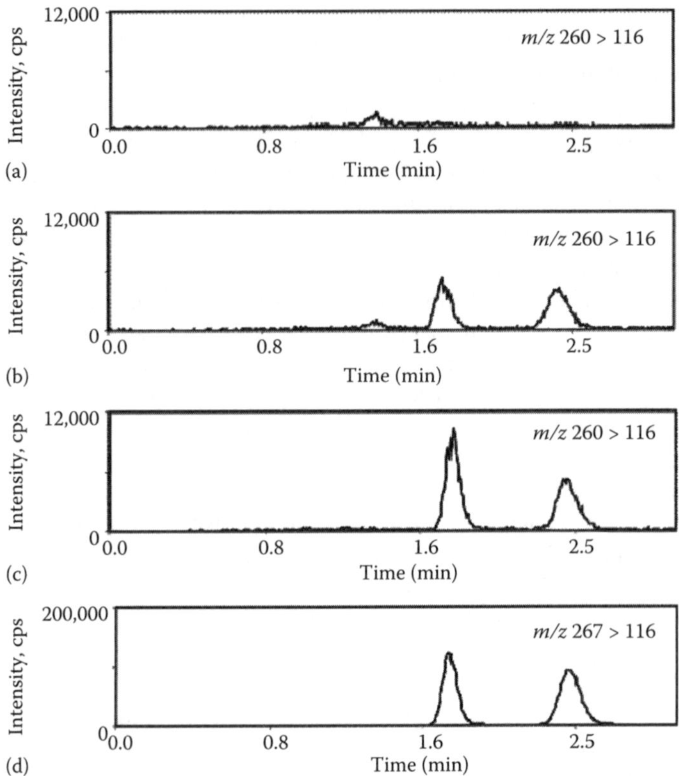

FIGURE 9.7 Reconstructed SFC–APCI/MS/MS chromatograms of racemic propranolol from (a) blank, (b) standard at 25 ng/mL, and (c) study mouse blood samples and the reconstructed SFC–APCI/MS/MS chromatogram of (d) D7-racemic propranolol from study mouse blood samples. (Adapted from Chen, J. et al., *Anal. Chem.*, 78, 1212, 2006. With permission.)

versatility, durability, and loading capacity [38]. The column normally employed in chiral HPLC may provide a higher column efficiency with SFC than with HPLC. In general, methanol as the starting modifier in SFC–MS is preferred over ethanol and isopropanol because it provides better ionization efficiency [45]. The chiral SFC retention is mainly governed by the polarity of the mobile phase. The additives to the modifier in the chiral SFC system were frequently suggested to improve the peak shape and enantioselectivity [45,46].

Conventionally, most chiral separation-based methods were developed using either UV or fluorescence detection. However, there is a trend toward interfacing enantioseparation technologies with more sensitive mass spectrometry–based methods without losing enantioselectivity [43]. Although the tandem mass spectrometric detection provides little selectivity for stereoisomeric drugs, it can provide complete resolution of the dosed drugs from endogenous materials and their metabolites. The primary concern of using APCI for the normal phase HPLC–MS system is the possible explosion hazard when a high flow of flammable solvents such as hexane is introduced into the ionization source at an elevated temperature in the presence of a corona discharge [47]. In contrast, SFC–APCI does not have this issue and can employ NP chromatography for analyses of chiral molecules because CO_2 is inert to heat and can quickly evaporate in a mass spectrometer ion chamber. By taking advantage of the inherent selectivity and sensitivity of mass spectrometric detection, implementing chiral SFC–MS method together with sample pooling technology allows one to simultaneously monitor samples containing multiple pairs of enantiomers with different masses within the same chromatographic run to further increase assay productivity [45]. The online coupling of chiral SFC to MS/MS detection was shown to be comparable with HPLC–MS/MS for most bioanalytical attributes such as specificity, linearity, accuracy, and ruggedness but advantageous for higher sample throughput and chromatographic resolution power [35,38,48]. With all of the aforementioned advantages, enantioselective SFC–MS assays for drug components are of increasing importance in the pharmaceutical arena.

9.5 ACHIRAL SFC–MS

SFC best emulates NP chromatography, but has the capability to separate a much broader range of analyte polarities amenable to SFC–MS if appropriate mobile-phase modifiers, additives, and columns are utilized [49–52]. As a general strategy, CO_2 combined with more polar solvents containing either acidic or basic additives as modifier to enhance the solvent strength of the mobile phase will allow for the elution of the analytes on a packed column with high efficiency. SFC–MS with two different columns (either a 2-ethyl-pyridine stationary phase, in a normal phase mode, or a pentafluorophenylpropyl stationary phase, in a RP mode) for the efficient separation of the contaminant quinocide from the drug primaquine diphosphate was reported to achieve a higher sample throughput than with either HPLC–MS or GC–MS [28]. In addition, the differential fragmentation of positional isomers at branching points with the ESI source was observed using the SFC–MS technique. Estrone, estradiol, estriol, and their metabolites are known to be potential biomarkers that have been linked to the possible development of breast cancer. Therefore, the ability to efficiently separate and quantify individual estrogen metabolites in hundreds of clinical samples is critical [51]. A packed column SFC–MS/MS approach was successfully developed by Xu et al. [51] for a broad analysis of 15 estrogen metabolites with around sevenfold reduction in assay times as compared to previous HPLC–MS/MS methods and without sacrificing the sensitivity achievable using HPLC–MS/MS (Figure 9.8a and b).

It is quite challenging to establish a reliable reversed-phase HPLC–MS method for very polar compounds showing little or no chromatographic retention prior to mass spectrometric detection. SFC is not limited to relatively nonpolar compounds but could be extended to allow the elution of polar and ionic compounds [16,53]. To illustrate the advantages of SFC–MS, Hsieh and his colleagues have applied it to quantitatively determine cytarabine (ara-C), an antineoplastic antimetabolite, in mouse plasma samples [16]. Here the separation of a very polar small molecule, ara-C, from endogenous compounds (sharing the same mass transition range as the analyte, as shown in Figure 9.9) as well as the elution of a nonpolar internal standard in

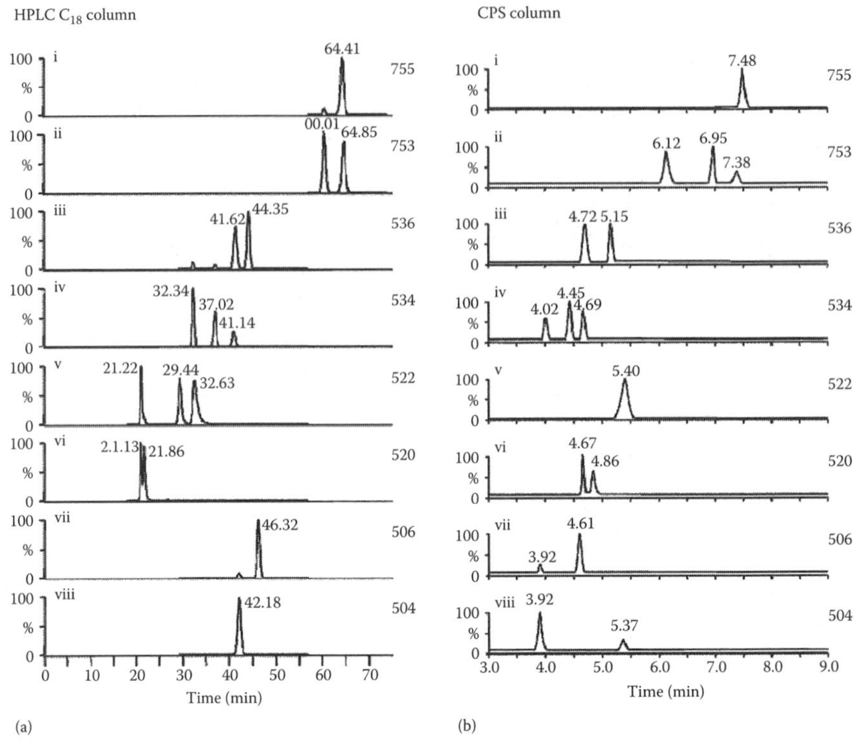

FIGURE 9.8 (a) HPLC separation of the 15 estrogen metabolites using a RP C-18 column and tandem mass spectrometry. (b) Packed column SFC separation of the 15 estrogen metabolites using a cyanopropyl column and tandem mass spectrometry. (Adapted from Xu, X. et al., *Anal. Chem.*, 78, 1553, 2006. With permission.)

mouse plasma extract was achieved by SFC using a bare silica stationary phase with an isocratic mobile phase composed of CO_2/methanol solvent containing ammonium acetate as additive [16]. As another example, for the elution of the most hydrophobic peptides, a 2-ethylpyridine bonded silica column using trifluoroacetic acid as an additive to a CO_2/methanol mobile phase was successfully employed to suppress deprotonation of peptide carboxylic acid groups and to protonate the peptides' amino groups prior to MS detection [53].

In HPLC, the direct serial coupling of achiral and chiral columns for the simultaneous achiral/chiral separation of a range of components in a complex mixture was not feasible due to either the increased backpressure or the different mobile-phase requirement. However, this concept is not restricted for SFC because the supercritical fluid not only has a significantly lower viscosity than a liquid but is also an effective eluent for both chiral and achiral stationary phases [52]. As an example, an SFC–MS approach using the coupled achiral/chiral column combination was reported to significantly enhance the separation efficiency for the reaction mixtures of cinnamonitrile and hydrocinnamonitrile intermediates to determine the diastereometric/enantiometric composition of the final product in one step [52]. Characterization of

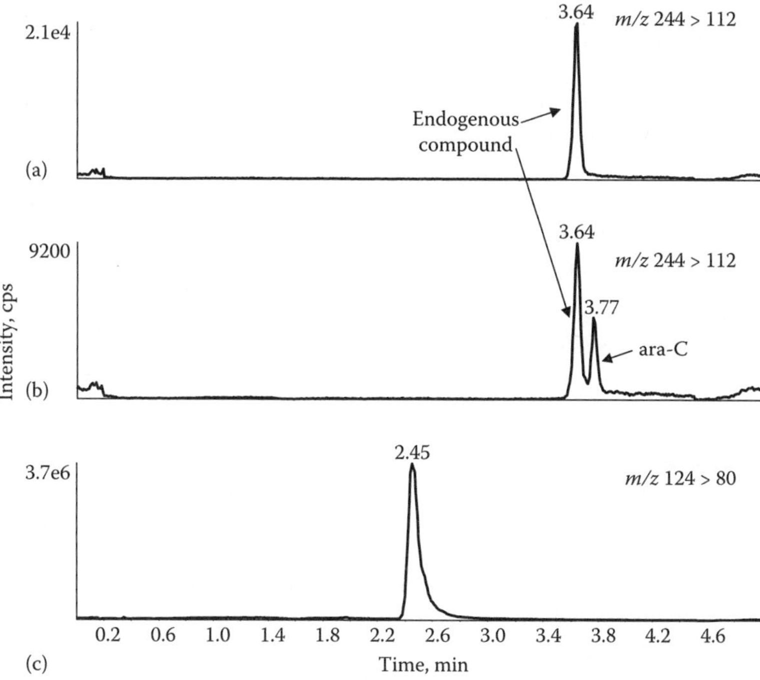

FIGURE 9.9 Reconstructed IP HPLC–MS/MS chromatograms of ara-C from (a) blank, (b) the spiked standard mouse plasma at 50 ng/mL, and the reconstructed IP HPLC–MS/MS chromatogram of nicotinic acid as the ISTD from (c) the study mouse plasma sample. (Adapted from Hsieh, Y. et al., *Anal. Chem.*, 79, 3856, 2007. With permission.)

low-molecular-weight polymeric materials is vital for elucidating their performance. A small difference in composition or structures of these polymers can largely impact performance [54]. The image data resulting from long column SFC–ESI/MS analysis of a commercial polymer are three-dimensional in nature (*m/z*, retention time, and intensity). The contour plots derived from SFC–MS data of complex mixtures are a compilation of at least hundreds of mass spectra and a useful tool in examining the separations of low-molecular weight polymers. The patterns generated by the contour plots of the SFC–MS analyses can be useful in assigning the composition of the polymer [54].

Metabolomics is the study of a comprehensive analysis of all the metabolites of an organism as an important research tool in postgenomic science. Metabolomics is normally carried out for a variety of small-molecule metabolite profiles and lipids as effectors for signal transduction [55]. Lipids are generally classified to be hydrophobic compounds. However, their structure diversity is high due to the complexity of a combination of hydrophobic acyl chain molecular species and their hydrophilicity increases in combination with hydrophilic molecules such as sugar and phosphoric acid, which requires efficient chromatographic techniques for separation of individual lipids in a complex condition. HPLC–MS is one of the major tools used to profile lipids [55]. As an example, a representative pharmaceutical application was performed by Mortuza et al. [56] who profiled phospholipids in rat urine after exposure to the phospholipidosis-inducing compound amiodarone. Bamba et al. [57] developed an SFC–MS system that enabled the rapid analysis of lipids with varied structures and polarities. The separation conditions for SFC and the detection conditions for mass spectrometry were optimized for simultaneous determination of phospholipids, glycolipids, neutral lipids, and sphingolipids. A cyanopropylated silica gel–packed column was used for the separation of all lipids within 15 min in support of lipid metabolomics as a fingerprinting method for the screening of diverse lipids but also for the profiling of individual components.

9.6 PARALLEL SFC–MS

Due to its inherent selectivity and sensitivity, MS-based methods allow the simultaneous determination of multiple components sharing the same retention times. An alternative approach for higher-throughput without sacrificing chromatographic integrity is the use of parallel chromatographic systems. For example, a threefold increase in sample throughput can be achieved by staggering injections onto the four columns, allowing the mass spectrometer to continuously monitor the chromatographic window of interest [58]. The parallel systems make use of unused chromatographic time windows where the analyst defines the acquisition windows placed into chromatographic timetable to provide a boost in throughput. There are two common ways in the serial mode to introduce multiple flows onto a mass spectrometric interface: with a valve selector or with parallel sprayers. The use of a commercial four-channel multiplex electrospray interface using the rotating aperture on a tandem mass spectrometer in support of *in vitro* and *in vivo* studies was demonstrated [59,60]. The mass data derived from each of the sprayer channels are separated to construct individual calibration curves to allow simultaneous method validation for the same analytes from different matrices.

The concept of using automated parallel HPLC–MS methods was further extended by Zeng et al. [61] to establish a parallel four-column SFC–MS system to perform higher-throughput enantioselective chromatographic method development and optimization. Chromatographic optimization of enantiomers was achieved on a single chiral column under several buffer conditions. This parallel SFC–MS approach allowing for simultaneous evaluation of multiple chiral stationary phases and/or evaluation of an array of solvent conditions, modifiers, and additives has resulted in a streamlined tool for enantioselective method development in support of drug discovery. The aforementioned parallel SFC–MS method was further enhanced by the same research group through an intelligent decision-making software program to assess enantiometric purity of several libraries containing compounds spanning a wide range of enantiomeric ratios with structurally diverse chemical scaffolds and stereogenic centers [31].

9.7 PREPARATIVE SFC–MS

As a result of the reduced viscosities and increased diffusivities of supercritical fluids, SFC using columns half the size in HPLC is often many times more productive, which makes large-scale separations more feasible. Preparative SFC–MS systems have the further advantage over preparative HPLC–MS systems for producing isolated fractions with minimal solvent waste, which is a key green chemistry concern. The evaporation of carbon dioxide during fraction collection allows the drying process of purified compounds in organic solvents more rapidly than the aqueous fractions eluted from RP preparative HPLC. For the preparative SFC–MS approach, the eluent from the column was split prior to fraction collector to allow a small portion of SFC to flow to the mass spectrometer [62]. Fraction collection is triggered by an ion signal exceeding an input threshold value detected from the mass spectrometer. The use of molecular weight responses to trigger fraction collection provides isolation of only a single component from a crude reaction mixture and validation of structure during the compound purification. Over the past few years, SFC–MS based purification process has experienced a rapid growth in the pharmaceutical and fine chemical industries [62–64].

For the preparative SFC–MS system, a make-up flow to the flow split to the mass spectrometer was recommended to avoid peak tailing and having a fluctuating baseline due to high mobile-phase flow rate. The addition of a small amount of basic organic compounds such as isopropylamine and ethyldimethylamine to the modifier was proven to be useful in improving both peak shape and loading capacity of pharmaceuticals containing basic functional groups. However, some of the basic additives such as isopropylamine might have a negative impact on mass spectrometric detection. For the ESI interface, a higher percentage of CO_2 in the mobile phase was reported to have a positive impact on ionization efficiency of the analytes in the positive mode but a deleterious effect in the negative mode [63]. Fraction collection using preparative SFC–MS does have some challenges. The supercritical fluid CO_2 undergoes rapid expansion to form an aerosol at the outlet of the pressure restrictor valve so that the standard low-pressure-rated valves on collection assembly can be frozen at the relatively high aerosol flow rates [62]. Also, the aerosol has a tendency to escape through the opening of the tube, leading to excessive sample losses and cross-contamination between fractions. To minimize the loss of sample, the use of a foil cover wrapped around the top of the collection tube, pressurized collection vessels, and cyclones were suggested to trap the samples [62].

9.8 MATRIX EFFECTS

A well-understood concern about assay reliability when developing new HPLC–MS methods is the possibility of encountering matrix ionization suppression problems [65–70]. For chiral SFC–MS applications, one aspect that might be overlooked is the matrix ionization suppression effect when the retention time window for the two enantiomers accounts for most of the total chromatographic run time. In general, matrix effect issues are associated with ionization sources, sample preparation procedures, or chromatographic conditions [68]. Continuous post-column infusion experiments are the standard method for monitoring ionization suppression [65,68]. The affected area of the chromatographic run is evaluated according to the differences in the infusion chromatograms between the mobile phase injection and the sample extract injection. The observed loss of API sensitivity is considered to be caused by matrix ion suppression effects due to sample extract constituents eluting from the analytical column.

Phospholipids are commonly found in mammalian blood and animal tissue as structure components in membranes and have been identified to be a major source of ion suppression [38,71]. Due to their hydrophilicity, phospholipids do not tend to be retained on RP HPLC. SFC may be an excellent solution if significant ion suppression of the analyte due to the co-elution of phospholipids is observed in the RP HPLC system. In addition, ion suppression from DMSO solvent was found to be a greater problem with the SFC–MS method than with the standard LC–MS method [21].

9.9 CONCLUSIONS

The integration of SFC with MS preserves chromatographic integrity and affords universal identification of chemicals. SFC–MS has been demonstrated as a complementary hyphenated technique to HPLC–MS. A direct comparison of HPLC–MS and SFC–MS highlights the effectiveness of the two techniques in terms of speed and resolution power of the assays. Some mixtures that are difficult to handle by HPLC may be susceptible to separation by SFC with greater productivity. Representative SFC–MS methods for chemical analysis are summarized in Table 9.1. A few examples have demonstrated the advantages of interfacing SFC with MS for quantitative and purification applications such as higher throughput, stereoisomeric selectivity, and reduced cost of operation. Although carbon dioxide has many practical advantages, including its near-ambient critical temperature and minimal interference with mass spectrometric detection, the use of alternative supercritical fluids, modifiers, and additives to carbon dioxide may further extend the applications of this technique. So far, the instrumental designs of SFC such as the stationary phases and API ionization source still rely on the same concepts used in HPLC–MS. A few investigations on a better understanding of fundamental processes of chromatographic retention and the API ionization mechanisms with extensive compounds under various SFC mobile-phase conditions are reported. All of these factors have restricted the future development of SFC–MS in a sustainable manner. Efforts toward the establishment of improved hardware for SFC–MS may elevate this technique to become the method of choice for high-throughput chemical analysis in both academic and industrial areas.

Preparative SFC offers unique advantages in chiral/achiral separation and purification of compound libraries, such as greater speed, faster evaporation time, and reduced disposal of organic solvent lending itself as a more environmentally friendly technique than HPLC. The scale-up of preparative SFC columns will allow handling separation of larger amount of chemicals. However, an increase in column diameter will result in the need for increasing the wall thickness of a stainless column which limits the scalability of constructing preparative SFC columns. To overcome the pressure and volume issues at larger scale, improvements in instrumental design such as using a parallel column SFC approach merits future exploration.

REFERENCES

1. Hsieh, Y. et al., Fast mass spectrometry-based methodologies for pharmaceutical analyses, *Comb. Chem. High Throughput Screen.*, 9(1), 3, 2006.
2. Hsieh, Y. and Korfmacher, W.A., Increasing speed and throughput when using HPLC-MS/MS systems for drug metabolism and pharmacokinetic screening, *Curr. Drug Metab.*, 7(5), 479, 2006.
3. Korfmacher, W., ed. *Using Mass Spectrometry for Drug Metabolism Studies*. 2005, CRC Press: Boca Raton, FL. p. 370.
4. Wainhaus, S. et al., Ultra fast liquid chromatography-MS/MS for pharmacokinetic and metabolic profiling within drug discovery, *Am. Drug Discov.*, 2(1), 6, 2007.
5. Korfmacher, W.A., Principles and applications of LC-MS in new drug discovery, *Drug Discov. Today*, 10(20), 1357, 2005.
6. Hsieh, Y. et al., Simultaneous fast HPLC-MS/MS analysis of drug candidates and hydroxyl metabolites in plasma, *J. Pharm. Biomed. Anal.*, 33(2), 251, 2003.
7. Hsieh, Y. et al., Simultaneous determination of a drug candidate and its metabolite in rat plasma samples using ultrafast monolithic column high-performance liquid chromatography/tandem mass spectrometry, *Rapid Commun. Mass Spectrom.*, 16(10), 944, 2002.
8. Wang, G. et al., Ultra-performance liquid chromatography/tandem mass spectrometric determination of testosterone and its metabolites in in vitro samples, *Rapid Commun. Mass Spectrom.*, 20(14), 2215, 2006.
9. Churchwell, M.I. et al., Improving LC-MS sensitivity through increases in chromatographic performance: Comparisons of UPLC-ES/MS/MS to HPLC-ES/MS/MS, *J. Chromatogr. B Analyt. Technol. Biomed. Life Sci.*, 825(2), 134, 2005.
10. Plumb, R.S. et al., A rapid simple approach to screening pharmaceutical products using ultra-performance LC coupled to time-of-flight mass spectrometry and pattern recognition, *J. Chromatogr. Sci.*, 46(3), 193, 2008.
11. Hoke, S.H., 2nd et al., Increasing bioanalytical throughput using pcSFC-MS/MS: 10 minutes per 96-well plate, *Anal. Chem.*, 73(13), 3083, 2001.
12. Lynam, K.G. and Nicolas, E.C., Chiral HPLC versus chiral SFC: Evaluation of long-term stability and selectivity of Chiralcel OD using various eluents, *J. Pharm. Biomed. Anal.*, 11(11–12), 1197, 1993.
13. Lee, M.L. and Markides, K.E., Chromatography with supercritical fluids, *Science*, 235(4794), 1342, 1987.
14. Wong, S.H., Supercritical fluid chromatography and microbore liquid chromatography for drug analysis, *Clin. Chem.*, 35(7), 1293, 1989.
15. Koski, I.J. et al., Analysis of prostaglandins in aqueous solutions by supercritical fluid extraction and chromatography, *J. Pharm. Biomed. Anal.*, 9(4), 281, 1991.
16. Hsieh, Y., Li, F., and Duncan, C.J., Supercritical fluid chromatography and high-performance liquid chromatography/tandem mass spectrometric methods for the determination of cytarabine in mouse plasma, *Anal. Chem.*, 79(10), 3856, 2007.
17. Zheng, J. et al., Study of the elution mechanism of sodium aryl sulfonates on bare silica and a cyano bonded phase with methanol-modified carbon dioxide containing an ionic additive, *J. Chromatogr. A*, 1090(1–2), 155, 2005.
18. Zheng, J. et al., Effect of ionic additives on the elution of sodium aryl sulfonates in supercritical fluid chromatography, *J. Chromatogr. A*, 1082(2), 220, 2005.
19. Hoffman, B.J. et al., Increasing UV detection sensitivity in the supercritical fluid chromatographic analysis of alcohol polyethers, *J. Chromatogr. A*, 1052(1–2), 161, 2004.
20. Kim, D.H., Lee, K.J., and Heo, G.S., Analysis of cholesterol and cholesteryl esters in human serum using capillary supercritical fluid chromatography, *J. Chromatogr. B Biomed. Appl.*, 655(1), 1, 1994.

21. Pinkston, J.D. et al., Comparison of LC/MS and SFC/MS for screening of a large and diverse library of pharmaceutically relevant compounds, *Anal. Chem.*, 78(21), 7467, 2006.
22. Pinkston, J.D., Advantages and drawbacks of popular supercritical fluid chromatography/mass interfacing approaches—a user's perspective, *Eur. J. Mass Spectrom. (Chichester, Eng)*, 11(2), 189, 2005.
23. Smith, R.D., Udseth, H.R., and Wright, B.W., Rapid and high resolution capillary supercritical fluid chromatography (SFC) and SFC/MS of trichothecene mycotoxins, *J. Chromatogr. Sci.*, 23(5), 192, 1985.
24. Bruchet, A. et al., Recent methods for the determination of volatile and non-volatile organic compounds in natural and purified drinking water, *J. Chromatogr.*, 562(1–2), 469, 1991.
25. Crowther, J.B. and Henion, J.D., Supercritical fluid chromatography of polar drugs using small-particle packed columns with mass spectrometric detection, *Anal. Chem.*, 57(13), 2711, 1985.
26. Edlund, P.O. and Henion, J.D., Packed-column supercritical fluid chromatography/mass spectrometry via a two-stage momentum separator, *J. Chromatogr. Sci.*, 27(6), 274, 1989.
27. Jones, D.C. et al., The analysis of beta-agonists by packed-column supercritical fluid chromatography with ultra-violet and atmospheric pressure chemical ionisation mass spectrometric detection, *Analyst*, 124(6), 827, 1999.
28. Brondz, I. et al., Nature of the main contaminant in the drug primaquine diphosphate: SFC and SFC-MS methods of analysis, *J. Pharm. Biomed. Anal.*, 43(3), 937, 2007.
29. Sjoberg, P.J. and Markides, K.E., Capillary column supercritical fluid chromatography-atmospheric pressure ionisation mass spectrometry interface performance of atmospheric pressure chemical ionisation and electrospray ionisation, *J. Chromatogr. A*, 855(1), 317, 1999.
30. Bolanos, B.J., Ventura, M.C., and Greig, M.J., Preserving the chromatographic integrity of high-speed supercritical fluid chromatography separations using time-of-flight mass spectrometry, *J. Comb. Chem.*, 5(4), 451, 2003.
31. Laskar, D.B. et al., Parallel SFC/MS-MUX screening to assess enantiomeric purity, *Chirality*, 20(8), 885, 2008.
32. Backstrom, B. et al., A preliminary study of the analysis of Cannabis by supercritical fluid chromatography with atmospheric pressure chemical ionisation mass spectroscopic detection, *Sci. Justice*, 37(2), 91, 1997.
33. Tuomola, M., Hakala, M., and Manninen, P., Determination of androstenone in pig fat using packed column supercritical fluid chromatography-mass spectrometry, *J. Chromatogr. B Biomed. Sci. Appl.*, 719(1–2), 25, 1998.
34. Pinkston, J.D. et al., Evaluation of capillary supercritical fluid chromatography with mass spectrometric detection for the analysis of a drug (mebeverine) in a dog plasma matrix, *J. Chromatogr.*, 622(2), 209, 1993.
35. Hoke, S.H., 2nd et al., Comparison of packed-column supercritical fluid chromatography–tandem mass spectrometry with liquid chromatography–tandem mass spectrometry for bioanalytical determination of (R)- and (S)-ketoprofen in human plasma following automated 96-well solid-phase extraction, *Anal. Chem.*, 72(17), 4235, 2000.
36. Baker, T. and Pinkston, J.D., Development and application of packed-column supercritical fluid chromatography/pneumatically assisted electrospray mass spectrometry, *J. Am. Soc. Mass Spectrom.*, 9(5), 498, 1998.
37. Pinkston, J.D. and Baker, T., Modified ionspray interface for supercritical fluid chromatography/mass spectrometry: Interface design and initial results, *Rapid Commun. Mass Spectrom.*, 9(12), 1087, 1995.

38. Chen, J. et al., Supercritical fluid chromatography-tandem mass spectrometry for the enantioselective determination of propranolol and pindolol in mouse blood by serial sampling, *Anal. Chem.*, 78(4), 1212, 2006.
39. Hsieh, Y. et al., Chiral supercritical fluid chromatography/tandem mass spectrometry for the simultaneous determination of pindolol and propranolol in metabolic stability samples, *Rapid Commun. Mass Spectrom.*, 19(21), 3037, 2005.
40. Hsieh, Y. et al., Supercritical fluid chromatography/tandem mass spectrometric method for analysis of pharmaceutical compounds in metabolic stability samples, *J. Pharm. Biomed. Anal.*, 40(3), 799, 2006.
41. Niessen, W.M., *Liquid Chromatography-Mass Spectrometry*, 3rd Edition. *Chromatographic Science Series*. Vol. 97. 2006, CRC Press: Boca Raton, FL.
42. Garzotti, M., Rovatti, L., and Hamdan, M., Coupling of a supercritical fluid chromatography system to a hybrid (Q-TOF 2) mass spectrometer: On-line accurate mass measurements, *Rapid Commun. Mass Spectrom.*, 15(14), 1187, 2001.
43. Garzotti, M. and Hamdan, M., Supercritical fluid chromatography coupled to electrospray mass spectrometry: A powerful tool for the analysis of chiral mixtures, *J. Chromatogr. B Analyt. Technol. Biomed. Life Sci.*, 770(1–2), 53, 2002.
44. Zhang, Y. et al., Enantioselective chromatography in drug discovery, *Drug Discov. Today*, 10(8), 571, 2005.
45. Zhao, Y. et al., Rapid method development for chiral separation in drug discovery using sample pooling and supercritical fluid chromatography-mass spectrometry, *J. Chromatogr. A*, 1003(1–2), 157, 2003.
46. Barnhart, W.W. et al., Supercritical fluid chromatography tandem-column method development in pharmaceutical sciences for a mixture of four stereoisomers, *J. Sep. Sci.*, 28(7), 619, 2005.
47. Hsieh, Y. and Wang, G., Integration of atmospheric pressure photoionization interfaces to HPLC-MS/MS for pharmaceutical analysis, *Amer. Pharm. Rev.*, 7, 88, 2004.
48. Coe, R.A., Rathe, J.O., and Lee, J.W., Supercritical fluid chromatography-tandem mass spectrometry for fast bioanalysis of R/S-warfarin in human plasma, *J. Pharm. Biomed. Anal.*, 42(5), 573, 2006.
49. Taylor, L.T., Supercritical fluid chromatography, *Anal. Chem.*, 80(12), 4285, 2008.
50. Cheng, X. and Hochlowski, J., Current application of mass spectrometry to combinatorial chemistry, *Anal. Chem.*, 74(12), 2679, 2002.
51. Xu, X. et al., Analysis of fifteen estrogen metabolites using packed column supercritical fluid chromatography-mass spectrometry, *Anal. Chem.*, 78(5), 1553, 2006.
52. Alexander, A.J. and Staab, A., Use of achiral/chiral SFC/MS for the profiling of isomeric cinnamonitrile/hydrocinnamonitrile products in chiral drug synthesis, *Anal. Chem.*, 78(11), 3835, 2006.
53. Zheng, J. et al., Feasibility of supercritical fluid chromatography/mass spectrometry of polypeptides with up to 40-mers, *Anal. Chem.*, 78(5), 1535, 2006.
54. Pinkston, J.D. et al., Characterization of low molecular weight alkoxylated polymers using long column SFC/MS and an image analysis based quantitation approach, *J. Am. Soc. Mass Spectrom.*, 13(10), 1195, 2002.
55. Bamba, T., Application of supercritical fluid chromatography to the analysis of hydrophobic metabolites, *J. Sep. Sci.*, 31(8), 1274, 2008.
56. Mortuza, G.B. et al., Characterisation of a potential biomarker of phospholipidosis from amiodarone-treated rats, *Biochim. Biophys. Acta.*, 1631(2), 136, 2003.
57. Bamba, T. et al., High throughput and exhaustive analysis of diverse lipids by using supercritical fluid chromatography-mass spectrometry for metabolomics, *J. Biosci. Bioeng.*, 105(5), 460, 2008.
58. Van Pelt, C.K. et al., A four-column parallel chromatography system for isocratic or gradient LC/MS analyses, *Anal. Chem.*, 73(3), 582, 2001.

59. Bayliss, M.K. et al., Parallel ultra-high flow rate liquid chromatography with mass spectrometric detection using a multiplex electrospray source for direct, sensitive determination of pharmaceuticals in plasma at extremely high throughput, *Rapid Commun. Mass Spectrom.*, 14(21), 2039, 2000.
60. Yang, L. et al., Evaluation of a four-channel multiplexed electrospray triple quadrupole mass spectrometer for the simultaneous validation of LC/MS/MS methods in four different preclinical matrixes, *Anal. Chem.*, 73(8), 1740, 2001.
61. Zeng, L. et al., Parallel supercritical fluid chromatography/mass spectrometry system for high-throughput enantioselective optimization and separation, *J. Chromatogr. A*, 1169(1–2), 193, 2007.
62. Wang, T. et al., Mass-directed fractionation and isolation of pharmaceutical compounds by packed-column supercritical fluid chromatography/mass spectrometry, *Rapid Commun. Mass Spectrom.*, 15(22), 2067, 2001.
63. Zhang, X. et al., Development of a mass-directed preparative supercritical fluid chromatography purification system, *J. Comb. Chem.*, 8(5), 705, 2006.
64. Searle, P.A., Glass, K.A., and Hochlowski, J.E., Comparison of preparative HPLC/MS and preparative SFC techniques for the high-throughput purification of compound libraries, *J. Comb. Chem.*, 6(2), 175, 2004.
65. Hsieh, Y. et al., Quantitative screening and matrix effect studies of drug discovery compounds in monkey plasma using fast-gradient liquid chromatography/tandem mass spectrometry, *Rapid Commun. Mass Spectrom.*, 15(24), 2481, 2001.
66. Matuszewski, B.K., Constanzer, M.L., and Chavez-Eng, C.M., Matrix effect in quantitative LC/MS/MS analyses of biological fluids: A method for determination of finasteride in human plasma at picogram per milliliter concentrations, *Anal. Chem.*, 70(5), 882, 1998.
67. Matuszewski, B.K., Constanzer, M.L., and Chavez-Eng, C.M., Strategies for the assessment of matrix effect in quantitative bioanalytical methods based on HPLC-MS/MS, *Anal. Chem.*, 75(13), 3019, 2003.
68. Mei, H., Matrix effects: Causes and solutions, in *Using Mass Spectrometry for Drug Metabolism Studies*, W. Korfmacher, ed. 2005, CRC Press: Boca Raton, FL. p. 103.
69. Mei, H. et al., Investigation of matrix effects in bioanalytical high-performance liquid chromatography/tandem mass spectrometric assays: Application to drug discovery, *Rapid Commun. Mass Spectrom.*, 17(1), 97, 2003.
70. Xu, X., Lan, J., and Korfmacher, W., Rapid LC/MS/MS method development for drug discovery, *Anal. Chem.*, 77(October 1), 389 A, 2005.
71. Chambers, E. et al., Systematic and comprehensive strategy for reducing matrix effects in LC/MS/MS analyses, *J. Chromatogr. B Analyt. Technol. Biomed. Life Sci.*, 852(1–2), 22, 2007.

10 Biomarkers of Efficacy and Toxicity: Discovery and Assay

Joanna R. Pols

CONTENTS

10.1 INTRODUCTION

The term "biomarker" refers to any small or large molecule or any measurement that is an indicator of a normal biological process, a disease state, or a pharmacological response to a drug treatment [1,2]. The various types of biomarkers are summarized in Table 10.1. Why is the pharmaceutical industry interested in biomarkers? In an elegant article, Kola and Landis [3] described a few key reasons due to which only one in nine compounds makes it through development and gets approved by regulatory agencies. The two top causes for a compound's attrition in the clinical phase of drug

TABLE 10.1
Types of Biomarkers

Role	Description	Examples
Disease biomarkers	Indicate the presence or likelihood of a particular disease in patients or in animal models.	High blood levels of proteins (troponin, C-reactive protein) associated with risk of stroke or heart disease, genotypes, and gene expression profiles linked with cancer.
Surrogate endpoints	Accepted by the FDA in support of drug approval as a substitute for desired clinical outcome, and often can be measured months or even years before meaningful clinical endpoints like mortality or morbidity.	Blood pressure, tumor shrinkage, psychometric testing, pain scales, CD4 blood count for HIV AIDS.
Efficacy or outcome biomarkers	Correlate with the desired effect of a treatment, but do not have as much validation as surrogate endpoints.	Pain scales, lung function tests, electrocardiogram readings, bone density, blood or urine levels of proteins and other analytes associated with disease.
Mechanism biomarkers	Suggest the drug effects on the desired pathway.	Evidence of downstream effects of the drug, such as activation or deactivation of enzymes and receptors.
Pharmacodynamic biomarkers	Used in early clinical trials to determine the dose that has the highest response. This dose is often below the maximum tolerated dose.	Blood or urine levels of proteins and other analytes associated with disease.
Target biomarkers	Show that a drug interacts with a particular target in *in vitro* studies, preclinical studies, and, increasingly, in *in vivo* imaging studies.	PET imaging studies to show residence time on a receptor.
Toxicity biomarkers	Indicate potentially harmful effects of a drug in cell-based, preclinical, or clinical studies.	Q-T prolongation, induction of cytochrome P-450.
Bridging or translational biomarkers	Can be used in both preclinical and clinical studies and may also be disease, efficacy, or toxicity biomarkers.	Any of the above examples that can be assessed in both humans and animal models of disease.

Source: Reprinted from Baker, M., *Nat. Biotechnol.*, 23, 297, 2005. With permission.

development in 2000 were lack of efficacy accounting for about 30% of failures and lack of safety accounting for another 30% of failures. This explains why biomarkers have been in the center of interest of pharmaceutical and diagnostic companies over the last five years. Currently, in the discovery stage, the majority if not all pharmaceutical companies focus on understanding first the molecular target and the molecular basis of a disease so that biomarkers of disease and efficacy can be proposed and eventually be used for demonstrating proof-of-concept in the clinical studies. In these early stages of drug development, biomarkers are used to optimize the selection of a lead compound based on its *in vitro* and *in vivo* preclinical activity profiles. As the compound is nominated for development, additional preclinical studies are initiated to better understand the safety and efficacy profile of a drug candidate. Biomarkers are used to study the correlation between the efficacy and PK/PD data to optimize clinical drug formulation as well as starting clinical dose and dosing regimen. Safety studies are conducted in preclinical animal species (typically rodent and non-rodent large animals) to characterize target organ toxicities and to define the maximum tolerated dose. At this stage opportunities arise to search for novel, more specific, and sensitive biomarkers of safety. Metabonomics and other "omics" can be used to study whether an observed toxicity is mechanism or non-mechanism based so that compounds with mechanism-based toxicity can be eliminated [3–5]. Once a compound enters the clinical phase of development, one of the biggest challenges is to evaluate and properly validate whether the biomarker findings from preclinical studies in animals translate into findings in humans. In the late clinical stage of drug development, biomarkers can improve the statistical power of clinical study designs so that fewer patients are enrolled and enable selection of patients who are more likely to respond to a treatment. Once a drug is marketed, biomarkers are used to diagnose a disease and can allow for a real-time monitoring of treatment safety and efficacy [2,5–7].

Although tremendous progress has been made in the area of small and large molecule biomarker research in the last decade, a question that needs to be asked is why so few novel biomarkers (whether safety or efficacy) are being discovered and applied in the pharmaceutical research and development arena. One of the primary reasons for this phenomenon is the fact that by the time a novel biomarker of a panel of biomarkers is discovered and validated, it is often too late for them to make an impact on decision-making [1]. Another problem is that novel biomarkers are not studied rigorously enough because of lack of time or resources before they can be used to influence decisions. Furthermore, for a biomarker to be used in regulatory decision-making, it needs to be validated. Cummings et al. [6] recently described in detail the biomarker validation process and Collings and Vaidya [8] discussed steps necessary for a successful development of a biomarker of safety from the moment it is first discovered until it can be successfully used in a clinical setting (Figure 10.1). Moreover, some argue that it is not worthwhile to search for new "unknown" biomarkers since there are a lot of "known" validated clinically relevant biomarkers described in the literature that can be applied. In addition, perhaps for the intellectual property reasons, the biomarker efforts within the pharmaceutical industry are quite extensive but the findings are not being made available for public knowledge.

The focus of this chapter is to review the progress that has been made in the area of mass spectrometry (MS)–based nontargeted metabonomics research in the context of

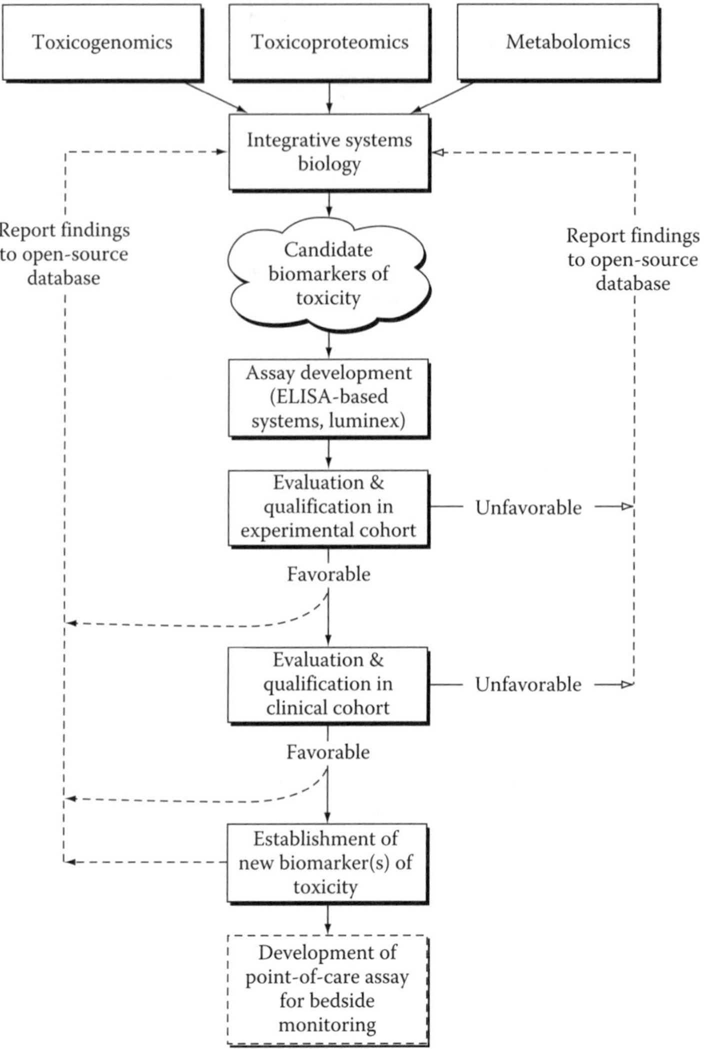

FIGURE 10.1 Biomarker discovery and evaluation process. (Reprinted from Collings, F.B. and Vaidya, V.S., *Toxicology*, 245, 167, 2008. With permission.)

pharmaceutical research and development. Examples of metabonomic biomarkers of safety and efficacy are presented and the necessary MS-based analytical procedures to generate such data are reviewed.

10.2 METABONOMICS VERSUS METABOLOMICS

In the last decade, the field of metabonomics or metabolomics has experienced tremendous growth and has been the subject of numerous reviews in the literature [9–12]. However, metabolite or metabolic profiling of low-molecular-weight compounds in biological samples is not new. In 1971 Horning [13] used the term "metabolic profiling" for the first time to describe a gas chromatography-mass spectrometry analysis of a patient sample. Lindon et al. [14] described the differences between the two terms "metabolomics" and "metabonomics." Metabolomics refers to the metabolome "which is defined as the metabolic composition of a cell and is analogous to the genome or proteome." Metabonomics, on the other hand, is the study of a whole organism with separate organs and multiple cell types. It evaluates not only the static cellular and biofluid concentrations of endogenous metabolites but also how those metabolites change over course of time [14]. Figure 10.2 illustrates how metabonomics relates to other "omics" sciences. Metabonomics is considered to be complementary to those other "omics" fields.

The field of metabonomics (or metabolomics), regardless of an analytical platform, involves an analysis of thousands of low-molecular-weight compounds, endogenous metabolites such as amino acids, sugars, vitamins, organic acids, or nucleotides, the end products of biochemical pathways [15,16]. The overall size of the metabonome remains debatable but it has been proposed that the number of small molecule metabolites ranges from a few thousand to tens of thousands. One of the greatest hopes for metabonomics is that it will eventually contribute to understanding the regulation of biochemical pathways and thus the interaction of proteins encoded by the genome [15]. Although many names have been used to describe this new field, including metabolomics, metabolic fingerprinting, and metabolic profiling, for the purpose of clarity, metabonomics is used throughout this chapter. It is important to point out that within the field of metabonomics, there are other more specific branches of metabonomics, such as lipidomics [17–21] or glycomics that specifically focus on lipid or sugar metabolites, respectively.

The applications of metabonomics techniques to pharmaceutical research and development have been extensively reviewed in the literature [9,10,12,14,22–33]. In metabonomics studies of pharmaceutical relevance, urine and plasma are the two biofluids most commonly profiled because they can be obtained noninvasively.

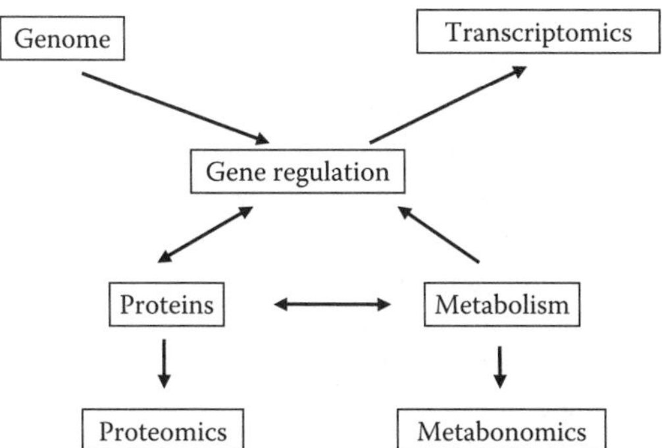

FIGURE 10.2 The relationships between the genome and the technologies for evaluating changes in gene expression (transcriptomics), protein levels (proteomics), and small-molecule metabolite effects (metabonomics). (Reprinted from Lindon, J.C. et al., *Anal. Chem.*, 75, 384A, 2003. With permission.)

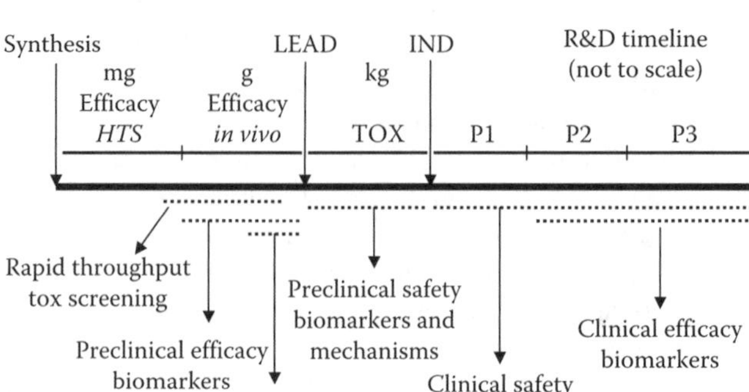

FIGURE 10.3 Schematic of typical drug-development timeline overlaid with points of entry and applications of metabonomics technology. Relative availability of bulk drug is indicated by mass designations (mg, g, kg quantities). (Reprinted from Robertson, D.G., *Toxicol. Sci.*, 85, 809, 2005. With permission.)

However, a number of other biofluids (e.g., bile, cerebrospinal fluid [CSF], or synovial fluid) and cell or tissue extracts can also be used [9,34–36]. In addition, it is generally considered that metabonomics offers an advantage over other "omics" technologies (genomics, transcriptomics, proteomics) because it can provide metabolic profile information on the whole-organism over time. A typical drug-development timeline together with potential opportunities for applying metabonomics technology within that timeline is presented in Figure 10.3 [11].

10.3 BIOMARKERS OF TOXICITY

From a drug safety perspective, there is a substantial interest within the pharmaceutical industry in the discovery of novel and more specific and sensitive markers of toxicity, in particular biomarkers of hepatotoxicity and nephrotoxicity. As mentioned above, the attrition of drug candidates in the development phase due to safety concerns is about 30% (combined preclinical and clinical) and as safety standards are raised, the drug development process has become more challenging [37]. Once a drug is approved and reaches a larger and broader patient population, other safety concerns such as adverse drug reactions arise. Adverse drug reactions are estimated to occur in ~5% of all treated patients. Thus, there is an increased need for more sensitive safety biomarker other than the traditional clinical pathology and hematology biomarkers (e.g., creatinine, urea, glucose) [37,38].

Ozer et al. [39] pointed out that even though all new drug products undergo an extensive preclinical and clinical testing prior to approval by regulatory agencies such as the Food and Drug Administration (FDA) or the European Medicines Agency (EMEA) and prior to being marketed, many drug candidates do not obtain the regulatory approval due to significant organ toxicity in preclinical species. Out of all of those preclinical drug candidates that are discontinued because of organ toxicity, up to half are due to liver toxicity. However, the correlation of the preclinical safety testing results to adverse events in human liver is only about 50%. Thus novel biomarkers of drug-induced liver injury are being explored in order to more accurately predict the potential for human liver toxicity and thus reduce the number of false negative results [39]. Ozer et al. [39] recently summarized current published large molecule biomarker candidates of hepatotoxicity and compared them with serum alanine animotransferase (ALT) activity level, the clinical chemistry gold standard of hepatotoxicity.

In addition to the high interest in the biomarkers of hepatotoxicity, there is also a great interest and need for more sensitive markers of nephrotoxicity because the traditional blood (creatinine, blood urea nitrogen) and urine markers (fractional excretion of sodium, urine osmolality) of kidney injury are insensitive and not specific enough to detect acute kidney injury. Vaidya et al. [40] and Ferguson [41] recently reviewed the most promising serum and urinary large molecule biomarkers for the early detection acute kidney injury (AKI) or acute renal failure (ARF). Some of the proposed biomarkers included clusterin, kidney injury molecule 1(KIM-1), microablumin, and interleukin-18 (IL-18). To date most studies have been focused on the discovery and characterization of several highly promising individual biomarkers. However, in order to translate these potential biomarkers into clinical application future research should include validation of those biomarkers individually and in a multiplexed manner to assess preclinical and clinical sensitivity and specificity [40,41].

One of the earliest hopes for metabonomics in drug safety testing was that it would provide an effective insight into understanding underlying mechanisms of toxicity of a given drug candidate. Being able to differentiate between mechanism-based and nonmechanism-based toxicity can be of great importance because in the early stages of drug development the latter can be potentially avoided by designing another more desirable analogue of a lead compound. To date, examples of this application of metabolomics in the most recent literature are rather limited most likely because of the proprietary nature of the pharmaceutical research. Nevertheless, the examples that have been described in the past indicate that metabonomics has consistently been shown to be a powerful tool in gaining insights on mechanisms of toxicity [4,11,12,42–47]. It is important to point out that metabonomics studies are typically an introduction to gaining insights on a potential mechanism of toxicity involved and are often followed by additional specific studies aimed at understanding the exact mechanisms involved.

Interestingly, the number of examples of toxicity biomarkers discovered using nuclear magnetic resonance (NMR)-based and MS-based metabonomics is quite impressive. In metabonomics studies in which a severe toxicity is observed, almost always the relative urinary concentrations of Krebs cycle intermediates are decreased and a concomitant decrease in urinary levels of hippuric acid is observed. These compounds have been proposed to be nonspecific markers of toxicity as they reflect a combination of complex changes in an organism [11,48–51]. Taurine has been known to be a specific marker for liver toxicity as its urinary levels are typically increased with necrosis and fatty liver. Several bile acids in the serum such as cholic, glycolic, and taurocholic acids have also been demonstrated to be sensitive markers of liver dysfunction [50]. In the literature, there are a few preclinical and clinical examples of using metabonomics in evaluating small molecule biomarkers of liver toxicity. In a very recent preclinical study, Sun et al. [52] investigated acute and chronic acetaminophen-induced hepatotoxicity using NMR- and ultrahigh-pressure liquid chromatography (UPLC)-MS-based metabonomics. A number of metabolites (energy-related metabolites, antioxidants, trigonelline, *S*-adenosyl-*L*-methionine) were depleted in urinary profiles of both groups of male Sprague-Dawley rats administered with acute and chronic doses of acetaminophen, indicating oxidative stress. In addition, results from histopathology and clinical chemistry studies well correlated with the metabolomics data. Based on similarities in patterns of metabolic changes, it was concluded that mechanisms for detoxification and recovery following acute and chronic administration of acetaminophen were also similar.

Yang et al. [53] investigated the usefulness of MS-based metabonomics in predicting liver toxicity in a clinical setting. The authors analyzed serum samples from 37 chronic hepatitis B patients and compared them with serum samples from 50 healthy individuals. A number of potential biomarkers for acute deterioration of liver function in chronic hepatitis B were identified including lysophosphatidylcholine (LPC) C18:0, LPC C16:0, LPC C18:1, LPC C18:2, and glycochenodeoxycholic acid. In another paper, Yang et al. [54] demonstrated that metabonomics methods were successfully used to differentiate between patients with hepatocirrhosis and hepatitis from those with liver cancer. Although the markers that allowed for that differentiation were not fully identified, such an approach showed promise for aiding in the correct and reliable diagnosis of the liver cancer.

To date, several small molecule biomarkers of renal toxicity have been described in the literature. Urine composition is considered to reflect kidney function and pathology. Thus, in the presence of proximal tubule damage, the relative concentrations of a number of metabolites (glucose, lactate, alanine, lysine, glutamine, glutamate, and valine) are markedly increased due to impaired reabsorption of those metabolites. An increase in the relative concentrations of those metabolites in urine is typically accompanied by a corresponding decrease in plasma levels [50,55].

An increase in urinary levels of trimethylamine *N*-oxide (TMAO), dimethylamine (DMA), methylamine (MA), and betain, with a decrease in dimethylglucine, citrate, and 2-oxoglutarate has been showed to be associated with renal papillary toxicity. A reduction in the urinary levels of products of tryptophan metabolism (kynurenic acid and xanthurenic acid) has been associated with perturbed peroxisomal function of proximal tubules [48–50,55,56].

10.4 BIOMARKERS OF EFFICACY

One of the reasons for an interest in metabonomics is the anticipation that global metabolite profiling of samples from diseased patients might lead to identification of new biomarkers of disease, help to understand the biochemical pathways leading to the disease, and thus allow for a more targeted design of new drugs. The idea to use metabolite profiling to diagnose human disease was brought about by studies of inborn errors of metabolism in infants. Millington and colleagues [57,58] pioneered the use of MS-based methods for monitoring fatty acid oxidation, as well as select organic and amino acids. Their work led to universal neonatal screening for metabolism disorders in the state of North Carolina. Vangala and Tonelli [37] pointed out that although in pediatric hospitals metabonomics has been successfully applied to screen and treat newborn babies with inborn errors of metabolism, the pharmaceutical industry has not yet embraced these metabonomics biomarkers for preclinical and clinical drug development.

Claudino et al. [59] recently reviewed current application of metabonomics in cancer research, specifically in clinical diagnosis of a disease, tumor biology characterization, monitoring a response to treatment with anticancer agents, and early prediction of toxicity from anticancer agents. The researchers currently attempt to integrate NMR-based metabonomic technology in their clinical breast cancer research projects, including profiling urine and blood for markers of cardiac toxicity in early breast cancer patients receiving an anthracycline-based adjuvant therapy, as well as searching for NMR-based metabolic profiles that might predict response to therapy in a population of advanced breast cancer patients.

Cancer biomarkers are currently most frequently detected by proteomics approaches in which plasma from cancer patients is analyzed for specific proteins or changes in their levels. In an effort to better understand the potential of using the human urine for the diagnosis of latent urologic malignancies, Kind et al. [60] performed a pilot study to discriminate between urinary samples from patients with renal cell carcinoma (RCC) and healthy individuals. Six urine samples randomly selected from patients with metastatic clear cell RCC and an equal number of randomly selected control urine samples were analyzed by various MS-based metabolomics techniques. Despite a small number of samples in this pilot study, the authors were able to clearly distinguish between RCC patients and normal individuals. Studies to validate those findings using a larger sample size are currently ongoing. In addition, Denkert et al. [61] applied MS-based metabolic profiling to study metabolite patterns in invasive ovarian carcinomas and ovarian borderline tumors, while Di Leo et al. [62] investigated metabolomics techniques as a novel strategy for identifying molecular markers to predict chemotherapy activity and toxicity in breast cancer.

Rozen et al. [63] demonstrated the applicability of metabonomics to detect metabolic signatures of motor neuron diseases (MNDs) such as amyotrophic lateral sclerosis (ALS) in a noninvasive manner. Blood plasma profiles of healthy human subjects were obtained and compared with those from patients with a MND on or off riluzole drug therapy. Further investigation of the signature molecules in ALS and other forms of MND would provide a better understanding of underlying biochemical pathways and perhaps also lead to a discovery of diagnostic markers and targets for drug design in this therapy area.

In another research article, Wang et al. [20] successfully applied LC–MS metabonomic techniques to the metabolite profiling of plasma phospholipids in type 2 diabetes mellitus (DM2) patients. Diabetes mellitus is associated with a metabolic disorder of lipid or fatty acid in phospholipids. The authors were not only able to differentiate between the samples from the DM2 patients and the healthy subjects but also identified a number of phospholipid molecular species that could be used as potential biomarkers for a differentiation of DM2 patients from the healthy individuals and perhaps even an early detection of DM2.

Lewis et al. [64] recently reviewed the application of metabolomics to cardiovascular biomarker and pathway discovery. The authors pointed out that although novel metabolomics techniques suffered from high signal-to-noise issues, the real challenge resided in human interindividual variability. This could be partially overcome by collecting serial samples in patients serving as their own biological controls. However, further testing in large cohorts would be necessary to detect more subtle metabolic changes and to provide robust statistical data on the utility of each marker. The authors provided a clinical study example [65] in which an MS-based metabolomics platform was applied to patients undergoing stress testing with myocardial perfusion imaging. Half of the patients had no evidence of ischemia (control group) while the other half had inducible ischemia (case group). Plasma samples obtained from all patients were profiled by LC–MS. Although the majority of detected metabolites displayed concordant changes in both patient groups, six metabolites yielded statistically significant changes ($p<0.0001$; c-statistic = 0.95). Lewis et al. [64] also pointed out that although plasma and serum offer the advantage of simple sample collection, they reflect the sum of all metabolic changes occurring throughout the body. Sampling specific tissues enables localization of metabolic changes and thus might be helpful in assessing the sensitivity and specificity of signature plasma metabolic profiles.

In an excellent mini-review, Griffin and Kauppinen [66] summarized the recent advances in the field of human brain tumor metabolomics research, in particular advances in high-resolution multinuclear nuclear magnetic resonance spectroscopy (MRS). They provided a list of metabolites including alanine, lactate, myo-inositol, choline-containing metabolites, nucleotides, CH_3 and CH_2 lipids, and polyunsaturated fatty acids that have been commonly identified as perturbed in brain tumors using tissue extract or *in vivo* MRS analysis. Others have made attempts to apply metabonomics techniques to study the etiology of multiple sclerosis in hopes of identifying new therapeutic targets [67], to monitor kidney transplant [67–70], to study aging and caloric restriction [71–73], to evaluate exercise-induced improvements in insulin sensitivity [74], or to study Parkinson's disease [75], but these and similar metabonomics efforts are still primarily in its infancy.

10.5 METABONOMICS ANALYTICAL METHODS

A typical workflow for an untargeted metabonomics analysis is presented in Figure 10.4 [76].

10.5.1 Factors Affecting Metabolic Profiling

When designing a metabonomic experiment it is critical to control as much as possible all the factors that affect the baseline of detected metabolites. These factors include but are not limited to species, strain, gender, diet, age, diurnal variation, changes in gut microflora, lifestyle, and time of sampling [34,50,77–82].

It is well understood that the majority of variation in metabonomic experiments comes from the biological rather than analytical variation [83]. Nevertheless, it is critical that the reproducibility of an analytical method is acceptable and that

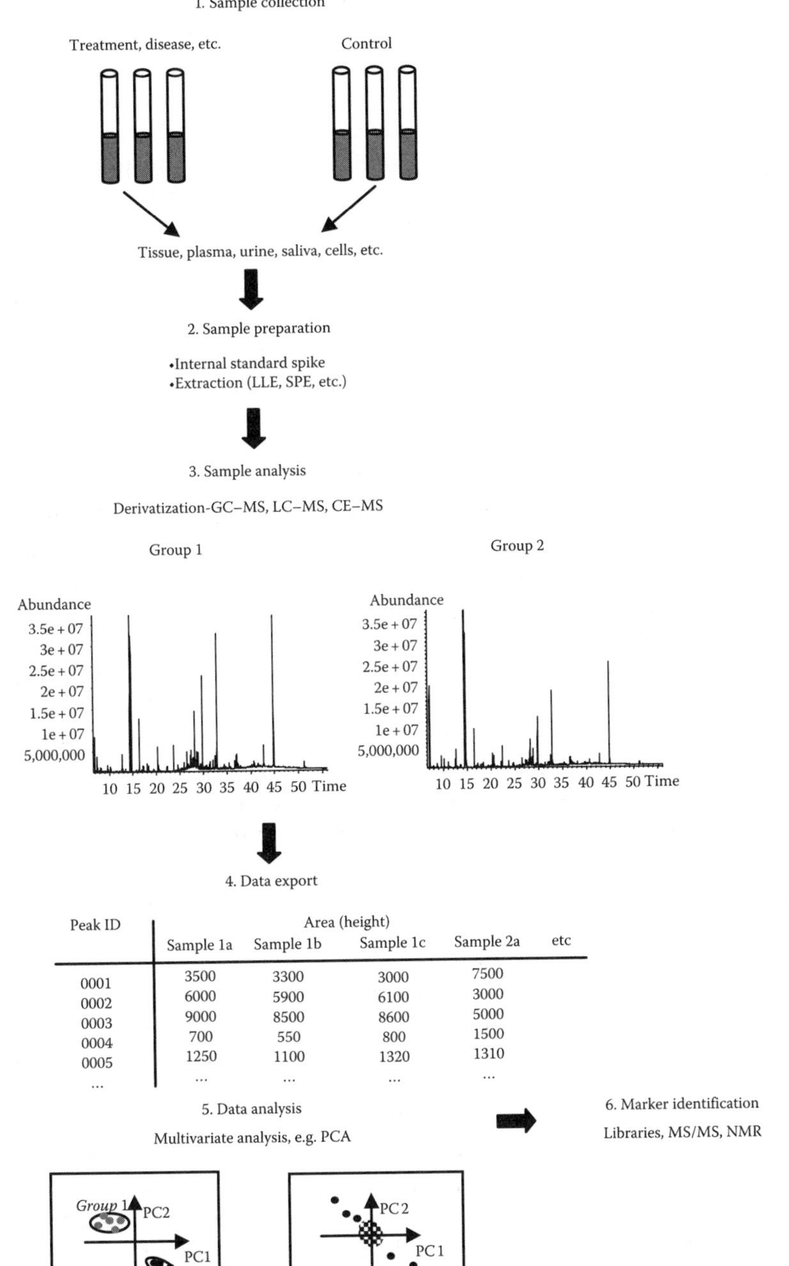

FIGURE 10.4 A typical workflow for an untargeted metabonomics analysis. (Reprinted from Dettmer, K. et al., *Mass Spectrom. Rev.*, 26, 51, 2007. With permission.)

sources of analytical variation such as changes in retention, mass accuracy, or signal intensity are minimized [84]. In addition, since metabonomic studies typically involve sample preparation prior to analysis, it is critical at a minimum to use quality control (QC) samples as much during the sample processing steps as during actual sample analysis. There are no clear guidelines on what should be used as QC samples in metabonomic studies. Some investigators choose to use pooled biological matrix samples as QC samples, as recently proposed by Sangster et al. [85]. In that metabonomic study involving LC–MS and gas chromatography (GC–MS) analyses, all plasma or urine samples were pooled per matrix and split into multiple QC samples, which were then randomly analyzed at the beginning, middle, and end of the entire experiment. An alternative approach to the pooled sample approach is to spike several pseudo-internal standards into samples of interest. Luc et al. [86] demonstrated the use of over 20 stable label isotope compounds with diverse chemistry as internal standards in metabonomics experiments. Whatever the QC samples used, they enable to monitor analytical processes and increase the credibility of metabonomic data sets [78,87].

The largest inherent biological variability is found in human studies because in preclinical studies factors such as strain, gender, or diet can be more easily controlled. Determining the normal variation is crucial for any study whether preclinical or clinical, regardless of the analytical technique used for sample analysis. Saude et al. [88] studied the variation of metabolites in normal human urine and pointed out that, conclusions from a metabonomic study heavily rely upon having a well-defined healthy group with well-characterized metabolite homeostasis. Currently, many Phase I clinical studies recruit healthy volunteers such that the total number of healthy control subjects is limited. Thus, metabolite measurements of the normal cohort are pooled and a certain level of homeostasis in normal urine is assumed to occur. Data generated in this manner for the control subjects are used for comparison with test subjects. A key finding of the study conducted by Saude et al. [88] was that the variance of the pooled metabolite concentrations within the normal population was high, as was the degree of variation of metabolites for individuals over time. Thus, before any comparisons can be made in any clinical or preclinical study between the test and the control groups, it is critical to first define the inter- and intra-individual metabolite variance within the control population. Otherwise, a simple single comparison of control versus test groups may generate misleading results.

When designing a metabonomics study, a great care must be taken so that rigorous protocols are employed to ensure reproducible sampling and that there is an adequate number of biological replicates per group. For example, in a preclinical study involving rodents, one would use 6–8 animals per dose group, while in a clinical study it is desirable to include as many as 20–30 individuals per dose group. In longitudinal studies or studies in which there is a limited number of controls (e.g., in Phase I clinical studies the number of control subjects is typically less than 10), each animal or subject will provide its own predose sample that can be used as its own control [78].

Another important aspect of conducting a metabonomics study, especially in a clinical setting, is to ensure strict control over samples between sample collection and sample receipt. In a clinical study, samples are typically collected from multiple locations all over the country or the world. Therefore, variation in time collection, time exposed to light prior to packaging or actual shipping time, is impossible to control. Shurubor et al. [83] using high-performance liquid chromatography coupled with electrochemical array detector (HPLC–ECD) assessed stability of human plasma metabolome under simulated shipping conditions up to 48 h. The authors demonstrated that most of the metabolites were sufficiently stable for use even after 48 h of exposure to shipping conditions. However, stability of some metabolites differed between individuals, suggesting that there were some biological factors involved that had influence on the stability of samples, a parameter that is typically considered as analytical.

Unlike other "omics" techniques, it is clear that metabonomics is very much an emerging field and new lessons on how to conduct a proper metabonomics study are learned each day. A very positive development is that in an effort to standardize the reporting standards for metabonomics studies, a first draft of a minimum requirement for the description of the biological materials and/or processes examined in a metabolomic study involving mammalian subjects was published in 2007 [89].

10.5.2 Sample Preparation

Sample preparation is one of the most critical steps in a successful metabonomic analysis. The composition of any biological (plasma, urine, bile) or an *in vitro* sample is very complex, and the quantity of metabolites detected depends to a large extent on how the sample is prepared for analysis. In a nontargeted metabolomics experiment, the purpose of sample preparation is to remove any large molecules (proteins, peptides) and salts, while retaining a wide range of small molecule metabolites. The more steps in sample preparation, the narrower will be the chemical diversity of compounds present in the final sample [90].

The most popular methods for extracting biological samples (in particular plasma, urine, and tissue) include solvent extraction with protein precipitation (PP), solid-phase extraction (SPE), and liquid–liquid extraction (LLE). While urine or bile can be directly injected for an LC–MS analysis, plasma must be first extracted to remove proteins. In some cases, it is desirable to clean up urine and bile samples and thus such samples are also most commonly subjected to SPE.

10.5.2.1 Plasma

Shipkova et al. [91] extensively investigated sample preparation approaches for nontargeted LC–MS metabonomic profiling of plasma. She concluded that traditional SPE approaches for extracting plasma lead to significant sample losses. Overall, the most promising procedure for processing plasma yielding the highest intensity of individual components was an extraction with three volumes of cold CH_3OH with 0.1% formic acid at (−20°C) to precipitate proteins. The solvent was evaporated under nitrogen and sample was reconstituted in 200 μL of H_2O:CH_3OH (95:5), vortexed, and centrifuged. A 5 μL aliquot was subjected to an LC–MS analysis. Alternatively, a modified C8 SPE approach could also be used to extract plasma samples. In this case, a plasma sample was first diluted 10-fold with 0.1% formic acid. A 500 μL diluted plasma sample was loaded onto a C8 cartridge and without washing was eluted with 500 μL CH_3OH. The eluent was evaporated under nitrogen and reconstituted in 200 μL of H_2O:CH_3OH (95:5), vortexed, and centrifuged. A 5 μL aliquot was injected for an LC–MS.

10.5.2.2 Urine

Traditionally, it has been established that urine samples should be diluted 1:4 with water prior to an LC–MS analysis. Plumb et al. [92,93] reported obtaining good results from LC–MS analysis for rat urine after a 1:4 dilution with water. Waybright et al. [94] investigated the applicability of this dilution method to human urine which is less concentrated than mouse or rat urine. He concluded that a 1:4 dilution of human urine is not optimum for the detection of lower abundance metabolites. Although, the chromatogram for neat untreated urine was more intense, injection of neat urine on an HPLC column without column washing between injections was not recommended because of contamination of the HPLC column.

Since urine is a complex biological matrix composed of hydrophobic and hydrophilic molecules, salts, proteins, and other compounds, it is useful to remove any of the interfering compounds that may interfere with chromatographic separation. Waybright et al. [94] demonstrated that C18 SPE cartridge was appropriate to extract the hydrophobic fraction of urine without any appreciable losses. A Waters Oasis HPLB (30 mg) cartridge was conditioned with 1 mL of CH_3OH and then with 1 mL of 10 mM ammonium acetate buffer adjusted to pH 4.01. A 500 μL urine sample was loaded onto a cartridge, washed with 1 mL of the buffer and the hydrophobic fraction was eluted with 1 mL of CH_3OH. The eluent was lyophilized at ambient temperature and reconstituted in 500 μL acetonitrile/0.1% formic acid in H_2O (1:1, v:v). A 5 μL aliquot was subjected to LC–MS analysis. Shipkova et al. [91] also demonstrated that SPE of urine showed overall higher recovery of both hydrophilic and hydrophobic fraction than the direct dilution approach. In this case, 50 μL of urine was first diluted 10-fold with 0.1% formic acid. A 500 μL diluted urine sample was loaded onto a C18 cartridge and without washing was eluted with 500 μL CH_3OH. The eluent was dried down under nitrogen and reconstituted in 200 μL of H_2O:CH_3OH (95:5), vortexed, and centrifuged. A 5 μL aliquot was injected for LC–MS analysis. In addition, if a urine sample is to be analyzed by GC–MS, it is recommended to remove excessive urea in urine samples by a urease treatment. High levels of urea can interfere with the analysis and cause an overload of a column and/or detector [76].

10.5.2.3 Tissue

Upon collection, tissues are typically frozen in liquid nitrogen and then homogenized by grinding the frozen sample to a fine powder. Extraction of tissue components is most commonly performed using LLE. Polar compounds are extracted with polar solvents including methanol, ethanol, isopropanol, acetonitrile, water, or polar solvent mixtures, while nonpolar compounds are extracted with ethyl acetate or chloroform. Such an LLE yields two fractions (polar and nonpolar) that can be analyzed separately [76,95,96].

10.5.3 Sample Stability

Saude and Sykes [97] studied the effects of preparation and storage on the stability of human urine for metabolomic studies. Urine samples were collected from healthy male and female subjects and prepared by four different procedures (raw, centrifuged, filtered, and containing the bacteriostatic preservative sodium azide) and analyzed by NMR. The samples were stored at three different conditions at room temperature (22°C), in a refrigerator (4°C), or in a deep-freezer (–80°C) and changes in concentrations of 55 metabolites were monitored every week for one month. The researchers concluded that samples should be stored in a deep freeze (–80°C) and the number of freeze–thaw cycles should be minimized as much as possible. Bacterial contamination of the urine had significant changes on urinary metabolites and could be controlled by either filtration (MS-based analyses) or by the addition of sodium azide (NMR-based analyses).

Gika et al. [98,99] investigated short-term sample stability for human urine by LC–MS. Urine samples were collected from normal female and male volunteers into containers containing one protease tablet and frozen at –80°C or –20°C for one month. Upon defrosting, samples were centrifuged at 13,000 rpm for 10 min and then diluted 1:4 with 0.1% formic acid for LC–MS analysis. All samples were kept at 4°C in an autosampler. LC–MS analyses showed that there were no differences observed between samples stored at –80°C or –20°C even after 1 month. Therefore, storing urine samples at –20°C for at least one month is acceptable. In addition, based on principal component analysis (PCA) the data obtained from samples that had been subjected from one to nine freeze–thaw cycles showed that overall stability of the samples was also acceptable.

The above examples illustrate that it is critical to establish short- and long-term stability, as well as freeze–thaw stability data for any matrix that is being investigated (animal or human plasma, urine, tissue extract, etc.) before any firm conclusions can be drawn from a metabolomics study, especially if the data are to be used for drug development decision-making.

10.5.4 GC–MS

Gas chromatography and more recently two-dimensional GC (GC×GC), when combined with MS detection, typically electron impact (EI) and more recently time-of-flight (TOF), is an excellent choice for nontargeted profiling of metabolites [78,87,100–106]. GC–MS provides very high chromatographic resolution, good sensitivity, and robustness. In addition, one of the reasons to use GC–MS rather than LC–MS for metabonomic analyses is the availability of commercial EI spectral libraries and structure databases that can be used for putative biomarker identification. However, the most practical limitation of GC-based metabolite profiling is the fact that only a limited number of compounds can be analyzed directly. Most samples following extraction need to be derivatized to reduce polarity and achieve thermal stability and volatility. This introduces more variability into an experiment, as well as makes the identification of the original molecule prior to derivatization very difficult [78,101]. The most widely used derivatization procedure involves dissolving a dry extract in pyridine, reacting it with methoxylamine hydrochloride for 90 min at 28°C followed by *N*-methyl-*N*(trimethylsilyl)-trifluoroacetamide (MSTFA) for 30 min at 37°C [60,107]. The first step ensures the conversion of carbonyl groups to oximes in order to stabilize sugars in an open-ring conformation and to prevent decarboxylation of α-ketoacids. In the second step, the active hydrogen in polar groups such as alcohols, amines, and carboxylic acids is replaced with a trimethylsilyl group, which decreases the polarity and increases the thermal stability and volatility of many compounds. In addition, there are a number of other derivatizing agents that can be used as summarized by Dettmer et al. [76]. Using ethyl chloroformate has recently gained popularity especially for the analysis of urine samples because it can not only be used in a nonaqueous medium but can also be used in an aqueous medium [87].

To illustrate an example of how a sample could be analyzed by GC–MS, Kind et al. [60] recently described a GC–MS procedure utilized in their laboratory. Briefly, following derivatization samples were analyzed by GC–MS using an Agilent 6890 N gas chromatograph interfaced with a TOF Pegasus III mass spectrometer. The system was fitted with both an Agilent injector and a Gerstel temperature-programmed injector. The Agilent injector temperature was maintained at 250°C. Injections of 1 μL were made in a split (1:5) mode and chromatography was performed using an Rtx-5Sil MS column (30 m × 0.25 mm i.d., 0.25 μm film thickness) and an Integra-Guard column. Helium gas flow was maintained at 1 mL/min. The GC oven temperature was programmed at 50°C and was ramped 20°C/min to a final temperature of 330°C. Mass spectra were acquired at 20 scans/s over a *m/z* 50–500 mass range.

Alternative GC–MS methods for metabonomic profiling are described in the literature [60,76,78,87,108]. An example of total ion chromatogram (TIC) of a human urine extract analyzed by GC–MS is shown in Figure 10.5 [76]. Figure 10.6 illustrates metabonomics approaches for biomarker screening in plasma, urine, and tissue samples using GC–MS [87].

As mentioned above, GC–MS metabonomic investigations traditionally have relied on GC–EI–MS and more recently TOF mass analyzers have become more popular. Analyses by GC–EI–MS rely primarily on quadrupole instruments, which provide high sensitivity and large dynamic range, but scan speed is low and only nominal mass accuracy can be obtained. The TOF mass analyzers on the other hand offer high mass accuracy and mass resolution in comparison to quadrupole instruments. They also allow for high scan speeds that are necessary for acquiring adequate number of data points across a high-resolution narrow (0.5–1 s wide) chromatographic peak. This is particularly important for two-dimensional GC analyses in which peak capacity is increased but the peak widths are significantly decreased [76,78,109].

10.5.5 LC–MS

LC–MS has only very recently started to be used extensively for metabonomics analyses and the amount of recently published reports in this area is extensive [110–116]. In many ways LC–MS is an ideal technique for metabolic profiling as many biological fluids such as urine and plasma upon PP can be directly injected onto an

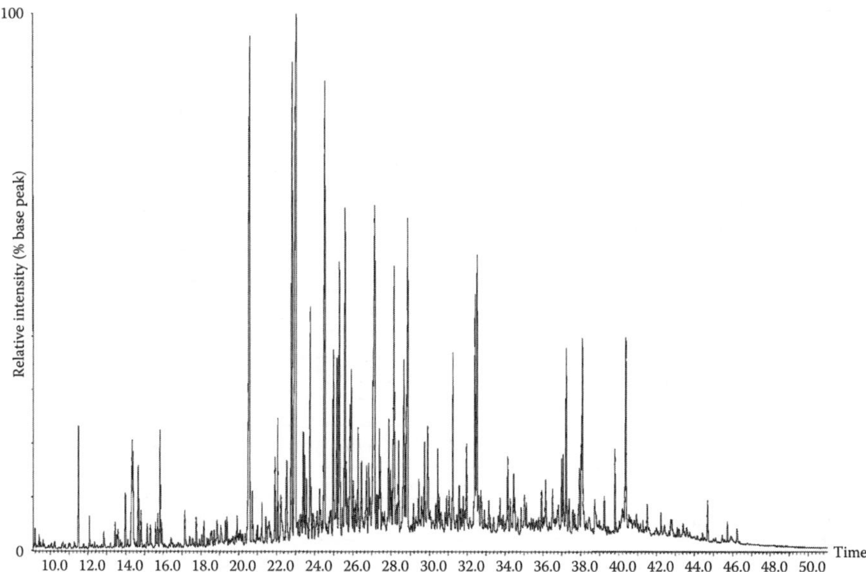

FIGURE 10.5 TIC of a human urine (1 mL) extract after silylation with MSTFA and GC–MS analysis. (Reprinted from Dettmer, K. et al., *Mass Spectrom. Rev.*, 26, 51, 2007. With permission.)

LC column. In addition, LC–MS offers good dynamic range, can be very sensitive depending on a class of compounds, and by generating structural data provides an opportunity to identify potential biomarkers [78].

One of the early most commonly used chromatographic methods for LC–MS-based metabonomics analysis was published by Plumb et al. in 2003 [77]. In this case, a 20 μL aliquot of diluted mouse urine was injected onto a Symmetry C18 column (2.1 × 100 mm, 3.5 μm). The mobile phase consisted of 0.1% formic acid in water (A) and acetonitrile (B). A 10 min linear gradient was used with 0%–20% B over 0.5–4 min, 20%–95% B over 4–8 min, 95% B held for 1 min and at 9 min returned to 100% A. The flow rate was 0.6 mL/min and the column effluent was split with 20% of the flow (120 μL) diverted to a mass spectrometer. Samples were analyzed using Waters Alliance 2795 HPLC system coupled to a Micromass QTof-Micro mass spectrometer equipped with an electrospray source and leucine enkephalin lock-spray. Data were acquired in a full-scan mode from *m/z* 100 to 1400 from 0 to 10 min, in either positive or negative mode. More details on analytical procedures can be found in the publication itself. Other similar LC–MS and UPLC–MS methods have been described in the literature [92–94,98,99,117–120].

Luc et al. [86] evaluated a number of different columns and gradients for LC–MS-based nontargeted metabonomics analyses of rat plasma and urine samples. The authors compared equivalent UPLC and HPLC columns of the same size and with the same packing material. The only difference was in the particle size, which was 1.7 and 3.5 μm for UPLC and HPLC columns, respectively. Luc et al. concluded that although the UPLC peaks had improved peak widths (FWHM) and overall intensity,

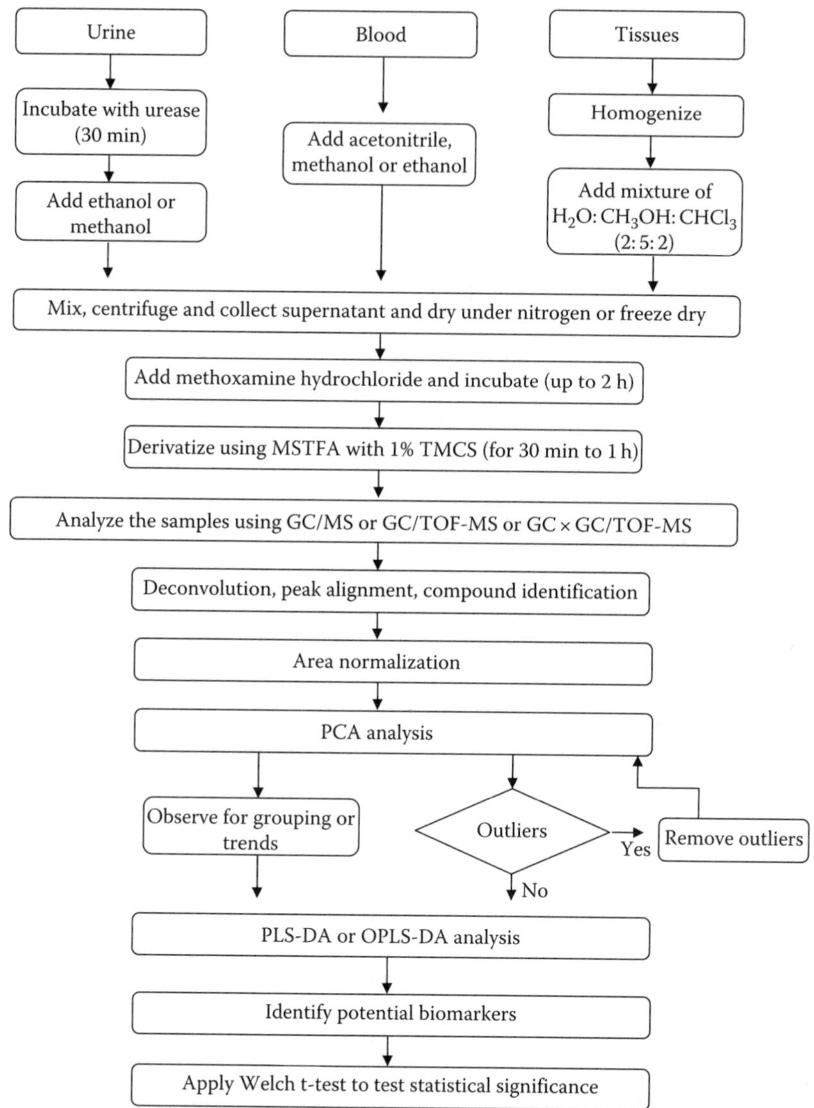

FIGURE 10.6 Metabonomic approaches for biomarker screening in plasma, urine, and tissue samples using GC–MS. (Reprinted from Pasikanti, K.K. et al., *J. Chromatogr. B Analyt. Technol. Biomed. Life Sci.*, 871, 202, 2008. With permission.)

this did not translate into a significant increase in the number of components detected by an LTQ-Orbitrap mass spectrometer operating at 15,000 mass resolution in *m/z* 85–1000 acquisition range. Moreover, step gradients provided much better separation of polar compounds than linear gradients. One of the most successful separation methods included a shallow 10 min gradient (100% A to 80% A in 6 min, 80% A to 5% A in 2 min, and 5% A to 100% A in 1.5 min), where mobile phase A consisted of 0.1% formic acid in water and mobile phase B consisted of 98:2 acetonitrile:water (0.1% formic acid). A Waters UPLC BEH C18 column (2.1 × 100 mm, 1.7 μ) was maintained at 45°C and the flow was adjusted to 0.6 mL/min. In a separate investigation, Waybright et al. [94] also reported that a step elution gradient rather than a linear gradient was the most efficient method for achieving optimal peak resolution when analyzing human urine samples (Figure 10.7).

Nordstrom et al. [121] also compared the differences between UPLC and HPLC for metabolomic profiling. This time a Waters Acquity system was coupled to Micromass Q-Tof-Micro mass spectrometer and data were acquired in the m/z 100–1000 range with an acquisition of ~2 spectra per second. About 20% more components were detected using UPLC versus HPLC and the length of the chromatographic separation was one of the most crucial parameters affecting the number of detected features. These examples demonstrate the importance of appropriate mass analyzers when utilizing UPLC. For more on UPLC see Chapter 8.

Wong et al. [122] published an excellent article on an approach to develop a method for untargeted urinary metabolite profiling using UPLC–QToF MS. The authors studied the effects of varying sample pretreatment and chromatographic

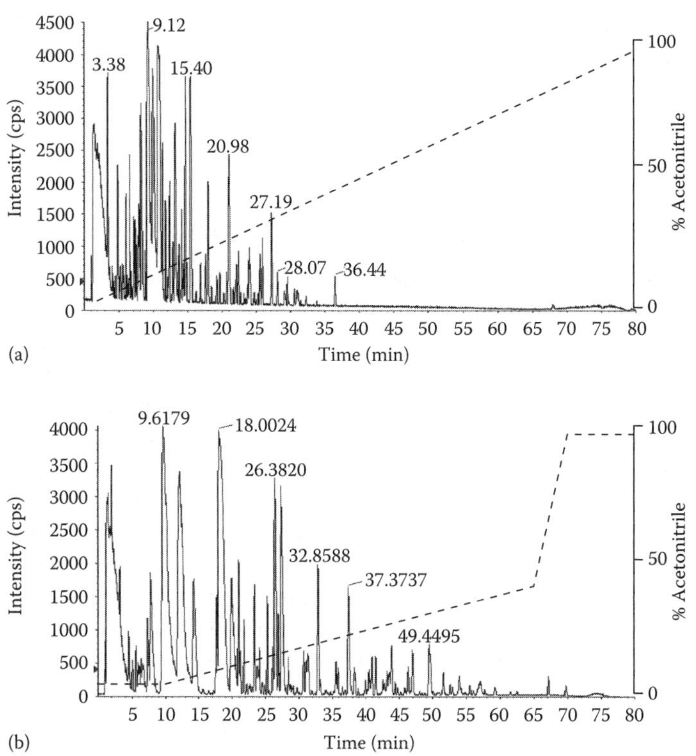

FIGURE 10.7 Chromatographic comparison mobile-phase gradients (dotted line) using a linear (a) vs. a step (b) gradient. (Reprinted from Waybright, T.J. et al., *J. Liquid Chromatogr. Related Technol.*, 29, 2475, 2006. With permission.)

conditions on metabolite profiles. As in the case of Luc's investigation [86], the step elution gradient yielded the highest number of components and the best chromatographic resolution. In addition, untreated filtered urine yielded the highest number of components but dilution with methanol provided more homogenous metabolic profiles with less variation in the number and intensity of the detected components.

Gika et al. [119] investigated the use of high-temperature UPLC–MS for metabonomics analysis. The studies showed that for rat urine, but not plasma, chromatography at elevated temperatures provided better results (higher peak capacity and better peak asymmetry) than conventional reverse-phase HPLC. Burton et al. [123] recently published a paper on the sources and appearance of artifacts that are inherent to the analytical technique in LC–MS metabolomics experiments and that cannot be easily avoided. They can however be detected using statistical tools such as multivariate analysis (MVA). Guy et al. [124] also in a very recent paper discussed the importance of method validation for metabolomic profiling of urine samples using UPLC–TOFMS.

An interesting approach to separation of highly polar compounds by hydrophilic interaction chromatography (HILIC) was described by Kind et al. [60]. Raw human urine samples were mixed with an equal amount of acetonitrile and centrifuged prior to injection onto a normal phase Luna NH_2 column (150 × 3 mm, 3 μm). Mobile phase A consisted of acetonitrile and mobile phase B consisted of 13 mM ammonium acetate adjusted to pH 9.1 with ammonium hydroxide. The separation gradient included holding 20% B for the first 5 min, increasing a gradient to 35% B over 20 min, followed by a gradient to 90% B over 30 min. The flow rate was 0.5 mL/ min and column was maintained at 40°C.

10.5.6 Ionization Techniques and Mass Analyzers

Once components of a biological sample are separated on a chromatographic column, they must be ionized for an analysis in a mass spectrometer. In GC, samples are vaporized and then ionized by EI or chemical ionization (CI). Want et al. [125] point out that the advantages of EI include good sensitivity and unique fragmentation. However, since often molecular ions cannot be detected because of fragmentation, this poses a problem for identifying unknown compounds. In addition, only compounds with relatively small molecular weights can be analyzed because of the requirement for thermal desorption. CI is a much softer ionization technique and thus molecular ions can be generated. Nevertheless, a compound still needs to be vaporized to be ionized.

For metabolic profiling using LC–MS, samples are ionized via atmospheric pressure ionization (API) techniques, most commonly electrospray ionization (ESI), although other ionization techniques such as atmospheric pressure chemical ionization (APCI) and atmospheric pressure photoionization (APPI) have also been reported but they are not widely used for metabonomic analyses. In LC–ESI–MS-based metabonomics experiments, samples are typically ionized in positive and negative ionization modes. Waybright et al. [94] reported that a greater number of chromatographic peaks were identified in the early portion of a typical chromatogram in the positive ionization mode, while the negative ionization mode provided more chromatographic peaks in the later portions of the chromatogram, indicating unique differences between compounds detected in the two ionization modes (Figure 10.8).

Along with advances in various ionization sources, significant improvements have been made in the area of mass analyzers. Mass analyzers can be differentiated based on several attributes such as scan speed, duty cycle, mass resolution, mass range, and cost [126]. The most common analyzers used for metabonomics analyses include the quadrupole and TOF-based analyzers [125–127]. Some other analyzers that have been reported for use in MS-based metabonomics analyses are the ion traps, Orbitraps, and Fourier transform mass spectrometers [128,129].

Quadrupole mass filters are one of the most common and cost effective mass analyzers. Although they have limited mass resolution (Table 10.2) and are less sensitive than other mass analyzers, they are durable and suitable for high-throughput analyses. To perform a tandem mass analysis (MS/MS), a triple quadrupole MS is used in which three quadrupoles are placed in series to select, fragment, and analyze ions of interest. Quadrupole ion-trapping devices such as linear 2D ion traps are

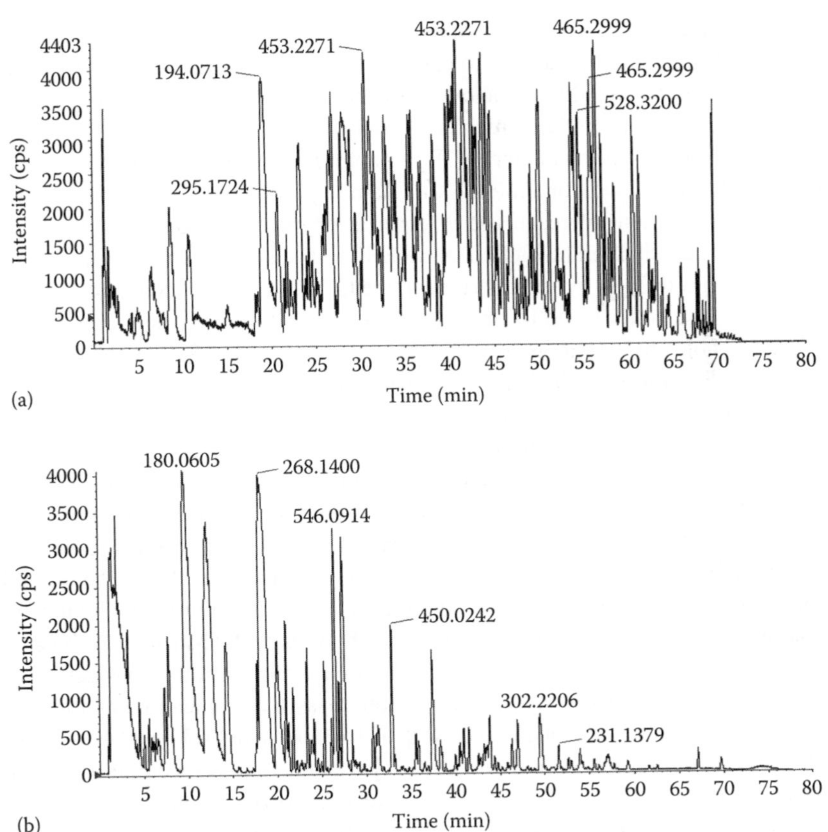

FIGURE 10.8 Raw urine chromatograms using a C18 column, step gradient, and negative (a) and positive (b) MS ionization modes. (Reprinted from Waybright, T.J. et al., *J. Liquid Chromatogr. Related Technol.*, 29, 2475, 2006. With permission.)

TABLE 10.2
Typical Values for Mass Resolving Power and Mass Accuracy of Commercially Available Analyzers

Analyzers	Mass Resolving Power[a]	Mass Accuracy
Triple quadrupoles and ion traps	~500	~0.3–1 Da
TOF devices	8,000–20,000	1–5 ppm[b]
LTQ-Orbitrap	Up to 100,000	<3 ppm
FTICR	Up to 1,000,000[c]	<1 ppm

Source: Reprinted from Werner, E. et al., *J. Chromatogr. B*, 871, 143, 2008. With permission.

[a] m/Δm, FWHM.

[b] When internal calibration or lock mass is used.

[c] Higher value can be achieved depending on the magnetic field intensity.

good alternatives to conventional quadrupole systems. They offer more sensitivity in the full-scan mode and the ability to carry out MS/MS analyses as with triple quadrupole instruments. In addition, a low mass cut-off can be now reduced by utilizing pulsed-Q-dissociation (PQD), which is a novel fragmentation mechanism available with the LTQ mass spectrometer. TOF is another mass analyzer most commonly used for MS-based metabonomics. As its name applies it sorts ions based on the time it takes for them to reach the detector after receiving equal kinetic energy. TOF offers relatively high resolution, extremely fast scanning capabilities, and mass accuracy (Table 10.2). A hybrid instrument, the Q-TOF combines a quadrupole mass filter for mass selection, a collision cell, and a TOF MS for analysis. This makes it an obvious choice for metabonomics analyses since accurate MS/MS spectra can be acquired [125]. However, Q-TOF instruments are limited by the properties of the time-to-digital converter detector that is only able to record one ion per dead time. Intense ion signals become saturated leading to distortions in the mass-peak shape and therefore deviations in mass accuracy. Recently some improvements have been made by extending dynamic range and introducing an online mass lock spray [90].

Fourier transform ion cyclotron mass spectrometry (FT-ICR-MS) instrumentation offers excellent sensitivity, accuracy (<1 ppm), and high mass resolution (>1,000,000) (Table 10.2). However, because of being too expensive, difficult to use, and not compatible with conventional HPLC columns and flow rates, FTMS has not been frequently used in pharmaceutical research. This changed with an introduction of a hybrid instrument consisting of a linear ion-trap mass spectrometer compatible with LC and an ion-cyclotron-resonance (ICR) detector. Such a hybrid instrument is compatible with conventional HPLC and allows for acquisition of accurate mass data-dependent MS^n spectra. Sanders et al. [128] recently reviewed the utility of hybrid LTQ-FTMS for drug metabolism and metabonomics applications while Brown et al. [129] reviewed the metabolomics applications of FT–ICR–MS.

The Orbitrap is a new FT–MS instrument that has a more modest performance in comparison to the FT–ICR–MS. It achieves a maximum >100,000 resolution and

<3 ppm mass accuracy (Table 10.2). It is very stable and provides accurate mass measurements (<2 ppm error) for over 24 h post-calibration [90,130]. Because ions are trapped in an electrostatic field rather than a superconducting magnet, the Orbitrap is also less expensive than FT–ICR–MS. A hybrid instrument that couples a linear ion trap to an Orbitrap was recently introduced. Makarov et al. [131] described its performance in detail. In a recent article Dunn et al. [132] reported the advantages of coupling a UPLC system with the LTQ-Orbitrap mass spectrometer for metabolite profiling of serum samples.

10.6 DATA PROCESSING AND ANALYSIS

In a GC–MS or LC–MS metabonomics experiment each file represents a single biological sample. The basic goal of data processing is to transform raw data files to a standard and uniform format so that statistical analyses can be carried out. A variety of instrument vendors utilize different proprietary data formats. Thus, the first step in data processing requires a conversion of such raw proprietary data into common raw data format such as ASCII text or binary netCDF. More recently, a universal mzXML format has become more popular. Many vendor instrument software packages contain scripts to allow for such a file conversion [76].

A typical data processing workflow includes data filtering, feature detection, data alignment, and normalization, but not always necessarily in this order depending on a software package used. Filtering methods allow for removal of effects such as chemical and random noise or a baseline. The purpose of feature detection is to identify all true ion signals and characterize them as so-called features that include *m/z*, retention time, and intensity for each ion. This step provides quantitative information on a given ion that can be then compared across all samples. Alignment is performed to correct retention time differences between runs and to combine data from different samples. Depending on the software used, alignment is sometimes performed prior to the feature detection step. Normalization is intended to remove undesirable systematic bias in ion intensities between measurements and at the same time retain the biological variation that is of interest. This is a very challenging task because of the number of chemically diverse molecules in a metabonomics experiment, differences in extraction recoveries for those compounds, as well as differences in their ionization efficiencies in a mass spectrometer. There are two major approaches to data normalization including using statistical models to derive optimal scaling factors for each samples (e.g., median of intensities) or using a number of internal or external standards (e.g., ion intensity or peak area) [76,133–137]. There are a number of commercial and freely available software packages for processing of MS-based metabonomics data [32,76,134–136,138–141]. Some of the most widely used software tools are summarized in Table 10.3.

An in-depth review of statistical methods for metabonomic data analysis is beyond the scope of this chapter. Briefly, there are a few main approaches to data analysis. Examples of multivariate data analyses include the so-called unsupervised analyses such as PCA, independent component analysis (ICA), and hierarchical clustering analysis (HCA), while partial least square differential analysis (PLS-DA) is

TABLE 10.3
Commercial and Free Software Tools for MS Metabonomics Data Processing

Name	Vendor or License Type
Commercial	
BlueFuse	BlueGnome, Cambridge, United Kingdom
Genedata Expressionist	Genedata, Basel, Switzerland
MarkerLynx	Waters, Milford, MA, United States
MarkerView	Applied Biosystems, Foster City, CA, United States
MassHunter	Agilent Technologies, Santa Clara, CA, United States
Metabolic Profiler	Bruker Daltonic & Bruker BioSpin, Billerica, MA, United States
metAlign	PlanResearch International B.V., Wageningen, the Netherlands
MS Resolver	Pattern Recognition Systems, Bergen, Norway
Rosetta Elucidator	Rosetta Biosoftware, Seattle, WA, United States
SIEVE	Thermo Fisher Scientific, Waltham, MA, United States
Free	
COMSPARI	GNU General Public License
MathDAMP	Free
MET-IDEA	Free
MZmine	GNU General Public License
SpecArray	GNU General Public License
XCMS	GNU General Public License

Source: Adapted from Katajamaa, M. and Oresic, M., *J. Chromatogr. A*, 1158, 318, 2007. With permission.

an example of one of the supervised methods. Typically, the unsupervised methods are used initially for sample classification or if sample identity is unknown. The supervised methods are used for searching for specific biomarkers, for example, biomarkers of disease, by comparing samples from healthy subjects and patients. Alternatively, when a purpose of a study is to simply discover general biomarkers common statistical methods such as ANOVA are often utilized [15,76,117,125,142–145]. An example of applying PLS-DA to an analysis of data obtained from UPLC–Q-TOF MS profiling of human urine samples from healthy, insulin-sensitive, and insulin-resistant subjects is presented in Figure 10.9 [146].

10.7 METABOLITE IDENTIFICATION AND STRUCTURE ELUCIDATION

Various approaches to metabolite identification and structure elucidation have been extensively described in the literature [78,90,101,107,109,125,127,146–151]. Fully identifying unknown metabolites in a metabonomics study remains a great challenge even today. For many metabonomics studies the most frequent goal is to understand biochemical pathway interactions or general metabolic changes and thus to achieve that goal it is critical to unequivocally characterize unknown putative diagnostic biomarkers. Otherwise, such studies are of no consequence. It is often perceived that mass accuracy is the most important parameter in this identification process

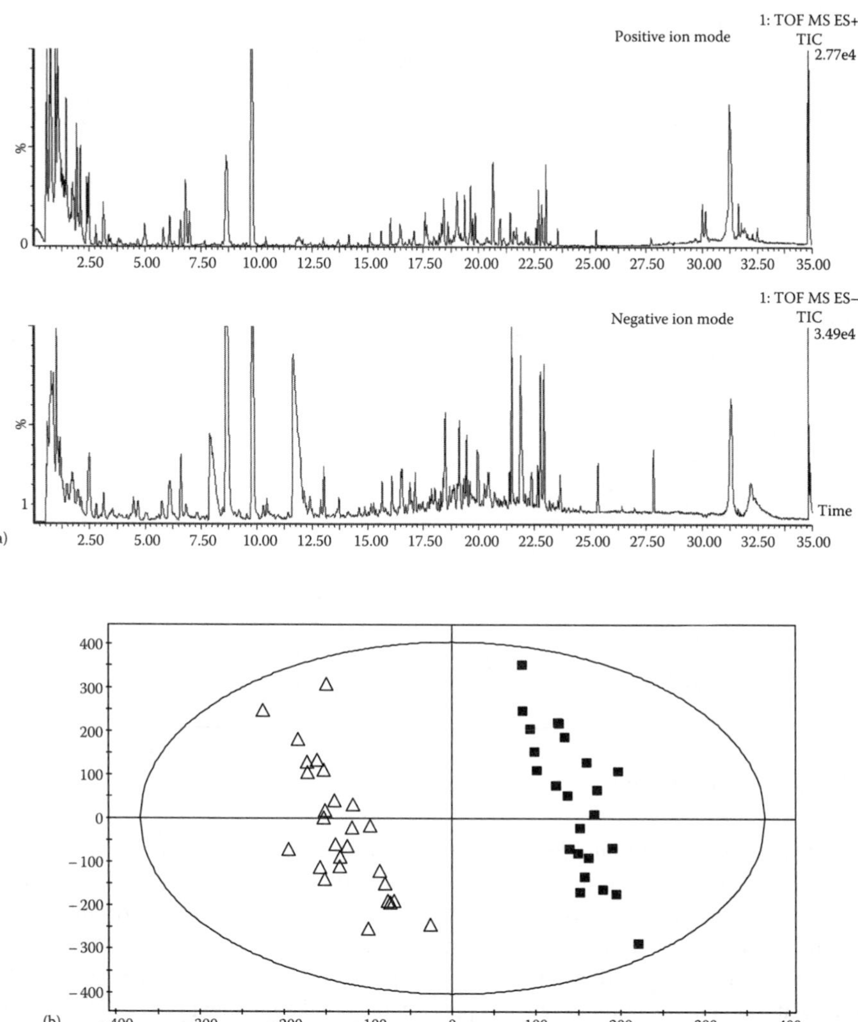

FIGURE 10.9 (a) Representative reversed-phase UPLC-Q-TOF MS TICs of a urine sample analyzed in positive and negative ion modes; (b) PLS-DA scores plot (OSC filtered) of healthy, insulin-sensitive subjects (Δ) and prediabetic, insulin-resistant individuals (■); (c) corresponding PLS-DA loading plot. The variables are labeled with *m/z*, and *m/z* 194.34 is labeled by an arrow; and (d) differences in the TIC peak height of *m/z* between insulin-sensitive subjects (Δ) and insulin-resistant individuals (■). $p < 0.05$ vs. insulin-resistant individuals analyzed by the Wilcoxon rank sum test. (Reprinted from Chen, J. et al., *Anal. Chem.*, 80, 1280, 2008. With permission.)

(*continued*)

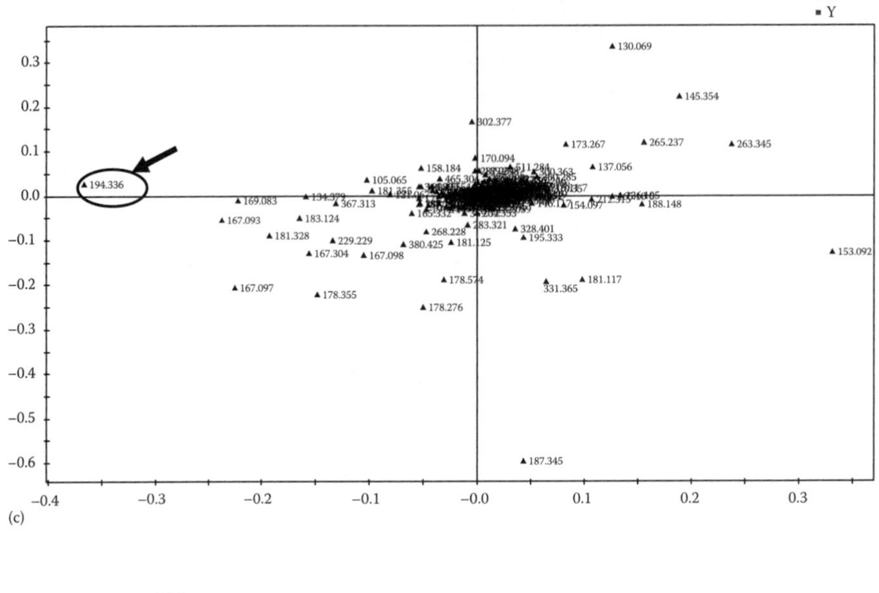

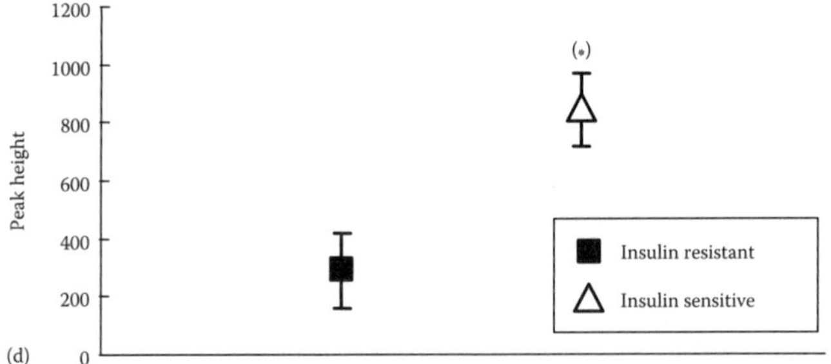

FIGURE 10.9 (continued)

because it can provide an elemental composition for an unknown. However, even with ultrahigh-resolution instruments, like the Orbitrap, there are still a number of elemental compositions possible for a given compound. The most important step in the molecular formula finding is applying constraints that reduce the number of possible elemental compositions. The two most popular constraints or rules include the nitrogen rule and the number of rings plus double bond equivalents (RDBE) rule. The first one states that odd nominal molecular mass compounds contain an odd number of nitrogen atoms. However, as Werner et al. [127] point out, this rule becomes unreliable for masses above 500 Da because of the large number of elements contributing to the mass and thus because of an increased mass defect (above 0.5 Da), it is possible to overestimate the integer mass by 1 Da. The RDBE rule allows one to calculate the number of rings and double bonds in a molecule from the number of C, H, and N atoms a molecule contains. One of the most important, but often overlooked tools to reduce the number of possible molecular formulas, is the analysis of isotopic patterns in a mass spectrum. Kind and Fiehn [148] showed that by using isotopic abundance patterns as the only constraint, it is possible to remove over 95% of false molecular formula candidates. For this reason, the use of instruments like the Orbitrap or FT–ICR becomes of great interest, because these high-resolution instruments allow for optimum resolution of isotope peaks. In addition, Kind and Fiehn [149] also proposed an algorithm for filtering molecular formulas. The algorithm is based on seven heuristic rules which include restriction for the number of elements, LEWIS and SENIOR chemical rules, isotopic patterns, hydrogen to carbon ratios, element ratio of N, O, P, and S versus C, element ratio probabilities, and presence of trimethylsilylated compounds. These rules have been implemented in an automated script within the Excel program.

Once a molecular formula is correctly assigned, tandem MS experiments can be used to determine the structure of unknown metabolites. As with determining of molecular formula for a molecular ion, high mass accuracy data for the fragment ions are often critical to efficiently characterize a metabolite. It is important to point out that mass spectrometry alone cannot distinguish between stereoisomers and thus in some cases additional spectroscopic analyses such as NMR or additional experiments (e.g., H/D exchange) may need to be carried out for complete identification of a compound. In addition, it is worthwhile mentioning that criteria for identifying nonnovel versus novel compounds differ significantly. Many compounds detected in a metabonomics experiment are nonnovel and thus can be identified based on co-characterization with an authentic standard. As Bedair and Sumner [109] point out a guideline for the identification of nonnovel metabolites has been recently proposed by the Chemical Analysis Working Group. The guideline states that a minimum of two independent and orthogonal data relative to an authentic standard analyzed under identical experimental conditions are needed for metabolite identification. Examples of such data include for example accurate mass and tandem MS, or mass and NMR spectra. Identifications performed without authentic standards based solely on comparisons to spectra available in spectral libraries are clearly insufficient to prove identity of a nonnovel metabolite. For novel metabolites, a standard array of physicochemical properties must be determined, including elemental analysis, accurate mass, UV/IR and/or NMR analysis, melting or boiling point, and optical rotation.

Moreover, there are a number of publicly available free and commercial spectral (e.g., MS- and NMR-based), chemical, and biochemical/metabolic pathway databases that can be useful in identification and characterization of metabolites, as well as interpretation of metabonomics data at a biochemical pathway level. Examples of such databases are listed in Table 10.4. In a very recent publication, Loftus et al. [150] provided an example in which unknown analytes in complex biological matrices can be detected and characterized using high mass accuracy ESI MS^n and formula prediction software, as well as by comparison to mass spectral databases, rather than by following the standard identification route via comparison to an authentic standard. This seems to be an attractive and a promising alternative for future profiling studies.

TABLE 10.4
Databases for Identification and Metabonomic Data Interpretation

Database	Source
MS-Based	
Fiehn Library	Fiehn Laboratory Univ California Davis
Golm Metabolome Database (GMDM@CSB.DB)	Max Planck Institute of Molecular Plant Physiology
Human Metabolome Database (HMDB)	Genome Alberta and Genome Canada
KNApSAcK (Comprehensive Species-Metabolite Relationship Database)	Nara Institute of Science and Technology (NAIST)
MassBank	Keio University, RIKEN PSC (Japan), and others
Metabolome of Tomato Database (MoTo DB)	Plant Research International
Metlin	The Scripps Research Institute
NIST/EPA/NIH Mass Spectral Library (NIST 0.5)	National Institute of Standards and Technology (NIST)
SpecInfo	Daresbury Laboratory
Spectral Database for Organic Compounds, SDBS	National Institute for Advanced Industrial Science and Technology (AIST)
Chemical	
Chemical Entities of Biological Interest (ChEBI)	European Bioinformatics Institute (United Kingdom)/European Molecular Biology Lab
Chemical Structure Lookup Service (CSLS)	CADD Lab, Med. Chem. NCI, NIH
ChemFinder	Cambridge Soft
Merck Index	John Wiley & Sons, Inc.
LipidBank	Japanese Conference on the Biochemistry of Lipids (Japan)
PubChem	National Institutes of Health (NIH)
SciFinder	Chemical Abstract Service (CAS)
Pathways	
Biochemical, Genetic and Genomic (BiGG) Database	University of California
BioCyc (HumanCyc, MetaCyc)	SRI International
Kyoto Encyclopedia of Genes and Genomes (KEGG)	Kyoto University/ Tokyo University
LIPID Metabolites and Pathway Strategy (LIPID MAPS)	LIPID MAPS Bioinformatics Core

Sources: Adapted from Moco, M. et al., *Trends Anal. Chem.*, 26, 855, 2007; Werner, E. et al., *J. Chromatogr. B*, 871, 143, 2008. With permission.

10.8 CONCLUSIONS

It is quite clear that the pharmaceutical industry has seriously embraced the concept of utilizing biomarkers to accelerate the drug discovery and development process.

MS-based metabonomics as well as other "omics" technologies have been explored for this purpose. However, the extent of these efforts varies widely depending on the focus of each pharmaceutical company. The primary reason for not utilizing the metabonomics technology, whether for discovering novel biomarkers of safety, efficacy, or target engagement, is the uncertainty of what the data mean and how they can be used to mitigate financial risk. In addition, metabonomics is a relatively young science with the analytical platform changing and improving rapidly. From a regulatory perspective, metabonomics together with other "omics" technologies have shown promise to reduce the development time of drugs by aiding in the selection of drug candidates or by altering the manner in which drugs are investigated and approved. However, the full potential of metabonomics and other associated technologies is yet to be realized [152]. It is likely that over time, metabonomics will evolve to find a wider application in the pharmaceutical research and development process.

ACKNOWLEDGMENT

The author thanks Dr. Swapan Chowdhury and Mr. Kevin Alton for reviewing this chapter and for their valuable comments.

REFERENCES

1. Baker, M., In biomarkers we trust? *Nat. Biotechnol.*, 23(3), 297, 2005.
2. Wagner, J.A., Strategic approach to fit-for-purpose biomarkers in drug development, *Annu. Rev. Pharmacol. Toxicol.*, 48, 631, 2008.
3. Kola, I. and Landis, J., Can the pharmaceutical industry reduce attrition rates?, *Nat. Rev. Drug Discov.*, 3(8), 711, 2004.
4. Mortishire-Smith, R.J. et al., Use of metabonomics to identify impaired fatty acid metabolism as the mechanism of a drug-induced toxicity, *Chem. Res. Toxicol.*, 17(2), 165, 2004.
5. Marrer, E. and Dieterle, F., Promises of biomarkers in drug development—a reality check, *Chem. Biol. Drug Des.*, 69(6), 381, 2007.
6. Cummings, J. et al., Biomarker method validation in anticancer drug development, *Br. J. Pharmacol.*, 153(4), 646, 2008.
7. Lock, E.A. and Bonventre, J.V., Biomarkers in translation: Past, present and future, *Toxicology*, 245(3), 163, 2008.
8. Collings, F.B. and Vaidya, V.S., Novel technologies for the discovery and quantitation of biomarkers of toxicity, *Toxicology*, 245(3), 167, 2008.
9. Lindon, J.C., Holmes, E., and Nicholson, J.K., Metabonomics techniques and applications to pharmaceutical research & development, *Pharm. Res.*, 23(6), 1075, 2006.
10. Nicholson, J.K. et al., Metabonomics: A platform for studying drug toxicity and gene function, *Nat. Rev. Drug Discov.*, 1(2), 153, 2002.
11. Robertson, D.G., Metabonomics in toxicology: A review, *Toxicol. Sci.*, 85(2), 809, 2005.
12. Robertson, D.G., Reily, M.D., and Baker, J.D., Metabonomics in preclinical drug development, *Expert Opin. Drug Metab. Toxicol.*, 1(3), 363, 2005.
13. Horning, E.C., Metabolic profiles: Gas-phase methods for the analysis of metabolites, *Clin. Chem.*, 17(8), 802, 1971.

14. Lindon, J.C. et al., Contemporary issues in toxicology the role of metabonomics in toxicology and its evaluation by the COMET project, *Toxicol. Appl. Pharmacol.*, 187(3), 137, 2003.
15. Kaddurah-Daouk, R., Kristal, B.S., and Weinshilboum, R.M., Metabolomics: A global biochemical approach to drug response and disease, *Annu. Rev. Pharmacol. Toxicol.*, 48, 653, 2008.
16. van Ravenzwaay, B. et al., The use of metabolomics for the discovery of new biomarkers of effect, *Toxicol. Lett.*, 172(1–2), 21, 2007.
17. Want, E.J. et al., Phospholipid capture combined with non-linear chromatographic correction for improved serum metabolite profiling, *Metabolomics*, 2(3), 145, 2006.
18. Yetukuri, L. et al., Bioinformatics strategies for lipidomics analysis: Characterization of obesity related hepatic steatosis, *BMC Syst. Biol.*, 1, 12, 2007.
19. Ogiso, H., Suzuki, T., and Taguchi, R., Development of a reverse-phase liquid chromatography electrospray ionization mass spectrometry method for lipidomics, improving detection of phosphatidic acid and phosphatidylserine, *Anal. Biochem.*, 375(1), 124, 2008.
20. Wang, C. et al., Plasma phospholipid metabolic profiling and biomarkers for type 2 diabetes mellitus based on high-performance liquid chromatography/electron mass spectrometry and multivariate statistical analysis, *Anal. Chem.*, 77(13), 4108, 2005.
21. Roberts, L.D. et al., A matter of fat: An introduction to lipidomic profiling methods, *J. Chromatogr. B Analyt. Technol. Biomed. Life Sci.*, 871(2), 174, 2008.
22. Lindon, J.C., Holmes, E., and Nicholson, J.K., So what's the deal with metabonomics? *Anal. Chem.*, 75(17), 384A, 2003.
23. Lindon, J.C., Holmes, E., and Nicholson, J.K., Metabonomics and its role in drug development and disease diagnosis, *Expert Rev. Mol. Diagn.*, 4(2), 189, 2004.
24. Lindon, J.C., Holmes, E., and Nicholson, J.K., Metabonomics in pharmaceutical R&D, *FEBS J.*, 274(5), 1140, 2007.
25. Lindon, J.C. and Nicholson, J.K., Analytical technologies for metabonomics and metabolomics, and multi-omic information recovery, *Trends Anal. Chem.*, 27(3), 194, 2008.
26. Griffin, J.L., The potential of metabonomics in drug safety and toxicology, *Drug Discov. Today Tech.*, 1(3), 285, 2004.
27. Griffin, J.L. and Bollard, M.E., Metabonomics: Its potential as a tool in toxicology for safety assessment and data integration, *Curr. Drug Metab.*, 5(5), 389, 2004.
28. Keun, H.C. and Athersuch, T.J., Application of metabonomics in drug development, *Pharmacogenomics*, 8(7), 731, 2007.
29. Kramer, J.A., Sagartz, J.E., and Morris, D.L., The application of discovery toxicology and pathology towards the design of safer pharmaceutical lead candidates, *Nat. Rev. Drug Discov.*, 6(8), 636, 2007.
30. Ackermann, B.L., Hale, J.E., and Duffin, K.L., The role of mass spectrometry in biomarker discovery and measurement, *Curr. Drug Metab.*, 7(5), 525, 2006.
31. Schlotterbeck, G. et al., Metabolic profiling technologies for biomarker discovery in biomedicine and drug development, *Pharmacogenomics*, 7(7), 1055, 2006.
32. Robertson, D.G., Reily, M.D., and Baker, J.D., Metabonomics in pharmaceutical discovery and development, *J. Proteome Res.*, 6(2), 526, 2007.
33. Dieterle, F. et al., Application of metabonomics in a compound ranking study in early drug development revealing drug-induced excretion of choline into urine, *Chem. Res. Toxicol.*, 19(9), 1175, 2006.
34. Ritchie, S., Comprehensive metabolomic profiling of serum and cerebrospinal fluid: Understanding disease, human variability, and toxicity, in *Surrogate Tissue Analysis*, 1st edition, Burczynski, M.E., and Rockett, J.C., Eds., CRC Press LLC, Boca Raton, FL, 2006, p. 165.

35. Wikoff, W.R. et al., Metabolomic analysis of the cerebrospinal fluid reveals changes in phospholipase expression in the CNS of SIV-infected macaques, *J. Clin. Invest.*, 118(7), 2661, 2008.
36. Bobeldijk, I. et al., Quantitative profiling of bile acids in biofluids and tissues based on accurate mass high resolution LC-FT-MS: Compound class targeting in a metabolomics workflow, *J. Chromatogr. B Analyt. Technol. Biomed. Life Sci.*, 871(2), 306, 2008.
37. Vangala, S. and Tonelli, A., Biomarkers, metabonomics, and drug development: Can inborn errors of metabolism help in understanding drug toxicity? *AAPS J.*, 9(3), E284, 2007.
38. Farkas, D. and Tannenbaum, S.R., In vitro methods to study chemically-induced hepatotoxicity: A literature review, *Curr. Drug Metab.*, 6(2), 111, 2005.
39. Ozer, J. et al., The current state of serum biomarkers of hepatotoxicity, *Toxicology*, 245(3), 194, 2008.
40. Vaidya, V.S., Ferguson, M.A., and Bonventre, J.V., Biomarkers of acute kidney injury, *Annu. Rev. Pharmacol. Toxicol.*, 48, 463, 2008.
41. Ferguson, M.A., Vaidya, V.S., and Bonventre, J.V., Biomarkers of nephrotoxic acute kidney injury, *Toxicology*, 245(3), 182, 2008.
42. Espina, J.R. et al., Detection of in vivo biomarkers of phospholipidosis using NMR-based metabonomic approaches, *Magn. Reson. Chem.*, 39(9), 559, 2001.
43. Slim, R.M. et al., Effect of dexamethasone on the metabonomics profile associated with phosphodiesterase inhibitor-induced vascular lesions in rats, *Toxicol. Appl. Pharmacol.*, 183(2), 108, 2002.
44. Clayton, T.A. et al., An hypothesis for a mechanism underlying hepatotoxin-induced hypercreatinuria, *Arch. Toxicol.*, 77, 208, 2003.
45. Robertson, D.G. et al., Metabonomic assessment of vasculitis in rats, *Cardiovasc. Toxicol.*, 1(1), 7, 2001.
46. Zhang, J. et al., Mechanisms and biomarkers of cardiovascular injury induced by phosphodiesterase inhibitor III SK&F 95654 in the spontaneously hypertensive rat, *Toxicol. Pathol.*, 34(2), 152, 2006.
47. Niemann, C.U. and Serkova, N.J., Biochemical mechanisms of nephrotoxicity: Application for metabolomics, *Expert Opin. Drug Metab. Toxicol.*, 3(4), 527, 2007.
48. Lenz, E.M. et al., Cyclosporin A-induced changes in endogenous metabolites in rat urine: A metabonomic investigation using high field 1H NMR spectroscopy, HPLC-TOF/MS and chemometrics, *J. Pharm. Biomed. Anal.*, 35(3), 599, 2004.
49. Lenz, E.M. et al., A metabonomic investigation of the biochemical effects of mercuric chloride in the rat using 1H NMR and HPLC-TOF/MS: Time dependent changes in the urinary profile of endogenous metabolites as a result of nephrotoxicity, *Analyst*, 129(6), 535, 2004.
50. Keun, H.C., Metabonomic modeling of drug toxicity, *Pharmacol. Ther.*, 109, 92, 2006.
51. Jia, L. et al., Serum metabolomics study of chronic renal failure by ultra performance liquid chromatography coupled with Q-TOF mass spectrometry, *Metabolomics*, 4(2), 183, 2008.
52. Sun, J. et al., Metabonomics evaluation of urine from rats given acute and chronic doses of acetaminophen using NMR and UPLC/MS, *J. Chromatogr. B Analyt. Technol. Biomed. Life Sci.*, 871(2), 328, 2008.
53. Yang, J. et al., High performance liquid chromatography-mass spectrometry for metabonomics: Potential biomarkers for acute deterioration of liver function in chronic hepatitis B, *J. Proteome Res.*, 5(3), 554, 2006.
54. Yang, J. et al., Diagnosis of liver cancer using HPLC-based metabonomics avoiding false-positive result from hepatitis and hepatocirrhosis diseases, *J. Chromatogr. B Analyt. Technol. Biomed. Life Sci.*, 813(1–2), 59, 2004.

55. Williams, R.E. et al., D-Serine-induced nephrotoxicity: A HPLC-TOF/MS-based metabonomics approach, *Toxicology*, 207(2), 179, 2005.
56. Lenz, E.M. et al., Metabonomics with 1H-NMR spectroscopy and liquid chromatography-mass spectrometry applied to the investigation of metabolic changes caused by gentamicin-induced nephrotoxicity in the rat, *Biomarkers*, 10(2–3), 173, 2005.
57. Frazier, D.M. et al., The tandem mass spectrometry newborn screening experience in North Carolina: 1997–2005, *J. Inherit. Metab. Dis.*, 29(1), 76, 2006.
58. Roe, C.R., Millington, D.S., and Maltby, D.A., Identification of 3-methylglutarylcarnitine. A new diagnostic metabolite of 3-hydroxy-3-methylglutaryl-coenzyme A lyase deficiency, *J. Clin. Invest.*, 77(4), 1391, 1986.
59. Claudino, W.M. et al., Metabolomics: Available results, current research projects in breast cancer, and future applications, *J. Clin. Oncol.*, 25(19), 2840, 2007.
60. Kind, T. et al., A comprehensive urinary metabolomic approach for identifying kidney cancer, *Anal. Biochem.*, 363(2), 185, 2007.
61. Denkert, C. et al., Mass spectrometry-based metabolic profiling reveals different metabolite patterns in invasive ovarian carcinomas and ovarian borderline tumors, *Cancer Res.*, 66(22), 10795, 2006.
62. Di Leo, A. et al., New strategies to identify molecular markers predicting chemotherapy activity and toxicity in breast cancer, *Ann. Oncol.*, 18(Suppl. 12), xii8, 2007.
63. Rozen, S. et al., Metabolomic analysis and signatures in motor neuron disease, *Metabolomics*, 1(2), 101, 2005.
64. Lewis, G.D., Asnani, A., and Gerszten, R.E., Application of metabolomics to cardiovascular biomarker and pathway discovery, *J. Am. Coll. Cardiol.*, 52(2), 117, 2008.
65. Sabatine, M.S. et al., Metabolomic identification of novel biomarkers of myocardial ischemia, *Circulation*, 112(25), 3868, 2005.
66. Griffin, J.L. and Kauppinen, R.A., A metabolomics perspective of human brain tumors, *Febs. J.*, 274(5), 1132, 2007.
67. Ibrahim, S.M. and Gold, R., Genomics, proteomics, metabolomics: What is in a word for multiple sclerosis? *Curr. Opin. Neurol.*, 18(3), 231, 2005.
68. Wishart, D.S., Metabolomics: The principles and potential applications to transplantation, *Am. J. Transplant.*, 5(12), 2814, 2005.
69. Wishart, D.S., Metabolomics in monitoring kidney transplants, *Curr. Opin. Nephrol. Hypertens.*, 15(6), 637, 2006.
70. Wishart, D.S., Metabolomics: A complementary tool in renal transplantation, *Contrib. Nephrol.*, 160, 76, 2008.
71. Kristal, B.S. and Shurubor, Y.I., Metabolomics: Opening another window into aging, *Sci. Aging Knowledge Environ.*, 26, 19, 2005.
72. Kristal, B.S. et al., Metabolomics in the study of aging and caloric restriction, *Methods Mol. Biol.*, 371, 393, 2007.
73. Kristal, B.S. et al., High-performance liquid chromatography separations coupled with coulometric electrode array detectors: A unique approach to metabolomics, *Methods Mol. Biol.*, 358, 159, 2007.
74. Kuhl, J. et al., Metabolomics as a tool to evaluate exercise-induced improvements in insulin sensitivity, *Metabolomics*, 4(3), 272, 2008.
75. Michell, A.W. et al., Metabolomic analysis of urine and serum in Parkinson's disease, *Metabolomics*, 4(3), 191, 2008.
76. Dettmer, K., Aronov, P.A., and Hammock, B.D., Mass spectrometry-based metabolomics, *Mass Spectrom. Rev.*, 26(1), 51, 2007.
77. Plumb, R. et al., Metabonomic analysis of mouse urine by liquid-chromatography-time of flight mass spectrometry (LC-TOFMS): Detection of strain, diurnal and gender differences, *Analyst*, 128(7), 819, 2003.

78. Lenz, E.M. and Wilson, I.D., Analytical strategies in metabonomics, *J. Proteome Res.*, 6(2), 443, 2007.
79. Granger, J.H. et al., A metabonomic study of strain- and age-related differences in the Zucker rat, *Rapid Commun. Mass Spectrom.*, 21(13), 2039, 2007.
80. Rezzi, S. et al., Human metabolic phenotypes link directly to specific dietary preferences in healthy individuals, *J. Proteome Res.*, 6(11), 4469, 2007.
81. Plumb, R.S. et al., A rapid screening approach to metabonomics using UPLC and oa-TOF mass spectrometry: Application to age, gender and diurnal variation in normal/Zucker obese rats and black, white and nude mice, *Analyst*, 130(6), 844, 2005.
82. Gu, H. et al., Monitoring diet effects via biofluids and their implication for metabolomics studies, *Anal. Chem.*, 79(1), 89, 2007.
83. Shurubor, Y.I. et al., Biological variability dominates and influences analytical variance in HPLC-ECD studies of the human plasma metabolome, *BMC Clin. Pathol.*, 7, 9, 2007.
84. Sangster, T.P. et al., Investigation of analytical variation in metabonomic analysis using liquid chromatography/mass spectrometry, *Rapid Commun. Mass Spectrom.*, 21(18), 2965, 2007.
85. Sangster, T. et al., A pragmatic and readily implemented quality control strategy for HPLC-MS and GC-MS-based metabonomic analysis, *Analyst*, 131(10), 1075, 2006.
86. Luc, C.E. et al., Evaluation of columns and gradients for LC/MS-based non-targeted metabonomics. *Proceedings of the 56th ASMS Conference*, Denver, CO, 2008.
87. Pasikanti, K.K., Ho, P.C., and Chan, E.C., Gas chromatography/mass spectrometry in metabolic profiling of biological fluids, *J. Chromatogr. B Analyt. Technol. Biomed. Life Sci.*, 871(2), 202, 2008.
88. Saude, E.J. et al., Variation of metabolites in normal human urine, *Metabolomics*, 3, 439, 2007.
89. Griffin, J.L. et al., Standard reporting requirements for biological samples in metabolomics experiments: Mammalian/in vivo experiments, *Metabolomics*, 3, 179, 2007.
90. Moco, M. et al., Metabolomics technologies and metabolite identification, *Trends Anal. Chem.*, 26(9), 855, 2007.
91. Shipkova, P.A. et al., Sample preparation approaches and data analysis for non-targeted LC/MS metabonomic profiling of plasma and urine. *Proceedings of the 55th ASMS Conference*, Indianapolis, IN, 2007.
92. Plumb, R.S. et al., Metabonomics: The use of electrospray mass spectrometry coupled to reversed-phase liquid chromatography shows potential for the screening of rat urine in drug development, *Rapid Commun. Mass Spectrom.*, 16(20), 1991, 2002.
93. Plumb, R.S. et al., Use of liquid chromatography/time-of-flight mass spectrometry and multivariate statistical analysis shows promise for the detection of drug metabolites in biological fluids, *Rapid Commun. Mass Spectrom.*, 17(23), 2632, 2003.
94. Waybright, T.J. et al., LC-MS in metabonomics: Optimization of experimental conditions for the analysis of metabolites in human urine, *J. Liquid Chromatogr. Related Technol.*, 29(17), 2475, 2006.
95. He, X. and Becker, C.H., A quick mass spectrometry based tissue metabolomics method: Multi-fraction preparation using automated homogenizer and liquid-liquid extraction. *Proceedings of the 56th ASMS Conference*, Denver, CO, 2008.
96. Wu, H. et al., High-throughput tissue extraction protocol for NMR- and MS-based metabolomics, *Anal. Biochem.*, 372(2), 204, 2008.
97. Saude, E.J. and Sykes, B.D., Urine stability for metabolomics studies: Effects of preparation and storage, *Metabolomics*, 3(1), 19, 2007.
98. Gika, H.G. et al., Within-day reproducibility of an HPLC-MS-based method for metabonomic analysis: Application to human urine, *J. Proteome Res.*, 6(8), 3291, 2007.

99. Gika, H.G., Theodoridis, G.A., and Wilson, I.D., Liquid chromatography and ultra-performance liquid chromatography-mass spectrometry fingerprinting of human urine: Sample stability under different handling and storage conditions for metabonomics studies, *J. Chromatogr. A*, 1189(1–2), 314, 2008.
100. Jonsson, P. et al., Extraction, interpretation and validation of information for comparing samples in metabolic LC/MS data sets, *Analyst*, 130(5), 701, 2005.
101. Kopka, J., Current challenges and development in GC-MS based metabolite profiling technology, *J. Biotech.*, 124, 312, 2006.
102. Kanani, H.H. and Klapa, M.I., Data correction strategy for metabolomics analysis using gas chromatography-mass spectrometry, *Metab. Eng.*, 9(1), 39, 2007.
103. Fiehn, O. and Kind, T., Metabolite profiling in blood plasma, *Methods Mol. Biol.*, 358, 3, 2007.
104. Fiehn, O., Extending the breadth of metabolite profiling by gas chromatography coupled to mass spectrometry, *Trends Analyt. Chem.*, 27(3), 261, 2008.
105. Kanani, H., Chrysanthopoulos, P.K., and Klapa, M.I., Standardizing GC-MS metabolomics, *J. Chromatogr. B Analyt. Technol. Biomed. Life Sci.*, 871(2), 191, 2008.
106. Strehmel, N. et al., Retention index thresholds for compound matching in GC-MS metabolite profiling, *J. Chromatogr. B Analyt. Technol. Biomed. Life Sci.*, 871(2), 182, 2008.
107. Fiehn, O. et al., Identification of uncommon plant metabolites based on calculation of elemental compositions using gas chromatography and quadrupole mass spectrometry, *Anal. Chem.*, 72(15), 3573, 2000.
108. Wiklund, S. et al., Visualization of GC/TOF-MS-based metabolomics data for identification of biochemically interesting compounds using OPLS class models, *Anal. Chem.*, 80(1), 115, 2008.
109. Bedair, M. and Sumner, L.W., Current and emerging mass-spectrometry technologies for metabolomics, *Trends Anal. Chem.*, 27(3), 238, 2008.
110. Wilson, I.D. et al., HPLC-MS-based methods for the study of metabonomics, *J. Chromatogr. B Analyt. Technol. Biomed. Life Sci.*, 817(1), 67, 2005.
111. Boernsen, K.O., Gatzek, S., and Imbert, G., Controlled protein precipitation in combination with chip-based nanospray infusion mass spectrometry. An approach for metabolomics profiling of plasma, *Anal. Chem.*, 77(22), 7255, 2005.
112. Tolstikov, V.V., Fiehn, O., and Tanaka, N., Application of liquid chromatography-mass spectrometry analysis in metabolomics: Reversed-phase monolithic capillary chromatography and hydrophilic chromatography coupled to electrospray ionization-mass spectrometry, *Methods Mol. Biol.*, 358, 141, 2007.
113. Nordstrom, A. et al., Multiple ionization mass spectrometry strategy used to reveal the complexity of metabolomics, *Anal. Chem.*, 80(2), 421, 2008.
114. Metz, T.O. et al., High-resolution separations and improved ion production and transmission in metabolomics, *Trends Anal. Chem.*, 27(3), 205, 2008.
115. Theodoridis, G., Gika, H.G., and Wilson, I.D., LC-MS based methodology for global metabolite profiling in metabonomics/metabolomics, *Trends Anal. Chem.*, 27(3), 251, 2008.
116. Lu, X. et al., LC-MS-based metabonomics analysis, *J. Chromatogr. B*, 866, 64, 2008.
117. Crockford, D.J. et al., Statistical search space reduction and two-dimensional data display approaches for UPLC-MS in biomarker discovery and pathway analysis, *Anal. Chem.*, 78(13), 4398, 2006.
118. Gika, H.G., Theodoridis, G.A., and Wilson, I.D., Hydrophilic interaction and reversed-phase ultra-performance liquid chromatography TOF-MS for metabonomic analysis of Zucker rat urine, *J. Sep. Sci.*, 31(9), 1598, 2008.
119. Gika, H.G. et al., High temperature-ultra performance liquid chromatography-mass spectrometry for the metabonomic analysis of Zucker rat urine, *J. Chromatogr. B Analyt. Technol. Biomed. Life Sci.*, 871(2), 279, 2008.

120. Gika, H.G. et al., Evaluation of the repeatability of ultra-performance liquid chromatography-TOF-MS for global metabolic profiling of human urine samples, *J. Chromatogr. B Analyt. Technol. Biomed. Life Sci.*, 871(2), 299, 2008.
121. Nordstrom, A. et al., Nonlinear data alignment for UPLC-MS and HPLC-MS based metabolomics: Quantitative analysis of endogenous and exogenous metabolites in human serum, *Anal. Chem.*, 78(10), 3289, 2006.
122. Wong, M.C. et al., An approach towards method development for untargeted urinary metabolite profiling in metabonomic research using UPLC/QToF MS, *J. Chromatogr. B Analyt. Technol. Biomed. Life Sci.*, 871(2), 341, 2008.
123. Burton, L. et al., Instrumental and experimental effects in LC-MS-based metabolomics, *J. Chromatogr. B Analyt. Technol. Biomed. Life Sci.*, 871(2), 227, 2008.
124. Guy, P.A. et al., Global metabolic profiling analysis on human urine by UPLC-TOFMS: Issues and method validation in nutritional metabolomics, *J. Chromatogr. B Analyt. Technol. Biomed. Life Sci.*, 871(2), 253, 2008.
125. Want, E.J. et al., From exogenous to endogenous: The inevitable imprint of mass spectrometry in metabolomics, *J. Proteome Res.*, 6(2), 459, 2007.
126. Ackermann, B.L. et al., Current applications of liquid chromatography/mass spectrometry in pharmaceutical discovery after a decade of innovation, *Annu. Rev. Anal. Chem.*, 1, 357, 2008.
127. Werner, E. et al., Mass spectrometry for the identification of the discriminating signals from metabolomics: Current status and future trends, *J. Chromatogr. B*, 871, 143, 2008.
128. Sanders, M. et al., Utility of the hybrid LTQ-FTMS for drug metabolism applications, *Curr. Drug Metab.*, 7(5), 547, 2006.
129. Brown, S.C., Kruppa, G., and Dasseux, J.L., Metabolomics applications of FT-ICR mass spectrometry, *Mass Spectrom. Rev.*, 24(2), 223, 2005.
130. Lim, H.K. et al., Metabolite identification by data-dependent accurate mass spectrometric analysis at resolving power of 60,000 in external calibration mode using an LTQ/Orbitrap, *Rapid Commun. Mass Spectrom.*, 21(12), 1821, 2007.
131. Makarov, A. et al., Performance evaluation of a hybrid linear ion trap/orbitrap mass spectrometer, *Anal. Chem.*, 78(7), 2113, 2006.
132. Dunn, W.B. et al., Metabolic profiling of serum using ultra performance liquid chromatography and the LTQ-Orbitrap mass spectrometry system, *J. Chromatogr. B Analyt. Technol. Biomed. Life Sci.*, 871(2), 288, 2008.
133. Shurubor, Y.I. et al., Analytical precision, biological variation, and mathematical normalization in high data density metabolomics, *Metabolomics*, 1(1), 75, 2005.
134. Katajamaa, M., Miettinen, J., and Oresic, M., MZmine: Toolbox for processing and visualization of mass spectrometry based molecular profile data, *Bioinformatics*, 22(5), 634, 2006.
135. Katajamaa, M. and Oresic, M., Processing methods for differential analysis of LC/MS profile data, *BMC Bioinform.*, 6, 179, 2005.
136. Katajamaa, M. and Oresic, M., Data processing for mass spectrometry-based metabolomics, *J. Chromatogr. A*, 1158(1–2), 318, 2007.
137. Sysi-Aho, M. et al., Normalization method for metabolomics data using optimal selection of multiple internal standards, *BMC Bioinform.*, 8, 93, 2007.
138. Smith, C.A. et al., XCMS: Processing mass spectrometry data for metabolite profiling using nonlinear peak alignment, matching, and identification, *Anal. Chem.*, 78(3), 779, 2006.
139. Styczynski, M.P. et al., Systematic identification of conserved metabolites in GC/MS data for metabolomics and biomarker discovery, *Anal. Chem.*, 79(3), 966, 2007.
140. Benton, H.P. et al., XCMS(2): Processing tandem mass spectrometry data for metabolite identification and structural characterization, *Anal. Chem.*, 80(16), 6382, 2008.

141. Lu, H. et al., Comparative evaluation of software for deconvolution of metabolomics data based on GC-TOF-MS, *Trends Anal. Chem.*, 27(3), 215, 2008.
142. Bijlsma, S. et al., Large-scale human metabolomics studies: A strategy for data (pre-) processing and validation, *Anal. Chem.*, 78(2), 567, 2006.
143. Jonsson, P. et al., A strategy for modelling dynamic responses in metabolic samples characterized by GC/MS, *Metabolomics*, 2(3), 135, 2006.
144. Wagner, S. et al., Metabonomics and biomarker discovery: LC-MS metabolic profiling and constant neutral loss scanning combined with multivariate data analysis for mercapturic acid analysis, *Anal. Chem.*, 78(4), 1296, 2006.
145. Westerhuis, J.A. et al., Assessment of PLSDA cross validation, *Metabolomics*, 4(1), 81, 2008.
146. Chen, J. et al., Practical approach for the identification and isomer elucidation of biomarkers detected in a metabonomic study for the discovery of individuals at risk for diabetes by integrating the chromatographic and mass spectrometric information, *Anal. Chem.*, 80(4), 1280, 2008.
147. Smith, C.A. et al., METLIN: A metabolite mass spectral database, *Ther. Drug Monit.*, 27(6), 747, 2005.
148. Kind, T. and Fiehn, O., Metabolomic database annotations via query of elemental compositions: Mass accuracy is insufficient even at less than 1 ppm, *BMC Bioinform.*, 7, 234, 2006.
149. Kind, T. and Fiehn, O., Seven Golden Rules for heuristic filtering of molecular formulas obtained by accurate mass spectrometry, *BMC Bioinform.*, 8, 105, 2007.
150. Loftus, N. et al., Profiling and biomarker identification in plasma from different Zucker rat strains via high mass accuracy multistage mass spectrometric analysis using liquid chromatography/mass spectrometry with a quadrupole ion trap-time of flight mass spectrometer, *Rapid Commun. Mass Spectrom.*, 22(16), 2547, 2008.
151. Wishart, D.S. et al., HMDB: The Human Metabolome Database, *Nucleic Acids Res.*, 35, D521, 2007.
152. Leighton, J.K., Application of emerging technologies in toxicology and safety assessment: Regulatory perspectives, *Int. J. Tox.*, 24, 153, 2005.

11 Imaging Mass Spectrometry for Small Molecules

Fangbiao Li

CONTENTS

11.1 INTRODUCTION

Mass spectrometry-based bioanalytical methods play an important role in the entire process of searching new medicines to improve the quality of human life. High-performance liquid chromatography coupled to a tandem mass spectrometer (HPLC–MS/MS) has become the standard tool for the determination of pharmaceuticals in various *in vitro* and *in vivo* samples stemming from drug discovery and development experiments [1–5]. However, the major disadvantage of the HPLC–MS/MS technique is that it does not readily provide information on the distribution of the administrated drug and its metabolites.

Alternatively, imaging mass spectrometry (IMS) using matrix-assisted laser desorption/ionization (MALDI) can be used to simultaneously map the distribution of pharmaceuticals in thin tissue sections to determine how a drug is distributed in animal tissues [6–9]. MALDI–IMS has been extensively employed to measure macromolecules such as peptides and proteins in tissue sections [10–13] (Figure 11.1). Although MALDI–IMS has been applied almost exclusively as an analytical tool for

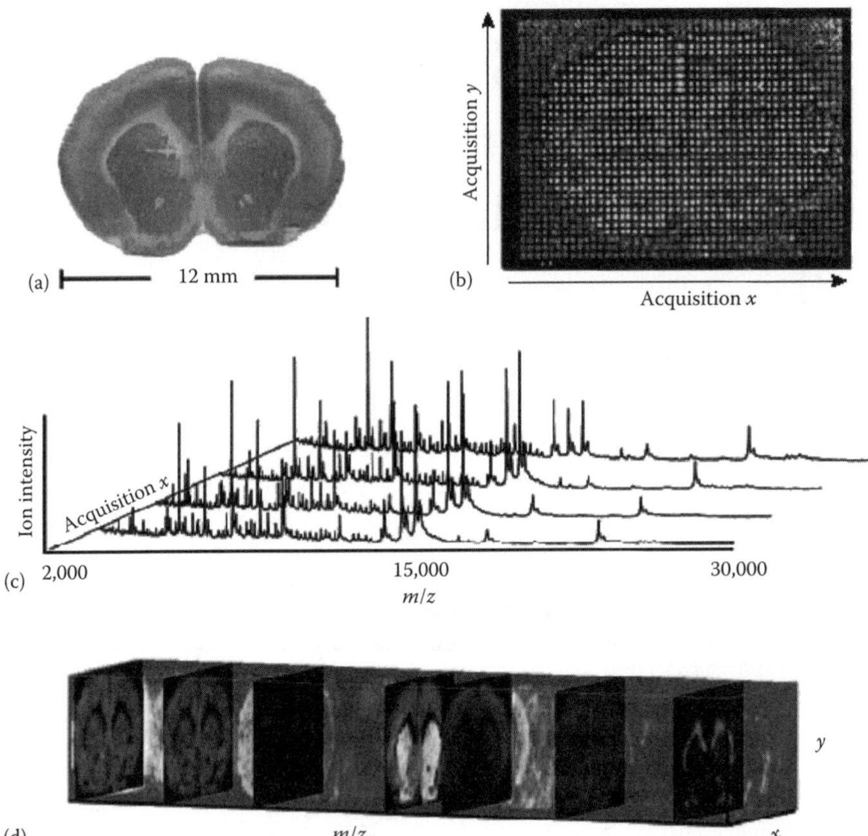

FIGURE 11.1 (See color insert following page 210.) General overview of MALDI–IMS. (a) Fresh section cut from sample tissue. (b) Mounted section after matrix application using a robotic picoliter volume spotter. (c) Partial series of mass spectra collected along one row of coordinates (x-axis). (d) Three-dimensional volumetric plot of complete dataset with selected m/z "slices" or ion images. Principal axes are x, y, and m/z. Color of each voxel is determined by ion intensity. (From Cornett, D. et al., *Nat. Methods*, 4, 828, 2007. With permission.)

biopolymers, the potential of using this technique for both qualitative and quantitative determination of small molecules (<1500 Da) such as enzyme subtracts [14], acetylcholine, 3,4-dihydroxyphenylalanine [15], vitamins [16], and others [17,18] has been demonstrated. Several teams of researchers have extended MALDI–MS imaging methodology to visualize the distribution of candidate drug compounds in different regions of biological tissue sections [19–24]. In this chapter, we describe the use of MALDI mass spectrometric imaging as a tool for the pharmaceutical analysis of drug compounds and their metabolites in the drug discovery and development process. This technique offers significant advantages over alternative technologies, such as autoradiography. For example, MALDI–MS imaging can distinguish intact drugs from their metabolites without the addition of an isotope label. This allows distribution studies of lead compounds to occur much earlier in the drug-discovery process and allows compounds with unfavorable distribution characteristics to be rapidly discarded. In addition, the reliance on radioactive isotopes can be reduced, which is advantageous from both an economical and environmental standpoint.

Distribution studies of drug candidates in animals are crucial in both drug discovery and development. These studies provide information about where the drug accumulates in the body. For example, it may be important to know if the test compound accumulates in the target organ or if it selectively accumulates in other organs or tissues (such as the brain, which could result in neurotoxicity or other unwanted side effects). It is also important to know how long a drug remains in the body and MALDI–IMS can help to measure that parameter as well. The accurate assessment of preclinical ADME (absorption/distribution/metabolism/excretion) and toxicological parameters of new chemical entities (NCEs) is essential for deciding whether or not to develop the NCE. Techniques currently in use for assessing the distribution of drug candidates in tissues will be discussed; these techniques include positron emission tomography (PET), magnetic resonance imaging (MRI), autoradiography, and secondary ion mass spectrometry (SIMS). Their strengths and limitations will be addressed, especially with regard to the spatial analysis of drugs and metabolites in tissues. The technique of MALDI–IMS will be discussed, and the examples of its use will be presented. Finally, the further development of the technique will be addressed, focusing on improvements and advance necessary to allow the technology to reach its full potential.

11.2 CURRENT IMAGING TECHNIQUES

Several analytical techniques exist to image the distribution of small molecules in the body. Generally, they can be divided into two groups: noninvasive *in vivo* techniques and *ex vivo/in vitro* techniques. *In vivo* techniques, such as PET [25–32] and MRI [32–36], allow the distribution of a drug to be evaluated over time in a living animal. In contrast, the *ex vivo/in vitro* techniques, such as autoradiography [29,30], require the removal of the tissue of interest and thus can only image the distribution of a drug at a fixed time.

All these techniques require labeled drugs. The visualization of PET requires that a drug contain a radioactive positron emitter, such as ^{11}C, ^{13}N, ^{15}O, or ^{18}F [32]. Similarly, a radioactive isotope, typically ^{14}C, ^{3}H, or ^{125}I, is necessary for a compound to be visualized with autoradiography. While MRI does not require a radioactive isotope, only isotopes with nuclear spin, such as ^{1}H, ^{13}C, and ^{19}F are visible with MRI [32,33]. Because the sensitivity of MRI depends on the magnetic properties of the monitored nucleus and its natural abundance, drugs typically need to contain ^{19}F or be enriched in ^{13}C in order to be effectively monitored in the body. Contrast agents, which include paramagnetic atoms such as gadolinium, iron, and manganese, are often used to increase the contrast in MRI images and are frequently used as models for drugs that cannot be imaged on their own [33,34]. The main limitation with using a label or tracer for imaging, especially in terms of drug metabolism, is that only the label and not the parent compound is imaged. Thus, the experimentally determined distribution of a drug in a tissue may be inaccurate if extensive metabolism has occurred.

In vivo imaging techniques have several advantages for drug-discovery applications, including the ability to examine a drug's actions directly in humans and the ability to re-measure the same individual under different conditions. However, there are significant limitations. The spatial resolution is limited with PET, and MRI-PET has a resolution of ~3–6 mm [26,32,33], while MRI has a resolution of ~2 mm [34,35], but can be improved to 700–800 μm [36] with some modifications to the instrument used. This limits the size regime in the body in which meaningful distribution data can be acquired. The use of a labeled compound typically occurs late in the drug development process due to the increased cost and time involved to synthesize the labeled compound. Additionally, there are time constraints with PET imaging due to the half-lives of the positron emitters. The half-lives of some commonly used positron emitters are ^{15}O, 2 min; ^{13}N, 10 min; ^{11}C, 20 min; and ^{18}F, 110 min. Thus a drug must be synthesized, administered, and monitored on the time-scale of the half-life of the radiotracer.

Autoradiography is one of the most common techniques used today in the pharmaceutical industry for examining the distribution of a drug candidate in the body *ex vivo* [37,38]. In quantitative whole-body autoradiography (QWBA), a radiolabeled drug is administered to animal and, after a specific time, the animal is sacrificed, flash-frozen, and sectioned. The radioactivity in the sections is then analyzed, and the result is an image showing where the drug has accumulated in various organs. Individual organs may be dissected and analyzed separately as well. This technique is also widely used in psychopharmacology for *in vitro* experiments, in which a radiolabeled compound is incubated with a section of brain tissue, and the resulting image is used to determine where in the brain the compound is binding in order to locate the specific receptors of the compound [38,39].

Autoradiography has significant advantages over the noninvasive imaging techniques, especially in terms of drug discovery. As it is typically performed on laboratory animals, it occurs earlier in the drug-discovery process. Additionally, the resolution is better that PET and MRI, ranging from a few micrometers for film to ~30–50 μm for the β-imager [38,39], ~70 mm for microchannel plates [40,41], and ~25–100 μm for commercial phosphor-imaging systems (Fujifilm). In contrast, the time required to acquire an autoradiographic image varies substantially and can be prohibitively long. It often takes weeks or months to acquire an image on film, while the same image may be acquired on the β-imager in 8–12 h [38,42]. Also, because a radiolabel is required, the extra cost and time required for additional synthesis are still limitations.

11.3 PRINCIPLES OF MALDI IMAGING MASS SPECTROMETRY

Due to its molecular specificity, there is a great deal of interest in using mass spectrometry as an imaging tool. Instruments used for IMS can be classified according to how ions are generated from the sample either by irradiation by pulsed laser or bombardment by energetic particles [43]. SIMS systems use particle bombardment with a continuous beam of highly focused, energetic ions such as Cs^+, Au_3^+, and C_{60}^+. Laser-based systems include MALDI and laser desorption ionization (LDI) instruments.

SIMS has been used for imaging surfaces for more than 40 years [44]. The primary ion beam has an energy of ~25 keV which desorbs ions from the surface of the sample. These secondary ions are predominantly elements, atomic clusters, and organic fragments that are typically analyzed by time-of-flight (TOF), quadrupole, or magnetic sector instruments. While protonated molecules can be formed, for example, several analyses have been reported involving the $[M + H]^+$ ions of cholesterol [45], benzodiazepines [46], and crystal violet [47], the secondary ions formed most abundantly include elemental ions (e.g., Na^+, K^+) and molecular fragment ions. Thus, its primary use has been for the localization of inorganic and atomic species [44].

Applications of SIMS imaging to biological/pharmaceutical analyses have been in two key areas: atomic imaging of drugs [48–50] and tissue mapping via molecular imaging of fragment ions [51–53]. In the first application, the drug of interest must be effectively labeled; it must contain an atom not natively found in cells or tissues, and only that atom is monitored. The other main application of SIMS to biological imaging has focused on tissue mapping. Compared to the traditional imaging techniques discussed earlier, the benefits of the SIMS imaging techniques include excellent spatial resolution (submicrometer to ~1 μm) and high sensitivity and molecular specificity. One of the main drawbacks, however, especially in regards to metabolite localization, is the low efficiency of molecular ionization [47]. Thus it is difficult to differentiate a compound and structurally similar metabolites if they share the same label or if they form the same fragment, because that is what is typically analyzed. While advances are being made toward increasing molecular ion yields, especially utilizing cluster ion beams such as C_{60}^+ and SF_5^+ [47,54], the full potential of this technology for pharmaceutical imaging has not been reached.

MALDI–IMS employs a matrix solution to mix with the analytes and to form cocrystals. These crystals absorb energy at the wavelength of the laser beam resulting in desorption and ionization of the analytes that were included in the crystals. The concept of MALDI–IMS was introduced in 1997 by Caprioli and coworkers [55] for rapid and direct profiling of the analytes within a tissue section or an organ. In this process, matrix is uniformly deposited over various tissue sections to extract analytes into the matrix and to produce matrix crystals (Figure 11.2) [56]. A raster of the organ or whole-body sections containing the compound of interest under a stationary laser beam is then performed over a predetermined two-dimensional array to generate ion plumes directly from the tissue sections in a MALDI plate array as depicted in Figure 11.3 [56]. For the small molecule application, commercially available MALDI source instruments have been utilized. These instruments utilize a large sample plate sufficient to contain whole tissue slices. The movement of the sample stages is automatically accomplished in the *x*- and *y*-directions to locate the edges of the tissue sample and to define the exact region of interest (Figure 11.2).

During an imaging mass spectrometric experiment, MALDI–MS signals of small molecules from tissue sections within a user-defined area are first obtained as a function of acquisition times which are associated with the location of an array of pixels as given in Figure 11.4 [56]. Thus, two-dimensional ion maps of biological tissues are reconstructed with drug signals of a given *m/z* value monitored in each spectrum from each pixel within a defined array to provide specific molecular images as demonstrated in Figure 11.5 [56]. In order to detect small molecules in complex

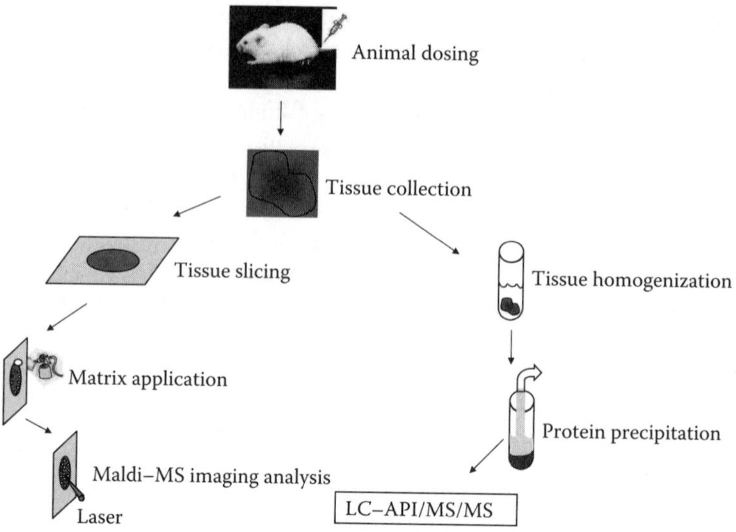

FIGURE 11.2 General procedure for MALDI–MS imaging analysis and conventional quantitative HPLC/MS analysis.

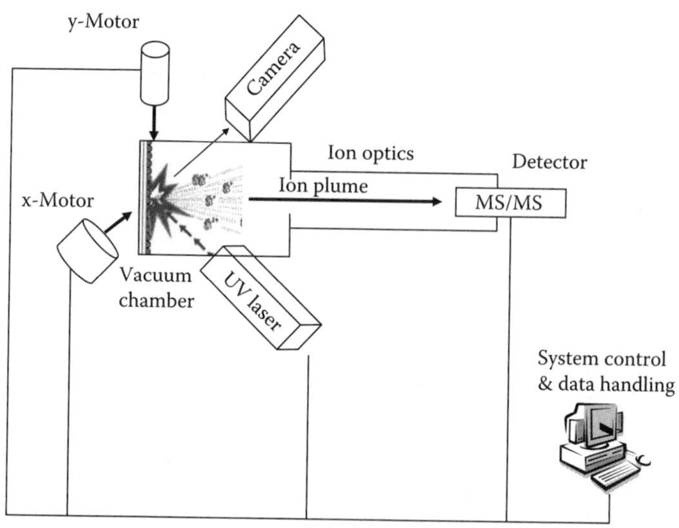

FIGURE 11.3 Diagram for an automated MALDI–IMS system used for direct analysis of pharmaceuticals in tissues. (From Hsieh, Y. et al., *J. Pharmacol. Toxicol. Methods.*, 55, 193, 2007. With permission.)

biological tissue sections using MALDI, a tandem mass spectrometer is required in order to separate the analyte ions from the background interference ions produced by the matrix [22]. In most MS/MS systems, two analyzers are connected together by a collision cell. A protonated molecule is generated in the ion source. Then the first analyzer selects the ion as the precursor ion for collisionally activated dissociation

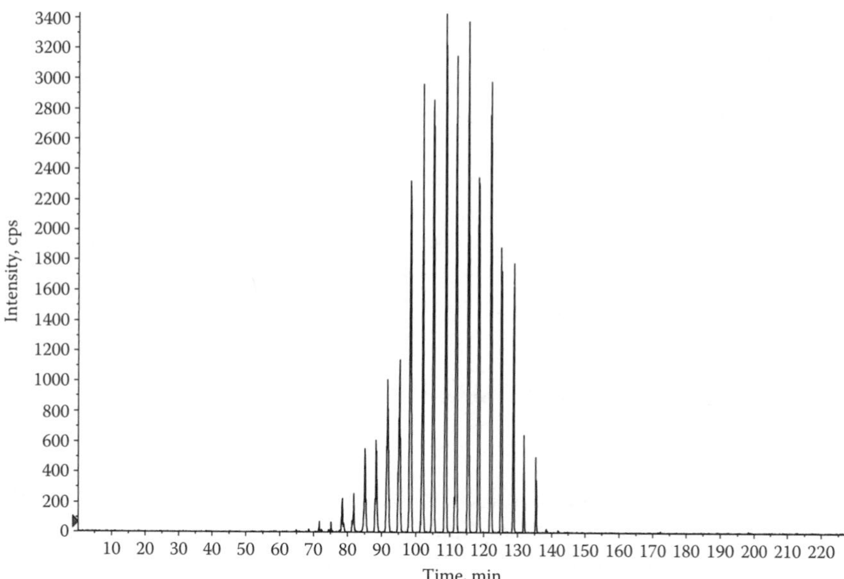

FIGURE 11.4 MALDI–MS signals from a droplet deposited in the rat brain tissue section containing a drug candidate are plotted against acquiring time which is directly related to the locations of each pixel. (From Hsieh, Y. et al., *J. Pharmacol. Toxicol. Methods.*, 55, 193, 2007. With permission.)

FIGURE 11.5 An ion image of a spot in the rat brain slice was obtained after reconstruction of MALDI–MS signals derived from the spot of chemical residue while "rastering" the tissue sample under the laser beam. (From Hsieh, Y. et al., *J. Pharmacol. Toxicol. Methods.*, 55, 193, 2007. With permission.)

(CAD). This precursor ion passes through a collision cell, which breaks the precursor ion into product ions. Analysis of the resulting product ions by the second analyzer generates a mass spectrum of the fragment ions and providing structure information. By monitoring the transition of a selected precursor ion to its dissociated product ions, tandem mass spectrometry systems give researchers a tool for distinguishing between precursor ions of similar mass-to-charge ratios, such as matrix ions and small molecule drug candidate compounds [22]. For example, a hybrid quadrupole-time-of-flight (QqTOF) analyzer was utilized in order to resolve some of the mass spectrometric interference from the matrix, as CAD fragmentation of protonated drug molecules could be performed [19,21,22].

11.4 FUNDAMENTALS OF IMAGING MASS SPECTROMETRY

11.4.1 Tissue Collection and Slicing

Tissue collection and slicing are critical to the spatial integrity of measured molecular distribution [57]. Inappropriately handling tissue samples in the sample preparation steps may cause delocalization and degradation of the analytes. A typical study may involve samples collected over a lengthy period of time, and standardized procedures are therefore required to minimize experimental variability over the time course of the study. Ideally, animal tissues are frozen immediately following collection to prevent drug dispersing in tissue. For the analysis, animal tissues are sliced to expose the area of interest. Cryostats are normally used as a standard tool for slicing frozen tissue to reducing sample contamination [56]. Contamination of the tissue surface due to the use of optimal cutting temperature (OCT) polymer as an embedding medium for stabilizing the organ specimens should be avoided because it would lead to ionization suppression in MALDI–IMS analysis [13,58]. Recently, it has been reported that distilled water can be used as embedding medium to overcome this problem [59]. There are several direct ways to transfer tissue slices to the sample plate [56]. First, the frozen tissue slice is gently positioned on the cold target plate (approximately −15°C) using an artist's brush and thaw-mounted onto the plate by quickly moving out of the cryostat chamber. Second, the frozen section adheres to the MALDI plate held at room temperature. Third, the section is stuck to a double-sided transparent tape which is then attached to the sample plate. In addition, blotting the tissue onto cellulose or carbon-conducting membranes is an indirect way to transfer the sections. Tissue containing drugs subject to photodegradation should be maintained in a dark container as much as possible during the processing steps.

11.4.2 MALDI Matrix Application

Once the section is mounted to the sample plate with the desired orientation, matrix solution is deposited on the tissue surface by electrospray deposition, aero spray, or using robotics to deposit small matrix droplets across the tissue surface before MALDI analysis [13]. The most common and least expensive devices available for applying matrix are hand-held aerosol sprayers or air brushes. The main advantage of these devices is that, with careful application, a dispersion of very small droplets

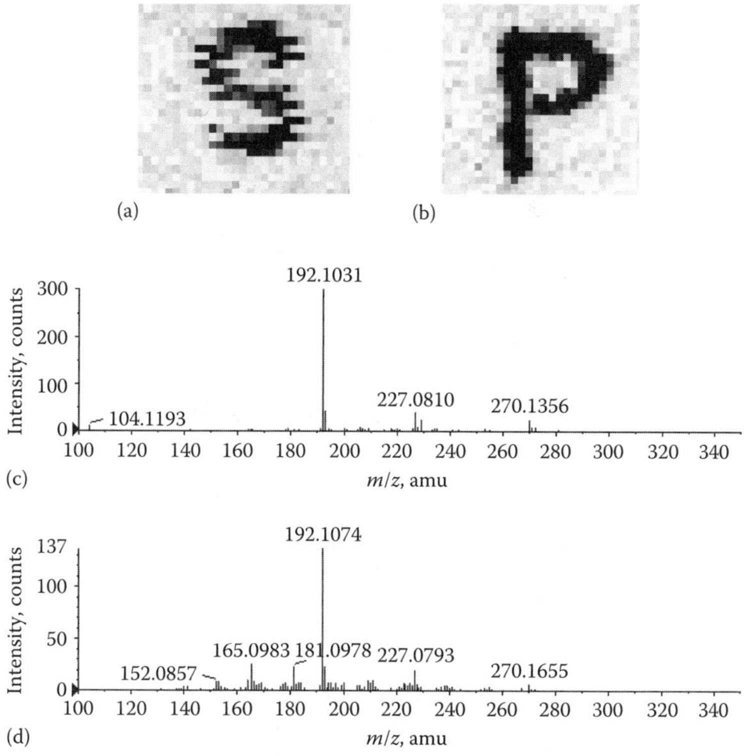

FIGURE 11.9 MALDI–MS images of letters (a) S and (b) P containing norclozapine and clozapine in rat brain sections, respectively. MALDI–MS/MS spectra of (c) clozapine and (d) norclozapine in rat brain sections. (From Hsieh, Y. et al., *Rapid Commun. Mass Spectrom.*, 20, 965, 2006. With permission.)

with a calculated protonated monoisotopic molecular weight that differed from the molecular weight of a sinapinic acid matrix cluster ion by less than 0.2 amu [21]. In this case, CAD was employed to fragment the drug ion at *m/z* 695 into a dominant fragment ion at *m/z* 228. The sinapinic acid cluster ion at *m/z* 695 fragmented into noninterfering ions at *m/z* 246 and *m/z* 471. Thus, MS/MS was used to localize the parent compound in a mouse tumor sample from a mouse dosed at 80 mg/kg.

MALDI–MS imaging was reliable and has been verified with autoradiography [21]. Another example using MALDI–MS/MS to image drugs in tissue was a study on clozapine levels in rat brain tissue in which results were compared with autoradiography studies. An autoradiograph taken from a rat brain following dosing with ^{3}H-labeled clozapine revealed that the drug was distributed throughout the rat brain, with the highest concentration in the lateral ventricle (Figure 11.10). A MALDI–MS/MS image of the clozapine fragment ion (*m/z* 327 → *m/z* 192) strongly correlated with the autoradiograph, showing the most intense signal in the ventricle area.

The MALDI instrument utilizes a relatively large target plate sufficient to contain a mouse whole-body tissue slice. Recently, some research teams had extended

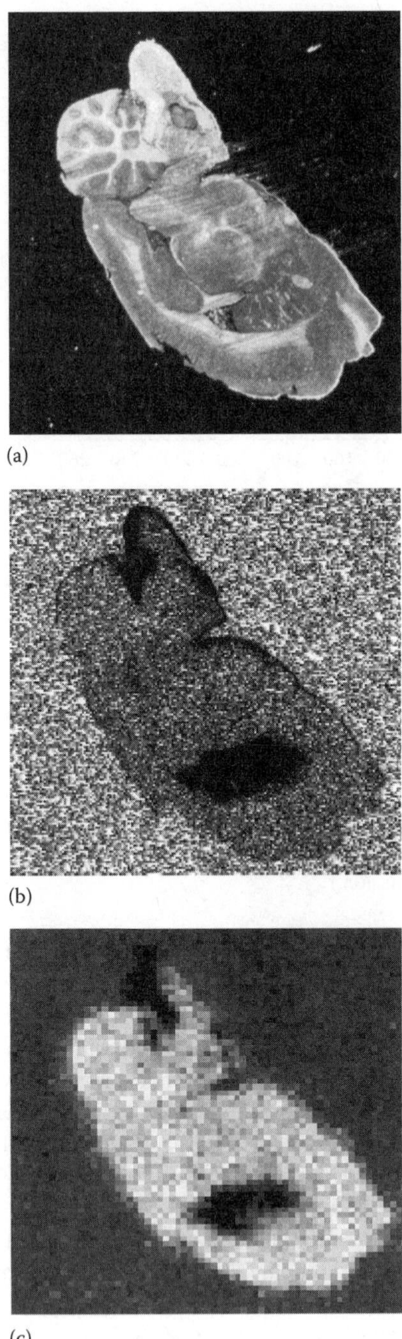

FIGURE 11.10 (a) Optical images, (b) radioautographic images, and (c) MALDI–MS/MS images from the study of rat brain tissue section. (From Hsieh, Y. et al., *Rapid Commun. Mass Spectrom.*, 20, 965, 2006. With permission.)

the applications of MALDI–IMS from an isolated organ tissue to whole-body sections [59]. Distinguishing drugs from metabolites has been demonstrated using MALDI–MS/MS imaging in whole-body rat sections. Olanzapine, an antipsychotic drug, was administered at a physiologically relevant dose to 10-week-old rats (given orally at 8 mg/kg). The animals were euthanized at 2 and 6 h postdose. Whole-body sections were obtained on acetate film tape and mounted onto MALDI plates using double-sided conductive tape. MALDI–MS/MS analyses were performed on the whole-body sections to detect simultaneously olanzapine (*m/z* 313 → *m/z* 256) and two metabolites, *N*-desmethyl-olanzapine (*m/z* 299 → *m/z* 256) and 2-hydroxymethyl-olanzapine (*m/z* 329 → *m/z* 272). As shown in Figure 11.11, all three compounds can be detected in almost all tissues at both 2 and 6 h postdose. Olanzapine was detected throughout the body at 2 h, and was present in the target organs, brain and spinal cord. However, by 6 h it was significantly decreased in intensity in the brain and spinal cord (66% less ion signal). As expected from the results of classical studies, neither metabolite was detected in the brain or spinal cord. Both metabolites were present to a high degree in the liver and bladder. These data were also consistent with

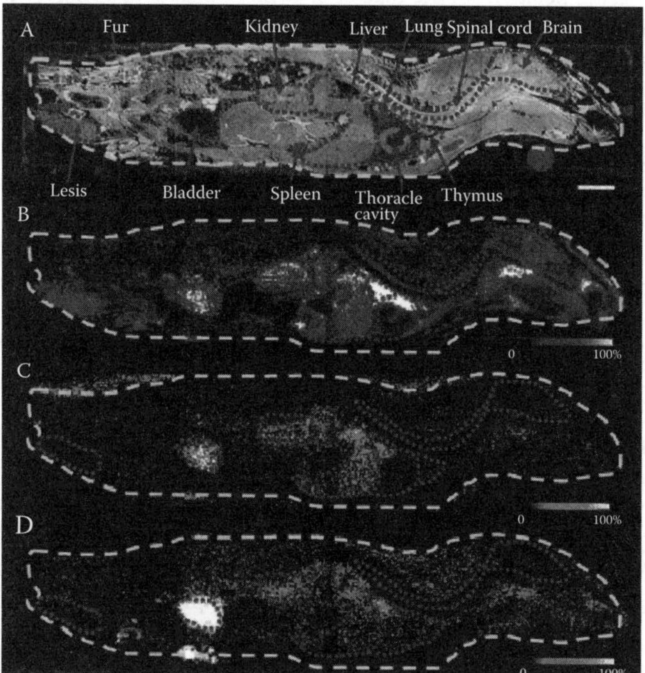

FIGURE 11.11 (See color insert following page 210.) Detection of drug and metabolite distribution at 2 h postdose in a whole rat sagittal tissue section by a single IMS analysis. (A) Optical image of a 2 h post-OLZ dosed rat tissue section across four gold MALDI target plates. (B) Organs outlined in red. Pink dot used as time point label. MS/MS ion image of OLZ (*m/z* 256). (C) MS/MS ion image of *N*-desmethyl metabolite (*m/z* 256). (D) MS/MS ion image of 2-hydroxymethyl metabolite (*m/z* 272). Bar, 1 cm. (From Khatib-Shahidi, S. et al., *Anal. Chem.*, 78, 6448, 2006. With permission.)

previous autoradiographic data and with the known metabolic pathways of rats [59]. This application highlights the ability of the technology to provide unambiguous localization information for several small molecules simultaneously.

There are also several successful examples of using MALDI–MS imaging for small-molecule analysis of the protonated molecule without CAD fragmentation [19,69]. One study used chlorisondamine, a neuronal nicotinic ganglionic blocker, and cocaine administered to rats [69]. Both compounds were detected in the brains of rats via MALDI–IMS using α-cyano-4-hydroxycinnamic acid (CHCA) and 2,6-dihydroxybenzoic acid (DHB) as the matrices for chlorisondamine and cocaine, respectively. However, MS/MS was employed to confirm the identity of the protonated ions. Another study was reported by Bunch et al. to map the distribution of a xenobiotic substance, ketoconazole, in skin [19]. In this case, the detection of the protonated drug compound (along with diagnostic ^{37}Cl isotope signals) enabled the drug to be localized on the tissue surface and also localized by depth following the analysis of a cross-section.

These studies illustrate that MALDI–MS imaging can be performed in several ways to detect small molecules. Care must be taken when using MS to ensure there are no interfering peaks from the matrix or from endogenous compounds present in the tissue. In addition, the specific fragmentation information obtained from MS/MS experiments adds a high degree of confidence in the identification of a signal as coming from the analyte of interest. However, the MS-only mode can give a multiplex advantage, in that many signals are detected at the same time from the same pixel. Sample preparation procedures, including the choice of matrix, extraction solvent, and detection method, must be optimized for each drug because of the wide variety of structures, solubilities, and physicochemical properties of drug compounds.

11.6 APPLICATION TO IMAGING OF PHOSPHOLIPIDS

Phospholipids played a very important role in many essential biological functions, including cell signaling, energy storage, and membrane structure and function. The analysis of lipids is challenging because of the wide molecular diversity of this class of compounds and their relative insolubility in aqueous systems. Using multiple stages of mass spectrometry [13,70,71], ion species may be unambiguously identified depending on how the lipid ion initially fragments. The use of ion mobility (IM) has also been reported in conjunction with MALDI–MS as a means to fractionate lipids [72–74] (Figure 11.12). In this technique, ions are first separated by IM and then analyzed by TOF mass spectrometry. The desorbed lipid ions fall on a trend line that is separate from those of oligonucleosides, peptides, proteins, and drugs and metabolites having the same nominal mass. This allows ions originating from lipids to be distinguished from other small molecules. In contrast with IMS, imaging IM–MS provides several unique advantages [72–74] including (1) the selective imaging of isobaric species or structural/conformational subpopulations of the same species on the basis of IM; (2) the separation/rejection of undesirable endogenous chemical noise; (3) the reduction of ion suppression effects in the source of the TOFMS, by temporal IM separation of analytes; and (4) the potential utility for nearly simultaneous IM–MS/MS of all analytes at a particular pixel coordinate.

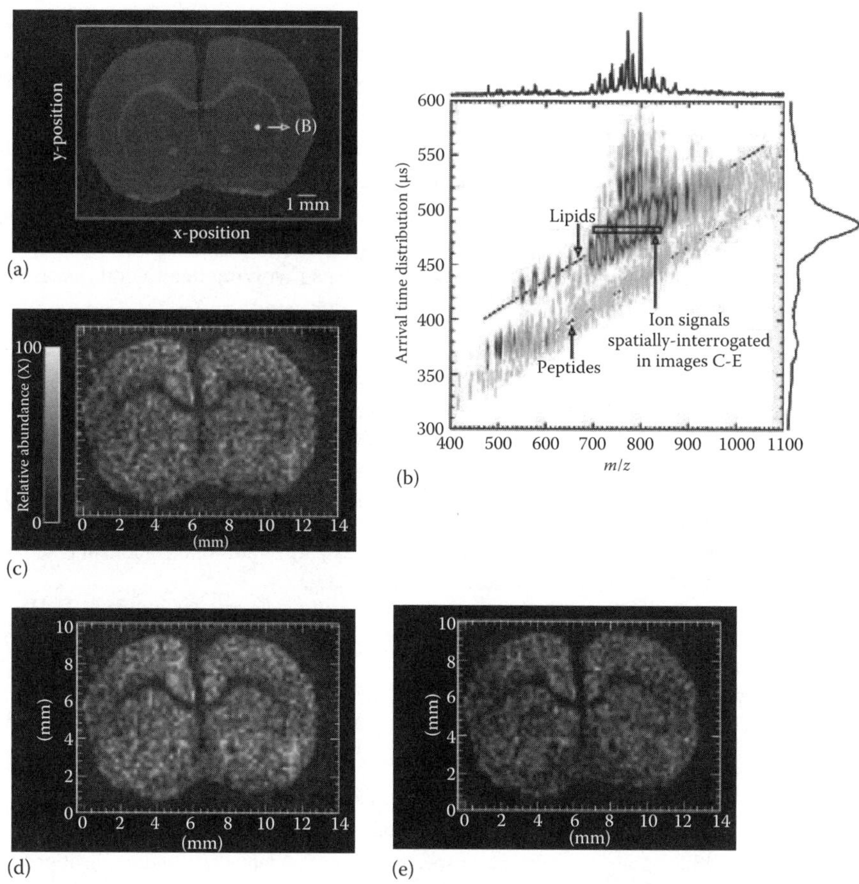

FIGURE 11.12 (See color insert following page 210.) Imaging IM–MS for mapping the spatial coordinates of analytes based on structure and *m/z*. (a) An optical image of a thin coronal tissue section (12 μm) of a rat brain adjacent to the section used for imaging IM–MS. The spatial coordinates of the white spot indicated in (a) were interrogated to yield an IM–MS molecular profile as indicated in (b). Lipid imaging of the analytes indicated by the rectangle about an IM arrival time distribution of 480–484 μs and 700–840 *m/z* is illustrated in (c). (d, e) Extracted ion density maps for two phospholipids at *m/z* 771–776 [PC 32:0 + K]$^+$ and 819–823 [PC 32:0 + K]$^+$, in (d) and (e), respectively. MALDI was performed at a laser repetition rate of 1000 Hz with a spatial resolution of 200 μm. *Note* that the images in c, d, and e represent raw data which have not been smoothed or interpolated. The relative abundance scale in c is the same for d and e. (From McLean, J.A. et al., *J. Mass Spectrom.*, 42, 1099, 2007. With permission.)

11.7 CONCLUSIONS

During the past several years, significant progress was made for the MALDI–MS imaging technique. It has been proven to be a valuable tool for studying drug distribution in animal tissues. MALDI–IMS is a simple, quick, and molecularly specific analytical platform that can distinguish small molecules from their metabolites on tissue sections. This technique is still in a very early stage of the development. There is a need for the further improvement in instrumentation and sample preparation. The use of new ion sources [75], new matrices [76], improved lasers [77], and advanced imaging software will improve our current ability to profile tissues by expanding the technological capabilities to decrease analysis time, and to increase sensitivity and versatility. The collaboration between analytical scientists and biological scientists will result in better experiment design and result interpretation. The potential of MALDI–IMS is great, and advances in the instrumentation and operating protocols will bring new applications and insights into molecular processes involving health and disease.

REFERENCES

1. Hsieh, Y. et al., High performance liquid chromatography–atmospheric pressure photoionization/tandem mass spectrometric analysis for small molecules in plasma, *Anal. Chem.*, 75, 3122, 2003.
2. Hsieh, Y. et al., Direct plasma analysis of drug compounds using monolithic column liquid chromatography and tandem mass spectrometry, *Anal. Chem.*, 75, 1812, 2003.
3. Korfmacher, W.A., Lead optimization strategies as part of a drug metabolism environment, *Curr. Opin. Drug Discov. Develop.*, 6, 481, 2003.
4. Korfmacher, W.A., Bioanalytical analysis in a drug discovery environment. In W.A. Korfmacher (Ed.), *Using Mass Spectrometry for Drug Metabolism Studies*, Boca Raton, FL: CRC Press, pp. 1–34, 2005.
5. Korfmacher, W.A., Principles and applications of LC-MS in new drug discovery, *Drug Discov. Today,* 10, 1357, 2005.
6. Chaurand, P. et al., Imaging mass spectrometry: Principles and potentials, *Toxico. Path.*, 33, 92, 2005.
7. Rohner, T.C., Staab, D., and Stoeckli, M., MALDI mass spectrometric imaging of biological tissue sections, *Mech. Ageing Dev.*, 126, 177, 2005.
8. Rubakhin, S.S. et al., Imaging mass spectrometry: Fundamentals and applications to drug discovery, *Drug Discov. Today*, 10, 823, 2005.
9. Rudin, M., Rausch, M., and Stoeckli, M., Molecular imaging in drug discovery and development: Potential and limitations of nonnuclear methods, *Molecular Imaging Biology,* 7, 5, 2005.
10. Chaurand, P., Schwartz, J., and Caprioli, R.M., Profiling and imaging proteins in tissue sections by MS, *Anal. Chem.*, 76, 87, 2004.
11. Stoeckli, M. et al., Imaging mass spectrometry: A new technology for the analysis of protein expression in mammalian tissues, *Nat. Med.*, 7(4), 493, 2001.
12. Todd, P.J. et al., Organic ion imaging of biological tissue with secondary ion mass spectrometry and matrix-assisted laser desorption/ionization, *J. Mass Spectrom.*, 36, 355, 2001.
13. Cornett, D.S. et al., MALDI imaging mass spectrometry: Molecular snapshots of biochemical systems, *Nat. Methods*, 4(10), 828, 2007.
14. Kang, M., Tholey, A., and Heinzle, E., Quantitation of low molecular mass substrates and products of enzyme catalyzed reactions using matrix-assisted laser desorption/ionization time-of-flight mass spectrometry, *Rapid Commun. Mass Spectrom.*, 14, 1972, 2000.
15. Duncan, M.W., Matanovic, G., and Cerpa-Poljak, A., Quantitative analysis of low molecular weight compounds of biological interest by matrix-assisted laser desorption ionization, *Rapid Commun. Mass Spectrom.*, 7, 1090, 1993.
16. Chen, Y. and Ling, Y., Detection of water-soluble vitamins by matrix-assisted laser desorption/ionization time-of-flight mass spectrometry using porphyrin matrices, *J. Mass Spectrom.*, 37, 716, 2002.

17. Donegan, M. et al., Sample preparation methods for MALDI analysis of small molecule metabolites, *Proceedings of the 51st ASMS Conference*, Montreal, Quebec, Canada, 2003.
18. Ayorinde, F.O. et al., Use of meso-tetrakis(pentafluorophenyl)porphyrin as a matrix for low molecular weight alkylphenol ethoxylates in laser desorption/ionization time-of-flight mass spectrometry, *Rapid Commun. Mass Spectrom.*, 13, 2474, 1999.
19. Bunch, J., Clench, M.R., and Richards, D.S., Determination of pharmaceutical compounds in skin by imaging matrix-assisted laser desorption/ionisation mass spectrometry, *Rapid Commun. Mass Spectrom.*, 18, 3051, 2004.
20. Crossman, L. et al., Investigation of the profiling depth in matrix-assisted laser desorption/ionization imaging mass spectrometry, *Rapid Commun. Mass Spectrom.*, 20, 284, 2006.
21. Hsieh, Y. et al., Matrix-assisted laser desorption/ionization imaging mass spectrometry for direct measurement of clozapine in rat brain tissue, *Rapid Commun. Mass Spectrom.*, 20, 965, 2006.
22. Reyzer, M.L. et al., Direct analysis of drug candidates in tissue by matrix-assisted laser desorption/ionization mass spectrometry, *J. Mass Spectrom.*, 38, 1081, 2003.
23. Troendle, F.J., Reddick, C.D., and Yost, R.A., Detection of pharmaceutical compounds in tissue by matrix-assisted laser desorption/ionization and laser desorption/chemical ionization tandem mass spectrometry with a quadrupole ion trap, *J. Am. Soc. Mass Spectrom.*, 10, 1315, 1999.
24. Hsieh, Y. et al., Mapping pharmaceuticals in tissues using MALDI imaging mass spectrometry, *J. Pharmacol. Toxicol. Methods*, 55, 193, 2006.
25. Boy, C. et al., Imaging dopamine D_4 receptors in the living primates brain: A positron emission tomography study using the novel D_1/D_4 antagonist [^{11}C]SDZ GLC 756, *Synapse*, 30, 341, 1998.
26. Ekesbo, A. et al., Effects of the substituted (S)-3-phenylpiperidine (-)-OSU6162 on PET measurements of [^{11}C]SCH23390 and [^{11}C]raclopride binding in primate brains, *Neuropharmacology*, 38, 331, 1999.
27. Saleem, A. et al., Modulation of fluorouracil tissue pharmacokinetics by eniluracil: In vivo imaging of drug action, *Lancet*, 355, 2125, 2000.
28. Moresco, R.M. et al., PET in psychopharmacology, *Pharmacol. Res.*, 44(3), 151, 2001.
29. Ishiwata, K. et al., Positron emission tomography and ex vivo and in vitro autoradiography studies on dopamine D_2-like receptor degeneration in the quinolinic acid-lesioned rat striatum: Comparison of [^{11}C]raclopride, [^{11}C]nemonapride, and [^{11}C]*n*-methylspiperone, *Nucl. Med. Biol.*, 29, 307, 2002.
30. Kassiou, M. et al., (+)-[^{76}Br]A-69024 : A non-benzazepine radioligand for studies of dopamine D_1 receptors using PET, *Nucl. Med. Biol.*, 29, 295, 2002.
31. Aboagye, E.O. and Price, P.M., Use of positron emission tomography in anticancer drug development, *Invest. New Drugs*, 21, 169, 2003.
32. Singh, M. and Walnuch, V., Physics and instrumentation for imaging in vivo drug distribution, *Adv. Drug Delivery Rev.*, 41, 7, 2000.
33. Port, R.E. and Wolf, W., Noninvasive methods to study drug distribution, *Invest. New Drugs*, 21, 157, 2003.
34. Faas, H. et al., Monitoring the intragastric distribution of a colloidal drug carrier model by magnetic resonance imaging, *Pharm. Res.*, 18(4), 460, 2001.
35. Kupriyanov, V.V. et al., The effects of drugs modulating K+ transport on Rb+ uptake and distribution in pig hearts following regional ischemia: 87Rb MRI study, *NMR Biomed.*, 15, 348, 2002.
36. Noworolski, S.M. et al., High spatial resolution ^{1}H-MRSI and segmented MRI of cortical gray matter and subcortical white matter in three regions of the human brain, *Magnet. Reson. Med.*, 41, 21, 1999.

37. Solon, E.G. and Kraus, L., Quantitative whole-body autoradiography in the pharmaceutical industry: Survey results on study design, methods, and regulatory compliance, *J. Pharmacol. Toxicol.*, 46, 73, 2002.
38. Coe, R.A.J., Quantitative whole-body autoradiography, *Regul. Toxicol. Pharm.*, 31, S1, 2000.
39. Langlois, X. et al., Use of the beta-imager for rapid ex vivo autoradiography exemplified with central nervous system penetrating neurokinin 3 antagonists, *J. Pharmacol. Exp. Ther.*, 299(2), 712, 2001.
40. Tribollet, E. et al., Localization and quantitation of tritiated compounds in tissue sections with a gaseous detector of beta particles: Comparison with film autoradiography, *Proc. Natl. Acad. Sci.*, 88, 1466, 1991.
41. Lees, J.E. et al., An MCP-based system for beta autoradiography, *IEEE Trans. Nucl. Sci.*, 46(3), 636, 1999.
42. Lees, J.E., Fraser, G.W., and Carthew, P., Microchannel plate detectors for ^{14}C autoradiography, *IEEE Trans. Nucl. Sci.*, 45(3), 1288, 1998.
43. McDonnell, L.A. and Heeren, R.M.A., Imaging mass spectrometry, *Mass Spectrom. Rev.*, 26, 606, 2007.
44. Pacholski, M.L. and Winograd, N., Imaging with mass spectrometry, *Chem. Rev.*, 99, 2977, 1999.
45. Pacholski, M.L. et al., Static time-of-flight secondary ion mass spectrometry imaging of freeze-fractured, frozen-hydrated biological membranes, *Rapid Commun. Mass Spectrom.*, 12, 1232, 1998.
46. Audinot, J.-N. et al., Detection and quantification of benzodiazepines in hair by TOF-SIMS: Preliminary results, *Appl. Surf. Sci.*, 203–204, 718, 2003.
47. Winograd, N., Prospects for imaging TOF-SIMS: From fundamentals to biotechnology, *Appl. Surf. Sci.*, 203–204, 13, 2003.
48. Guerquin-kern, J.-L. et al., Complementary advantages of fluorescence and SIMS microscopies in the study of cellular localization of two new antitumor drugs, *Microsc. Res. Techniq.*, 36, 287, 1997.
49. Chandra, S., Lorey II, D.R., and Smith, D.R., Quantitative subcellular secondary ion mass spectrometry (SIMS) imaging of boron-10 and boron-11 isotopes in the same cell delivered by two combined BNCT drugs: In vitro studies on human gliblastoma T98G cells, *Radiat. Res.*, 157, 700, 2002.
50. Fragu, P. and Kahn, E., Secondary ion mass spectrometry (SIMS) microscopy: A new tool for pharmacological studies in humans, *Microsc. Res. Techniq.*, 36, 296, 1997.
51. Todd, P.J. et al., Organic ion imaging of biological tissue with secondary ion mass spectrometry and matrix-assisted laser desorption/ionization, *J. Mass Spectrom.*, 36, 355, 2001.
52. McCandlish, C.A., McMahon, J.M., and Todd, P.J., Secondary ion images of the rodent brain, *J. Am. Soc. Mass Spectrom.*, 11, 191, 2000.
53. Todd, P.J. et al., Organic SIMS of biologic tissue, *Anal. Chem.*, 69, 529A, 1997.
54. Wong, S.C.C. et al., Development of a $C60^+$ ion gun for static SIMS and chemical imaging, *Appl. Surf. Sci.*, 203–204, 219, 2003.
55. Caprioli, R.W., Farmer, T.B., and Gile, J., Molecular imaging of biological samples: Localization of peptides and proteins using MALDI-TOF MS, *Anal. Chem.*, 69, 4751, 1997.
56. Hsieh, Y., Chen, J., and Korfmacher, W.A., Mapping pharmaceuticals in tissues using MALDI imaging mass spectrometry, *J. Pharmacol. Toxicol. Methods*, 55(2), 193, 2007.
57. Schwartz, S.A., Reyzer, M.L., and Caprioli, R.M., Direct tissue analysis using matrix assisted laser desorption/ionization mass spectrometry: Practical aspects of sample preparation, *J. Mass Spectrom.*, 38, 699, 2003.

58. Chaurand, P. et al., New developments in profiling and imaging of proteins from tissue sections by MALDI mass spectrometry, *J. Proteome Res.*, 5, 2889, 2006.
59. Khatib-Shahidi, S. et al., Direct molecular analysis of whole-body animal tissue sections by imaging MALDI mass spectrometry, *Anal. Chem.*, 78, 6448, 2006.
60. Shimma, S. et al., Direct MS/MS analysis in mammalian tissue sections using MALDI-QIT-TOFMS and chemical inkjet technology, *Surf. Interface Anal.*, 38, 1712, 2006.
61. Aerni, H.R. et al., Automated acoustic matrix deposition for MALDI sample preparation, *Anal. Chem.*, 78, 827, 2006.
62. Ohkawa, T. and Bunch, J., MALDI MS imaging to reveal distribution of benzodiazepine drug and metabolite molecules in rat brain, *Proceedings of the 55th ASMS Conference*, Indianapolis, IN, 2007.
63. Cohen, S.L. and Chait, B.T., Influence of matrix solution conditions on the MALDI-MS analysis of peptides and proteins, *Anal. Chem.*, 68, 31, 1996.
64. Dreisewerd, K.L. et al., Influence of the laser intensity and spot size on the desorption of molecules and ions in matrix-assisted laser desorption/ionization with a uniform beam profile, *Int. J. Mass Spectrom.*, 141, 127, 1995.
65. Stoeckli, M. et al., Compound and metabolite distribution measured by MALDI mass spectrometric imaging in whole-body tissue sections, *Int. J. Mass Spectrom.*, 260, 195, 2007.
66. Reyzer, M.L. and Caprioli, R.M., MALDI-MS-based imaging of small molecules and proteins in tissues, *Curr. Opin. Chem. Biol.*, 11, 29, 2007.
67. Pevsner, P.H. et al., Direct identification of proteins from T47D cells and murine brain tissue by matrix-assisted laser desorption/ionization postsource decay/collision-induced dissociation, *Rapid Commun. Mass Spectrom.*, 21, 429, 2007.
68. Atkinson, S.J. et al., Examination of the distribution of the bioreductive drug AQ4N and its active metabolite AQ4 in solid tumours by imaging matrix-assisted laser desorption ionization mass spectrometry, *Rapid Commun. Mass Spectrom.*, 21, 1271, 2007.
69. Wang, H.-Y.J. et al., Localization and analyses of small drug molecules in rat brain tissue sections, *Anal. Chem.*, 77, 6682, 2005.
70. Garrett, T.J. et al., Analysis of intact tissue by intermediate-pressure MALDI on a linear ion trap mass spectrometer, *Anal. Chem.*, 78, 2465, 2006.
71. Cha, S.W. et al., Colloidal graphite-assisted laser desorption/ionization mass spectrometry and MS^n of small molecules. 1. Imaging of cerebrosides directly from rat brain tissue, *Anal. Chem.*, 79, 2373, 2007.
72. Woods, A.S. et al., IR-MALDI-LDI combined with ion mobility orthogonal time-of-flight mass spectrometry, *J. Preoteome Res.*, 5, 1484, 2006.
73. Jackson, S.N. et al., Direct tissue analysis of phospholipids in rat brain using MALDI-TOFMS and MALDI-ion mobility-TOFM, *J. Am. Soc. Mass Spectrom.*, 16, 133, 2005.
74. McLean, J.A. et al., Profiling and imaging of tissues by imaging ion mobility-mass spectrometry, *J. Mass Spectrom.*, 42, 1099, 2007.
75. Wiseman, J.M. et al., Mass spectrometric profiling of intact biological tissue by using desorption electrospray ionization, *Angew Chem. Int. Ed. Engl.*, 44, 7094, 2005.
76. Lemaire, R. et al., Solid ionic matrixes for direct tissue analysis and MALDI imaging, *Anal. Chem.*, 78, 809, 2006.
77. Holle, A. et al., Optimizing UV laser focus profiles for improved MALDI performance, *J. Mass Spectrom.*, 41, 705, 2006.

12 MALDI IMS for Proteins and Biomarkers

Michelle L. Reyzer and Richard M. Caprioli

CONTENTS

12.1 INTRODUCTION

The introduction of matrix-assisted laser desorption/ionization (MALDI) [1–3] in the late 1980s ushered in a new era of biological research, allowing large, intact biomolecules, primarily proteins, to be analyzed by mass spectrometry (MS) for the first time. Building upon this innovation, for the past decade MALDI MS has been applied to the direct analysis of proteins and peptides from thin tissue sections [4–6]. This type of analysis is advantageous for several reasons, including minimal sample preparation requirements, relatively fast analysis, and the analysis of the active proteome complement that is present at a given location and time. One primary advantage of this technology is the ability to spatially map, or image, the distribution of any protein or peptide across the entire tissue section.

This ability to combine spatial information with high molecular specificity is unique to MS, and over the past decade MALDI imaging has flourished. The goal of this chapter is to briefly summarize the MALDI imaging mass spectrometry (IMS) experiment for proteins and biomarkers, to describe seminal research illustrating the strengths and versatility of the technique, and to elaborate on current advances, examples, and challenges relating to direct protein imaging by MALDI MS. While this technology is currently widely available and new applications and advances are continuing to be reported by many researchers, this chapter highlights examples from the author's laboratory.

12.2 MALDI IMS: A BRIEF HISTORY

As was first described in 1997 [4], the application of MALDI MS to the direct analysis of proteins in tissue sections is illustrated in Figure 12.1. The general procedure is to obtain a thin section of a sample of interest, apply matrix (which is necessary for the MALDI process), and then acquire mass spectra at discrete *x*-, *y*-coordinates over the entire sample section. Each individual mass spectrum contains signals corresponding to proteins (and other endogenous compounds, including peptides, lipids, and metabolites) present in the sample at a unique set of coordinates. The intensities of any given molecular species may then be plotted as a function of position on the sample surface.

In the imaging approach, the technology is a powerful discovery tool, highlighting areas of the tissue where certain proteins are abundant or absent, where proteins may colocalize with other proteins, and where proteins may localize in histologically well-defined areas. In addition, the signals obtained primarily represent the proteins as they are found in that region of tissue at that moment of time, including all the processing and/or posttranslational modifications they have undergone. Indeed, the first applications of this technology in the imaging mode showed localizations of proteins in well-defined and symmetric compartments of mouse brain tissue, as shown in Figure 12.2 [6]. Similarly, differential localizations of several proteins, including thymosin β4, β actin, and S100A4, were shown in different regions of a section of mouse xenograft glioblastoma tumor.

Another way of implementing this technology is in the profiling mode, which can simply be thought of as a low-resolution imaging experiment with high morphological

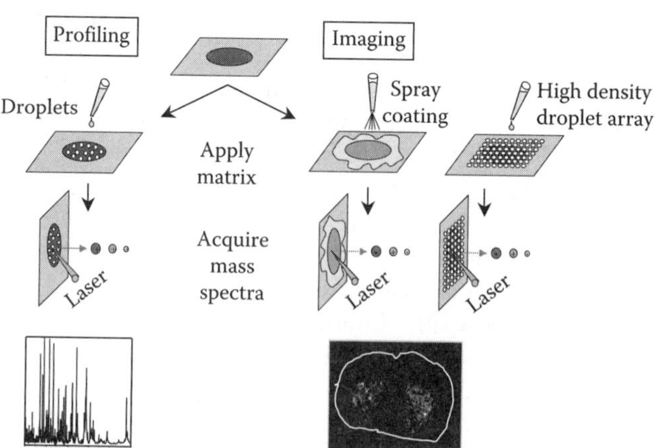

FIGURE 12.1 (See color insert following page 210.) General procedure for MALDI MS imaging experiments in profiling (left side) and imaging (right side) modes.

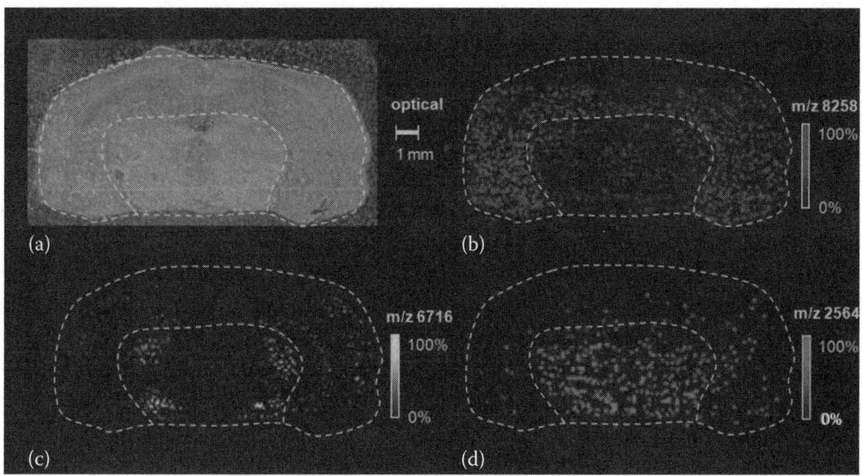

FIGURE 12.2 (See color insert following page 210.) One of the first examples of mass spectrometric images of proteins in a mouse brain section. (a) Optical image of the tissue section mounted on a gold-coated plate. (b) *m/z* 8528 in the regions of the cerebral cortex and the hippocampus. (c) *m/z* 6716 in the regions of the substantia nigra and medial geniculate nucleus. (d) *m/z* 2564 in the midbrain. (Reproduced from Stoeckli, M. et al., *Nat. Med.*, 7, 493, 2001. With permission.)

specificity. In this mode, matrix is deposited in discrete locations upon the tissue surface corresponding to areas that are highly enriched in a cell population of interest (normal epithelial cells, areas of surrounding stroma, invasive cancerous cells, etc.). Mass spectra are acquired at each spot (so that the spectra are still linked to a location on the tissue surface), but the analysis is more typically a biostatistical analysis where protein profiles or signatures are compared between and among different regions, often with the goal of discovering biomarkers that can be used in diagnostic and/or prognostic analyses. In many cases, both imaging and profiling experiments are undertaken on the same samples, thus generating complementary data that, for example, visually show the location of statistically significant biomarkers.

The first applications of the profiling approach involved human lung cancer [7] and human brain cancer (glioma) [8,9] samples. In both cases, groups of signals obtained from cancerous tissues were found to be significantly different from signals obtained from noncancerous tissues, and different statistical algorithms were utilized to successfully distinguish cancer from noncancer and to classify subtypes of tumors.

12.3 TECHNOLOGY DEVELOPMENTS: DATA ACQUISITION AND PROCESSING

In the decade since its inception, many significant advances have occurred to make the MALDI imaging process more widely available, more automated, and more robust. These improvements involved more sophisticated sample preparation methods as well as commercial instrumental and software improvements. In general, MALDI imaging may be performed on any mass spectrometer equipped with a MALDI source, but realistically some type of software improvements are necessary to facilitate data acquisition. Indeed, the first images were generated via manual control of the sample stage where data acquisition took ~1 min/pixel. Figure 12.3a shows an example of an "image" of rat pancreas tissue taken before any image generation or processing software was available. It shows a graphical representation of the location of the signal intensity at *m/z* 5802 (insulin) along a linear coordinate.

Thus, the first improvement was a software program written in-house (MS Image Tool) to automate instrument control, data sampling, and data processing [10]. This improved data acquisition times to ~1 s/pixel or better. Currently, software to automate image acquisition and perform some image processing is commercially available from several MS instrument vendors, including FlexImaging from Bruker Daltonics, ImageQuest from Thermo Scientific, MassLynx from Waters, and oMALDI server [11] and TissueView from Applied Biosystems/MDS Sciex. These tools allow the user to specify instrumental parameters such as spatial resolution, laser shots per spot, and pixel coordinates, as well as provide some data processing and image generation capabilities. Along with higher frequency lasers and faster electronics, these software packages allow for imaging times on the order of 10–100 ms/pixel (depending on desired resolution and sensitivity). An example of an image acquired using current technology of similar tissue (human pancreas) is shown in Figure 12.3b. Clearly, advances in the imaging process are visible, as the expression of insulin (*m/z* 5808) is more accurately visualized as part of the tissue section.

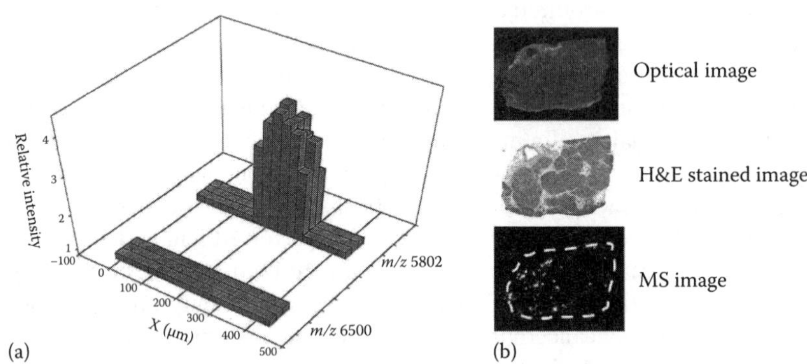

FIGURE 12.3 (See color insert following page 210.) Technological developments have facilitated the development and processing of MALDI image data. (a) Rudimentary "image" of an area of rat pancreas tissue, 450 × 75 μm. The reconstructed ion image at *m/z* 5802 shows the presence of insulin in an islet while the image at *m/z* 6500 serves as a control. (Reproduced from Caprioli, R.M. et al., *Anal. Chem.*, 69, 4751, 1997. With permission.) (b) Software developments allowed this image of human normal pancreas tissue (adjacent to tumor) to be obtained. From top to bottom: optical image of the tissue section, H&E stained section, and MALDI MS image of insulin (*m/z* 5808) visualized with Biomap software obtained at 150 × 150 μm resolution. Insulin is clearly visualized within many islet cells.

In addition, several software programs are freely available to assist with the imaging process, including BioMap from Novartis [12] (which is an image processing program for use on any data set that has been transformed into the Analyze file format), *CreateTarget* and *Analyze This!* [13] (which work with Bruker instruments to create custom geometry files and to export data into the Analyze format for further analysis with BioMap), and Axima2Analyze (converts data from Shimadzu Axima QIT instruments to Analyze format). These software packages primarily assist with image generation and data processing, and are currently available for download from the Mass Spectrometry Imaging interest group website (http://www.maldi-msi.org).

In the interest of generating a "universal" tool that can be used to analyze MS data acquired from any one of the various MS instruments commonly used for peptide and protein analysis (often with their own proprietary data file structure), the Human Proteome Organization (HUPO) has proposed creating a standard format for MS data, called mzML (building on a previous version, mzXML) [14]. A new imaging software package has been developed to take advantage of the XML format, MITICS [15], which serves to both control image data acquisition on Applied Biosystems MALDI-TOF Voyager series instruments (MITICS Control) and provide for data processing and image reconstruction in the XML format (MITICS Image). An extension of the standard MS data format, called i-mzML (http://www.i-mzml.org), for imaging data has also been proposed. This trend toward making all imaging data available in one standard format will assist in image evaluation and validation between and among different acquisition platforms and should allow further development of the technology to progress more rapidly.

12.4 TECHNOLOGY DEVELOPMENTS: AUTOMATION

In addition to the increased availability of image-specific software packages, there are a number of tools now commercially available that are designed to automate various aspects of sample preparation. The ultimate goal for imaging applications is to acquire images that accurately show the distribution of proteins while maintaining spatial integrity. The requisite addition of a matrix to the tissue section, typically as a solution, introduces the possibility of analyte delocalization as analytes may be solubilized by the matrix solution and migrate over the surface of the tissue section. Two instruments are now commercially available that enable automated spray-coating of matrix on tissue sections. Bruker Daltonics manufactures the ImagePrep system, which nebulizes small amounts of matrix solution and allows the "mist" to fall onto a tissue section mounted on a glass slide. By monitoring the transmission of light through the glass slide, the instrument can estimate the relative "wetness" of the matrix coating, and optimize the timing for future coats of matrix. The TM-Sprayer by LEAP Technologies [16] utilizes a spray nozzle to create a fine mist of matrix, which is deposited on the sample plate as it is moved in an x, y pattern under the spray nozzle. One recent report also describes the use of a commercial desktop inkjet printer for automated matrix deposition [17]. This approach provides a less expensive alternative to the commercially available robotic sprayers and accommodates the possibility of using multiple solvents.

The ultimate goal of these robotic instruments, as with manual spray-coating, is to generate a thin, homogeneous coating of matrix crystals over the tissue surface while maintaining the spatial integrity of the solubilized proteins. Automating the process should make it more reproducible and thus improve the quality of the images produced.

12.5 TECHNOLOGY DEVELOPMENTS: SAMPLE PREPARATION

Apart from automating the matrix application process, it is critical to evaluate the resulting matrix crystals and coating for analyte extraction, localization, and effect on tissue architecture. Several reports have shown that even for standard analytes mixed with matrices, there is an uneven distribution of the analyte within the MALDI crystals and that sample preparation influences the resulting distribution [18–20]. Thus, other active areas of research are focused on optimizing matrix application and sample preparation protocols.

For example, one way to avoid solvent-induced analyte delocalization is to apply matrix without any solvent. Two groups have explored this approach, one using sublimation to apply matrix to tissue sections [21], and the other using finely ground powdered matrix applied through a sieve [22]. Both of these approaches were evaluated on mouse brain tissue and were found to produce homogeneous coatings consisting of fine microcrystals (<20 μm). Relatively high-resolution images (~30–100 μm) were acquired from various phospholipids present in the brain tissues illustrating high degrees of analyte localization within various substructures. An example of the DHB crystals obtained after dry-coating DHB on brain tissue is shown in Figure 12.4a, along with representative images of phospholipids obtained from spray-coating and dry-coating mouse brain sections (Figure 12.4b). While these results are very promising for lipids, the applicability of these techniques to larger, more polar analytes, including peptides and proteins, is yet to be fully examined.

Another approach to matrix application is to robotically deposit small (pL to nL) droplets of matrix on discrete areas of the sample surface. When the droplets are deposited as arrays over the entire tissue surface, mass spectra can be acquired at each matrix spot and reconstructed into an image. In this case, potential analyte delocalization is limited to the area under the matrix spot, typically ~100–250 μm with commercially available spotters (e.g., Portrait 630 by Labcyte [23], ChIP by Shimadzu, TM iD by LEAP Technologies). This approach is fundamentally a trade-off between spatial integrity and resolution, because what limits resolution in most cases is the diameter of the laser beam, which is typically much smaller than the diameter of the matrix spot (on the order of 30–50 μm). Samples prepared by spray-coating, either manually or robotically, may be imaged at the diameter of the laser beam, but may not have maintained analyte localization throughout the spray process.

Another goal of matrix application, important to diagnostic samples, is the preservation of tissue integrity for subsequent histological analysis. Classical histological techniques usually involve immersing the tissue in a solution, such as formalin or ethanol, to fix the proteins while maintaining cellular structure. Formalin works by crosslinking proteins, and thus is not readily compatible with MALDI MS analysis,

FIGURE 12.4 (See color insert following page 210.) Developments in sample preparation have led to matrix coating procedures that induce minimal delocalization. (a) Scanned optical image of a dry-coated sagittal mouse brain section on a gold-coated MALDI plate, with a 4× magnification of the DHB layer shown on the right. (b) Comparison of phospholipid images obtained from a sagittal mouse brain either spray-coated (middle) or dry-coated (right) with DHB. (Adapted from Puolitaival, S.M. et al., *J. Am. Soc. Mass Spectrom.*, 19, 882, 2008. With permission.)

although some work has been done with samples fixed in this way using alternative approaches [24]. (This will be discussed in more detail in the application section.) Implementing solvent rinses as a means to improve peptide and protein sensitivity for MALDI IMS presupposes that the solvent will rinse away compounds on the tissue surface, primarily salts and lipids, that compete for protons or otherwise inhibit protein cocrystallization and/or ionization. However, any time the tissue is immersed in a liquid, the possibility exists first that analytes of interest may be soluble in the rinsing solvent and thus be extracted and lost, and second that soluble analytes may be extracted and redeposited elsewhere on the tissue surface, thus inducing delocalization. Methodology development must incorporate a full investigation of these effects prior to the implementation of any washing protocol, and it must be kept in mind that the washes may behave differently on different tissue surfaces (i.e., liver vs. brain vs. tumor). Nonetheless, different rinses and fixation protocols have been described by several groups and by and large have been shown to improve MALDI sensitivity for peptides and proteins while maintaining tissue architecture and preserving the tissues over time [25–29].

First reported as a method to improve the crystal coverage and crystal formation of α-cyano-4-hydroxycinnamic acid (CHCA) on tissues, largely due to the removal of surface salts and lipids [29], further studies have systematically investigated the tissue fixation properties of various organic solvent rinses [26,27]. Lemaire et al. [27] developed a tissue-washing procedure based on organic solvents typically used for lipid extraction. Using rat brain sections as the standard tissue, they compared MALDI signals in the protein range with and without rinsing the tissue with different organic solvents, including chloroform, hexane, acetone, toluene, and xylene. They found a 44% increase in the number of signals detected (m/z >3000) in tissue rinsed with xylene and a 34% increase in tissue rinsed with chloroform compared to control (untreated) tissue, as well as an increase in overall signal intensity. A decrease in the standard deviation was observed for all solvents tested. Tissue cellular structure was determined with optical microscopy and immunohistochemistry to be intact after rinsing with no notable delocalization of proteins.

Imaging experiments in both the peptide and protein range were undertaken with rat brain sections at least 6 months old prior to analysis [27]. Two adjacent sections were obtained, one was left untreated and the other was rinsed with chloroform. Sinapinic acid was applied as the matrix for protein imaging. For older tissues it was difficult to recognize brain structures without any treatment, but after rinsing the structure became apparent. The signal obtained for a relatively high mass protein at *m/z* 14,306 showed an increase in intensity after chloroform rinsing compared to the untreated section. Visualizing the spatial localization of the protein was thus enhanced after chloroform rinsing. The authors concluded that treatments with these organic solvents improved the detection and sensitivity of proteins via lipid removal without generating any peptide or protein delocalization. They also noted however, that there was not a reduction of salt adducts as has been shown with alcohol rinses.

Seeley et al. also systematically evaluated a series of organic solvents for the effects on protein signals from tissue using rat liver tissue as a model [26]. Alcohol treatments (ethanol, methanol, isopropanol) were evaluated in addition to lipid removing solvents (chloroform, xylene, toluene, hexane). In this case the goal was to determine the overall effect on protein signal after rinsing (number of signals, signal intensity) as well as tissue preservation, that is, how long after sectioning/rinsing/matrix application could the tissue be analyzed with minimal disruption to the signals. Similar to Lemaire, they found increases in both the number of protein signals and the signal reproducibility after most washes (except acetone) compared to the untreated control. They also reported that overall signal intensity and quality was maintained with no significant deterioration after washing with isopropanol for up to 1 week. Ethanol and isopropanol were also found to leave the tissues virtually indistinguishable from untreated tissues after histological staining, thus indicating no disruption of tissue architecture [26].

Pathologists use staining to enhance contrast between cellular regions on tissue sections to diagnose clinical specimens, with hematoxylin and eosin (H&E) being a common staining protocol. Combining this histological information with MALDI analyses of protein expression is quite powerful for biomarker discovery, because the cellular location of specific biomarkers can be assessed. However, this is typically accomplished using serial sections of a given tissue specimen because common staining protocols, such as H&E, have been found to compromise the MALDI signal [30]. Alternative staining protocols were thus evaluated in order to develop a protocol that could be used both for histological assessment and for MALDI MS. Toluidine blue, nuclear fast red, methylene blue, Terry's polychrome, and cresyl violet were applied to mouse liver sections in either water- or ethanol-based solvents [28]. Excess stain was removed with ethanol washing, and the resulting stained tissues were evaluated both for optical detection of cellular features and for the resulting MALDI spectra after matrix application. In this case, MALDI signals from methylene blue and cresyl violet were similar to control (ethanol washed) tissue, in terms of number of signals and signal intensity. It was also shown that cresyl violet offered high nuclear to cytoplasm contrast necessary for adequate histological analysis of cancer tissue [28]. This analysis allows for MALDI IMS and histological analysis to be performed on the same tissue section instead of serial sections, which can eliminate any errors associated with sectioning and is valuable for very small clinical diagnostic specimens, such as needle biopsies.

Agar et al. also investigated histology-compatible tissue preparation protocols for MALDI IMS [25]. In this case, they modified several established immunohistochemistry protocols to include dissolved MALDI matrices in the solvent rinses, and they examined the resulting tissue architecture and MALDI protein signals using mouse brain tissue as a model. They found an ethanol/methanol/acetonitrile/0.1% trifluoroacetic acid (in water) mix in a 2:2:1:1 ratio containing 25–30 mg/mL of matrix resulted in the best MALDI signal (in terms of number of signals and S/N of the resulting signals) while minimizing diffusion of proteins and tissue deformation. Again, this protocol would allow one single tissue section to be stained, visually examined for histological features, and then subjected to MALDI IMS analysis where the resulting protein images could be directly correlated with tissue architecture.

12.6 EXAMPLES AND APPLICATIONS

MALDI MS offers the ability to discretely analyze the proteomic composition of many well-defined substructures in various tissues, without the need for dissection, extraction, or homogenization. It has thus been used as a tool for the study of normal development [31] and of many varied pathological processes, used in a targeted or nontargeted approach. A wide variety of tissues, both animal and human, including brain [6,12,25,28,31,32], liver [26,28], kidney [33,34], eye [35], pituitary [36], pancreas [4], prostate [37], colon [5], and breast [38] have been analyzed using this methodology. Analyses of whole-body animal sections, rat or mouse, have also been described [39–41]. Pathological conditions, such as cancer [6–9,24,26,38,42–47], Parkinson's disease [48,49], Alzheimer's disease [12], diabetes, tuberculosis, preterm labor, and heart disorders have been examined. These lists are not meant to be inclusive, and certainly with time the kinds of tissues and conditions examined will expand. Rather, this chapter aims to highlight several applications that illustrate the unique capabilities and versatility of the MALDI IMS experiment.

12.6.1 Applications to Development and Disorders

The study of Parkinson's disease is often done in animal models where one side of the brain is lesioned, generally via administration of a chemical, such as 1-methyl-4-phenyl-1,2,3,6-tetrahydropyridine (MPTP) or 6-hydroxydopamine (6-OHDA), that disrupts the dopaminergic pathway. Differences in protein expression between the lesioned and unlesioned sides of the brain have been analyzed via MALDI MS. One study revealed significantly decreased levels of Purkinje cell protein 4 (PCP-4 or PEP-19) in the striatum of MPTP-lesioned mice compared to control [48]. This result was in agreement with data acquired from a variety of other analytical techniques, including liquid chromatography–tandem mass spectrometry, high-resolution MS, and *in situ* hybridization. In a more recent study, the same group used a targeted approach to evaluate the distribution and relative quantitation of the immunophilin FKBP-12 in the striatum of 6-OHDA-lesioned rats [49]. MALDI MS profiling revealed increased levels of intact FKBP-12 (*m/z* 11,791) in the dorsal and middle part of the dopamine-depleted striatum, but no difference in the level of FKBP-12 in the ventral part compared to the striatum in the intact hemisphere of the brain. These results were also supported by data from Western blotting and two-dimensional (2D) gel electrophoresis.

MALDI IMS was recently utilized to examine embryo implantation in mice [50]. A comprehensive understanding of the spatial and temporal changes that occur in the uterine environment prior to and during embryo implantation has yet to be realized. Successful implantations are necessary for successful pregnancies, so understanding the reasons for implantation failure has important ramifications for understanding and potentially treating some causes of infertility. However, because the biochemical events surrounding implantation are quite complex, with the roles and expressions of a large number of gene products remaining poorly understood during this process, this is an analytically challenging task. In addition, these implantation sites are approximately 2 mm in diameter, necessitating an analytical technique capable of high-resolution analysis. MALDI IMS allowed the authors to analyze hundreds of proteins involved in various biochemical processes during the peri-implantation period, and thus provide unique proteomic snapshots of implantation and interimplantation sites.

Implantation and interimplantation sites were excised from dissected mouse uteri on several days postimplantation and prepared for MALDI MS analysis. Profiling experiments were undertaken where matrix was deposited on discrete morphological areas of the tissue, and proteomic signatures were recorded. It was determined that 50 signals were significantly changed in the uterus due to the presence and proximity of the embryo [50]. Imaging experiments were subsequently undertaken, and the resulting images for several of the statistically significant proteins, including calcyclin and ubiquitin, are shown in Figure 12.5. As shown, both the location and expression of the proteins changes as a result of time (days 5, 6, or 8 after implantation). The imaging results are in good agreement with the profiling results and mRNA localizations as determined by *in situ* hybridization (Figure 12.5).

MALDI IMS was further used to examine the proteomic profiles of wild type and *Pla2g4a* null mice, which have known defects in implantation and experience implantation delays. The results of these studies suggest that proteome signatures are different between wild type and *Pla2g4a* null mice, which may help distinguish which proteins are critical for on-time implantation, which leads to more positive pregnancy outcomes.

The evolution of protein expression during normal mouse prostate development was examined with MALDI IMS technology [37]. In this case, protein expression and spatial localization was followed as a function of lobe specificity and age: during organ development (1–5 weeks), at sexual maturation (6 weeks), and as adult (10, 15, or 40 weeks). In addition, the protein signatures from normal development

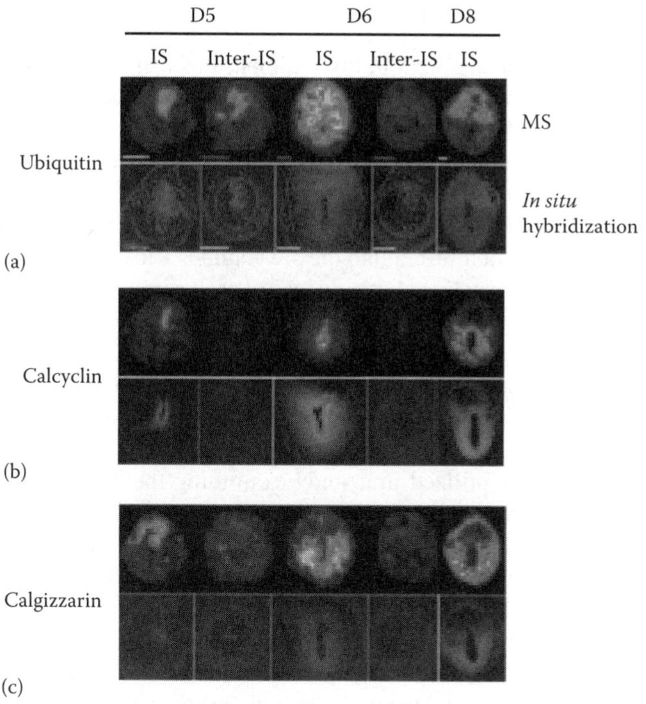

FIGURE 12.5 (See color insert following page 210.) Imaging MS shows spatial localization for selected proteins of interest on days 5, 6, and 8 of pregnancy. Spatial localization of (a) ubiquitin, (b) calcyclin, and (c) calgizzarin by imaging MS (upper panels) and respective mRNAs by *in situ* hybridization (lower panels) in the implantation site (IS) or interimplantation site (inter-IS) of the mouse uterus. Bar, 700 μm. (Adapted from Burnum, K.E. et al., *Endocrinology*, 149, 3274, 2008. With permission.)

were compared to 15 week prostate tumors. Several protein signals were found to be uniquely expressed during the very early time points, such as *m/z* 4937 and *m/z* 4965 (consistent with the molecular weights of thymosin β 10 and 4, respectively), suggesting these proteins may play a role in early prostate development. In contrast, signals identified as probasin and spermine-binding protein (SBP) were increasingly detected as the prostate sexually matured. Most proteins showing this trend were absent in the tumors, indicating that normal protein synthesis was interrupted as the cells dedifferentiated. In addition, numerous signals were found to be lobe-specific.

One of the strengths of this technique is the ability to identify posttranslationally modified proteins by following a change in mass. In this study, the authors reported a change in the proportion of α-*N*-acetylated (*m/z* 17,882) and nonacetylated (*m/z* 17,840) cyclophilin A (Cyp A) between the normal prostate and prostate tumors. (The identity of Cyp A was confirmed by HPLC separation, MALDI MS peptide fingerprinting, and MS/MS analysis). These results were confirmed by Western blotting and immunohistochemical analyses [37].

12.6.2 Three-Dimensional Imaging

The ability to localize biomarkers in three dimensions using MALDI imaging MS is an emerging approach for the analysis of protein profiles in organs. The initial proof-of-principle study optically registered 264 coronal mouse brain sections (20 μm thick) spanning the area of the corpus callosum to a stereotaxic coordinate system to produce a three-dimensional (3D) volume [32]. MALDI protein profiles were obtained on 10 of the 264 sections, which were approximately 400–500 μm apart, and the protein data inserted into the 3D volume. Thus, the distribution of any protein signal could be visualized not only within each coronal brain section but also over the entire 3D structure. Additional studies of 3D imaging have recently been reported for the rat brain model. The colocalization of not only proteins but also peptides or small endogenous or exogenous compounds were assessed in a 3D space.

One of the strengths of the MALDI MS technique for biomarker discovery is that many different types of compounds can be analyzed on a single analytical platform. This unique feature was utilized in a study examining the distribution of proteins and peptides in the same 3D volume from the substantia nigra and interpeduncular nucleus [51]. For this analysis, two series of adjacent sections were cut every 200 μm throughout the rostrocaudal axis of the rat midbrain. One section was coated with sinapinic acid as matrix for protein analysis and the other with DHB for peptide analysis. The matrix was spotted using a robotic matrix dispenser in an array with 200 μm spacing in both the *x*- and *y*-directions. As the sections were acquired every 200 μm (the *z*-direction), a single voxel consisted of the intensity of any *m/z* signal at one 200 × 200 × 200 μm point in 3D space. The small distance (12 μm) between adjacent sections taken for protein and peptide analyses was determined to be roughly the size of one medium-sized neuron, and thus negligible in terms of shifting the morphology in the regions, compared to the large 200 μm distance between points in 3D space. Similar to the initial experiment, a 3D volume of the analyzed brain area was constructed based on registering anatomical features, and the protein and peptide data were inserted into the 3D volume. Figure 12.6 illustrates the process from intrasection registration of the MALDI images with the matrix spots (Figure 12.6a), to creation of the 3D volume (Figure 12.6b), to the visualization of unique peptides and proteins in the same 3D space (Figure 12.6c). The results showed significant overlap for the peptide substance P (*m/z* 1347.8, shown in green) and the protein PEP-19 (*m/z* 6717, shown in purple) in the substantia nigra, while differential localization was observed in the interpeduncular nucleus [51].

Extending the visualization of proteins in three dimensions, another recent study attempted to integrate the 3D representation of MALDI MS protein data with 3D data acquired from another imaging modality, namely, magnetic resonance imaging [44]. This study utilized a mouse with a brain tumor, where the whole mouse head was subjected to MRI imaging on a 7T magnet, after which the mouse was sacrificed and frozen. Sequential cryosectioning for subsequent MALDI analysis was then performed, with high-resolution digital images acquired for each blockface section. In this case, the 3D volume recreated from the blockface images was coregistered to the magnetic resonance volume of the mouse's head. Due to the fact that the brain is surrounded by bone, minimal deformations occurred to the brain

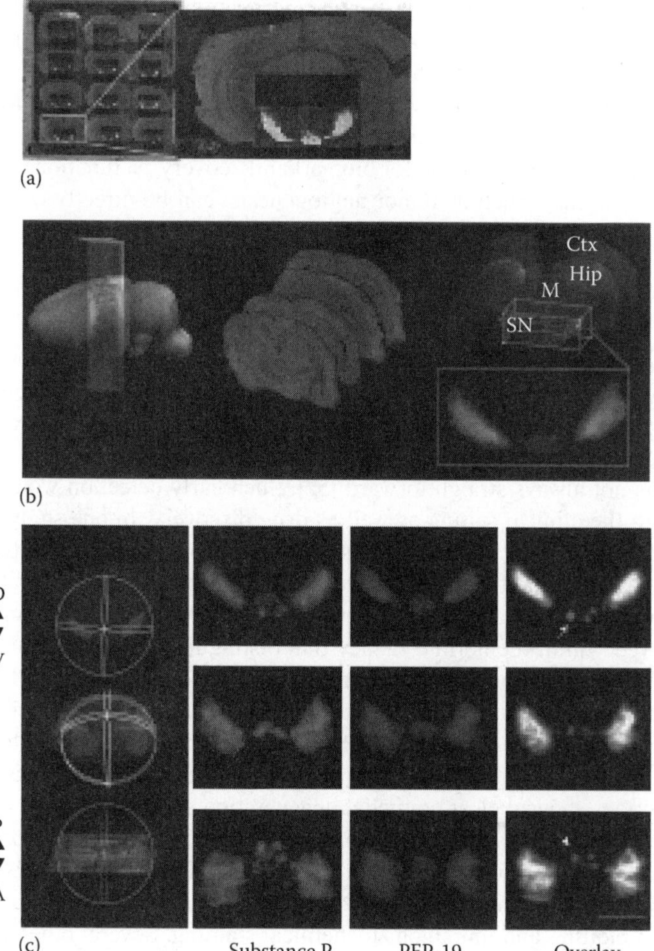

FIGURE 12.6 (See color insert following page 210.) 3D volume reconstructions of one peptide and one protein as detected by MALDI IMS of the rat ventral midbrain. (a) Intrasection registration of pixelated IMS images is performed so that each individual pixel matches its matrix deposit. (b) The intersection registration uses cresyl violet stained images to create a *z*-stack for alignment using Amira software. The 3D surface reconstructions of cortex (Ctx), hippocampus (Hip), midbrain (M), and substantia nigra (SN) visualize the alignment. The box outlines the volume analyzed by MALDI IMS. (c) MALDI IMS images of substance P (*m/z* 1347.8 $[M + H]^+$) and PEP-19 (*m/z* 6717 $[M + H]^+$) were registered using the same factors obtained for intra- and intersection registration. Virtual *z*-stacks were created in Image J software and 3D volume reconstructions of substance P and PEP-19 were rotated for frontal and dorsal views (top and bottom rows, respectively). Overlap of substance P and PEP-19 localization was assessed by combining *z*-stacks. Differential localization was observed in the interpeduncular nucleus. Anterior (A), posterior (P), ventral (V), dorsal (D). Scale bars, 2 mm. (Reproduced from Andersson, M. et al., *Nat. Methods*, 5, 101, 2008. With permission.)

during processing, so a relatively simple registration was required. Ion volumes were created for two proteins, the astrocytic phosphoprotein Pea 15 (*m/z* 15,035) and fatty acid-binding protein 5 (Fabp5, *m/z* 15,076) obtained from the MALDI data and rendered against the MRI data. These results showed the feasibility of correlating *in vivo* anatomical imaging with spatially resolved *ex vivo* molecular imaging. This is quite promising for biomarker discovery, as functional information from *in vivo* studies, such as tumor angiogenesis, can be directly correlated with molecular signatures from these areas.

12.6.3 Markers of Toxicity and Infection

MALDI MS has been utilized as part of a new strategy to investigate potential markers of toxicity. The kidney is often the major site of drug toxicity because it serves to filter and excrete exogenous compounds from the body. Nephrotoxicity is usually reversible if detected early so the toxicant can be discontinued, however early detection is not always straightforward [52]. Such early detection would have benefits in both the clinical setting as well as drug discovery. In one study, the known nephrotoxicant gentamicin was administered to rats for seven consecutive days [33]. Sections were acquired from kidneys from both control and gentamicin treated rats, and proteomic profiles and images were acquired. Differential protein expression was observed within the normal kidney that distinguished the substructures of the kidney, with unique signals expressed in the cortex, medulla, and papilla.

The gentamicin-treated kidneys showed differential expression of a number of signals compared to controls, mostly in the cortex, consistent with the known action of gentamicin. This correlated well with H&E stained sections, where the treated sections showed clear epithelial degeneration and regeneration in the proximal tubules. The most significant signal at *m/z* 12,959 was highly upregulated in the treated cortex but not detectable in the control (Figure 12.7). This signal was identified as transthyretin, a blood transporter protein. This result was confirmed by Western blotting and immunohistochemical staining. Several other signals were found to be differentially expressed in treated vs. control kidneys, and the use of IMS enabled the localization of the affected proteins. For example, a signal at *m/z* 8242 was found to be present throughout the cortex and medulla in the control kidney, but depleted only in the cortex after gentamicin treatment. Further studies are required to discover whether transthyretin is a general marker for kidney toxicity, a marker for proximal tubule damage, or is specific for gentamicin or aminoglycoside-induced toxicity. Nonetheless, this result is promising and provides additional information not obtainable by conventional 2D gel electrophoresis analysis of kidney homogenates [53] or plasma samples [54].

Understanding the molecular events underlying the immune response to infection may lead to the design of better antimicrobial agents, a task that grows more urgent as more bacteria become resistant to current treatments. MALDI MS was utilized to examine host factors that limit microbial growth in abscesses in a mouse model of *Staphylococcus aureus* infection [34]. An initial imaging experiment comparing kidneys from control mice and mice infected with *S. aureus* revealed high signals from a protein at *m/z* 10,165 that was localized to the abscesses. This signal was

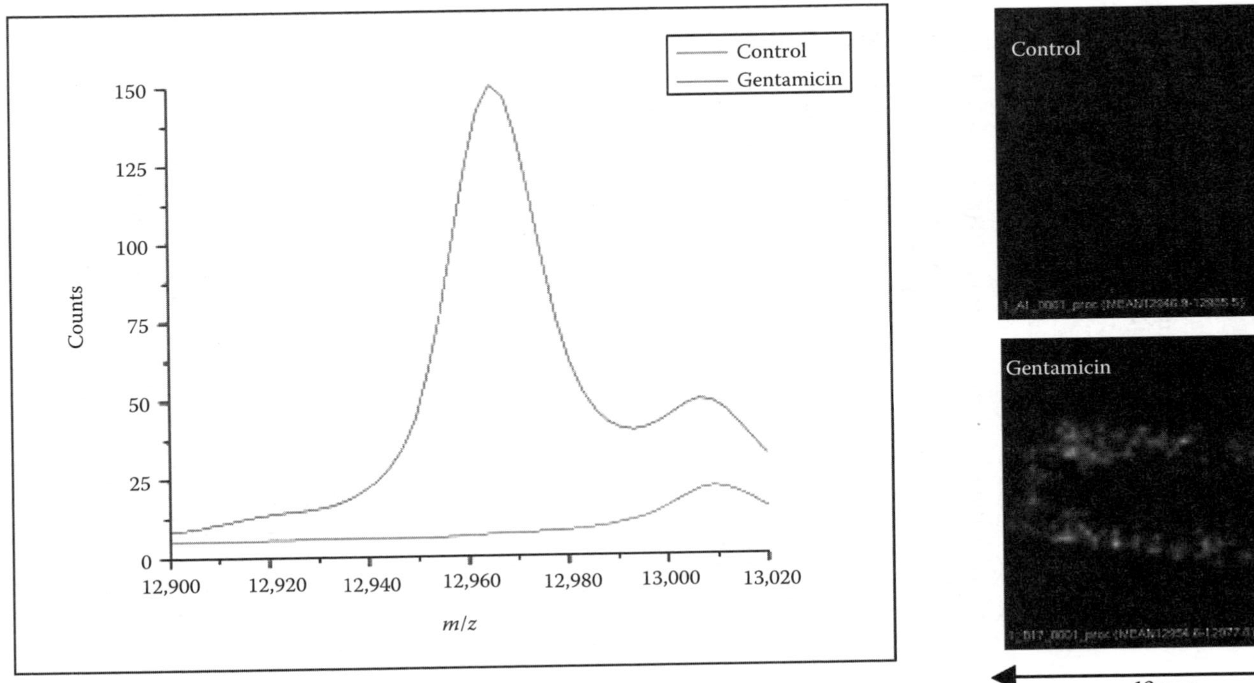

FIGURE 12.7 (See color insert following page 210.) Differential imaging analysis of a control vs. a gentamicin-treated rat kidney section. Average spectrum (cortex) and intensity distribution of m/z 12,959 on a control and a treated kidney section. (Adapted from Meistermann, H. et al., *Mol. Cell. Proteomics*, 5, 1876, 2006. With permission.)

subsequently identified as S100A8 (calgranulin A). Other imaging experiments also revealed a signal at *m/z* 12,976 that colocalized with S100A8 in the abscesses of infected kidneys. This protein was identified as S100A9 (calgranulin B). These two proteins make a functional heterodimer, calprotectin, which is a calcium-binding protein known to inhibit the growth of many bacterial pathogens. It was further shown that the proteins are only present in the abscesses in immunocompetent (neutrophil-replete) mice. A more thorough investigation of how calprotectin exerts its antibacterial activity determined that calprotectin likely starves the bacteria of the metal ions (Ca^{2+}, Mn^{2+}, Zn^{2+}) that *S. aureus* needs to survive [34].

12.7 NEW APPROACHES

A limitation of protein technologies is the relatively small subset of proteins that are readily amenable to detection (typically small proteins <50,000 Da, soluble proteins, and relatively abundant proteins). One approach to making this technology more suitable for other types of compounds, such as membrane proteins or mRNA, is the use of reporter tags. This strategy can be used to analyze compounds not generally amenable to MALDI, to multiplex the analysis of such compounds [55] (i.e., label several compounds with reporters of different mass so they may all be analyzed in one experiment), or to amplify the signal obtained from a low abundance analyte [56].

A recent report describes the use of MALDI profiling (untargeted) followed by a targeted imaging approach, to determine biomarkers for ovarian cancer [46]. In this study 25 late-stage ovary carcinomas and 23 benign ovaries were analyzed using MALDI MS. One putative biomarker was found in ~80% of the ovarian cancer samples at *m/z* 9744. This signal was subsequently identified as a fragment of the 11S proteasome activator complex, Reg-alpha. Western blotting and immunocytochemical studies were carried out to validate the specific presence of this protein in cancer samples compared to benign ovaries. Finally, a targeted imaging experiment was carried out, with a primary antibody to Reg-alpha applied to the carcinoma tissue. The primary antibody was associated with a secondary antibody that had been modified with a photocleavable linker and a reporter peptide. MALDI analysis of this tissue section detected the reporter peptide at the sites that were occupied by the anti-Reg-alpha primary antibody. These results correlated well with the untargeted MALDI imaging experiment showing the localization of the fragment at *m/z* 9744. Taken together, these findings support the 11S proteasome activator complex as a potential biomarker of ovarian cancer.

Another approach is to digest proteins directly on tissue sections and then analyze the resulting peptides. This technique can make peptides from insoluble membrane proteins or high molecular weight, low abundance proteins available for analysis, and it can provide a built-in validation, as all peptides from the same protein should have the same distribution. This was recently demonstrated for proteins in rat brain tissue [57]. For example, tryptic fragments from the higher molecular weight proteins actin (41 kDa), tubulin (55 kDa), and synapsin-1 (74 kDa) were all identified using *in situ* tryptic digestion on a single tissue section. The distribution of one tryptic peptide from synapsin-1 (*m/z* 1496.75) is shown in Figure 12.8. In addition, this approach can also be a means to profile formalin-fixed tissue specimens whose native proteins have

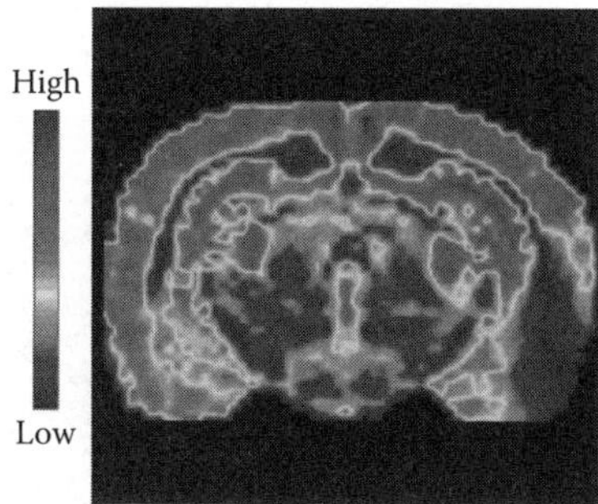

FIGURE 12.8 (See color insert following page 210.) On-tissue digestion aids the detection of high-molecular-weight species. Distribution of ion detected at *m/z* 1496.75 corresponding to a tryptic peptide of the 71 kDa protein, synapsin I. (Reproduced from Groseclose, M.R. et al., *J. Mass Spectrom.*, 42, 254, 2007. With permission.)

been crosslinked by the formalin fixation. This approach has tremendous potential, given the vast stores of archived samples in tissue banks that have been fixed in this manner. There are several reports describing the use of immunohistochemical antigen retrieval methods for more general proteomic profiling studies [58–60]. These results suggest that it is possible to extract proteomic identifications that are similar to those obtained from fresh frozen tissues.

One example of using on-tissue digestion for formalin-fixed tissue analysis was recently reported [24]. In this application, tissue microarrays (TMA) were created by embedding formalin-fixed needle core biopsies from lung tumor patients in paraffin blocks. Analysis of TMAs allows for hundreds or thousands of samples to be processed together in a high-throughput approach. Analyzing TMAs by MALDI IMS offers the additional advantage of molecular and spatial specificity. The TMA described in this paper contained 100 duplicate needle core biopsies from 50 patients diagnosed with nonsmall-cell lung cancer (NSCLC) and 10 adjacent normal lung tissue samples.

On-tissue digestion was performed on a section of the TMA to generate diagnostic peptides from the formalin-crosslinked proteins. After digestion, matrix was applied and spectra were acquired from all biopsy samples in one MS run. A serial section of the TMA was acquired for H&E staining and histological examination by a pathologist. Spectra were classified according to the histological area they were obtained from, and statistical analyses were carried out to develop classification algorithms to predict which biopsy correlated to which diagnosis (specifically adenocarcinoma or squamous cell carcinoma). The analysis is illustrated in Figure 12.9. As shown, the statistical classification made by the support vector machine (SVM) of four biopsies is in agreement with the diagnosis made by the pathologist. This model classified adenocarcinoma spectra with an accuracy of 97.9% and squamous cell carcinoma with an accuracy of 98.6%. This approach combines histology-directed protein profiling with a visualization of the statistical results on hundreds of samples at once. While much work still needs to be done in terms of validation and analysis of fresh frozen samples, the potential of using this

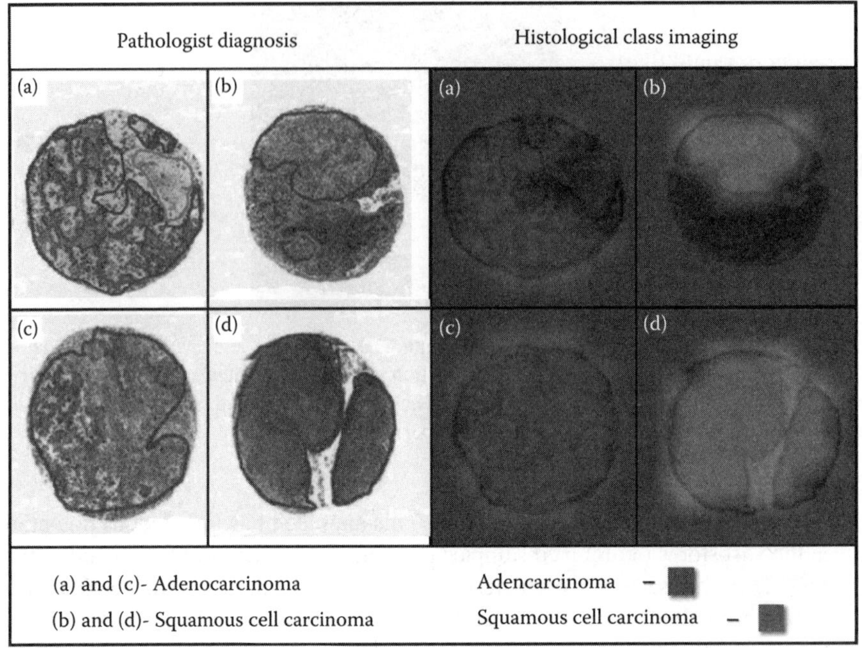

FIGURE 12.9 (See color insert following page 210.) On-tissue digestion of lung tumor TMA allows for high-throughput diagnostic analyses. Shown is a visual representation of the statistical classification of four biopsies compared to the marking and diagnosis based on histology. (Reproduced from Groseclose, M.R. et al., *Proteomics*, 8, 3715, 2008. With permission.)

technology as a fast, high-throughput, objective tool for the rapid diagnosis of lung tumor biopsies is enormous.

12.8 CONCLUSIONS

The use of MALDI IMS for protein analysis and biomarker discovery has been described. In the relatively short time that this technology has been around, it has enjoyed rapid and widespread use. Unparalleled as a discovery tool, MALDI IMS can simultaneously show the location and state of hundreds of proteins in a given tissue section at one time. This capability has been demonstrated in applications involving understanding normal human development and pathological conditions from Parkinson's disease to cancer. Used in conjunction with histological analysis, protein signatures can be correlated with tissue structures and probed using various statistical algorithms to generate lists for biomarker discovery. As biomarkers already discovered are validated and developed into clinically relevant tools, MALDI IMS will become an invaluable tool in the clinical setting. And as technological innovations continue to make the imaging process better, increasing resolution and sensitivity and decreasing analysis time, MALDI IMS will no doubt continue to be applied to new biological challenges well into the foreseeable future.

ACKNOWLEDGMENT

The authors would like to thank Nipun Merchant for the human pancreas tissue and NIH/NIGMS grant R01 GM58008 for financial support.

REFERENCES

1. Karas, M. et al., Matrix-assisted ultraviolet laser desorption of non-volatile compounds, *Int. J. Mass Spectrom. Ion Processes*, 78, 53, 1987.
2. Karas, M. and Hillenkamp, F., Laser desorption ionization of proteins with molecular masses exceeding 10,000 daltons, *Anal. Chem.*, 60(20), 2299, 1988.
3. Tanaka, K. et al., Protein and polymer analyses up to m/z 100 000 by laser ionization time-of-flight mass spectrometry, *Rapid Commun. Mass Spectrom.*, 2(8), 151, 1988.
4. Caprioli, R.M., Farmer, T.B., and Gile, J., Molecular imaging of biological samples: Localization of peptides and proteins using MALDI-TOF MS, *Anal. Chem.*, 69(23), 4751, 1997.
5. Chaurand, P., Stoeckli, M., and Caprioli, R.M., Direct profiling of proteins in biological tissue sections by MALDI mass spectrometry, *Anal. Chem.*, 71(23), 5263, 1999.
6. Stoeckli, M. et al., Imaging mass spectrometry: A new technology for the analysis of protein expression in mammalian tissues, *Nat. Medicine*, 7(4), 493, 2001.
7. Yanagisawa, K. et al., Proteomic patterns of tumour subsets in non-small-cell lung cancer, *Lancet*, 362, 433, 2003.
8. Schwartz, S.A. et al., Protein profiling in brain tumors using mass spectrometry: Feasibility of a new technique for the analysis of protein expression, *Clin. Cancer Res.*, 10, 981, 2004.
9. Schwartz, S.A. et al., Proteomic-based prognosis of brain tumor patients using direct-tissue matrix-assisted laser desorption ionization mass spectrometry, *Cancer Res.*, 65(17), 7674, 2005.
10. Stoeckli, M., Farmer, T.B., and Caprioli, R.M., Automated mass spectrometry imaging with a matrix-assisted laser desorption ionization time-of-flight instrument, *J. Am. Soc. Mass Spectrom.*, 10, 67, 1999.
11. Zhao, J.Y. et al., Application of MALDI-MS/MS molecular imaging software for small molecule profiling in tissue samples, *Proceedings of the 51st ASMS Conference*, Montreal, Quebec, Canada, 2003.
12. Stoeckli, M. et al., Molecular imaging of amyloid beta peptides in mouse brain sections using mass spectrometry, *Anal. Biochem.*, 311, 33, 2002.
13. Clerens, S., Ceuppens, R., and Arckens, L., *CreateTarget* and *Analyze This!*: New software assisting imaging mass spectrometry on Bruker Reflex IV and Ultraflex II instruments, *Rapid Commun. Mass Spectrom.*, 20(20), 3061, 2006.
14. Pedrioli, P.G.A. et al., A common open representation of mass spectrometry data and its application to proteomics research, *Nat. Biotech.*, 22(11), 1459, 2004.
15. Jardin-Mathe, O. et al., MITICS (MALDI Imaging Team Imaging Computing System): A new open source mass spectrometry imaging software, *J. Proteomics*, 71, 332, 2008.
16. Nye, G.J., Norris, J.L., and Nickerson, S., Optimization of the performance of automated matrix spraying for MALDI imaging mass spectrometry, *Proceedings of the. 54th ASMS Conference*, Seattle, WA, 2006.
17. Baluya, D.L., Garrett, T.J., and Yost, R.A., Automated MALDI matrix deposition method with inkjet printing for imaging mass spectrometry, *Anal. Chem.*, 79(17), 6862, 2007.
18. Dai, Y., Whittal, R.M., and Li, L., Confocal fluorescence microscopic imaging for investigating the analyte distribution in MALDI matrices, *Anal. Chem.*, 68(15), 2494, 1996.

19. Garden, R.W. and Sweedler, J.V., Heterogeneity within MALDI samples as revealed by mass spectrometric imaging, *Anal. Chem.*, 72(1), 30, 2000.
20. Bouschen, W. and Spengler, B., Artifacts of MALDI sample preparation investigated by high-resolution scanning microprobe matrix-assisted laser desorption/ionization (SMALDI) imaging mass spectrometry, *Int. J. Mass Spectrom.*, 266(1–3), 129, 2007.
21. Hankin, J.A., Barkley, R.M., and Murphy, R.C., Sublimation as a method of matrix application for mass spectrometric imaging, *J. Am. Soc. Mass Spectrom.*, 18(9), 1646, 2007.
22. Puolitaival, S.M. et al., Solvent-free matrix dry-coating for MALDI imaging of phospholipids, *J. Am. Soc. Mass Spectrom.*, 19(6), 882, 2008.
23. Aerni, H.-R., Cornett, D.S., and Caprioli, R.M., Automated acoustic matrix deposition for MALDI sample preparation, *Anal. Chem.*, 78(3), 827, 2006.
24. Groseclose, M.R. et al., High-throughput proteomic analysis of formalin-fixed paraffin-embedded tissue microarrays using MALDI imaging mass spectrometry, *Proteomics*, 8(18), 3715, 2008.
25. Agar, N.Y.R. et al., Matrix solution fixation: Histology-compatible tissue preparation for MALDI mass spectrometry imaging, *Anal. Chem.*, 79(19), 7416, 2007.
26. Seeley, E.H. et al., Enhancement of protein sensitivity for MALDI imaging mass spectrometry after chemical treatment of tissue sections, *J. Am. Soc. Mass Spectrom.*, 19(8), 1069, 2008.
27. Lemaire, R. et al., MALDI-MS direct tissue analysis of proteins: Improving signal sensitivity using organic treatments, *Anal. Chem.*, 78(20), 7145, 2006.
28. Chaurand, P. et al., Integrating histology and imaging mass spectrometry, *Anal. Chem.*, 76, 1145, 2004.
29. Schwartz, S.A., Reyzer, M.L., and Caprioli, R.M., Direct tissue analysis using matrix-assisted laser desorption/ionization mass spectrometry: Practical aspects of sample preparation, *J. Mass Spectrom.*, 38, 699, 2003.
30. Xu, B.J. et al., Direct analysis of laser capture microdissected cells by MALDI mass spectrometry, *J. Am. Soc. Mass Spectrom.*, 13(11), 1292, 2002.
31. Laurent, C. et al., Direct profiling of the cerebellum by matrix-assisted laser desorption/ionization time-of-flight mass spectrometry: A methodological study in postnatal and adult mouse, *J. Neurosci. Res.*, 81, 613, 2005.
32. Crecelius, A.C. et al., Three-dimensional visualization of protein expression in mouse brain structures using imaging mass spectrometry, *J. Am. Soc. Mass Spectrom.*, 16, 1093, 2005.
33. Meistermann, H. et al., Biomarker discovery by imaging mass spectrometry: Transthyretin is a biomarker for gentamicin-induced nephrotoxicity in rat, *Mol. Cell. Proteomics*, 5, 1876, 2006.
34. Corbin, B.D. et al., Metal chelation and inhibition of bacterial growth in tissue abscesses, *Science*, 319, 962, 2008.
35. Thibault, D.B. et al., MALDI tissue profiling of integral membrane proteins from ocular tissues, *J. Am. Soc. Mass Spectrom.*, 19(6), 814, 2008.
36. Altelaar, A.F.M. et al., High-resolution MALDI imaging mass spectrometry allows localization of peptide distributions at cellular length scales in pituitary tissue sections, *Int. J. Mass Spectrom.*, 260(2–3), 203, 2007.
37. Chaurand, P. et al., Monitoring mouse prostate development by profiling and imaging mass spectrometry, *Mol. Cell. Proteomics*, 7(2), 411, 2008.
38. Cornett, D.S. et al., A novel histology directed strategy for MALDI-MS tissue profiling that improves throughput and cellular specificity in human breast cancer, *Mol. Cell. Proteomics*, 5(10), 1975, 2006.
39. Stoeckli, M. et al., Imaging of a [beta]-peptide distribution in whole-body mice sections by MALDI mass spectrometry, *J. Am. Soc. Mass Spectrom.*, 18(11), 1921, 2007.

40. Rohner, T.C., Staab, D., and Stoeckli, M., MALDI mass spectrometric imaging of biological tissue sections, *Mech. Ageing Dev.*, 126(1), 177, 2005.
41. Khatib-Shahidi, S. et al., Direct molecular analysis of whole-body animal tissue sections by imaging MALDI mass spectrometry, *Anal. Chem.*, 78(18), 6448, 2006.
42. Yang, E.C.C. et al., Protein expression profiling of endometrial malignancies reveals a new tumor marker: Chaperonin 10, *J. Proteome Res.*, 3, 636, 2004.
43. Taguchi, F. et al., Mass spectrometry to classify non-small-cell lung cancer patients for clinical outcome after treatment with epidermal growth factor receptor tyrosine kinase inhibitors: A multicohort cross-institutional study, *J. Natl. Cancer Inst.*, 99(11), 838, 2007.
44. Sinha, T.K. et al., Integrating spatially resolved three-dimensional MALDI IMS with in vivo magnetic resonance imaging, *Nat. Methods*, 5(1), 57, 2008.
45. Reyzer, M.L. et al., Early changes in protein expression detected by mass spectrometry predict tumor response to molecular therapeutics, *Cancer Res.*, 64, 9093, 2004.
46. Lemaire, R. et al., Specific MALDI imaging and profiling for biomarker hunting and validation: Fragment of the 11S proteasome activator complex, Reg alpha fragment, is a new potential ovary cancer biomarker, *J. Proteome Res.*, 6, 4127, 2007.
47. Amann, J.M. et al., Selective profiling of proteins in lung cancer cells from fine-needle aspirates by matrix-assisted laser desorption ionization time-of-flight mass spectrometry, *Clin. Cancer Res.*, 12(17), 5142, 2006.
48. Skold, K. et al., Decreased striatal levels of PEP-19 following MPTP lesion in the mouse, *J. Proteome Res.*, 5, 262, 2006.
49. Nilsson, A. et al., Increased striatal mRNA and protein levels of the immunophilin FKBP-12 in experimental Parkinson's disease and identification of FKBP-12-binding proteins, *J. Proteome Res.*, 6, 3952, 2007.
50. Burnum, K.E. et al., Imaging mass spectrometry reveals unique protein profiles during embryo implantation, *Endocrinology*, 149(7), 3274, 2008.
51. Andersson, M. et al., Imaging mass spectrometry of proteins and peptides: 3D volume reconstruction, *Nat. Methods*, 5(1), 101, 2008.
52. Schetz, M. et al., Drug-induced acute kidney injury, *Curr. Opin. Critical Care*, 11, 555, 2005.
53. Charlwood, J. et al., Proteomic analysis of rat kidney cortex following treatment with gentamicin, *J. Proteome Res.*, 1, 73, 2002.
54. Bandara, L.R. et al., A correlation between a proteomic evaluation and conventional measurements in the assessment of renal proximal tubular toxicity, *Toxicol. Sci.*, 73, 195, 2003.
55. Thiery, G. et al., Multiplex target protein imaging in tissue sections by mass spectrometry—TAMSIM, *Rapid Commun. Mass Spectrom.*, 21(6), 823, 2007.
56. Yang, J. et al., Towards LDI imaging mass spectrometry via molecular interaction and photo labile tags, *Proceedings of the 55th ASMS Conference*, Indianapolis, IN 2007.
57. Groseclose, M.R. et al., Identification of proteins directly from tissue: *In situ* tryptic digestions coupled with imaging mass spectrometry, *J. Mass Spectrom.*, 42(2), 254, 2007.
58. Xu, H. et al., Antigen retrieval for proteomic characterization of formalin-fixed and paraffin-embedded tissues, *J. Proteome Res.*, 7, 1098, 2008.
59. Shi, S.-R. et al., Protein extraction from formalin-fixed, paraffin-embedded tissue sections: Quality evaluation by mass spectrometry, *J. Histochem. Cytochem.*, 54(6), 739, 2006.
60. Ronci, M. et al., Protein unlocking procedures of formalin-fixed paraffin-embedded tissues: Application to MALDI-TOF imaging MS investigations, *Proteomics*, 8, 3702, 2008.

13 Ambient Ionization Methods and Their Early Applications in ADME Studies

Jing-Tao Wu

CONTENTS

13.1 INTRODUCTION

One of the most significant developments in mass spectrometry in the recent years is the introduction of a new class of ionization methods where samples in either solid or liquid state can be directly ionized in their native environment under ambient conditions (rather than inside a mass spectrometer) without any sample preparation. This new class of ionization methods is often referred to as ambient ionization methods [1,2]. Because these methods generally ionize analytes on the surface or near the surface of the samples at atmospheric pressure, they have also been called atmospheric pressure surface sampling/ionization methods or direct/open air ionization methods [3]. Since the first reports on ambient ionization with desorption electrospray ionization (DESI) [4] and direct analysis in real time (DART) [5], numerous reports have been published on the applications of these new ionization methods as well as the introduction of many related ambient ionization methods such as desorption atmospheric pressure chemical ionization (DAPCI) [6], atmospheric solid analysis probe (ASAP) [7], and electrospray-assisted laser desorption/ionization (ELDI) [8]. Recently, two reviews of the various established and emerging ambient ionization methods have been published [2,3].

While the applications of ambient ionization mass spectrometry have been widely reported in areas such as biomedical, homeland security/forensics, and pharmaceutical process monitoring, the evaluation and the use of these techniques for absorption, distribution, metabolism, and excretion (ADME) studies on drugs or drug metabolites have emerged more slowly mostly due to two reasons. First, samples for ADME studies are generally in complex biological matrixes, and, therefore, ion suppression is a concern if samples are directly subject to ambient ionization methods without any sample preparation. Second, most of the ADME applications are quantitative in nature. Unlike conventional liquid chromatography–tandem mass spectrometry (LC–MS/MS)-based techniques where samples with a well-defined volume are injected for analysis, sampling reproducibility can be a challenge for ambient ionization methods since there is no defined sampling volume. In spite of the challenges, the unique and attractive features of ambient ionization methods such as real-time analysis without sample preparation or vacuum constraints have driven the research interest in this area and publications in this field are emerging.

In this chapter, we first provide an introduction to DESI and DART as the two most established ambient ionization methods. We then review some details on the early applications of DESI and DART in ADME studies. We close this chapter with some perspectives on the future role of ambient ionization methods in ADME studies.

13.2 TWO MOST ESTABLISHED AMBIENT IONIZATION METHODS: DESI AND DART

Instead of introducing a defined volume of samples in liquid state onto a mass spectrometer as in LC–MS/MS-based techniques, there has always been a desire and interest in analyzing samples in solid state using a desorption or ablation method. Matrix-assisted laser desorption ionization (MALDI) [9] and secondary ion mass spectrometry (SIMS) [10] are the two most established ionization methods for these types of applications where samples are ionized directly inside the vacuum chamber of a mass spectrometer. The introduction of atmospheric pressure (AP) MALDI [11] allowed samples to be placed outside the enclosure of a mass spectrometer although extensive sample preparation is still required prior to analysis. The first report on ambient ionization mass spectrometry, where samples were analyzed under ambient conditions without any sample preparation, was published in 2004 by Cooks et al. using the DESI technique [4]. A schematic diagram of this DESI technique is shown in Figure 13.1. In this method, an electrospray probe carrying the solvent is sprayed toward the surface of the sample at an angle. The charged solvent droplets are believed to first prewet the sample surface and dissolve or collect analytes on the sample surface. The subsequently arriving charged droplets impact this surface solvent layer and create off-spring droplets containing the analyte. The analyte in these droplets are ionized through the same mechanism as electrospray before being sampled into a mass spectrometer. DESI applies to both small and large molecules and the resulting mass spectrum resembles that of electrospray.

Shortly after the report on the DESI technique, DART was reported as another ambient ionization method [5]. In this method, a gas with a high ionization potential such as helium or nitrogen flows through a probe that contains multiple chambers. The gas first flows into a discharge chamber where electrical discharge occurs by applying a potential of several thousand volts across a cathode and an anode. The

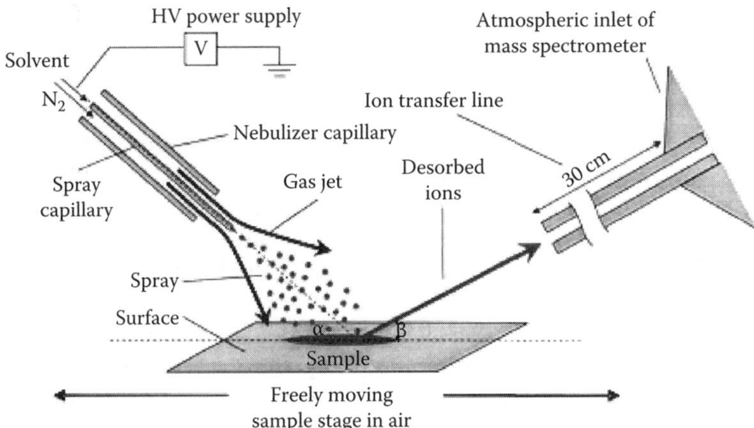

FIGURE 13.1 Schematic diagram of DESI. (Reprinted from Takats, Z. et al., *Science*, 306, 471, 2004. With permission.)

discharge generates ions, electrons, and electronic or vibronic excited-state species. As this mixture moves through the rest of the chambers, it is heated and ions are removed from the mixture. The gas flow coming out of the probe can be aimed directly at the orifice of the mass spectrometer with the sample in between or the gas flow can be reflected off a sample surface into the mass spectrometer. It is believed that the excited-state species (metastable helium atoms or nitrogen molecules) are mostly responsible for the desorption and ionization of the compounds off the sample surface. DART generally applies to small molecules and does not ionize large molecules such as proteins and peptides.

After the introduction of DESI and DART techniques, they have found wide applications across different fields. DESI has been used to characterize the active ingredients in formulated tablets, gels, and ointments [12,13]. It has also been evaluated in pharmaceutical cleaning validations [14], rapid analysis of controlled substances [15], explosives [16], and as a chromatographic detector [17]. DART has been used for reaction monitoring in drug discovery [18], counterfeit drug identification [19], a chromatographic detector [20], flavors and fragrances analysis [21], forensic science [22], bacterial fatty acid profiling [23], self-assembled monolayer analysis [24], and so on. The literature on the applications of ambient ionization methods are quickly expanding.

13.3 EARLY APPLICATIONS OF AMBIENT IONIZATION METHODS IN ADME STUDIES

Although still at a very early stage, ambient ionization methods, mostly DESI and DART, have been evaluated and used in the ADME studies for drug candidates. The fact that ambient ionization methods can analyze samples without sample preparation seems extremely attractive for bioanalysis as sample preparation is widely considered as the key bottleneck in bioanalysis. Also, the fact that samples can be analyzed intact under ambient conditions generated interest in using ambient ionization mass spectrometry as an imaging tool to study drug distribution in tissue sections.

The two key challenges in bioanalysis with ambient ionization methods are reproducibility and matrix effects [25]. Overall, good sample-to-sample reproducibility is not as readily achieved in ambient ionization/sampling methods as with conventional LC–MS/MS-based methods where samples with a defined volume can be injected onto the system with high precision. An initial key question during the evaluation of ambient ionization methods for quantitative applications was whether reproducibility could be largely improved through a more refined sample introduction mechanism, or was it a more fundamental limitation of the technique itself. Recently Yu et al. [26] and Zhao et al. [27] independently demonstrated that at least for samples with homogenous matrices such as plasma, excellent sample-to-sample reproducibility can be achieved with DART using an improved automated sample introduction device. An example of a reproducibility test is shown in Figure 13.2 where repeated sample introduction was made on the DART/MS/MS system with untreated rat plasma containing 1 μM of benzoylecgonine. The %CV of the peak height for nine consecutive sample injections was 3.1%, which is comparable to that of conventional LC–MS/MS methods. These reports demonstrated that the reproducibility of DART, and potentially other ambient sampling methods can be optimized to meet the requirements for quantitative applications.

Matrix effects are another key aspect that needs to be carefully considered for quantitative applications with ambient ionization methods. Currently an extensive comparison of the matrix effects among various ambient mass spectrometric methods is not available. DESI has been evaluated for salt tolerance, and it was found that DESI is more tolerant of the presence of salts than is electrospray [28]. On the other

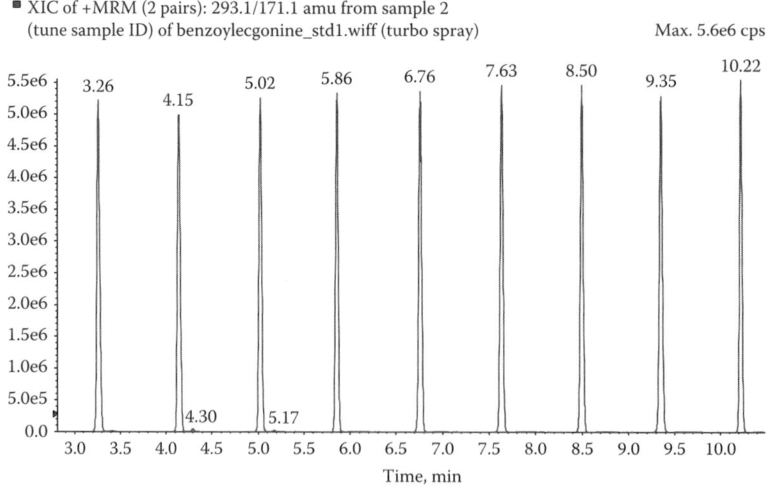

FIGURE 13.2 Reproducibility test of repeated injections of a rat plasma sample containing 1 μM of benzoylecgonine on the DART/MS/MS system. The %CV of the peak height is 3.1%. (Reprinted from Yu, S. et al., *Anal. Chem.*, 81, 193, 2009. With permission.)

hand, based on the mass range, DART appears to offer some advantage in handling matrix suppression from biological samples because it generally only ionizes small molecules and not the large molecules in the sample matrix such as proteins and peptides. Also, one of the key ionization mechanisms in DART largely resembles APCI, which involves proton transfer from ionized water cluster to the analyte in the gas phase [29]. Therefore, DART may be expected to be potentially less prone to ion suppression than DESI, which has an ionization mechanism that mimics electrospray ionization. The recent work by both Yu et al. and Zhao et al. has evaluated the matrix effects of various compounds in different matrices. One set of these results are summarized in Figure 13.3, where the matrix effects of methacin, verapamil, alprazolam, proparacaine, and a proprietary compound in untreated rat plasma were measured. It was found that all of these compounds could be directly detected from untreated rat plasma using DART, and as expected, the matrix effects were highly compound dependent. The level of matrix effects ranged from about 5%–70% indicating that signal suppression up to 20-fold could occur with untreated plasma, which could result in an elevated quantitation limit. The matrix effects of verapamil in untreated rat whole blood and bile, and in rat liver and brain homogenates, are also shown in the figure. The results indicate that, with some compromise in sensitivity, it is possible to analyze samples directly from tissue homogenates with DART. Zhao et al. evaluated the limits of detection (LOD) in plasma for 39 commercially available drugs on the DART/MS/MS system and found the LOD to range from 0.25 to 20 ng/mL [27]. Since these LODs are generally higher than those obtained on a conventional LC–MS/MS system, they can meet the needs for many of the bioanalytical applications in drug discovery, but may not be suitable for applications that demands for the highest sensitivity, particularly with compounds that have high matrix effects.

In addition to reproducibility, matrix effects, and sensitivity, DART has been evaluated for dynamic range, precision, and accuracy [26]. In general, DART/MS/MS on a triple-quadrupole instrument such as an API-4000 has showed a dynamic range across a concentration range of three to four orders of magnitude. The precision and accuracy can also meet the commonly accepted ±15% criteria for quality control samples and back-calculated standards. It is interesting to note that, according to the work by Yu et al., satisfactory results in precision and accuracy can be achieved without the use of an internal standard for plasma samples. This allowed for the further simplification of the assay procedures as there was no need to add the internal standards. As a result, the DART/MS/MS was successfully used to analyze samples from a limited number of discovery PK studies and *in vitro* metabolic stability studies and the results were broadly similar to those generated using conventional LC–MS/MS systems.

A major limitation of the DART/MS/MS method in bioanalysis is the potential metabolic interference resulting from ion source fragmentation. The helium atom at the excited state carries an internal energy of 19.8 eV, which is higher than the bond energy in many labile metabolites. Since no chromatographic separation is used in the DART/MS/MS method, metabolites with labile chemical bond, such as a glucuronides or *N*-oxides, may break down in the ion source and interfere with the assay of the parent compound [27]. This limitation is believed to be more of a

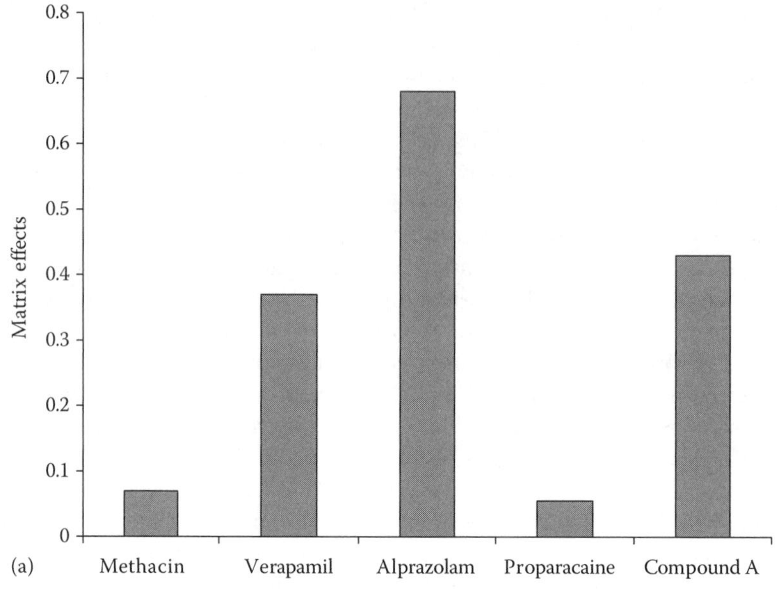

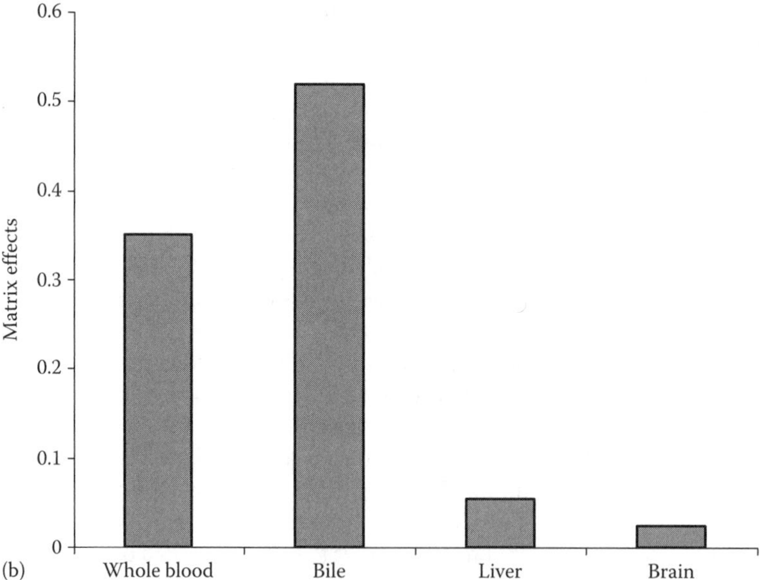

FIGURE 13.3 Matrix effects with DART ionization. (a) Matrix effects of test compounds in plasma. (b) Matrix effects of verapamil in different tissue homogenates. Matrix effect is defined as the ratio of analyte peak height in matrix over that in neat solvent. A matrix effects value of 1 is no matrix effect. (Reprinted from Yu, S. et al., *Anal. Chem.*, 81, 193, 2009. With permission.)

concern with DART than with DESI, which is a softer ionization method. Because of this limitation, the application of DART/MS/MS in bioanalysis for pharmacokinetic studies should be carried out with caution. Some suitable applications include either early-stage screening studies, where any positive results will be confirmed by LC–MS/MS, or studies on late-stage discovery compounds whose major metabolic pathways are known.

Besides the early work in bioanalysis using DART, the feasibility of using DESI to profile drug metabolites in urine has been successfully demonstrated by Cooks et al. although no quantitative results have been reported [30–32]. The same group also evaluated the quantitative analysis of small molecules in neat solvent spotted on porous polytetrafluorethylene surface, and it was found that DESI can be employed for quantitative analysis by using this porous surface [33]. Recently Takats et al. demonstrated the use of DESI coupled with solid-phase extraction for bioanalysis [34]. In this work, the eluent from the solid-phase extraction was evaporated on the closing frit of the extraction cartridge and then was subjected to DESI analysis. Significant improvement in sensitivity was reported compared to conventional solid-phase extraction as a result of the drying and concentrating process at the frit of the extraction cartridge. However, to date, the capability of full quantification of small molecules in blood or plasma using DESI without sample preparation has not yet been reported.

In addition to some early applications in bioanalysis, ambient ionization mass spectrometry has been used as an imaging tool to study drug distribution in tissue sections. Most of the work reported so far involved the use of DESI as the ambient ionization method. Compared to other mass spectrometry-based tissue imaging techniques such as MALDI and SIMS, DESI allows tissue samples to be analyzed under ambient conditions without sample preparation, which simplifies the procedure and prevents the redistribution of analytes during matrix deposition. A major drawback of DESI as an imaging tool is its relatively low spatial resolution (typically 250 μm) and therefore cannot be used for cellular or subcellular imaging.

Wiseman et al. recently reported the procedures of using DESI for ambient molecular imaging of tissue sections [35]. Since DESI did not require sample preparation and a vacuum environment for analysis, the overall procedure was very simple. Examples in their work include the imaging of the distribution of clozapine in the sagittal section of the brain tissue in the rats that received a 250 μg dose of clozapine via an intracerebral ventricular injection. The authors claimed that the procedures should also apply to other types of tissues.

A most comprehensive DESI-based imaging study on drug tissue distribution was recently reported by Kertesz et al. [36,37]. In this work, they used DESI/MS/MS to obtain the drug distribution images from whole-body thin tissue sections of rats dosed with propranolol. They also compared the results generated by DESI/MS/MS with those from whole body autoradiography. Figure 13.4 shows the comparison of the images generated by these two methods. As we can see, the images look qualitatively similar for propranolol tissue distribution. Tissue sections showing the most radioactivity were also excised and analyzed by HPLC to differentiate the parent from the metabolites. After correction for metabolites, the relative signal of propranolol calculated from radiography and measured by DESI/MS/MS are shown

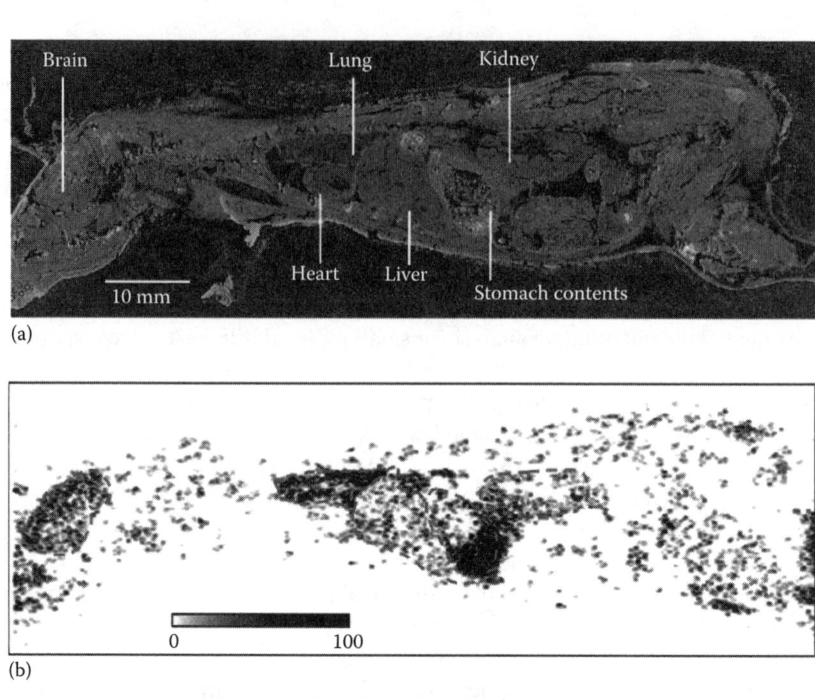

(b)

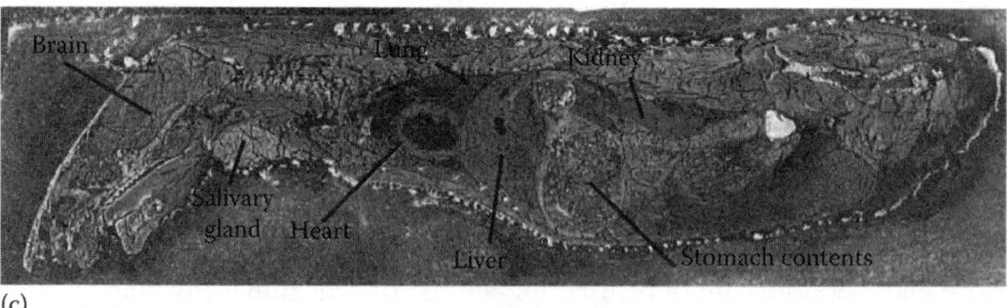

(c)

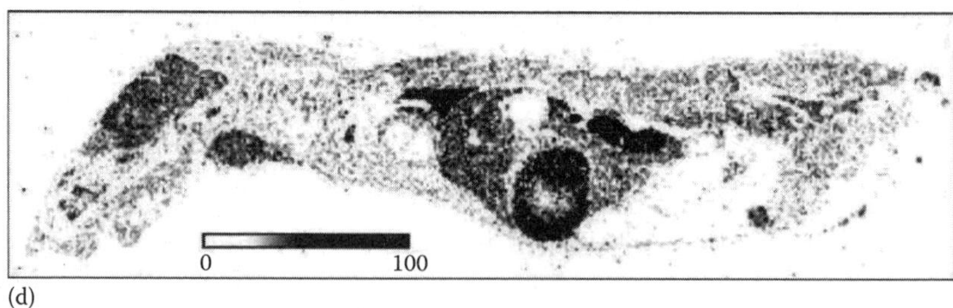

(d)

FIGURE 13.4 Comparison of distribution images of propranolol from whole body thin tissue sections obtained using DESI and autoradiography: (a) Scanned optical image of a whole body tissue section of a mouse dosed with propranolol; (b) Images measured by DESI/MS/MS in the tissue section in (a); (c) Scanned optical image of a whole body tissue section of a mouse dosed with [^{3}H] propranolol; (d) Autoradioluminograph of [^{3}H] propranolol-related material in the tissue section in (c). (Reprinted from Kertesz, V. et al., *Anal. Chem.*, 80, 5168, 2008. With permission.)

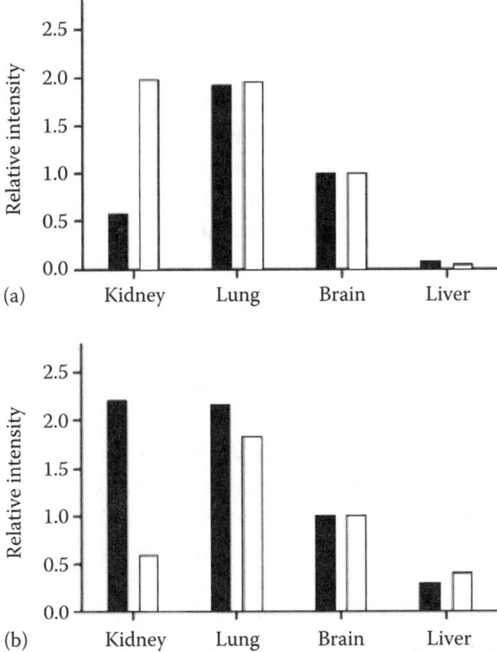

FIGURE 13.5 Comparison of relative signal of propranolol from whole body autoradiography (after correction with radio-profiling results) (filled bars) and DESI/MS/MS (empty bars) for kidney, lung, brain, and liver: (a) 20 min after drug administration and (b) 60 min after drug administration. Signals were normalized in all cases to the propranolol signal in the brain. (Reprinted from Kertesz, V. et al., *Anal. Chem.*, 80, 5168, 2008. With permission.)

in Figure 13.5. Nominal agreement was achieved between the two techniques for the amount of propranolol in the brain, lung, and liver. However, the two techniques showed a marked discrepancy on propranolol amount in the kidney, which was unexplained. Also, a major metabolite (hydroxypropranolol glucuronide) known to be present in the kidney and liver was not detected by DESI/MS/MS.

Although often considered beyond the scope of ADME studies, it is worth pointing out that DESI has been evaluated and used for tissue profiling and imaging [38]. One of these examples was the direct tissue profiling of human liver adenocarcinoma using DESI. Differences in tissue profiles, including elevated levels of certain phospholipids in the tumor region, were observed compared to the nontumor region. This experiment demonstrated the utility and feasibility of using DESI as a potential real-time diagnostic tool in surgery.

13.4 CONCLUSIONS AND FUTURE DIRECTIONS

Since the first publication in 2004, ambient ionization mass spectrometry has become one of the fastest growing areas in mass spectrometry. Because of the diversity of mechanisms for ionization under ambient conditions, a large variety of ambient ionization techniques has been developed. Among these techniques, DESI and DART are the two most established methods that have been evaluated and used in different applications including ADME studies. Some initial proof-of-concept work has been done with either DART or DESI in bioanalysis and tissue imaging. DART has showed good reproducibility, wide dynamic range, and manageable matrix effects with untreated plasma samples in bioanalysis. With the current instrumentation of the DART/MS/MS system, the detection limits are generally about one order of magnitude higher than those with conventional LC–MS/MS. A key limitation with the DART/MS/MS system in bioanalysis is the potential metabolite interference resulting from ion source fragmentation in the absence of chromatographic separations. DESI has showed to be a potentially useful imaging tool for studying drug distribution in tissues without vacuum constraints and the need for sample preparation. The largest limitation of tissue imaging with DESI is the relatively low spatial resolution.

Compared to the dominant role of LC–MS/MS in bioanalysis, the use of ambient ionization mass spectrometry in bioanalysis is still at its infancy. It is not expected that these ambient ionization techniques will replace LC–MS/MS in drug discovery and development due to some of the limitations described above. However, these techniques will enhance our current bioanalytical capability to address different needs in drug discovery and development. First, ambient ionization may be a solution for screening PK or *in vitro* screening studies where a quick turnaround of the data is critical for decision-making. By eliminating sample preparation, ambient ionization methods offer the potential for close to real-time turnaround. The recent work by Yu et al. and Zhao et al. on DART has fully demonstrated this potential [26,27]. Spooner et al. recently reported the use of dried blood spots on paper as a sample collection technique for the determination of pharmacokinetics in the clinic [39]. If this technique can be combined with ambient ionization methods, it may significantly simplify the blood collection procedures in the clinic by dramatically reducing blood draw volume, employing less invasive sampling (finger or heel prick vs. conventional venous cannula), and streamlining sample storage and shipping. This kind of dry spot sampling methods, once enabled by direct analysis with ambient ionization methods, may change the way we do bioanalysis for certain types of studies. The second area where ambient ionization methods may find its unique advantage is in clinical diagnostics. Because of its simplicity in sample handling, it may potentially be deployed routinely into clinical labs or even at bedside, particularly when combined with a portable mass spectrometer. It can be used to monitor drugs, endogenous compounds, or biomarkers in real time to make timely decisions on a therapy. In the initial report on DESI, Takats et al. demonstrated the feasibility of sampling the finger of a person who had taken 10 mg of the over-the-counter antihistamine loratadine [4]. Another interesting example of the application in this area was the use of DART to analyze hydrocarbons in a live fly [40]. In this work, a small steel probe was used to carefully depress the abdomen of restrained but awake fly. Cuticular hydrocarbons were directly sampled through the probe by DART/MS. A third area of potential applications, which is yet to be explored, is to take advantage of the intact sample to study binding and cellular partitioning of drugs. All of our conventional bioanalytical procedures require sample preparation, which will largely change the noncovalent binding and cellular distribution of drugs. By eliminating sample preparation with ambient ionization methods, it is potentially possible to study these properties particularly when a soft ambient ionization method is used. Zheng et al. recently reported the development of a low-temperature plasma probe for ambient desorption ionization [41]. This method provides soft ionization without damage to the sample or causing perceptible heating. This method or other to be developed "low impact" ambient ionization methods may provide utility in this type of applications.

Like its use in bioanalysis, ambient mass spectrometry is also at an early stage in tissue imaging. The distinct advantages of eliminating the need for sample preparation and the introduction into a vacuum system make it unique over other mass spectrometry-based imaging tools. Because of its relatively low spatial resolution, ambient ionization methods cannot replace MALDI or SIMS for imaging of tissue sections in many cases. However, the unique strength with ambient ionization is the potential applications in real-time or even *in vivo* imaging, which is highly desirable in many cases such as in surgery and pathology, but has been largely viewed as the major limitation of any mass spectrometry-based imaging tools.

Overall, ambient ionization methods have showed to be a highly promising tool for ADME studies. Although most of the early work to date has been focused on the throughput advantage, the future role of ambient ionization methods in the pharmaceutical industry and particularly for the ADME studies will largely depend on how their unique features can be utilized to solve problems that are not readily solvable with conventional LC–MS/MS methods.

REFERENCES

1. Cooks, R.G. et al., Ambient mass spectrometry, *Science*, 311, 1566, 2006.
2. Venter, A., Nefliu, M., and Cooks, R.G., Ambient desorption ionization mass spectrometry, *Trends Anal. Chem.*, 27(4), 284, 2008.
3. Van Berkel, G.J., Pasilis, S.P., and Ovchinnikova, O., Established and emerging atmospheric pressure surface sampling/ionization techniques for mass spectrometry, *J. Mass Spectrom.*, 43, 1161, 2008.
4. Takats, Z. et al., Mass spectrometry sampling under ambient conditions with desorption electrospray ionization, *Science*, 306, 471, 2004.
5. Cody, R.B., Laramee, J.A., and Durst, H.D., Versatile new ion source for the analysis of materials in open air under ambient conditions, *Anal. Chem.*, 77(8), 2297, 2005.
6. Williams, J.P. and Scrivens J.H., Rapid and accurate mass desorption electrospray ionization tandem mass spectrometry of pharmaceutical samples, *Rapid Commun. Mass Spectrom.*, 19, 3643, 2005.
7. McEwen, C.N., McKay, R.G., and Larsen, B.S., Analysis of solids, liquids, and biological tissues using solid probe introduction at atmospheric pressure on commercial LC/MS instruments, *Anal. Chem.*, 77(23), 7826, 2005.
8. Shiea, J. et al., Electrospray-assisted laser desorption/ionization mass spectrometry for direct ambient analysis of solids, *Rapid Commun. Mass Spectrom.*, 19, 3701, 2005.
9. Hillenkamp, F. and Karas, M., MALDI MS, in *A Practical Guide to Instrumentation, Methods and Applications*, Hillenkamp, F. and Peter-Katalinic, J. (eds.). Wiley-VCH Verlag GmbH & Co. KgaA, Weinheim, Germany, 2007.
10. Benninghoven, A., Hagenhoff, B., and Niehuis E., Surface MS: Probing real-world samples, *Anal. Chem.*, 65(14), 630A, 1993.

11. Moyer, S.C. and Cotter, R.J., Atmospheric pressure MALDI, *Anal. Chem.*, 74(17), 469A, 2002.
12. Chen, H. et al., Desorption electrospray ionization mass spectrometry for high-throughput analysis of pharmaceutical samples in the ambient environment, *Anal. Chem.*, 77(21), 6915, 2005.
13. Leuthold, L.A. et al., Desorption electrospray ionization mass spectrometry: Direct toxicological screening and analysis of illicit ecstasy tablets, *Rapid Commun. Mass Spectrom.*, 20, 103, 2006.
14. Soparawalla, S. et al., Pharmaceutical cleaning validation using non-proximate large-area desorption electrospray ionization mass spectrometry, *Rapid Commun. Mass Spectrom.*, 23, 131, 2005.
15. Rodriguez-Cruz, S.E., Rapid analysis of controlled substances using desorption electrospray ionization mass spectrometry, *Rapid Commun. Mass Spectrom.*, 20, 53, 2006.
16. Takats, Z. et al., Direct, trace level detection of explosives on ambient surfaces by desorption electrospray ionization mass spectrometry, *Chem. Commun.*, 15, 1950, 2005.
17. Kauppila, T.J. et al., New surfaces for desorption electrospray ionization mass spectrometry: Porous silicon and ultra-thin layer chromatography plates, *Rapid Commun. Mass Spectrom.*, 20, 2143, 2006.
18. Petucci, C. et al., Direct analysis in real time for reaction monitoring in drug discovery, *Anal. Chem.*, 79(13), 5064, 2007.
19. Fernández, F.M. et al., Characterization of solid counterfeit drug samples by desorption electrospray ionization and direct-analysis-in-real-time coupled to time-of-flight mass spectrometry, *Chem. Med. Chem.*, 1(7), 702, 2006.
20. Morlock, G. and Ueda Y., New coupling of planar chromatography with direct analysis in real time mass spectrometry, *J. Chromatogr. A*, 1143(1–2), 243, 2007.
21. Haefliger, O.P. and Jeckelmann, N., Direct mass spectrometric analysis of flavors and fragrances in real applications using DART, *Rapid Commun. Mass Spectrom.*, 21(8), 1361, 2007.
22. Laramee, J.A. et al., Forensic applications of DART mass spectrometry, in *Forensic Analysis on the Cutting Edge*, Blackledge, R.D., (ed.). John Wiley and Sons, Inc, Hoboken, NJ, p. 175, 2007.
23. Pierce, C.Y. et al., Ambient generation of fatty acid methyl ester ions from bacterial whole cells by direct analysis in real time (DART) mass spectrometry, *Chem. Commun.*, (8), 807, 2007.
24. Kpegba, K. et al., Analysis of self-assembled monolayers on gold surfaces using direct analysis in real time mass spectrometry, *Anal. Chem.*, 79(14), 5479, 2007.
25. Matuszewski, B.K., Costanzer, M.L., and Chavez-Eng, C.M., Matrix effect in quantitative LC/MS/MS analyses of biological fluids: A method for determination of finasteride in human plasma at picogram per milliliter concentrations. *Anal. Chem.*, 70, 882, 1999.
26. Yu, S. et al., Bioanalysis without sample cleanup or chromatography: The evaluation and initial implementation of direct analysis in real time ionization mass spectrometry for the quantification of drugs in biological matrixes, *Anal. Chem.*, 81(1), 193, 2009.
27. Zhao, Y. et al., Quantification of small molecules in plasma with direct analysis in real time tandem mass spectrometry, without sample preparation and liquid chromatographic separation, *Rapid Commun. Mass Spectrom.*, 22(20), 3217, 2008.
28. Jackson, A.U. et al., Salt tolerance of desorption electrospray ionization, *J. Am. Soc. Mass Spectrom.*, 18, 2218, 2007.
29. Cody, R.B., Observation of molecular ions and analysis of nonpolar compounds with the direct analysis in real time ion source, *Anal. Chem.*, 81(3), 1101, 2009.
30. Kauppila, T.J. et al., Desorption electrospray ionization mass spectrometry for the analysis of pharmaceuticals and metabolites, *Rapid Commun. Mass Spectrom.*, 20(3), 387, 2006.

31. Kauppila, T.J. et al., Rapid analysis f metabolites and drugs of abuse from urine samples by desorption electrospray ionization-mass spectrometry, *Analyst*, 132, 868, 2007.
32. Pan, Z. et al., Principal component analysis of urine metabolites detected by NMR and DESI-MS in patients with inborn errors of metabolism, *Anal. Bioanal. Chem.*, 387, 539, 2006.
33. Ifa, D.R. et al., Quantitative analysis of small molecules by desorption electrospray ionization mass spectrometry from polytetrafluoroethylene surfaces, *Rapid Commun. Mass Spectrom.*, 22(4), 503, 2008.
34. Dénes, J. et al., Analysis of biological fluids by direct combination of solid phase extraction and desorption electrospray ionization mass spectrometry, *Anal. Chem.*, 81(4), 1669, 2009.
35. Wiseman, J.M. et al., Ambient molecular imaging by desorption electrospray ionization mass spectrometry, *Nat. Protocols*, 3(3), 517, 2008.
36. Kertesz, V. et al., Comparison of drug distribution images from whole-body thin tissue sections obtained using desorption electrospray ionization tandem mass spectrometry and autoradiography, *Anal. Chem.*, 80(13), 5168, 2008.
37. Kertesz, V. and Van Berkel, G.J., Scanning and surface alignment considerations in chemical imaging with desorption electrospray mass spectrometry, *Anal. Chem.*, 80(4), 1027, 2008.
38. Wiseman, J.M. et al., Mass spectrometric profiling of intact biological tissue by using desorption electrospray ionization, *Angew. Chem. Int. Ed.*, 44, 7094, 2005.
39. Spooner, N., Lad, R., and Barfield, M., Dried blood spots as a sample collection technique for the determination of pharmacokinetics in clinical studies: Considerations for the validation of a quantitative bioanalytical method, *Anal. Chem.*, 81(4), 1557, 2009.
40. Yew, J.Y., Cody, R.B., and Kravitz, E.A., Cuticular hydrocarbon analysis of an awake behaving fly using direct analysis in real-time time-of-flight mass spectrometry, *Proc. Natl. Acad. Sci. USA*, 105(20), 7135, 2008.
41. Harper, J.D. et al., Low-temperature plasma probe for ambient desorption ionization, *Anal. Chem.*, 80(23), 9097, 2008.

14 Pharmaceutical Applications of Accelerator Mass Spectrometry

Lan Gao and Swapan Chowdhury

CONTENTS

14.1 INTRODUCTION

Drug development is a long and expensive process. Unfortunately, around 75% of drug development cost is associated with drug candidates that failed in the early stages of development and did not make it to the market [1,2]. One reason for the high failure rate in drug development is the unreliable prediction of absorption, distribution, metabolism, and excretion (ADME) in humans based on data generated in preclinical species. Up to 40% of new drugs under development are dropped during Phase I even though promising ADME data were obtained in animals [2]. Current methods of investigating drug metabolism and pharmacokinetics (PK) rely on animal, *in vitro*, and *in silico* models. There is always a concern that human metabolism pathways and PK might differ significantly from those predicted from the animal models. Thus, there is a critical need to confirm metabolism and PK parameters in humans as early as possible, prior to large investment in clinical trials.

A new approach to address the issue that has been gaining a lot of interest in recent years is to generate human PK data early in clinical development with microdosing. In microdosing studies, subpharmacological doses of drugs are administrated to human subjects to obtain metabolism profiles and PK parameters such as exposure, clearance, volume of distribution, bioavailability, and half-life. Current guidance for human microdosing allows for administrating less than 1/100th of the therapeutic dose predicted from animal and *in vitro* models, and must be less than 100 μg. Ultrasensitive analytical techniques are required for quantitative measurement of the drug and metabolites after microdosing. Both accelerator mass spectrometry (AMS) [3,4] and HPLC–MS/MS [5] have been used for biological sample analysis after microdosing. However, HPLC–MS/MS sensitivity is severely limited by the MS sensitivity of analytes and high background interference for many microdosing studies. An AMS is capable of achieving sensitivity in the range of 10^{-21} to 10^{-18} mol by directly detecting ^{14}C atoms [6–8] and is adequate for microdosing studies. Usually, the drug candidate is ^{14}C-labeled at a metabolically stable position for microdosing studies. Other stable isotopes, such as ^{3}H, ^{41}Ca, and ^{26}Al, have been evaluated as well.

The ultrahigh sensitivity of AMS enables it to follow trace amounts of radioisotope-labeled drugs in animals and humans postadministration. In addition to microdose PK studies, AMS has been utilized in many aspects of pharmaceutical research, such as absolute bioavailability determination, mass balance and metabolite profiling, and identifications of DNA and protein adducts.

14.2 ACCELERATOR MASS SPECTROMETRY

14.2.1 History of Accelerator Mass Spectrometry

Accelerator mass spectrometry (AMS) is a technique developed in the United States during the mid-1970s for measuring long-lived radionuclides that occur naturally in the environment. In AMS, ions are accelerated at very high energy (often millions of electron volts) and separated based on their energy, charge, and momentum. Initially AMS was mainly used for archaeological studies for detecting ^{14}C and radiocarbon dating purposes [9]. Living entities maintain a natural level of ^{14}C due to relative constant isotope production in the upper atmosphere. This amount, referred to as "Modern" carbon level, corresponds to 97.6 attomole of ^{14}C per milligram (mg) of total carbon, which is the background level of ^{14}C in the atmosphere in 1950 before the widespread atmospheric testing of atomic weapons (Figure 14.1). Upon death, the ^{14}C abundance starts to decrease at a constant rate due to β-decay of ^{14}C and permits age determination based on the $^{14}C/^{12}C$ ratio. Before AMS was available, the decay counting without isotope enrichment was possible for samples less than 60,000 years of age and required 7 g of sample and four days of counting. AMS extended the limit to 100,000 years with less than 5 mg of sample and a counting time of 2 h.

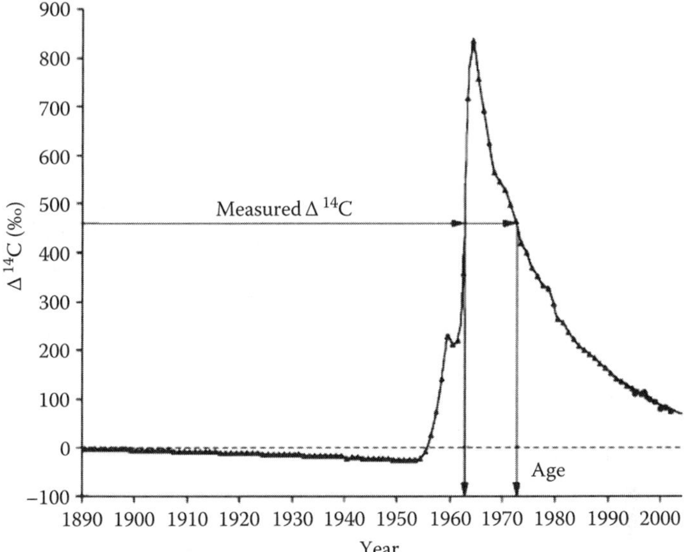

FIGURE 14.1 Atmospheric ^{14}C concentration as reflected in tree rings and the "bomb curve" arising due to atmospheric nuclear weapon tests in the 1950s and 1960s. (Adapted from Palmblad, M. et al., *J. Mass Spectrom.*, 40, 154, 2005. With permission.)

Using AMS to follow ^{14}C and other radioisotope tracers for biomedical studies was suggested in the 1980s [10,11]. In the early studies, AMS was used to examine the effect of chemical carcinogens at the DNA level and PK of xenobiotics in animals [12,13]. AMS applications to trace other long-lived isotopes of interest in biomedicine include ^{3}H, ^{26}Al, ^{41}Ca, ^{36}Cl, ^{129}I, and several others [11]. The success of AMS measurement relies on (1) the generation of stable negative ions, (2) an elemental isotope with a half-life of more than 10 years, (3) more than one stable isotope of the element, and (4) the isotope of interest should be of low natural abundance. Considering these factors, ^{14}C is the most suitable isotope for biomedical applications using AMS as a detector for measuring trace-level analytes.

14.2.2 Instrumentation

AMS does not determine the absolute concentration of isotopes; rather an isotope ratio of a rare isotope to an abundant stable isotope of the same element is measured. There is no commercially available AMS instrument that measures all isotopes, such that most instruments are built to detect one or two elements only. Highly specialized, complex, and time-consuming sample preparation is often required prior to AMS analysis. AMS instruments for various radioisotope analyses have been reviewed in a recent publication [14]. Here we would only discuss AMS instrument for ^{14}C analysis, the most important radioisotope for biomedical applications. For ^{14}C measurement, an AMS instrument usually consists of an ion source to generate negative ions, a low magnetic field for initial ion separation, a high-voltage accelerator to separate out isobaric ions, and to strip electrons from the negatively charged ions to form positive ions, an electrostatic mass analyzer and ion detectors (Figure 14.2) [15–20].

Biological samples are graphitized by a series of reactions into reduced carbon and loaded into a metal holder in the ion source [21]. Usually a cesium ion sputter is used in the ion source to generate the negatively charged ions (C^-, C^{2-}, CH^-, and CH^{2-}). These negative ions are first accelerated by passing through a preacceleration tube to energies as high as 80 keV. Ions are initially separated by their kinetic energy and momentum. However, ions of similar masses are not distinguished here.

Upon leaving the preacceleration tube, ions are guided into the high-voltage Pelletron accelerator (two-stage accelerator). A 45° electrostatic analyzer is usually used to sort out ions according to predetermined kinetic energies and a 90° injecting magnet is used to guide the ions of selected momentum into the accelerator. By discrete adjustment of the injecting magnet current, one can select only $^{12}C^-$, $^{13}C^-$, $^{14}C^-$ ions, and their corresponding isobaric ions, to pass into the accelerator.

The Pelletron accelerator contains up to 6 atm of SF_6 gas in a sealed tank to insulate the terminal, which allows it to attain several megavolts (typical ranges are from 1 to 10 MV) without discharging. Inside the accelerator, the acceleration tube is

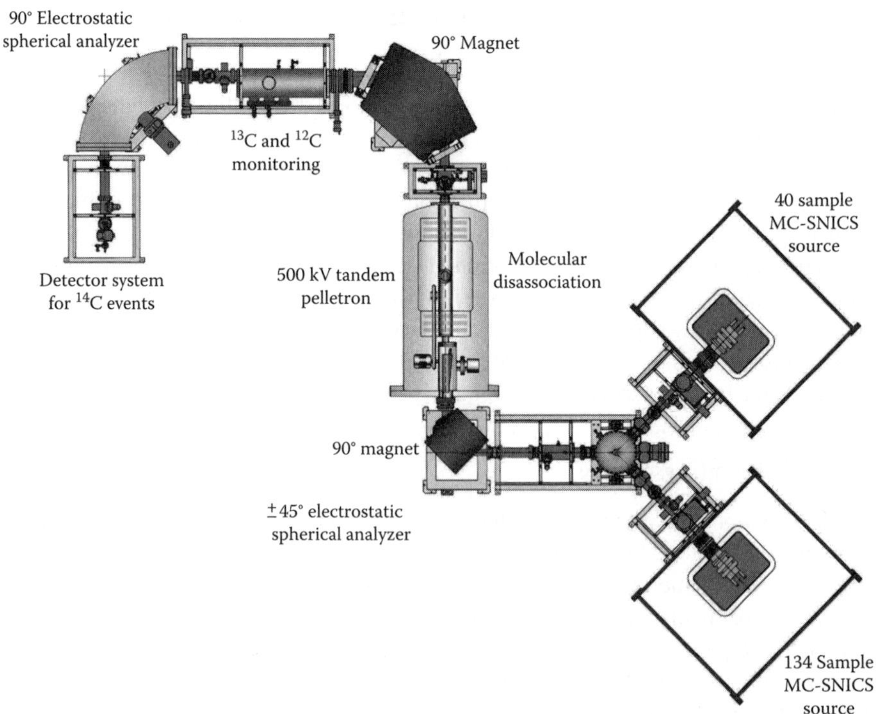

FIGURE 14.2 Schematic layout of a ^{14}C accelerator mass spectrometer with 500 kV double stage Pelletron accelerators. (Courtesy of National Electrostatics Corporation, Middleton, WI.)

maintained at ultrahigh vacuum (10^{-7} torr) and the end terminals at ground potential. The terminal at the midpoint of the acceleration tube is maintained at positive high voltage. A thin carbon foil or gas stripper, usually Argon, is used inside the midpoint terminal. Electrons are removed from the high-velocity negative ions when they pass through the stripper. Positive ions with varying charge states are formed and interfering molecular isobaric ions are dissociated and separated. The charge state depends on the ion velocity. With a 5–7 MV terminal, most C^- ions form C^{4+} ions, while mainly C^+ ions are formed in 250 kV accelerators. Positive ions are then further accelerated to ground potential in the second half of the accelerator.

Posterior to the Pelletron accelerator is a high-energy 90° analyzing magnet, which is used to separate the rare ions (^{14}C) from the abundant ions (^{12}C, ^{13}C). The ions are then measured in an energy-sensitive semiconductor or gas-filled ionization chamber. Rare ions (^{14}C) are measured by single particle counting in the gas ionization detector, while abundant ions are measured by electric current generated in the electron multipliers (Faraday cup). AMS does not determine absolute concentrations of radioisotopes. Each measurement is an isotope ratio, the ratio of rare isotope to a stable isotope of the same element ($^{14}C/^{12}C$ and/or $^{14}C/^{13}C$). The $^{14}C/^{12}C$ ratio, R, is presented in units of Modern, or percentage of Modern Carbon (pMC). One Modern corresponds to 98 attomole ^{14}C/mg ^{12}C or 6.11 fCi ^{14}C/mg ^{12}C, ^{14}C in the background atmosphere in 1950, prior to widespread atmospheric testing of atomic weapons.

While the majority of AMS instruments have a double-stage Pelletron accelerator, there is a trend to build a new generation of AMS instruments that can run at a lower voltage and with a smaller footprint. The earlier AMS instruments used 1–10 MV double-stage Pelletron accelerators [18] and required a footprint of 260 m^2. More recently AMS instruments with double-stage Pelletron accelerators are designed to operate at a lower energy, for example, 500 kV [17]. This high voltage requires a sealed pressure tank filled with SF_6 up to 6 atm to insulate from discharging. Any repair or maintenance on the Pelletron accelerator demands emptying the pressure tank and an expensive refill. Recently, an AMS instrument with a single-stage 250 kV accelerator is available commercially from National Electrostatics Corporation (Middleton, WI, United States) [22–24]. After dissociation of the accelerated negative ions, there is no second-stage acceleration. Because of the relatively low voltage of the accelerator, open-air insulation is allowed. A low cost solid-state silicon barrier detector has replaced the gas ionization detector in the instrument to reduce the cost. Although there is some reduction of the absolute sensitivity, the single-stage AMS is comparable to a double-stage 5 MV AMS instrument for pharmaceutical applications, because the sensitivity is limited by the background interference for biological samples (Figure 14.3). The biggest advantage of the single-stage AMS instrument is that it reduces the footprint greatly, from the 260 m^2 required for a typical 5 MV double-stage Pelletron AMS to around 56 m^2, which is much easier to fit into a typical analytical laboratory. With less complexity and no high-pressure insulation tank, the operation and maintenance are less challenging. Because of many useful utilities of AMS systems and the availability of small foot-print systems, it is expected that more pharmaceutical research organizations will develop in-house AMS capabilities in the very near future.

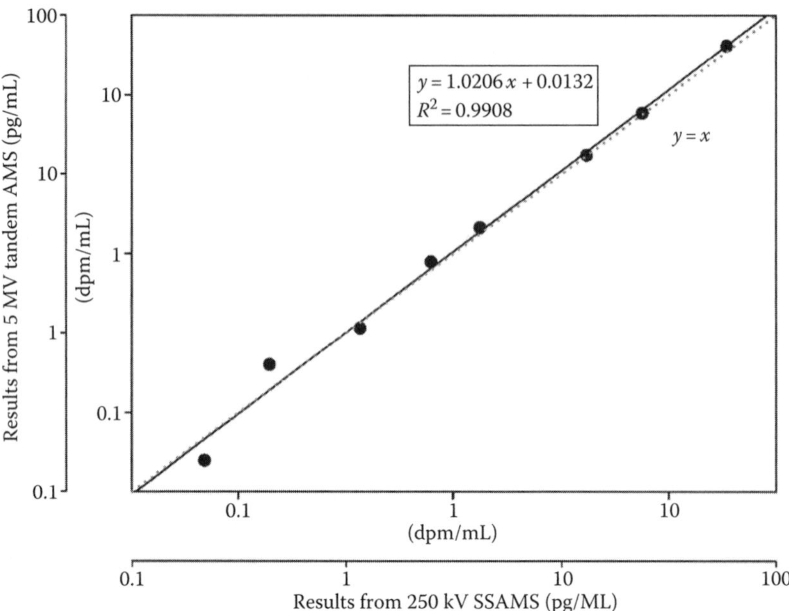

FIGURE 14.3 Comparison of mean data obtained on a 5MV double-stage AMS and a 250kV single-stage AMS from graphites prepared from human plasma spiked with radiocarbon over nominal range of 0.1–20dpm/mL. The conversion to pictogram per milliliter was based on the specific activity of 5.3MBq/mg for the radiocarbon source used in the experiment. Solid line: best fit regression line; dotted line: $y = x$. (Adapted from Young, G.C. et al., *Rapid Commun. Mass Spectrom.*, 22, 4035, 2008. With permission.)

14.2.3 AMS Sample Preparation

The most important consideration in preparing samples for AMS analysis is preventing contamination and knowing the source and amount of isotope (^{14}C) introduced during the process. Numerous precautions are taken throughout the procedures to ensure that the amount of isotope present is within the dynamic range of the spectrometer and the isotope detected is associated with the labeled sample under investigation. Disposable labware, a dedicated low-background working area and specialized equipment are often required to avoid sample contamination.

Most ^{14}C samples are measured as graphite. Alternatively, the ^{14}C samples can be analyzed as a gas [25–27], such as CO_2. Compared to gaseous forms, solid samples emit up to 10× the ions and make the most efficient use of the spectrometer. However, a gaseous source requires less amount of sample [28], can be operated continuously, and is compatible with gas chromatography (GC) as demonstrated in a recent report [25]. The solid-state sample is preferred due to mature standardized procedures, lower risk of cross-contamination in the ion source, shorter memory effects when a "hot" sample is encountered, higher sample throughput due to more efficient ionization, etc. A direct interface for gaseous ionization will have more applications as the technology becomes more mature.

The standard procedure (Figure 14.4) to produce a graphite ^{14}C sample involves oxidation of the biological sample to a gaseous mixture of CO_2 and water, cryogenically extracting the CO_2 gas, and reducing it to form filamentous graphite [21]. In most cases, the compound of interest from biological samples is isolated from the crude sample using HPLC and transferred to a prebaked quartz tube containing CuO powder (Figure 14.5). The quartz tube is vacuum-sealed using a high-temperature torch. The sample is then oxidized by heating in an oven at 900°C for 2 h or longer to be converted to CO_2, H_2O, and N_2. After cooling slowly to room temperature, the quartz tube is transferred into one end of a Y-shaped vacuum device called the combustion chamber. The other end of the Y-shaped device is a reaction chamber

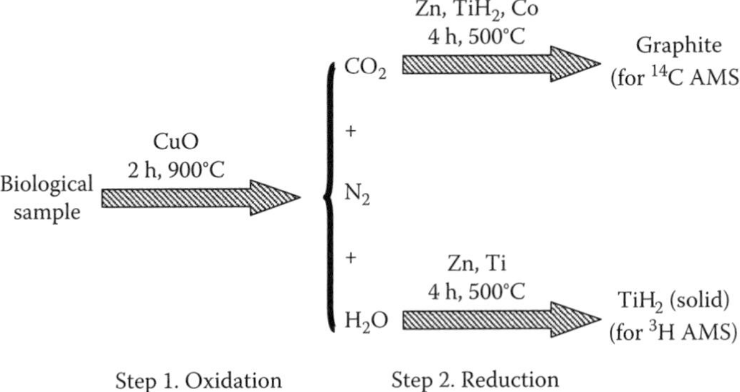

FIGURE 14.4 Stages of sample preparation for ^{14}C and ^3H AMS analysis.

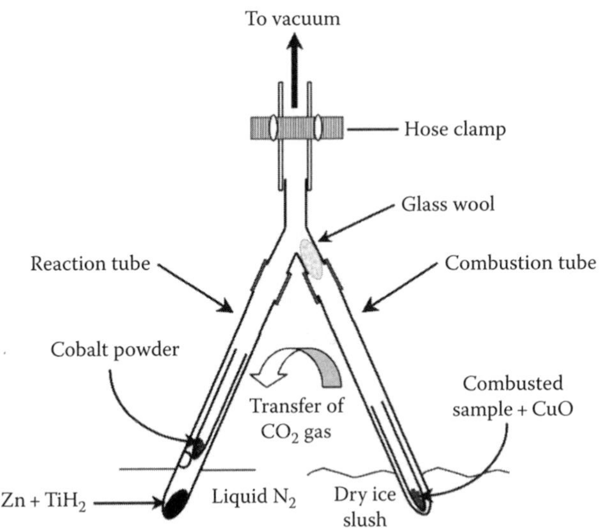

FIGURE 14.5 Sample manifold to transfer CO_2 from the combustion tube to the reaction tube. (Adapted from Vogel, J.S., *Radiocarbon*, 34, 344, 1992. With permission.)

containing reducing agents, such as titanium hydride, zinc, and cobalt powder. The gases in the reaction chamber are first pumped out followed by clamping off. The combustion chamber is placed into a dry-ice/isopropanol bath to condense water. The end of the combustion tube is broken to allow the transfer of CO_2 over to the reaction chamber, which is placed in liquid N_2 to trap the CO_2. Finally, the clamp is removed to release the N_2 gas and the reaction tube is sealed. The sealed reaction tube is placed in a furnace for ~4 h at ~500°C. The CO_2 is reduced to graphite by zinc and titanium hydride and deposited on the cobalt catalyst contained in the smaller inner tube of the reaction tube.

Sample preparation is the rate-limiting step in AMS studies. A number of modifications have been designed to improve the efficiency and to reduce the sample amount, including preparing graphite from CO_2 in a septa-sealed vial [29], miniaturizing the reactor size [7], and direct plasma/urine sample analysis by depositing on a CuO reaction bed heated by a laser [30]. Directly analyzing components from GC eluent by passing through a CuO reactor using AMS was reported recently [25]. In the foreseeable future, direct HPLC/AMS analysis may become a reality.

With the sample preparation method described earlier, the optimal amount of total carbon for AMS analysis is around 1.0 mg. When the total carbon level is far below that level, a precise amount of suitable carrier carbon is added. Petroleum-based tributyrin is often used as the carrier because it is a nonvolatile liquid, contains depleted levels of ^{14}C and ^{3}H, and its measurement is highly reproducible [18].

14.2.4 Data Processing

AMS measures the isotope ratio of ^{14}C:^{13}C:^{12}C and expresses the ratio as the amount of ^{14}C per mass of carbon. The isotope ratio is interpreted by knowing the sources of carbon in the studied sample, which is the sum of isotope abundance in natural and an appropriate added carrier carbon. This ratio can be converted to a drug concentration according to the equation from Vogel and Lover [20] and Salehpour et al. [7]:

$$K = (R_{\text{measure}} - R_{\text{natural}})\psi(W/L) \tag{14.1}$$

where

- K is the concentration of analyte
- R_{measure} is the isotope ratio of the sample containing the ^{14}C-analyte
- R_{natural} is the natural background isotope ratio of the sample (or the controlled background of the sample without ^{14}C-analyte)
- ψ is the carbon mass fraction in the sample (the percentage of carbon in the sample)
- W is the molecular weight of the analyte
- L is the specific molar activity of the analyte

AMS data only provide the total ^{14}C content from a collected sample. To determine the concentration of any individual analyte present in a mixture of analyte, the compound of interest must be separated from others by a chromatographic technique, usually HPLC, prior to AMS analysis.

For studies with a well-characterized drug or known metabolites, a calibration curve can be constructed by spiking a series of sample with predefined amounts of ^{14}C standard along with a known and equal amount of nonradiolabeled analyte as the internal standard [31]. The drug or metabolite concentration in the sample can be calculated directly from the working calibration curve by comparing the AMS measurement. This procedure avoids complicated unit conversion and compensating for compound loss during sample preparation.

14.3 APPLICATIONS OF AMS FOR PHARMACEUTICAL STUDIES

The great sensitivity of AMS gives it a number of advantages over other analytical techniques for detecting long-lived radioisotopes. The AMS technique is therefore ideally suited for studies in which only low doses of chemical or radioactivity are allowed. The long half-lives of radionuclides suitable for AMS analysis undergo negligible decay during the time course of biological *in vitro* and *in vivo* studies. In recent years, the AMS applications for microdosing PK studies, as well as metabolite profiling, absolute bioavailability, mass balance, and DNA/protein modification studies have grown rapidly. Some of the compounds that have been studied recently using AMS for their DNA/protein binding, bioavailability, and PK are shown in Figure 14.6.

14.3.1 Validation of AMS Instrument for Pharmaceutical Applications

In an effort to validate AMS for pharmaceutical applications, multiple studies have been conducted. Using ^{14}C standards, a precision of 0.5% has been achieved on a two-stage 500 kV AMS instrument at two different ^{14}C levels from replicate analysis over a 3 day period [17]. The linear response of AMS to ^{14}C concentration has been

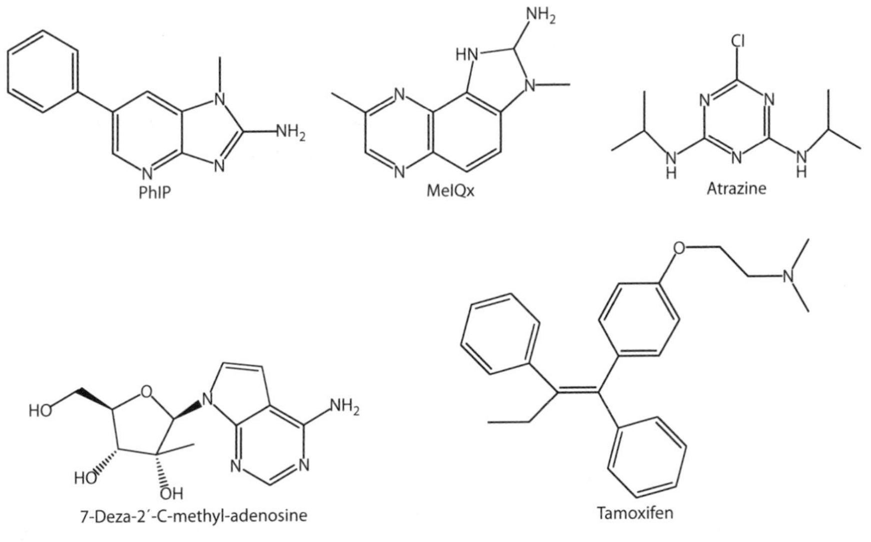

FIGURE 14.6 Structures of some compounds that have been studied using AMS.

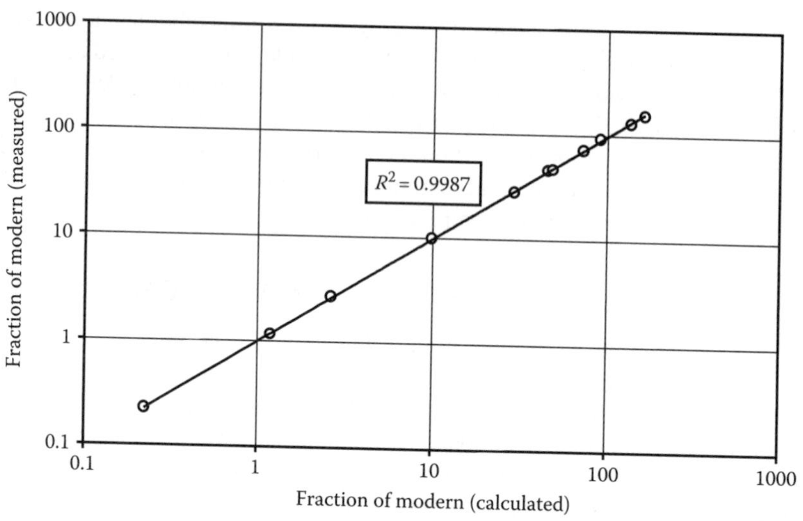

FIGURE 14.7 Linear response curve of ^{14}C AMS measurement to spiked ^{14}C concentration in human plasma between 0.1 and 150 Modern. (Adapted from Zoppi, U. et al., *Radiocarbon*, 49, 171, 2007. With permission.)

demonstrated (Figure 14.7) in standard samples by spiking different amount of ^{14}C malonic acid in tributyrin in the range from 0.1 to 150 Modern (correlation coefficient 0.9987).

Direct comparison of AMS with liquid scintillation counting (LSC) of biological samples has been demonstrated for analysis of plasma, urine, and feces. During a PK study with ^{14}C-labeled fluticasone in male Han Wistar rats, blood samples were analyzed using a liquid scintillation counter and diluted several thousand folds for AMS analysis. The two methods agreed with each other remarkably ($R^2 = 0.999$) [32]. In a similar comparison study, 10 human subjects were dermally exposed to the ^{14}C-labeled herbicide atrazine over 24 h at two different levels (0.167 mg, 6.45 μCi and 1.98 mg, 24.7 μCi) [33]. Urine and feces were collected over a 7 day period and analyzed using both LSC and AMS. Urine samples were diluted 10× with distilled water for AMS analysis and 1.23 mg of carbon carrier tributyrin was added to every sample to minimize the variability of carbon content among samples. Figure 14.8 shows the correlation of data for AMS and LSC analysis with replicate ($N \geq 4$) analysis of the same samples from a single subject at the same dose. The correlation is 0.999 and AMS has better precision at the low concentrations. Day-to-day AMS analysis of the same sample over a five-day period is very reproducible with a relative standard deviation of 3.9%. The correlation between two methods for all 10 subjects ranged from 0.992 to 0.998 for the entire sample set.

14.3.2 Microdose in Pharmacokinetics Studies

Drug development continues to be a long, complex, and expensive process. Over the past decade, the research and development cost for new drugs has increased

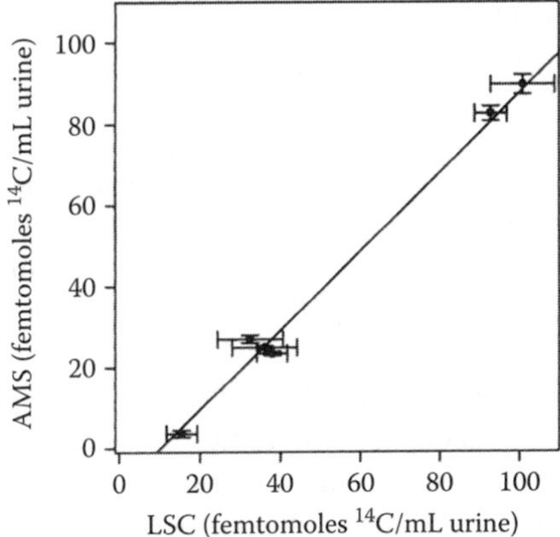

FIGURE 14.8 Comparison of data acquired by AMS and LSC from urine of a single subject collected after dermal dose of ^{14}C-atrazine. Error bars represent the standard deviation for replicate measurements of each sample ($N \geq 4$). (Adapted from Gilman, S.D. et al., *Anal. Chem.*, 70, 3463, 1998. With permission.)

exponentially, while the new chemical entities (NCEs) registered for marketing is either static or declining [34]. While safety, efficacy, and toxicology issues dominate the termination of drug development programs, up to 40% of the investigational drug candidates are still discontinued for further development after the first in-human study due to unacceptable human PK, even though promising ADME data were obtained in animals [2]. Regulatory authorities are concerned that excessive development costs as a result of high failure rates are preventing new life-saving medicines from reaching patients at an affordable price [35]. Conventional methods of drug development rely on animal, *in vitro*, and *in silico* models to predict human metabolism and PK. There is always a great concern that drug metabolism pathways and PK in humans might differ significantly from those derived from models. A recent approach to address these issues is to obtain human metabolism and PK through administration of subtherapeutic or subpharmacological doses (microdose) of investigational drugs in so-called Phase 0 studies. To detect drugs at very low doses, trace amount of ^{14}C are administrated with the microdose and the drug concentration is measured by AMS to determine the PK of the drug.

The microdosing strategy to quickly and safely obtain human PK data was first proposed around 2000 [1,4,8,18,32,34,36,37]. In 2004, the European agency for Evaluation of Medicinal Products (EMEA) published a position paper to encourage the microdosing strategy [38], while the FDA issued a similar guidance in 2006 [39]. Both regulatory authorities define a microdose as the administration of 1/100th of the pharmacological dose determined from animal models or *in vitro* systems, which must not exceed 100 μg. The guidance recommends that the sponsor should demonstrate a large multiple (100×) of the proposed human dose that does not induce adverse effects in the experimental animals, and the scaling from animals to humans should be based on body surface area.

The ability to conduct a microdose study relies on ultrasensitive analytical techniques that are capable of measuring drug and metabolite concentrations in the low picogram to femtogram per milliliter range. AMS, which has sensitivity to trace ^{14}C compounds in the zeptomole to attomole range [7,8], is the most suitable technique for such studies when ^{14}C-labeled drug candidate is dosed. Positron emission tomography (PET) has excellent sensitivity [40,41]. However, the radioisotope, ^{11}C, used for PET studies has a very short half-life, only 20 min. This limits the technique in that PK data can only be obtained for around 2 h after drug administration. Close proximity of the radiosynthesis laboratory, the facility to administer the drug to volunteer patients, and the PET analytical laboratory is also required. PET is more suitable for pharmacodynamics studies than for PK studies. When a drug candidate has suitable mass spectrometry sensitivity and can be easily extracted from plasma, high-performance liquid chromatography tandem mass spectrometry (HPLC–MS/MS) can be used for microdose studies as well [5,42,43]. The advantage of HPLC–MS/MS is that it does not require radioisotope-labeled materials, the isolation of parent drug from other drug-derived materials, and the instrument is readily available in many analytical laboratories. However, endogeneous matrix species could interfere with the detection for drug analysis in the picogram per milliliter concentration range or below. For AMS and PET, drugs must be labeled at metabolically stable sites. With the long half-life of ^{14}C, AMS is currently the preferred analytical technique for microdosing studies.

The success of a microdose study largely depends on establishing near-linear kinetics between a non therapeutic microdose and a pharmacological dose. As a first approximation, PK difference at both doses would be minimized if drug concentrations are well below K_m for the metabolizing enzymes. If a pharmacological dose achieves drug concentrations well above K_m in the liver or other sites of metabolism, nonlinear kinetics could be observed [34,44]. This problem was first addressed in a microdose study in dogs with an antiviral drug candidate (7-deaza-2′-C-methyladenosine) [3]. During this study, dogs were dosed orally at 0.02 mg/kg, 1.0 mg/kg, and intravenously at 0.02 mg/kg using ^{14}C-labeled drug. Blood samples were collected over 80 h and the total ^{14}C was analyzed using AMS. In addition, dogs were administrated using unlabeled drug (0.4 mg/kg) intravenously, followed by 1.0 mg/kg orally 2 weeks later. In this group blood samples were collected over 72 h and analyzed using HPLC–MS/MS. It is usually not a good practice to measure only the total ^{14}C in blood or plasma samples, for it represents the sum of parent and metabolites. However, in this case, a comparison of AMS and HPLC–MS/MS data showed that total ^{14}C determinations were very similar to parent drug, probably due to very low levels of circulating metabolites. PK between 0.02 mg/kg (Figure 14.9) and 1.0 mg/kg (Figure 14.10) doses were shown to be linear with respect to AUC (area under curve) on average. After dose-normalization, the oral AUCs (1.0 mg/kg) were 77% of that achieved from the 0.02 mg/kg IV dose, while oral C_{max} was 76% of the IV microdose.

In the Consortium for Resourcing and Evaluation AMS Microdosing (CREAM) trial, five compounds (diazepam, midazolam, warfarin, erythromycin, and ZK253) were selected for study and the results were published in 2006 [45]. Of the five compounds, PK between microdoses (100 μg in all cases) and pharmacological doses of

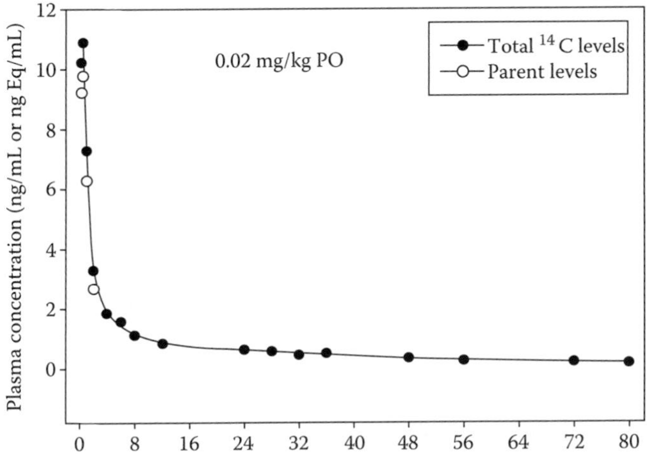

FIGURE 14.9 Plasma concentration of 7-deaza-2′-C-[^{14}C]methyl-adenosine (Compound A) (ng/mL; open circle) and total ^{14}C levels (ng Eq/mL; filled circle) after oral administration of 0.02 mg/kg (~100 nCi) ^{14}C-Compound A to dogs ($n = 2$). (Adapted from Sandhu, P. et al., *Drug Metab. Dispos.*, 32, 1254, 2004. With permission.)

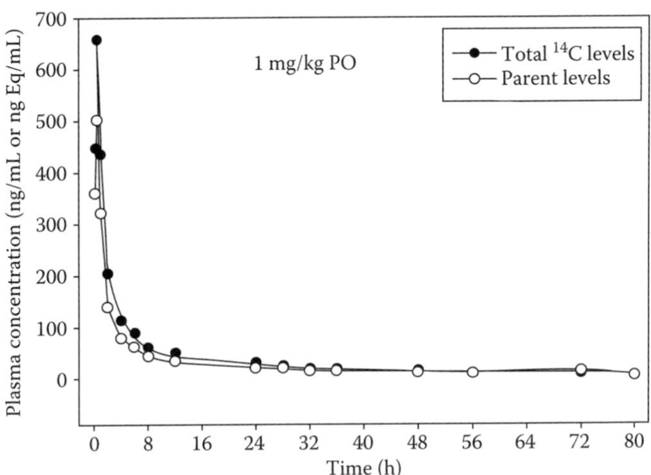

FIGURE 14.10 Plasma concentration of 7-deaza-2′-C-[^{14}C]methyl-adenosine (Compound A) (ng/mL; open circle) and total ^{14}C levels (ng Eq/mL; filled circle) after oral administration of 1 mg/kg (~100 nCi) ^{14}C-Compound A to dogs ($n = 2$). (Adapted from Sandhu, P. et al., *Drug Metab. Dispos.*, 32, 1254, 2004. With permission.)

diazepam (10 mg), midazolam (7.5 mg), and ZK253 (50 mg) were essentially linear within a factor of two. For example, diazepam was administrated intravenously at microdose (100 μg) and therapeutic dose (10 mg) over a period of 30 min. Plasma diazepam concentration from this microdose study predicted human PK using a therapeutic dose (Figure 14.11).

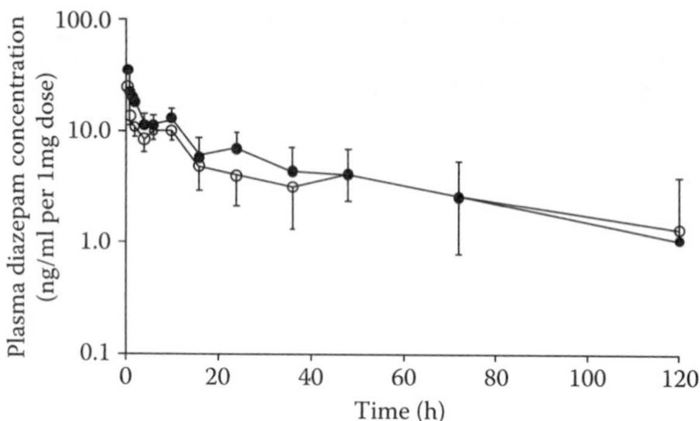

FIGURE 14.11 Semilog plot of plasma diazepam concentrations (geometric mean, SE) normalized to 1 mg dose vs. time after intravenous administration of 100 µg (open circle) and 10 mg (filled circle) doses over a period of 30 min, in which a linear PK was demonstrated over 100-fold concentration range. (Adapted from Lappin, G. et al., *Clin. Pharmcol. Ther.*, 80, 203, 2006. With permission.)

Warfarin showed nonlinear PK. There was no overlap in the dose-normalized concentrations except for those at the initial sampling time (1 h) and last sampling time (120 h) (Figure 14.12). The half-life of a therapeutic dose was around 37 h, while it was estimated to be 528 h by microdose. The clearance was estimated at 0.17 L/h using a microdose with complete absorption assumed, which was similar to that 0.19 L/h with a therapeutic dose. These findings suggested that distribution was the major source of nonlinearity in warfarin PK. The volume of distribution was 67 L by microdose versus 10 L with the therapeutic dose. It is extremely unlikely that the difference in distribution is caused by protein binding, since warfarin binding to albumin is not saturated at a therapeutic dose. Evidence from studies in rats suggested

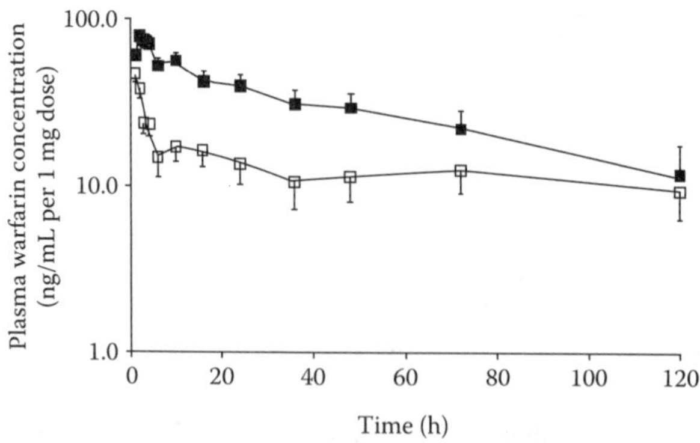

FIGURE 14.12 Semilog plot of plasma warfarin concentrations (geometric mean, SE) normalized to 1 mg dose vs. time after oral administration of 100 µg (open square) and 5 mg (solid square). (Adapted from Lappin, G. et al., *Clin. Pharmcol. Ther.*, 80, 203, 2006. With permission.)

that low-capacity, high-affinity warfarin binding to vitamin K-2,3-epoxide reductase in liver might be the cause. The other compound that did not exhibit linear PK is erythromycin. Erythromycin is extremely acid labile, and it was likely that most of the oral microdose degraded rapidly in the acidic environment of the stomach. Plasma erythromycin was undetectable after oral microdosing (100 μg). Therefore, no meaningful comparison of PK between doses could be made.

In a recent review [46] of 18 drugs studied using both microdoses and pharmacological doses, 15 of these 18 drugs demonstrated linear PK within a factor of 2. The three drugs that showed nonlinearity included warfarin in the CREAM trial [45], MLNX, and metoprolol. MLNX, a compound synthesized and studied in Millennium Pharmaceuticals, Inc., was orally administrated to rats at 0.01, 0.1, 1, and 10 mg/kg dose levels [42]. The compound showed linear PK between 0.01 and 1 mg/kg dose levels. However, nonlinearity was observed at the highest dose of 10 mg/kg. The relevance of this finding compared to a human therapeutic dose is unknown. In spite of the small number of studies reported, the microdose predictability is around 83%. The dose ranges established for linear PK have been from several hundred to over 10,000-fold.

An AMS-based microdosing study provides several advantages over conventional Phase I studies: (1) less than 100 g of GMP manufactured active pharmaceutical ingredient (API) is required, (2) a minimal preclinical toxicology package is required, (3) data can be generated in 4–6 months from commencement of the preclinical drug safety to obtaining human PK data, in contrast to around 12–18 months for a conventional Phase I study, and (4) the cost of conducting a microdose study is only a fraction of that required in full Phase I study programs. However, it should be noted that a microdose study does not provide information on human toxicity or efficacy. A microdose study can assist the candidate selection process, determine the first dose for the subsequent Phase I study and the likely pharmacological dose, calculate the cost of goods for a drug that is expensive to manufacture, and guide lead compound optimization if the microdosing study is carried out in an iterative manner. Probably the greatest advantage of a microdosing study is to terminate further development of a drug candidate if poor PK is observed.

14.3.3 Absolute Bioavailability

Absolute bioavailability is a measure of the true extent of systemic absorption of an extravascularly administrated drug. Along with clearance and volume of distribution, absolute bioavailability is one of the important parameters to characterize PK. Low bioavailability of a drug can be caused by incomplete dissolution when administrated as a solid, inability to permeate membranes, and metabolic instability (first-pass metabolism). Despite the importance of absolute bioavailability, it is not routinely assessed due to the cost and toxicology requirements for such a study in a conventional study design, which requires an intravenous reference. Safety issues may arise due to solubility limitation and toxicity associated with C_{max} effect. As a result, it is necessary to conduct a preclinical toxicological study with an IV formulation to ensure adequate human safety and potential problem. Bioavailability determined from animal models is not always predictive of that in human.

A conventional absolute bioavailability study is carried out by administrating an extravascular and an intravenous dose of a drug in human volunteers in a cross-over design. After the drug is administrated by one route, for example orally, plasma samples are taken at appropriate times and parent drug concentrations are determined. Following a suitable washout period, the same human volunteers are administrated the drug intravenously, and plasma samples are taken over time and parent drug concentrations are similarly measured. The total AUC is calculated for each dose route [47–50]. For intravenous administration, 100% of the drug is assumed to be in the systemic circulation. However, only portions of the dose might be absorbed into systematic circulation by extravascular route of administration. The absolute bioavailability (F) can be expressed by following equation if the clearance is constant for both dose routes:

$$F = \left(\frac{\mathrm{AUC_{ev}}}{\mathrm{AUC_{iv}}}\right) \cdot \left(\frac{\mathrm{DOSE_{iv}}}{\mathrm{DOSE_{ev}}}\right) \tag{14.2}$$

where

$\mathrm{AUC_{ev}}$ is the AUC calculated for the extravascular administration

$\mathrm{AUC_{iv}}$ is for the intravenous administration

$\mathrm{DOSE_{iv}}$ and $\mathrm{DOSE_{ev}}$ are the amounts of drug by intravenous and extravascular routes, respectively

If the clearance is plasma drug concentration-dependent, and either intravenous exposure is limited by blood stream solubility or the drug has very low bioavailability, there could be an error in the estimation of absolute bioavailability. For more on these PK parameters, see Chapter 3. Additionally, extravascular and intravenous administrations are performed at two different time periods; any changes of metabolism in the study subject may also affect the calculated absolute bioavailability.

An alternative method to study bioavailability is by concurrent intravenous administration of isotopically labeled drug and extravascular administration of unlabeled drug. Stable isotopes, such as ^{13}C, ^{15}N, and ^{18}O, have been used in absolute bioavailability studies [51–53] in an effort to distinguish drug in circulation following an oral dose from that originating from an IV dose. Therefore, HPLC–MS/MS can be used for the quantification of the drug separately from two different dosing routes. However, due to the high natural abundance of these stable isotopes, for example, 1.1% for ^{13}C and 0.37% for ^{15}N, analytical sensitivity is often compromised. Therefore, it is necessary to dose similar amount of stable isotope-labeled drug and the unlabeled compound, and/or incorporate a sufficient number of isotopically stable atoms into the molecule so that it is sufficiently differentiated during HPLC–MS/MS measurement from the unlabeled drug. Adequate solubility of drug in the IV formulation may become an issue when the dose approaches that administrated orally.

The list of AMS applications in pharmaceutical research continues to grow since its introduction in 1990s. With subattomole sensitivity, AMS is an ideal analytical technique for bioavailability studies using the isotope methodology with ^{14}C-labeled drug, since it has extremely low natural abundance (10^{-11}%) [54,55]. When AMS is used to measure plasma levels of ^{14}C-labeled parent drug, less than 100 μg drug is required for intravenous administration, typically with 0.1 μCi or less radioactivity. These low radioactive doses minimize radiosafety considerations and limit regulatory requirement. Solubility issues also become less of a problem at such low doses. As specified in the guidance for a microdosing study by the regulatory authorities [38,39], the oral toxicology package is often enough to support the safety of the intended intravenous microdose. It is also possible to combine an absolute bioavailability study with the conventional Phase I oral protocol by including the simultaneous intravenous microdosing alongside one of the oral doses. As a consequence, the cost of determining absolute bioavailability drops significantly with an AMS-based microdosing study, since no intravenous toxicology package is required, and the blood samples are collected according to standard procedures.

Nelfinavir is a marketed medicine to treat HIV infections. The absolute bioavailability of nelfinavir has been studied using AMS following microdosing with ^{14}C-labeled drug [56]. In a multiple dose study in healthy male adults, volunteers received oral doses (1250 mg) of nelfinavir twice daily from days 2 to 10, but only the morning oral dose on days 1 and 11. On days 1 and 11 after taking the morning oral dose, each volunteer received a single intravenous dose of 1 mg (~ 50 nCi) of ^{14}C-labeled nelfinavir. Total nelfinavir was analyzed by HPLC/UV, while ^{14}C-labeled nelfinavir and its main active metabolite AG1402 were analyzed using AMS after HPLC fraction collection. Nelfinavir bioavailability dropped from 0.88 to 0.47 from day 1 to day 11. The PK results from this study revealed slow oral absorption and rapid clearance of nelfinavir, but also evidence for metabolic enzyme induction after repeated oral dosing.

In cases where drug is applied locally or topically, parent drug concentrations can be so low that routine analytical methods may not have enough sensitivity to determine the concentration in plasma. AMS is the most suitable analytical tool for such cases. Hexaminolevulinate (HAL) is a diagnostic agent to visualize tumor bearing tissue in the bladder. It is administrated intravesically via a catheter. Ideally, the drug should be confined to the bladder only to minimize systemic exposure. The absolute bioavailability of HAL was determined by intravesical delivery of ^{14}C-labeled drug into the bladder and intravenously on two separate occasions in a cross-over study [57]. The patients were given 100 mg (0.3 μCi) intravesically, and approximately 1 mg (0.1 μCi) intravenously after a 2 week washout period. The plasma drug concentrations were determined by AMS. In order to avoid artifacts arising from dose-dependent kinetics, the doses were carefully selected to maintain similar plasma drug concentrations (within a factor of 1.5). As expected, a very low absolute bioavailability ~7% was obtained. In another study, the dermal exposure to the herbicide atrazine was initially investigated using the conventional liquid-scintillation counting. Absorption through human skin was much lower than expected from preliminary animal studies. However, the archived samples were later analyzed using AMS to successfully obtain an estimate for absolute bioavailability [58].

Absolute bioavailability is one of the important PK parameters that can provide knowledge about the uptake mechanism, extent of first pass effect, and potential to enhance bioavailability by formulation optimization. However, absolute bioavailability is not routinely measured in the clinical study, since by conventional means it requires an extensive toxicology package for intravenous administration, IV formulation development and is costly to implement in a traditional cross-over design. The combination of AMS with its exquisite sensitivity and the cost efficiency of a simpler study design that includes a safe and easy to manufacture IV microdose is a powerful incentive for more routinely conducting absolute bioavailability studies in the clinic.

14.3.4 Mass Balance and Metabolite Profiling

Radiolabeled mass balance studies are used to investigate the preferred clearance pathways and the extent of drug and drug-derived materials that are excreted from the body. Approximately 50~100 μCi of ^{14}C-labeled drug is used to trace the fate of a drug in a conventional human mass balance studies [59,60]. In addition to the total drug-derived radioactivity excreted in urine and feces, parent drug and metabolites are measured in urine, feces, and plasma. AMS has 10^3–10^9 times higher sensitivity than conventional LSC for ^{14}C detection. This enhanced sensitivity is sufficient to conduct a mass balance study with less than 100 nCi ^{14}C-labeled drug. This allows for conducting mass balance studies when higher level of radioactivity is not permitted for administration, or when absorption of the drug is poor or the drug is administrated locally by inhalation or topical administration.

The application of AMS for mass balance was first demonstrated in male Han Wistar rats using ^{14}C-fluticasone, a chemical for treatment of asthma and rhinitis [32]. Samples of blood and excreta were analyzed by LSC, then diluted several thousand times and analyzed by AMS. In a study involving GI181771, an investigational drug candidate, only 3.3 nCi of ^{14}C-labeled drug (2.7 mg) was orally administrated to healthy male volunteers [61]. Following analysis of excreta using AMS, more than 92% of the dose could be accounted for in the feces over 5 days and less than 6.5% in the urine. In the mass balance study of R115777, an anticancer drug candidate, 50 mg (~34 nCi) of ^{14}C-labeled drug was orally administrated in healthy male volunteers [36]. Plasma, urine, and feces samples were collected over 7 days. The samples were analyzed with HPLC and collected with 1 min fractions. The HPLC fractions were converted to graphite following the procedure described previously [21]. The cumulative radioactivity measured was 13.7% and 79.8% of administrated dose, respectively, in urine and feces.

In some studies, it is only possible to administer a low amount of a ^{14}C tracer [62]. Ixabepilone is an anticancer drug, which is very unstable when labeled with ^{14}C at high specific activity. Therefore, to ensure radiochemical stability, it is required to use low specific activity radiocarbon and ultrasensitive AMS for detection. In a single dose intravenous study, 80 nCi (70 mg) of ^{14}C-ixabepilone was administrated to humans. Blood, urine, and feces samples were collected over 7 days. Total radioactivity was analyzed using AMS, and the parent drug was analyzed using HPLC–MS/MS (Figure 14.13). More than 77% of the drug-derived material was excreted over 7 days with 52% in feces and 25% in urine.

Although AMS has exquisite sensitivity, it is not routinely used for metabolite profiling. AMS is not available for coupling in-line with HPLC analysis, and no structural information is obtained when samples are graphitized for AMS analysis. Metabolites detected by AMS require characterization using other analytical

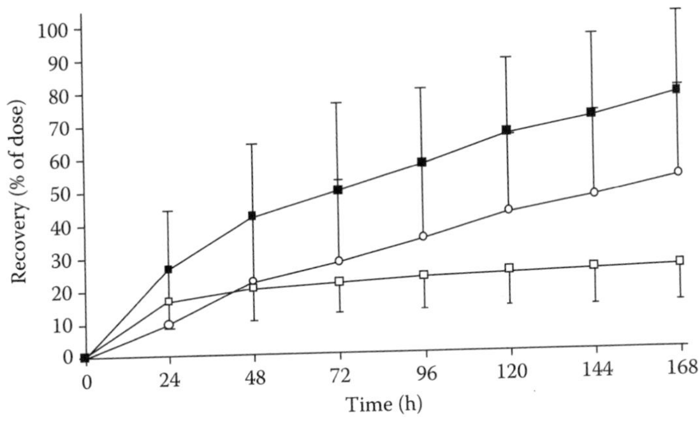

FIGURE 14.13 Mean urinary (□), fecal (○), and total (■) cumulative excretion (168 h) of ^{14}C-ixabepilone derived from eight patients after intravenous administration of 70 mg (80 nCi) ^{14}C-ixabepilone over 3 h. Error bars represents the standard deviation. (Adapted from Beumer, J.H. et al., *Invest. New Drugs*, 25, 327, 2007. With permission.)

techniques, such as HPLC–MS/MS. Nevertheless, when the need arises, metabolites can be detected by AMS with several thousand folds lower radioactive dose. In a study [36] involving R115777, 1 min HPLC fractions were collected for the entire HPLC 60 min run time and then analyzed using AMS to produce a metabolite profile. The radioactivity in all HPLC fractions was also measured after treatment with β-glucuronidase to assist with the metabolite identification. After β-glucuronidase treatment, the major polar early eluting HPLC peak decreased in abundance and shifted mainly to the position of the parent drug R115777 (Figure 14.14). The metabolite was later confirmed with authentic reference and HPLC–MS/MS analysis to be a glucuronide of R115777. The herbicide ^{14}C-atrazine has been studied using AMS for 7 days after dermal exposure [58]. A dose of less than 5 nCi to humans was sufficient to trace the metabolic profile by HPLC with fraction collection and AMS analysis. Atrazine mercapturate was identified as the major urinary metabolite and confirmed with an authentic reference standard. Using the same methodology, GI181771 fecal metabolites have been identified after an oral dose (2.7 mg, 3.3 nCi) of ^{14}C-labeled drug [61].

In addition to xenobiotic metabolism, AMS has been used to study the metabolism of endogenous compounds in humans. Such studies are generally difficult because the amount of radiotracer that can be administered is severely limited due to potentially long residence time in the body. Using AMS, it was possible to conduct the experiment with microdosing. The PK of β-carotene and its contribution to vitamin A synthesis in humans was investigated using AMS [63]. A single dose (930 ng, 60.4 nCi) of ^{14}C-labeled β-carotene was intravenously administrated to healthy human subjects. Blood samples were collected up to 22 h postdose; urine and feces were collected for 10 days. The study found that 1.0 mole of IV administrated β-carotene provided 0.19 moles of vitamin A. Folic acid metabolism has also been

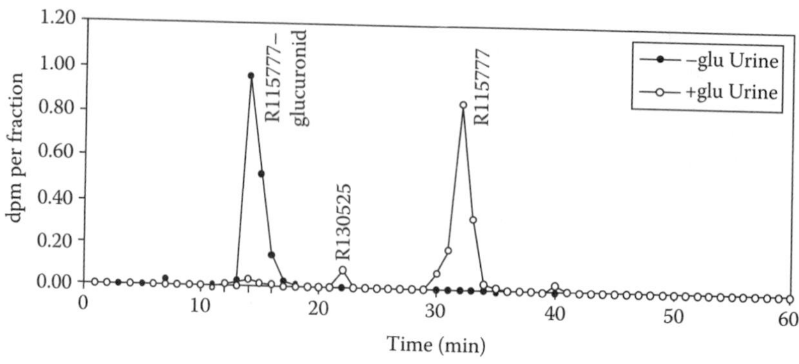

FIGURE 14.14 HPLC metabolite profile from humans receiving a single oral dose of 50 mg (34.35 nCi) ^{14}C-R115777 determined by AMS analysis of fractions from a combined pool of 0–24 h urine samples before (filled circle) and after (open circle) β-glucuronidase treatment. (Adapted from Garner, R.C. et al., *Drug Metab. Dispos.*, 30, 823, 2002. With permission.)

studied in humans using AMS following an oral dose (35 μg, 100 nCi) of ^{14}C-labeled material [18]. It was shown that the dose of ^{14}C-folic acid took 350 days to be eliminated from the body.

14.3.5 DNA and Protein Binding

The earliest success of AMS applied to biomedical research involved identification of DNA and protein binding of ^{14}C-labeled carcinogens and drugs in animals and humans [12,64]. It is important to understand the level of exposure, extent of adduct formation, and their biological effects. Formation of a covalent DNA adduct is considered an initial step in some chemically induced cancers. Protein adducts can serve as a surrogate biomarker for DNA binding, which may be present at higher levels than DNA adducts. Measuring DNA and protein adducts requires very sensitive analytical techniques, especially for environmental or therapeutic exposure in humans. Various techniques have been developed for this purpose, including ^{32}P-postlabeling, immunoassay, mass spectrometry, and fluorescence. AMS is the most sensitive technique to measure adducts after environmental exposure to ^{14}C-radiolabeled compounds.

Heterocyclic aromatic amines MeIQx (2-amino-3,8-dimethylimidazo[4,5-*f*]-quinoxaline) and PhIP (2-amino-1-methyl-6-phenylimidazo[4,5-*b*]pyridine) are formed at parts per billion (ppb) levels in meat and fish during cooking. Both compounds have been linked to colon, prostate, and breast cancers. Mouse liver levels of MeIQx DNA adducts were studied using oral doses ranging from 500 pg/kg to 50 mg/kg body weight [64]. A linear relationship between MeIQx DNA adducts and dose amount was established across the entire dosing range. ^{14}C-labeled MeIQx has been orally dosed to colon cancer patients (21.3 μg, 4.3 μCi) 3.5 ~ 6 h before surgery [13]. DNA was extracted from normal colon tissue surrounding the tumor and from the tumor tissue itself. Samples analyzed using AMS showed a similar level of DNA damage. The extent of DNA adduct formation in human colon was higher than that detected in mice, suggesting that human colon might be more sensitive to the genotoxic effects of MeIQx. In a similar clinical study, the colon cancer patients were orally dosed with ^{14}C-labeled PhIP (70 μg, 17 μCi) [13,65,66]. Again a higher extent of DNA adduct formation was found in human colon tissue than mice. Collectively, these data suggest that humans have a greater risk for colon cancer from these compounds than rodents. The ultrasensitivity of AMS made it possible to conduct such studies in humans with very low dose and limited safety risk.

Benzene is a human leukemogen, such that chronic exposure to the environmental pollutant benzene results in an increased risk of anemia and leukemia. In a mouse study [67], tissues, DNA, and proteins were analyzed using AMS over a 48 h period after intraperitoneal injection of ^{14}C-benzene at 5 μg/kg body weight (specific activity 50.3 mCi/mmol). With this dose that closely mimicked human environmental exposures, liver DNA adduct levels peaked at 0.5 h and bone marrow DNA adducts peaked 12–24 h postdose. The integrated DNA adduct level was found to be much higher in the bone marrow, the target tissue for benzene toxicity, than that observed in the liver. In addition, substantial protein modification was observed. Protein adduct levels were found to be 9 to 43 times higher than DNA adducts. Protein and associate radiolabeled protein adducts from mice liver and bone marrow following ^{14}C-benzene intraperitoneal (155 μg/kg body weight, specific activity 58.2 mCi/mmol) treatment were isolated using 1D and 2D gel electrophoresis, then excised from the gel for AMS analysis [68]. A number of protein and protein adducts were identified by in-gel digestion followed by conventional matrix-assisted laser desorption ionization–mass spectrometry (MALDI–MS) or LC–MS methods. The study found that the binding of ^{14}C-benzene and/or its metabolites to proteins was largely proportional to the abundance of protein without selective targets. This implies that benzene and its metabolites are highly reactive to most proteins and the core modification of benzene causing the carcinogenicity remains unknown.

Tamoxifen is used widely for treatment of breast cancer, and has been approved for the prevention of this disease in healthy women. However, long-term treatment with tamoxifen may increase the risk of endometrial cancer [69]. Using AMS, low levels of DNA damage were detected in rats that received a single oral dose of ^{14}C-labeled tamoxifen (90.3 mg/kg body weight, 15 μCi) [70]. In a follow-up study, women who were about to undergo hysterectomies were given a single oral therapeutic dose of ^{14}C-labeled tamoxifen (20 mg, 50 μCi), 18 h before surgery [71]. DNA was extracted from the uterus and analyzed using AMS. A low level of DNA damage was observed, and it was concluded that the damage alone was not enough to result in the formation of uterine neoplasias.

When measuring covalent DNA or protein adducts using AMS, the chemical reactants are often ^{14}C-labeled. AMS measures only the ratio of radioisotope to an abundant isotope (^{13}C or ^{12}C) without differentiating the signal origin. Rigorous purification is always employed to remove noncovalently bound compounds, in order to ensure that the detected AMS signal is arising from DNA or protein adducts. In many cases, gel electrophoresis [68] or HPLC separations [12,13,65,70] are used before AMS analysis. However, in one study, the formation of DNA adducts was directly confirmed by enzymatic digestion [72].

14.4 OTHER AMS APPLICATIONS

In addition to extensive application of AMS in pharmaceutical research using ^{14}C-labeled chemicals, other long half-life radioisotopes, such as ^{3}H, ^{41}Ca, ^{26}Al, and ^{36}Cl, have also been used [73] (Table 14.1).

^{41}Ca is a radioisotope with a half-life of 104,000 years and decays by emitting Auger electrons and soft x-rays. AMS provides about 10^8-fold better sensitivity for ^{41}Ca detection than LSC [74]. For AMS analysis [75,76], calcium is converted to either CaH_2, or CaF_2, and is usually detected as $^{41}Ca^{8+}$. In order to avoid potential interference from isobaric ^{41}K from biological samples, free calcium is often precipitated as the oxalate salt before converting to CaF_2. Since calcium is an important constituent for bones and cell signaling, its uptake and metabolism have been studied using AMS with ^{41}Ca-labeling to provide insight into Osteoporosis disease and neural disorders in both human [77,78] and animal models [76,79]. Calcium exchanges very slowly from skeletal tissues into the systemic circulation. Once administrated to humans, ^{41}Ca could last for lifetime, so that the dose amount is strictly limited. AMS is the only analytical technique available with enough sensitivity to measure ^{41}Ca in humans at a safe dose level ~5 nCi. For human studies, calcium uptake is determined by comparing the difference between ^{41}Ca in dietary intake and urinary/feces excretion. Direct measurement from bone biopsy would be too invasive for routine measurement.

Aluminum apparently plays an important role in Alzheimer's disease and other neural disorders. There are concerns about human exposure to aluminum in the drinking water, medical vaccines (hepatitis B), antiperspirants, and environmental contamination. However, research on aluminum uptake and metabolism has been greatly constrained due to lack of adequate analytical techniques. ^{26}Al is a radioisotope with a half-life of 710,000 years, and is rarely determined using scintillation

TABLE 14.1
Long-Lived Radioisotopes Suitable for AMS Measurement

Element	Isotope	Half-Life (Year) [91]	Natural Abundance[a]
Hydrogen	^{3}H	12.33	1×10^{-16} [14]
Beryllium	^{10}Be	1.51 M	7×10^{-15} [10]
Carbon	^{14}C	5730	1.2×10^{-12} [18]
Aluminum	^{26}Al	710 K	1×10^{-14} [10]
Chlorine	^{36}Cl	301 K	7×10^{-16} [10]
Calcium	^{41}Ca	104 K	1×10^{-15} [10]
Iodine	^{129}I	17.0 M	1×10^{-12} [10]

M, million; K, kilo.

[a] Natural abundance, or the lowest measurement from natural samples.

counting due to its extremely slow decay. When analyzed by AMS, biological samples must be processed to remove potential interferences from isobaric ^{26}Mg and $^{25}MgH^-$ before aluminum is converted to Al_2O_3 as a solid [80,81]. Very low doses of ^{26}Al can be administrated to humans, typically ~3–5 nCi in the form of $AlCl_3$. The absorption, distribution, and excretion of Al in animal models [82] and the differences between patients with renal failure [83–85] have been studied by administering ^{26}Al. Using AMS and microdialysis, the absorption and transport of aluminum across the blood–brain barrier have been investigated [84].

^3H is probably one of the most extensively used radioisotopes for pharmaceutical research. AMS has been used to measure ^3H in biological samples but to a much less extent compared to ^{14}C. As in the case of ^{14}C measurement using AMS, ^3H in the biological samples is ultimately converted to a solid in the form of titanium hydride (TiH_2) in a two-step process (Figure 14.4) [86]. Samples are heated at 900°C for 2 h in a quartz tube in the presence of CuO to form water and CO_2, the same oxidation conditions that are used for ^{14}C sample preparation for AMS analysis. The water is reduced to hydrogen gas with zinc, followed by reaction with titanium powder to form titanium hydride. The AMS sensitivity to ^3H is about 100–1000 times greater than direct scintillation counting. One potential application of ^3H AMS is dual-isotope labeling for experiments in conjunction with ^{14}C AMS. $^{14}CO_2$ and ^3H_2O can be extracted simultaneously from the same biological sample and separated cryogenically. In one such study, two compounds ^3H-PhIP and ^{14}C-MeIQx were dosed separately and simultaneously in rats and examined using AMS [87,88]. The study demonstrated the feasibility of studying complex mixtures using dual isotope labeling followed by AMS analysis.

14.5 FUTURE TRENDS

Since the first application with biological samples around 1990 [64,89,90], AMS has been shown to be a powerful tool for pharmaceutical research. The unique selectivity and ultrasensitivity of AMS allow study designs using very low doses of drugs with minimum exposure to radioactivity. The microdosing strategy for pharmaceutical drug development has accelerated the routine application of AMS. In the past there were a limited number of AMS facilities capable of conducting biological sample analysis. The large physical area required for the instrument, and the time-consuming sample preparation procedures limited the application of AMS in pharmaceutical and biomedical investigations, where throughput is critical. With the recent introduction of commercially available smaller footprint instrument [23], rapid growth of AMS to solve analytical problems in the pharmaceutical industry is expected.

Sample preparation for AMS analysis is probably the most time-consuming and rate-limiting step for biological sample analysis. The graphite conversion step is not compatible with real-time HPLC analysis, which is extensively used for separation of analytes and compound characterization. A large number of HPLC fractions are typically collected for metabolite profiling, however, no structural information can be obtained from AMS analysis. The recent development of an interface for direct GC ^{14}C AMS analysis using a gaseous ion source is encouraging [25,27]. Nonetheless, the overriding benefit of AMS is its exquisite sensitivity in particular to the presence of ^{14}C. This creates new opportunities for safe study designs that employ vanishing small amount of radioactivity to investigate target engagement, mechanism of action, and PK. It is easy to predict that the AMS applications in pharmaceutical and biomedical research will continue to grow in the ensuing years.

REFERENCES

1. Wilding, I.R. and Bell, J.A., Improved early clinical development through human microdosing studies, *Drug Discov. Today*, 10(13), 890, 2005.
2. DiMasi, J.A., Risks in new drug development: Approval success rates for investigational drugs, *Clin. Pharmacol. Ther.*, 69, 297, 2001.
3. Sandhu, P. et al., Evaluation of microdosing strategies for studies in preclinical drug development: Demonstration of linear pharmacokinetics in dogs of a nucleoside analog over 50-fold dose range, *Drug Metab. Dispos.*, 32(11), 1254, 2004.
4. Lappin, G. and Garner, R.C., Current perspectives of ^{14}C-isotope measurement in biomedical accelerator mass spectrometry, *Anal. Bioanal. Chem.*, 378, 356, 2004.
5. Yamane, N. et al., Microdose clinical trial: Quantitative determination of fexofenadine in human plasma using liquid chromatography/electrospray ionization tandem mass spectrometry, *J. Chromatogr. B*, 858, 118, 2007.
6. Vogel, J.S. et al., Attomole quantitation of protein separations with accelerator mass spectrometry, *Electrophoresis (Weinheim, Fed. Repub. Ger.)*, 22, 2037, 2001.
7. Salehpour, M., Possnert, G., and Bryhni, H., Subattomole sensitivity in biological accelerator mass spectrometry, *Anal. Chem.*, 80(10), 3515, 2008.
8. Barker, J. and Garner, R.C., Biomedical applications of accelerator mass spectrometry-isotope measurements at the level of the atom, *Rapid Commun. Mass Spectrom.*, 13, 285, 1999.
9. Muller, R.A., Radio isotope dating with a cyclotron, *Science*, 196(4289), 489, 1977.
10. Litherland, A.E., Ultrasensitive mass spectrometry with accelerators, *Ann. Rev. Nucl. Part. Sci.*, 30, 437, 1980.
11. Elmore, D. and Phillips, F.M., Accelerator mass spectrometry for measurement of long-lived radioisotope, *Science*, 236(4801), 543, 1987.
12. Turteltaub, K.W. and Dingley, K.H., Application of accelerator mass spectrometry (AMS) in DNA adduct quantification and identification, *Toxicol. Lett.*, 102–103, 435, 1998.
13. Mauthe, R.J. et al., Comparison of DNA-adduct and tissue-available dose levels of MeIQx in human and rodent colon following administration of a very low dose, *Int. J. Cancer*, 80, 539, 1999.
14. Hellborg, R. and Skog, G., Accelerator mass spectrometry, *Mass Spectrom. Rev.*, 27, 398, 2008.
15. Brown, K., Tompkins, E.M., and White, I.N.H., Applications of accelerator mass spectrometry for pharmacological and toxicological research, *Mass Spectrom. Rev.*, 25, 127, 2006.
16. Brown, K., Dingley, K.H., and Turteltaub, K.W., Accelerator mass spectrometry for biomedical research, *Methods Enzymol.*, 402, 423, 2005.
17. Zoppi, U. et al., Performance evaluation of the new AMS system at Accium Biosciences, *Radiocarbon*, 49(1), 171, 2007.
18. Turteltaub, K.W. and Vogel, J.S., Bioanalytical applications of accelerator mass spectrometry for pharmaceutical research, *Curr. Pharm. Design*, 6, 991, 2000.
19. Vogel, J.S., Accelerator mass spectrometry for quantitative *in vivo* tracing, *Biotechniques* Suppl., 25, 2005.

20. Vogel, J.S. and Lover, A.H., Quantitating isotopic molecular labels with accelerator mass spectrometry, *Methods Enzymol.*, 402, 402, 2005.
21. Vogel, J.S., Rapid production of graphite without contamination for biomedical AMS, *Radiocarbon*, 34(3), 344, 1992.
22. Klody, G.M. et al., New results for single stage low energy carbon AMS, *Nucl. Instrum. Methods Phys. Res. Sect. B*, 240, 463, 2005.
23. Young, G.C. et al., Comparison of a 250 kV single-stagte accelerator mass spectrometer with a 5 MV tandem accelerator mass spectrometer—fitness for purpose in bioanalysis, *Rapid Commun. Mass Spectrom.*, 22(24), 4035, 2008.
24. Synal, H.A., Jacob, S., and Suter, M., New concepts for radiocarbon systems, *Nucl. Instrum. Methods Phys. Res. Sect. B*, 161–163, 29, 2000.
25. Flarakos, J. et al., Integration of continuous-flow accelerator mass spectrometry with chromatography and mass-selective detection, *Anal. Chem.*, 80(13), 5079, 2008.
26. Bronk, C.R. and Hedges, R.E.M., A gaseous ion source for routine AMS radiocarbon dating, *Nucl. Instrum. Methods Phys. Res. Sect. B*, 52, 322, 1990.
27. Skipper, P.L. et al., Bringing AMS into the bioanalytical lab, *Nucl. Instrum. Methods Phys. Res. Sect. B*, 223–224, 740, 2004.
28. Ramsey, C.B. and Hedges, R.E.M., Hybrid ion sources: Radiocarbon measurement from microgram to milligram, *Nucl. Instrum. Methods Phys. Res. Sect. B*, 123, 539, 1997.
29. Ognibene, T.J. et al., A high-throughput method for the conversion of CO_2 obtained from biochemical samples to graphite in septa-sealed vials for quantification of ^{14}C via accelerator mass spectrometry, *Anal. Chem.*, 75(9), 2192, 2003.
30. Liberman, R.G. et al., An interface for direct analysis of ^{14}C in nonvolatile samples by accelerator mass spectrometry, *Anal. Chem.*, 76(2), 328, 2004.
31. Lappin, G. et al., High-performance liquid chromatography accelerator mass spectrometry: Correcting for losses during analysis by internal standardization, *Anal. Biochem.*, 378(1), 93, 2008.
32. Garner, R.C., Accelerator mass spectrometry in pharmaceutical research and development—A new ultrasensitive analytical method for isotope measurement, *Curr. Drug Metab.*, 1(2), 205, 2000.
33. Gilman, S.D. et al., Analytical performance of accelerator mass spectrometry and liquid scintillation counting for detection of ^{14}C-labeled atrazine metabolites in human urine, *Anal. Chem.*, 70(16), 3463, 1998.
34. Lappin, G. and Garner, R.C., Big physics, small doses: The use of AMS and PET in human microdosing of development drugs, *Nat. Rev. Drug Disc.*, 2 (March), 233, 2003.
35. USA Food and Drug Administration, Innovation or stagnation. Challenge and opportunity on the critical path to new medical products, 2004.
36. Garner, R.C. et al., Evaluation of accelerator mass spectrometry in human mass balance and pharmacokinetic study-experience with ^{14}C-labeled (*R*)-6-[amino(4-chlorophenyl)(1-methyl-1*H*-imidazol-5-yl)methyl]-4-(3-chlorophenyl)-1-methyl-2(1*H*)-quinolinone (R115777), a farnesyl transferase inhibitor, *Drug Metab. Dispos.*, 30(7), 823, 2002.
37. Garner, R.C., Less is more: The human microdosing concept, *Drug Discov. Today*, 10(7), 449, 2005.
38. European Medicines Agency, Position paper on non-clinical safety studies to support clinical trials with a single microdose, 2004.
39. U.S. Food and Drug Administration, Guidance for industry, investigators, and reviewers. Exploratory IND studies, 2006.
40. Bauer, M. et al., A positron emission tomography microdosing study with a potential antimyloid drug in healthy volunteers and patients with Alzheimer's disease, *Clin. Pharmacol. Ther.*, 80(3), 216, 2006.

41. Aboagye, E.O., Price, P.M., and Jones, T., *In vivo* pharmacokinetics and pharmacodynamics in drug development using positron-emission tomography, *Drug Discov. Today*, 6, 293, 2001.
42. Balani, S.K. et al., Evaluation of microdosing to assess pharmacokinetic linearity in rats using liquid chromatography-tandem mass spectrometry, *Drug Metab. Dispos.*, 34(3), 384, 2006.
43. Ni, J. et al., Microdosing assessment to evaluate pharmacokinetics and drug metabolism in rat using liquid chromatography-tandem mass spectrometry, *Pharm. Res.*, 25(7), 1572, 2008.
44. Garner, R.C. and Lappin, G., The phase 0 microdosing concept, *Br. J. Clin. Pharmacol.*, 61(4), 367, 2006.
45. Lappin, G. et al., Use of microdosing to predict pharmacokinetics at the therapeutic dose: Experience with 5 drugs, *Clin. Pharmacol. Ther.*, 80(3), 203, 2006.
46. Lappin, G. and Garner, R.C., The utility of microdosing over the past 5 years, *Expert Opin. Drug Metab. Toxicol.*, 4(12), 1499, 2008.
47. Azzaro, A.J. et al., Pharmacokinetics and absolute bioavailability of selegiline following treatment of healthy subjects with the selegiline transdermal system (6 mg/24 h): A comparison with oral selegiline capsules, *J. Clin. Pharmacol.*, 47, 1256, 2007.
48. Wade, M. et al., Absolute bioavailability and pharmacokinetics of treprostinil sodium administered by acute subcutaneous infusion, *J. Clin. Pharmacol.*, 44, 83, 2004.
49. Smith, P.F. et al., Absolute bioavailability and disposition of (−) and (+) 2′-deoxy-3′-oxa-4′-thiocytidine (dOTC) following single intravenous and oral doses of racemic dOTC in humans, *Antimicrob. Agents Chemother.*, 44, 1609, 2000.
50. Zhang, X. et al., Determination of 25-OH-PPD in rat plasma by high-performance liquid chromatography-mass spectrometry and its application in rat pharmacokinetic studies, *J. Chromatogr. B*, 858, 65, 2007.
51. Strong, J.M. et al., Absolute bioavailability in man of N-acetylprocainamide determined by a novel stable isotope method, *Clin. Pharmacol. Ther.*, 18(5 Pt 1), 613, 1975.
52. Sun, J.X. et al., Comparative pharmacokinetics and bioavailability of nitroglycerin and its metabolism from Transderm-Nitro, Nitrodisc, and Nitro-Dur II systems using a stable-isotope technique, *J. Clin. Pharmacol.*, 35(4), 390, 1995.
53. Brown, T.R., Stable isotopes in clinical pharmacokinetic investigations: Advantages and disadvantages, *Clin. Pharmacokinet.*, 18(6), 423, 1990.
54. Lappin, G. and Garner, R.C., AMS for measuring absolute bioavailability of new drugs, *Drug Discov.* (March), 42, 2006.
55. Lappin, G., Rowland, M., and Garner, R.C., The use of isotope in the determination of absolute bioavailability of drugs in humans, *Expert Opin. Drug Metab. Toxicol.*, 2(3), 419, 2006.
56. Sarapa, N. et al., The application of accelerator mass spectrometry to absolute bioavailability studies in humans: Simultaneous administration of an intravenous microdose of ^{14}C-nelfinavir mesylate solution and oral nelfinavir to healthy volunteers, *J. Clin. Pharmacol.*, 45, 1198, 2005.
57. Klem, B. et al., Determination of the bioavailability of [^{14}C]-hexaminolevulinate using accelerator mass spectrometry after intravesical administration to human volunteers, *J. Clin. Pharmacol.*, 46(4), 456, 2006.
58. Buchholz, B.A. et al., HPLC-accelerator MS measurement of atrazine metabolites in human urine after dermal exposure, *Anal. Chem.*, 71(16), 3519, 1999.
59. Beumer, J.H., Beijnen, J.H., and Schellens, J.H., Mass balance studies, with a focus on anticancer drugs, *Clin. Pharmacokinet.*, 45, 33, 2006.
60. Dain, J.G., Nicoletti, J., and Ballard, F., Biotransformation of clozapine in humans, *Drug Metab. Dispos.*, 25(5), 603, 1997.

61. Young, G. et al., Accelerator mass spectrometry (AMS): Recent experience of its use in a clinical study and potential future of the technique, *Xenobiotica*, 31(8/9), 619, 2001.
62. Beumer, J.H. et al., Human mass balance study of the novel anticancer agent ixabepilone using accelerator mass spectrometry, *Invest. New Drugs*, 25(4), 327, 2007.
63. Dueker, S.R. et al., Disposition of ^{14}C-beta-carotene following delivery with autologous triacyclglyceride-rich lipoproteins, *Nucl. Instrum. Methods Phys. Res. Sect. B*, 259, 767, 2007.
64. Turteltaub, K.W. et al., Accelerator mass spectrometry in biomedical dosimetry: Relationship between low-level exposure and covalent binding of heterocyclic amine carcinogens to DNA, *Proc. Natl. Acad. Sci. USA*, 87, 5288, 1990.
65. Turteltaub, K.W. et al., Macromolecular adduct formation and metabolism of heterocyclic amines in humans and rodents at low doses, *Cancer Lett. (Shannon, Irel.)*, 143, 149, 1999.
66. Garner, R.C. et al., Comparative biotransformation studies of MeIQx and PhIP in animal models and humans, *Cancer Lett. (Shannon, Irel.)*, 143, 161, 1999.
67. Creek, M.R. et al., Tissue distribution and macromolecular binding of extremely low doses of [^{14}C]-benzene in $B_6C_3F_1$ mice, *Carcinogenesis*, 18(12), 2421, 1997.
68. William, K.E. et al., Attomole detection of *in vivo* protein targets of benzene in mice: Evidence for a highly reactive metabolite, *Mol. Cell. Proteomics*, 1, 885, 2002.
69. Fisher, B. et al., Tamoxifen for prevention of breast cancer: Report of the National Surgical Adjuvant Breast and Bowel Project P-1 Study, *J. Natl. Cancer Inst.*, 90(18), 1371, 1998.
70. White, I.N.H. et al., Comparison of the binding of [^{14}C] radiolabelled tamoxifen or toremifene to rat DNA using accelerator mass spectrometry, *Chem. Biol. Interact.*, 106(2), 149, 1997.
71. Martin, E.A. et al., Tamoxifen DNA damage detected in human endometrium using accelerator mass spectrometry, *Cancer Res.*, 63(23), 8461, 2003.
72. Turteltaub, K.W. et al., MeIQx-DNA adduct formation in rodent and human tissues at low doses, *Mutat. Res.*, 376, 243, 1997.
73. Palmblad, M. et al., Neuroscience and accelerator mass spectrometry, *J. Mass Spectrom.*, 40(2), 154, 2005.
74. Weaver, C.M., Use of calcium tracers and biomarkers to determine calcium kinetics and bone turnover, *Bone*, 22, 103S, 1998.
75. Freeman, S.P.H.T. et al., Biological sample preparation and ^{41}Ca AMS measurement at LLNL, *Nucl. Instrum. Methods Phys. Res. Sect. B*, 99, 557, 1995.
76. Jiang, S. et al., The measurement of ^{41}Ca and its application for the cellular Ca^{2+} concentration fluctuation caused by carcinogenic substances, *Nucl. Instrum. Methods Phys. Res. Sect. B*, 223–234, 750, 2004.
77. Freeman, S.P.H.T. et al., The study of skeletal calcium metabolism with ^{41}Ca and ^{45}Ca, *Nucl. Instrum. Methods Phys. Res. Sect. B*, 172, 930, 2000.
78. Freeman, S.P.H.T. et al., Human calcium metabolism including bone resorption measured with ^{41}Ca tracer, *Nucl. Instrum. Methods Phys. Res. Sect. B*, 123, 266, 1997.
79. Mi, S. et al., Study on calcium absorptivity in calcium supplements evaluated by ^{41}Ca labeled calcium pool of osteoporotic rats, *Acta Nutrim. Sin.*, 30(1), 39, 2008.
80. Ittel, T.H. et al., Ultrasensitive analysis of intestinal absorption and compartmentalization of aluminium in uraemic rats: A ^{26}Al tracer study employing accelerator mass spectrometry, *Nephrol. Dial. Transplant.*, 12, 1369, 1997.
81. Hohl, C. et al., Medical application of ^{26}Al, *Nucl. Instrum. Methods Phys. Res. Sect. B*, 92, 478, 1994.
82. Yokel, R.A. and Florence, R.L., Aluminum bioavailability from the approved food additive leavening agent acidic sodium aluminum phosphate, incorporated into a baked good, is lower than from water, *Toxicology*, 227(1–2), 86, 2006.

83. Yokel, R.A. et al., Entry, half-life, and desferrioxamine-accelerated clearance of brain aluminum after a single ^{26}Al exposure, *Toxicol. Sci.*, 64, 77, 2001.
84. Yokel, R.A., Brain uptake, retention, and reflux of aluminum and manganese, *Environ. Health Perspect.*, 110(Suppl.), 699, 2002.
85. Steinhausen, C. et al., Investigation of the aluminium biokinetics in humans: A ^{26}Al tracer study, *Food Chem. Toxicol.*, 42(3), 363, 2004.
86. Chiarappa-Zucca, M.L. et al., Sample preparation for quantitation of tritium by accelerator mass spectrometry, *Anal. Chem.*, 74(24), 6285, 2002.
87. Dingley, K.H. et al., Attomole detection of ^{3}H in biological samples using accelerator mass spectrometry: Application in low-dose, dual-isotope tracer studies in conjunction with ^{14}C accelerator mass spectrometry, *Chem. Res. Toxicol.*, 11, 1217, 1998.
88. Dingley, K.H. et al., Effect of dietary constituents with chemopreventive potential on adduct formation of a low dose of the heterocyclic amines PhIP and IQ and phase II hepatic enzymes, *Nutr. Cancer*, 46, 212, 2003.
89. Vogel, J.S. et al., Application of AMS to the biomedical sciences, *Nucl. Instrum. Methods Phys. Res. Sect. B*, 52, 524, 1990.
90. Felton, J.S. et al., Accelerator mass spectrometry in the biomedical sciences: Applications in low-exposure biomedical and environmental dosimetry, *Nucl. Instrum. Methods Phys. Res. Sect. B*, 52, 517, 1990.
91. Kutschera, W., Progress in isotope analysis at ultra-trace level by AMS, *Int. J. Mass Spectrom.*, 242, 145, 2005.

Index

D

E

F

G

H

N

O

T

U

V

W

X

Z